Ion Exchange and Solvent Extraction

Ion Exchange and Solvent Extraction

A Series of Advances

Volume 15

edited by

Yizhak Marcus
The Hebrew University of Jerusalem
Jerusalem, Israel

Arup K. SenGupta
Lehigh University
Bethlehem, Pennsylvania

Jacob A. Marinsky
Founding Editor

CRC Press
Taylor & Francis Group
Boca Raton London New York

CRC Press is an imprint of the
Taylor & Francis Group, an **informa** business

First published 2002 by Marcel Dekker, Inc.

Published 2018 by CRC Press
Taylor & Francis Group
6000 Broken Sound Parkway NW, Suite 300
Boca Raton, FL 33487-2742

© 2002 by Taylor & Francis Group, LLC
CRC Press is an imprint of Taylor & Francis Group, an Informa business

No claim to original U.S. Government works

ISBN 13: 978-0-8247-0601-2 (hbk)

Visit the Taylor & Francis Web site at
http://www.taylorandfrancis.com

and the CRC Press Web site at
http://www.crcpress.com

Preface

The fifteenth volume of *Ion Exchange and Solvent Extraction* is concerned with the advances in solvent extraction, a field that also was covered in volume 13. This book was conceived at ISEC 1999, the International Solvent Extraction Conference, held in Barcelona, Spain. Several authors responded to the call by the editors to contribute comprehensive review chapters on subjects in which they are experts and in which important advances have been noted in recent years. In such diverse industrial fields as hydrometallurgy, pharmaceutical chemistry, organic fine chemicals, and biotechnology, there are new applications of solvent extraction. These are based on research conducted in laboratories all over the world, leading to the development of viable and profitable processes. Both basic research and process development are the keys to success in this field, and this volume addresses the need for persons active in the field of solvent extraction to learn of the advances made in it. These often occur rapidly, and it is expected that further advances, based partly on those presented here, will be discussed at the next International Solvent Extraction Conference, ISEC 2002, to be held in Cape Town, Republic of South Africa.

Chapter 1, dealing with an integrated method for the development and scaling up of extraction processes, is actually a manual for practitioners, both novices and experienced. It is written in a conversational manner, presenting sound procedures to be followed and cautions against pitfalls; it also provides detailed examples of actual processes that have been developed. Several sections include the important advice that in some cases the developer should tell the person who commissioned the work that solvent extraction is *not* the method that should be used for the separation job. In many other cases, however, solvent extraction can be used profitably, if the proper solvent and equipment are selected, on the basis of laboratory, bench-scale piloting, and full-scale piloting experiments, based on and backed-up by the proper simulation work. These are fully described in Chapter 1.

Chapter 2 is concerned with the design of pulsed extraction columns, one of the options discussed in Chapter 1. This design requires time-consuming and expensive laboratory-scale or pilot-plant experiments, in particular if detailed information on the drop-size distribution and/or the holdup profile of the dispersed phase is needed. The sensitivity of the mass transfer to even low concentrations of impurities mandates experiments with the real mixture in order to obtain reliable data for designing a technical column. Since the design cannot proceed from first principles, a "standard apparatus" was developed that allowed the measurement of the required data, based on a population-balance model. It permitted the determination of the rising velocity of single drops, the breakup of drops, the coalescence of drops, and the mass transfer coefficient of a drop under the conditions to be expected for the technical column. The excellent agreement between simulated and experimental data confirms the suitability of the design concept.

An implicit implementation of the development and scaling up of extraction processes, as outlined in Chapter 1, is seen in Chapter 3, on the purification of nickel by solvent extraction. Chapter 3 shows that industry has adopted solvent extraction as the major production method for nickel, replacing pyrometallurgical processes, at least in the stages concerned with purification of nickel from other base metals. This advance was made possible by the recent availability of commercial quantities of selective extractants as well as the demonstrated successful operation of large-scale solvent extraction plants. Modern flow sheets thus permit profitable recovery of highly pure nickel even from low grade ores, as well as the capture of high-purity cobalt as a by-product.

Another application of solvent extraction on an industrial scale is the clean up of contaminated soils and sludges, described in Chapter 4. The application of this technology to environmental problems appears to be gaining more widespread acceptance. This acceptance comes in spite of the justified criticism that inappropriate use of solvents in industry, including their use in solvent extraction, adds to the contamination of the environment. When a proper choice of solvents is made, this technology becomes viable for the cleaning up the environment. The processes that have been proposed and tested on a commercial scale are reviewed, and their acceptance by the regulatory authorities is discussed.

As explained in Chapter 1, the choice of the appropriate solvent for a given separation job by solvent extraction is the key problem in this technology. For the short term, commercially available solvents must be employed, even if their properties are not optimal. However, in the long run, in particular for highly specialized separation jobs, it is important to design new solvents with optimized properties for a given separation. How to design solvents with their ultimate use in mind is fully discussed in Chapter 5. In one sense, this chapter supplements the procedures recommended in Chapter 1 by providing the needed screening criteria. In another sense, the design of new solvents is a challenge met by the methodology described in Chapter 5. Computer-aided molecular design is invoked, including interactive, combinatorial, construct-and-test, and mathematical programming methods. The evolutionary approach to designer solvents is then discussed in detail, showing all its advantages, and is illustrated by the design of a solvent for the separation of phenols from neutral oils.

The emphasis in developments in solvent extraction has shifted in recent years from the more venerable fields of nuclear fuel reprocessing and hydro-metallurgy to the separation of organic substances. A tough problem in this field, of particular relevance for pharmaceutical chemistry and biotechnology, is the separation of optical isomers, only one type of enantiomer is biologically active. The difficulty arises because the differences in the extractabilities of the species formed with suitable extractants are minute. The attempts that have been made to solve this problem are described in Chapter 6. If a solvent ex-traction method for the separation of the enantiomers is to be developed the chiral recognition mechanisms must be understood. The design of a multistage process appears to be necessary, and the solutions proposed so far for this problem are discussed, the absence of the use of extraction columns being conspicuous in this respect.

Most of the known solvent extraction processes take place between an aqueous feed and an organic solvent/extractant that, to ensure low miscibility with the aqueous solution, is necessarily of low polarity. An exception, discussed fully in Volume 13 of this series, is the use of an aqueous solution of poly-ethylene glycols opposing a concentrated aqueous electrolyte solution in a bi-phasic aqueous extraction system. Water-miscible polar organic solvents, com-monly used in electrochemistry or analytical chemistry, such as the lower alcohols, N,N-dimethylformamide, acetonitrile, and dimethyl sulfoxide, have therefore been excluded from consideration in solvent extraction systems. This situation is now reversed, as shown in Chapter 7. It is shown that such solvents or their aqueous solutions can be successfully used when opposed to a nonpolar hydrocarbon to separate hydrophobic solutes by liquid–liquid distribution. The solutes include higher alcohols, long-chain amines, quaternary ammonium salts, nitroalkanes and nitriles, phosphoric acid esters, and terpenoid derivatives. The regularities governing such systems are explored, to enable the prediction of the best choice of solvent for a given separation problem.

Finally, Chapter 8 is devoted to a solvent extraction concept completely different from that discussed in the earlier chapters (mainly 1–3)—that is, nondispersive, membrane-based liquid–liquid distribution systems. In certain situations, mixer-settlers or columns, based on the dispersion of the two immiscible liquid phases in each other before being separated again, have not been sufficiently successful, and an alternative has been sought. Since the advent of commercially available hollow-fiber modules, among other porous-membrane-based devices, nondispersive methodologies have become attractive for use in such situations. Laboratory- and pilot-plant-scale studies of membrane-based liquid–liquid extraction systems are described, as are the modeling of such systems and the evaluation of the mass transfer coefficients. Some specialized applications, already in an advanced stage of implementation, include procedures used by the beverage and semiconductor industries.

The wide gamut of topics covered in these eight chapters attests to the diversity of the solvent extraction methodology and its applicability to separation technology in many industries. This volume is also evidence of the continuing advances in this field being contributed by laboratories and industrial research organizations all over the globe. Few readers will fail to find information or insight that will make their effort worthwhile.

Yizhak Marcus
Arup K. SenGupta

Contributors to Volume 15

Richard J. Ayen   Neptune Consulting, Wakefield, Rhode Island

Peter M. Cole   Hydrometallurgy Consultant, Bryanston, South Africa

André B. de Haan   Faculty of Chemical Technology, Twente University, Enschede, The Netherlands

Baruch Grinbaum   IMI (TAMI) Institute for Research and Development Ltd., Haifa, Israel

Hartmut Haverland   Gebrüder Lödige Maschinenbau GmbH, Paderborn, Germany

Sergey M. Leschev   Analytical Chemistry Department, Belarusian State University, Minsk, Republic of Belarus

James D. Navratil   Environmental Engineering and Science, Clemson University, Anderson, South Carolina

Izak Nieuwoudt   Department of Chemical Engineering, Institute for Thermal Separation Technology, University of Stellenbosch, Stellenbosch, South Africa

Anil Kumar Pabby   PREFRE Plant, Nuclear Recycle Group, Bhabha Atomic Research Centre, Tarapur, Maharashtra, India

Ana-Maria Sastre   Departament d'Enginyeria Quimica, Universitat Politècnica de Catalunya, Barcelona, Spain

Béla Simándi   Department of Chemical Engineering, Budapest University of Technology and Economics, Budapest, Hungary

Kathryn C. Sole   Anglo American Research Laboratories (Pty) Ltd, Johannesburg, South Africa

Braam van Dyk   Department of Chemical Engineering, Institute for Thermal Separation Technology, University of Stellenbosch, Stellenbosch, South Africa

Alfons Vogelpohl   Mass Transfer Laboratory, Technical University of Clausthal, Clausthal-Zellerfeld, Germany

Contents

Contents

Contents of Other Volumes

xiii

1

An Integrated Method for Development and Scaling Up of Extraction Processes

Baruch Grinbaum

IMI (TAMI) Institute for Research and Development Ltd., Haifa, Israel

I. INTRODUCTION

The development and scaling up of solvent extraction processes is long and complicated. Although solvent extraction (SX) has been used for over 100 years as an industrial unit operation, no general manual, which describes the steps in process development, has so far been published.

A. Genesis

". . . And then, after the client finished describing the problem, came the crucial question: How will you do it by solvent extraction?" This is how the development of most SX processes starts.

Like other phase–equilibrium separation processes, SX is usually just a purification stage in the process downstream, to recover the desired components. There are exceptions, of course: the IMI processes for manufacturing KNO_3, H_3PO_4, and $CaBr_2$ are stand-alone SX processes. The development of such a process starts when somebody awakes in the morning and asks the key question: Can we make this product entirely by SX, instead of the existing processes?

1

B. Integrated Approach to the Development of Solvent Extraction Flowsheets

The general objective of the integrated approach is to develop an optimal process with minimum resources and within certain time limits. This approach enables one, through limiting tests, to eliminate inferior options in the earliest possible stage, and go on to the advanced—and more expensive—stages of development only with the most promising ones. Blumberg [1] has described this approach fully.

The development procedure is similar for all SX processes:

1. Define exactly your input information and the final target of the process. Rule out SX if a better (i.e., cheaper) route is available.

2. Check the literature to find recommended solvents, equilibrium data, and process data for your solutes. Remember that *one can always save a day of literature survey by wasting two months on experimental work that has already been carried out and reported.* The data you need may not be available, but you will always find *something* that enables you to draw an analogy to your process and to start with a few feasible solvents.

3. Test the feasibility of the solvents that seem to be suitable to your process by testing their chemical (liquid–liquid equilibrium) and hydrodynamic (phase separation) properties.

4. For the solvent (or solvents) that seem to be optimal, prepare the equilibrium isotherms, and derive from them the correlation for the distribution coefficient.

5. Using the isotherms, carry out computerized simulation and determine the recommended flow sheet of the process. The simulation enables the design of an optimum flow sheet for a given solvent. If several solvents are considered, the parametric runs enable us to pick the best one. For multicomponent separation, the simulator also helps us to find the optimum order of extraction, configuration, composition, and amount of reflux streams. The results of such an optimization are more accurate than any experimental procedure and usually can be learned with less manpower. It should be emphasized that in the integrated approach, *the use of computerized simulation is of crucial importance: this is the stage at which maximum savings in time and manpower can be achieved.*

6. Carry out a full set of hydrodynamic and kinetic experiments for the solvents that were found optimal on the basis of the simulation. This test includes the measurement of density, viscosity, and interfacial tension for the various streams, and the determination of the phase separation ratio, the kinetics, and the mass transfer at both dispersions.

7. Based on these results, recommend the final composition of the solvent, the dispersion type, the flow sheet, and the equipment to be used. Decide

whether columns or mixer-settlers are to be used, and select the appropriate type for each. Remember that every battery may need different equipment, and there is no such thing as "the best equipment for solvent extraction." If in doubt, consider testing in the pilot plant both dispersions, two different types of equipment, and/or two or three compositions of the solvent. This is expensive, but cheaper than a wrong choice.

       8.   Test the recommended equipment(s) on bench scale, pilot plant scale, or both, to estimate the mass transfer, entrainment, and phase separation, and, optionally, accumulation phenomena (this last point can be checked only by running a pilot plant interconnected to the real plant over long periods).

Remember

In almost every development project, one has to work within tight timetables and budget limitations. The integrated approach means the allocation of enough resources to every stage of the project, so that the work is done efficiently in the first iteration, without any need of repetitions.

      If in doubt about when to switch from one stage to another, follow these simple rules:

       1.   Do not start any experimental work until the project has been fully defined and you have all the data from the client, the literature, and—for an existing process—also from the plant.

       2.   Do not do in the pilot plant what can be done in the laboratory.

       3.   Never test in a pilot plant what can be calculated by simulation (e.g., phase ratios or the number of theoretical stages).

       4.   Remember that every problem not solved in the pilot plant will reappear bigger in the industrial plant.

       5.   Use your intuition, because there are *never* enough data for full-proof process development and design.

       6.   The successful design of a plant is based on reliable data, not on timetables. Time spent in obtaining the correct and relevant data will always pay dividends [2]. Let nobody rush you to premature, and probably wrong, conclusions only because a deadline has passed. Remember Murphy's law for development (There is never enough time and money to do the work well but there always are plenty of resources to repeat it) and avoid its consequences.

      This chapter tries to draw the guidelines to every step in the process development.

II. PROJECT DEFINITION AND ANALYSIS

Exact definition of the project is essential to successful process development.

      The final product of the process development procedure is a report, which defines the flow sheet and the equipment that, using the recommended solvent and operating conditions, will profitably manufacture the product(s) at the given specifications of recovery, quality, and quantity.

Remember

Your output is a bunch of papers. If well done, you enable the client to erect and run a profitable industrial plant. But if done wrong, you cause damage far exceeding your lifetime salary.

It is important to realize that solvent extraction is not just a unit operation, like distillation. Because of the necessity to develop or fit a suitable solvent, SX is far more complicated and it shares a great many features of unit processes.

A. Definitions

Before going on, let us define a few basic concepts, which are used throughout this chapter. Most of them are standard SX nomenclature, but to avoid ambiguities, here they are

Solute: The components that undergo extraction are called the *distribuends*, but most researchers refer to them just as the *solutes*, and this is the name used in this chapter.

Stage: For the sake of simplicity, "stage" in this text is synonymous with transfer unit—that is, equivalent to one theoretical stage, whether physical stages (i.e., mixer-settlers) or columns are used.

Battery: The part of an SX plant that carries out a well-defined function (e.g., extraction, stripping). Each battery has at least two inputs and two outputs.

Flow Sheet: The *schematic* layout of the plant, namely, the number of batteries, number of stages in each battery, and full configuration of input and output streams, including flow rates and compositions of each stream. Note that the flow sheet tells you nothing about the type of the equipment that should be used.

A general flow sheet (Fig. 1) includes three batteries: extraction, purification/ scrubbing, and reextraction (stripping). The number of batteries may vary from one (extraction only) to many (if multiple species are to be separated).

Reflux: A stream that is an outlet of one battery and an input to another one.

Note

In extraction there is no reflux within a battery; that is, a stream leaving the battery should never be refluxed to another stage in that battery.

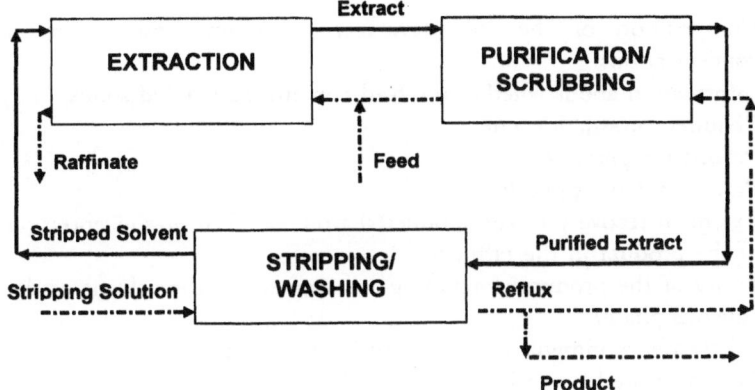

Figure 1 Extraction plant: basic configuration.

Internal Recycle: A stream that circulates within the same stage (i.e., from the settler to the mixer) to keep the desired local phase ratio in the mixer. *This is not a reflux.*

Extraction: Transfer of the solute from the feed stream (known in hydrometallurgy as pregnant liquor) to the solvent.

Reextraction: Transfer of the solutes from the loaded solvent to the product stream. Reextraction may be done by *washing* (i.e., reextraction with water) or by *stripping* with a chemical reagent.

Purification/Scrubbing: Removal of impurities from the extract. If the operation is done by using a part of the product, it is called *purification.* Removal by an external solution is called *scrubbing.*

Type of Dispersion: When the aqueous phase is dispersed in the solvent, the dispersion is called water-in-oil, or W/O. When the aqueous phase is the continuous one, the dispersion is called oil-in-water, or O/W. If no aqueous phase is present (e.g., extraction of BTX), "water" refers to the feed/product phase and "oil" to the solvent.

Client: For our purpose, the one who gave you the project and to whom you report: it may be your boss, a colleague from another department, or a representative from an external company that ordered the research.

B. Project Input

Two problems should be identified and eliminated before you start working: missing data and excessively high constraints.

Some important data are often missing in the initial problem definition. Before you start to analyze the project, be sure to have *all* the following data:

1. Composition of the feed (average, minimum, and maximum concentrations)
2. Information about whether the feed contains suspended solids
3. Required production rate
4. Process temperature
5. Process pH (if applicable)
6. Extent of recovery of the product(s) from the feed *or* concentration of the product in the raffinate
7. Purity of the products (including entrainment of the solvent or the aqueous phase)
8. Allowed entrainment of the solvent in the raffinate
9. Environmental limitations

Sometimes the answer for some of items 4–6 is: find the best, turning them into output variables. This strategy is fully legitimate (and increases significantly the scope of the project), but to avoid troubles in the future, be sure that you have written and signed authorization.

Sometimes the initial constraints of the process are impractical. Do not start working before clarifying this point with the client. The client, who often is not an SX expert, may not realize that a request for the level of recovery, product purity, or entrainment cannot be met without extreme (and unjustified) cost. Clarify this point *before* you start working.

C. Project Output

When the project is finished, the following parameters must be determined:

1. Composition of the solvent (see Section II.D)
2. Process temperature and pH for each battery
3. Composition of all outlet streams (including entrainment)
4. Detailed flow sheet
5. Phase ratio in each battery
6. The flux in each battery
7. The dispersion in each battery
8. The type of equipment to be used in the industrial plant for each battery and its *approximate* size
9. Auxiliary operations: stabilization of the solvent, sidestream treatment for removal of contaminants from the solvent, separation of entrained solvent, treatment of crud.

All these properties are linked. After the extractant has been chosen, the following links must be considered:

1. Phase Ratio

Through this chapter, "phase ratio" is given as O:A and defined as the ratio of the oil phase to the aqueous phase. Low and high phase ratios mean small flow of solvent and vice versa. The minimum phase ratio, to ensure the requested recovery, is obtained at limiting conditions (i.e., at an infinite number of transfer units). Lower phase ratios mean smaller inventories of solvent and lower total flows. So, we are interested in decreasing the phase ratio as much as possible. But, for a real process, some of the following penalties are paid for decreasing the phase ratio:

    a.   Increased concentration of the extractant in the solvent (if feasible)
    b.   At given recovery, increased number of transfer units
    c.   At given number of stages, decreased recovery

2. Number of Stages

A smaller number of stages diminishes the size and cost of the plant, including decreased solvent inventory. But (other factors remaining constant) it decreases the recovery. For a given recovery and number of stages, increased phase ratio is needed.

3. Concentration of Extractant

A high concentration of the extractant decreases the phase ratio; that is, smaller solvent inventory and smaller plant will suffice. On the other hand it may harm the phase separation, that is, it may lower the flux and require a larger plant. This also makes the solvent, per volume unit, more expensive, increasing the cost of entrainment losses.

4. The Type of Dispersion

The type of dispersion—water-in-oil (W/O) or oil-in-water (O/W)—may usually be determined by us. In many processes it has a significant influence on the mass transfer or phase separation properties, and so affects the size of the plant and the total cost of the operation. See the detailed discussion in Section VI.

    The project definition methodology may be demonstrated by the following examples:

Example 1
Uranium is to be extracted at 40° C from a feed that contains on the average 400 ppm of uranium. It may vary between 100 and 800 ppm, depending on the processed ore. The raffinate that goes for further downstream processing must contain less than 1 ppm uranium (ecological limit) and up to 50 ppm of solvent entrainment

(downstream process limit). The uranium should be stripped so that the final product of "yellow cake" is readily obtained. To avoid precipitation, the pH in the stripping should be below 5. The flow rate of the feed stream is 500 m³/h. The feed contains up to 1000 ppm suspended solids. No pH control for the extraction is needed.

Before any experiment is done, we already know the following:

1. The client's wishes. The problem is well defined, and you must meet the process constraints.
2. The concentration of the uranium in the feed varies over a very wide range. Consult the client to determine whether you must meet the limit of 1 ppm in raffinate *on average* (i.e., for a feed of 400 ppm uranium) or *always* (i.e., even for a feed of 800 ppm uranium).
3. The extract must be stripped with ammonia solution (to yield the yellow cake). Do not look for any alternative. To ensure full stripping without precipitation of the yellow cake the stripping should be pH-controlled. (Precipitation is done as a downstream process. If it happens in the stripping battery, the equipment will be blocked.)
4. The expected amount of mass transfer is very low (concentrations <0.1%). Low concentration of extractant in the diluent and low O:A phase ratio is recommended.
5. The extractant must be able to extract quantitatively the uranium (yield ~ 99.9% is needed).
6. The extremely low level of solvent entrainment implies W/O dispersion, and the use of either columns or mixer-settlers with entrainment recovery (entrainment from a mixer-settler usually >50 ppm).

Extraction of uranium is a well-known problem. A short literature survey will give you the solvents that are in use, some equilibrium data, and even recommended flow sheets. You merely adjust the compositions of the solvents to your process, check them, and choose the right equipment for the industrial plant.

Example 2

A new aromatic, water-soluble compound, with boiling point about 200°C, is to be extracted from the aqueous feed. Its concentration in the feed is 500 g/L. The raffinate should contain less than 1 g/L of the organic compound (economical constraint). Some isomers of the product, which were generated during the synthesis, could be extracted just as well. The product will be removed from the solvent and the contaminants by distillation. The maximum quantity of the solvent (assuming that it is a relatively harmless compound) in the final product is 50 ppm. Production rate of the product is 2000 tons/year. There is no ecological constraint on the raffinate.

In this case you know the following:

1. The solvent must be volatile, so that it can be distilled efficiently from the product. Consult the distillation experts about any solvent that you consider suitable.
2. Huge mass transfer is expected. A very efficient extractant is necessary, since the product must be extracted with a yield of 99.7%; and yet a high concentration of the solute in the solvent is essential, to decrease the cost of the distillation. Probably many equilibrium stages will be needed. The equipment must be designed to enable this mass transfer.
3. The phase ratio in the process is variable, since 50% of the aqueous phase is extracted to the solvent. The highest O:A ratio is near the raffinate outlet, and the lowest near the extract outlet.
4. The selectivity of the solvent toward the contaminants is irrelevant because they will be separated in the distillation. Treat it as a single-solute problem.
5. You will have to find the best extraction temperature. The pH is not an experimental parameter for the extraction process.
6. The extraction and the distillation should be developed by one team, given the obvious interaction in this case.
7. Both O/W and W/O dispersions may be considered.

The literature may yield a great deal of information about extraction of aromatic compounds, which will help you to choose the solvents that meet the process constraints. But you will have to use your intuition, and you will have to test a few of the candidates. This is a more difficult project than the case of uranium extraction.

Example 3
We would like to consider purification of 20,000 tons/year of toluene, which is contaminated with 5–10% of benzene and 2–3% of xylenes. The toluene may contain no more than 100 ppm contaminants.

DO NOT WASTE YOUR TIME. Prepare a polite answer (the request came probably from somebody whom you do not want to insult, e.g., your boss) and simply forget it. SX is an efficient unit operation, but it has its limits. For liquids with reasonable relative volatility, distillation is always cheaper and more efficient, and purification of a mixture of aromatic solvents by distillation is a well-known procedure. Use extraction when distillation fails, or at least staggers.

D. Choice of the Solvent

Once you have the definition of the problem, the very first stage is to find the suitable solvent. *This is the heart of the problem.* Unless an efficient solvent is found, there will be no SX process.

A rigorous treatment of the various types of solvent and the mechanism that applies to each appears in the book by Marcus and Kertes, Ion Exchange and Solvent Extraction of Metal Complexes [3]. Simplified descriptions of various types of solvent are given by Ashbrook and Ritcey [4], Blumberg [1], and others.

For engineering purposes, the solvents may be divided into two main categories:

Physical Extraction: The existing compounds are transferred from one phase to the other, by means of simple distribution or solvation.

Chemical Extraction: The extraction is a chemical process, causing the creation of new molecules, either by ion pairing or complexation.

Remarks

1. According to the IUPAC convention, chemical extraction may be referred to as "reactive extraction." Yet this term may be misleading and should rather be kept for processes in which extraction is carried out in parallel to a chemical reaction in one of the phases, in a manner similar to reactive distillation.

2. There is a small group of processes that does not fit this simplified classification. These are the processes involving composite solvents that contain both anionic and cationic extractants and extract salts. Their extraction mechanism is clearly chemical, yet they do not have maximum loading, and according to our definition should be treated as physical extractants.

The distribution curves of a solute for chemical and physical extraction are shown in Fig. 2.

1. What is a Good Solvent?

A good solvent should have as many as possible of the following properties:

    a.   High loading (high distribution coefficient)
    b.   High selectivity
    c.   Easy retrieval of the solute
    d.   Rapid phase separation
    e.   Rapid kinetics
    f.   Minimum solubility in the feed
    g.   Chemical stability under process conditions
    h.   Availability at reasonable price from commercial sources
    i.   Safety and environmental friendliness.

These properties deserve some explanation.

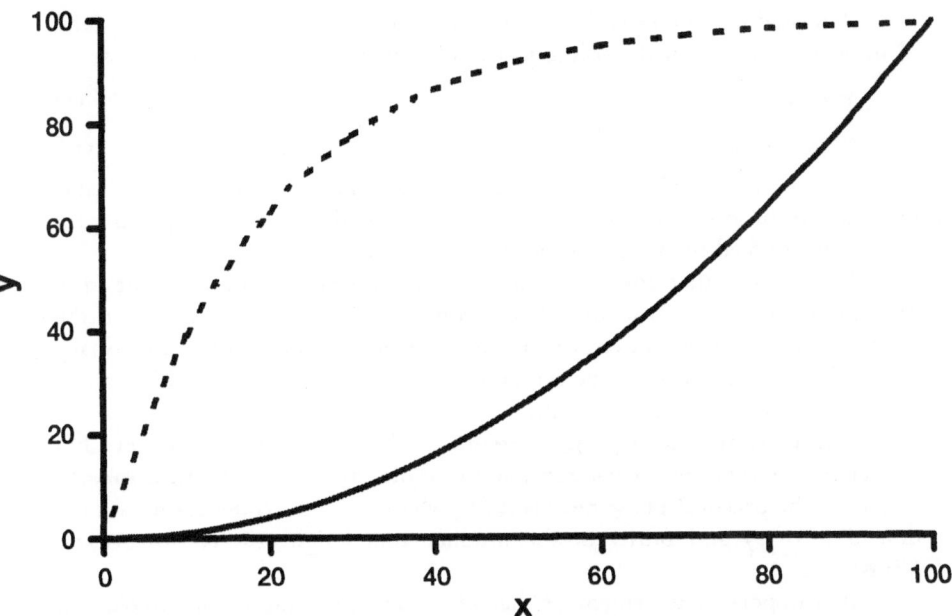

Figure 2 Solvent loading: dashed curve, chemical extraction; solid curve, physical extraction.

a. High Loading. The definition of *loading* in the SX literature is not unique. In this chapter it is simply a synonym for y—the concentration of the solute in the solvent—which is the really important parameter for process development and is expressed either as grams per liter or weight percent.

In chemical extraction the *maximum solvent loading capacity* y_{max} is defined as the concentration of the solute in the solvent that utilizes all the molecules of the extractant. When the chemistry of the process is known, it may be calculated from stoichiometric considerations. Otherwise it is determined by measuring the loading under *limiting conditions* (see Section III.A). For this case the *relative loading* may be defined as the actual concentration of the solute in the solvent divided by the maximum one.

For physical extraction there is no theoretical upper limit for the concentration of the solute in the solvent (i.e., no maximum loading): the concentration of the solute in the solvent goes up with its concentration increasing in the feed, according to Eq. (1). So y_{max} is the concentration of the solute in the solvent in equilibrium with the feed, that is, at limiting conditions (see Section III.A).

A related parameter is the distribution coefficient D, defined as the ratio of the concentration of the solute in the solvent, y, and the aqueous phase, x.

$$y = Dx \tag{1}$$

$$y_{max} = D_{lc} x_{feed} \tag{1'}$$

where D_{lc} is the distribution coefficient at the limiting conditions. Its value may vary from less than one to hundreds. It should be high enough to keep the solvent-to-feed phase ratio in a reasonable range.

In a solvent that contains a diluent, the maximum loading in the ideal case should be a linear function of the concentration of the extractant in the solution. In reality we often see deviations from ideality, which are usually expressed by the decrease in the actual maximum loading capacity with increasing concentration of the extractant.

The maximum loading y_{max} determines the phase ratio. To achieve full extraction, *the minimum phase ratio*, at an infinite number of stages, *is equal to* x_{feed}/y_{max}. The practical phase ratio is a/D_{lc}, where $a = 1.2–2$, depending on the desired recovery and purity of the product and design criteria (see Section VII.A).

It is important to remember that $D < 1$ in extraction is undesirable, but definitely not a no-go factor. This just means that the flow of the solvent is higher than the flow of the feed. In purification and reextraction, the lower the value of D, the better. Remember that in our nomenclature D is always y/x, which need not be a constant but may vary with x.

b. High Selectivity. This parameter applies to *selective extraction*, that is, separation between a few solutes and extraction of only the desired ones.

Higher selectivity implies a smaller purification stream. Note that in selective extraction, the selectivity is similar to the relative volatility in distillation. We usually use a part of the product as a purification stream, so that the purification stream is practically a *reflux*. The ratio of the reflux to product stream is sometimes called the *reflux ratio*.

c. Easy Retrieval of the Solute. Retrieval of the solute can be done either by distillation or reextraction. When the solute is an organic compound, distillation is the natural choice. In this case, we must see that the relative volatility between the solvent and the solute is as large as possible. Sometimes it is worthwhile to reject an excellent solvent, having a very high distribution coefficient but low relative volatility, in favor of a less efficient solvent that yields lower loading but is easy to distill.

In the case of a water-soluble solute, like a salt or an acid, the best solvent is the one from which the solute can be reextracted by water (i.e., *washed*). In most cases of chemical extraction, the product must be *stripped*, using a suitable chemical reagent.

High loading and easy stripping usually contradict one another. This is logical: high loading means high affinity of the solute to the extractant. The resulting complex has a lower Gibbs energy than the initial system, so the process cannot be fully reversible, and the reextraction must involve dilution, input of energy, or addition of a chemical agent for stripping.

For the comparison of solvents it is convenient to refer to the *differential loading*, that is, the difference between the concentration of the solute in the extract and in the stripped solvent,

$$\Delta y = y_{extr} - y_{ss} \tag{2}$$

where y_{extr} is the loading in the extract and y_{ss} the loading in the stripped solvent.

Example 4

Extraction of H_2SO_4 from waste streams can be done quantitatively, in one or two equilibrium stages, by using a linear long chain aliphatic tertiary amine, such as Alamine 336 [5]. But the resulting extract cannot be washed with water, even at elevated temperatures. When a branched chain aliphatic tertiary amine like tris(2-ethylhexyl) amine (TEHA) is used, the extraction is less efficient and seven equilibrium stages are needed to extract about 80% of the acid. But the solute can be reextracted with hot water, to yield a pure solution of H_2SO_4 in water, only slightly more dilute than in the feed solution. It goes without saying that TEHA is the recommended extractant for this application.

d. Rapid Phase Separation.   The rate of phase separation is determined by three physical properties: difference in density between the phases, interfacial tension, and viscosity of both phases.

i.   Difference in Density of the Phases: Usually the larger the better. In the rather rare case of very dense liquids, like dibromomethane, which has a density above 2 g/cm^3, extra mixing is needed to ensure good phase contact. The minimum recommended value of the difference in density is 0.05 g/cm^3.

ii.   Interfacial Tension: Any value between 5 and 25 mN/m should give good drop breakage and coalescence. With higher interfacial tension, the drops tend to be stable and not to break. With lower interfacial tension they do not coalesce. The minimum recommended interfacial tension for mixer-settlers or columns is 2 mN/m [6].

iii.   Viscosity: The lower the viscosity of both phases, the more rapid the phase separation, due to the decrease in the drag force on the droplets of the dispersed phase. The viscosity of the continuous phase is the most important one. Therefore, the more viscous phase should be the dispersed one, if possible.

If the interfacial tension and the difference in densities are below the minimum values, change the diluent and/or add a modifier to increase the too low parameter. If such a modification is not feasible, it is strongly recommended to look for another solvent. Otherwise you must rule out mixer-settlers or columns and look for centrifugal extractors.

There is a trade-off between these properties. For example, if the viscosity of both phases is high (>10 mPa·s), the minimum interfacial tension or the difference in densities should be at least double the minimum recommended value.

e. Rapid Kinetics. This is usually not a problem. All physical and most of the chemical extractants reach equilibrium within 1 minute—good enough for any practical purpose. Yet some extractants are slower, especially in hydrometallurgy. They may need up to 3 minutes to reach equilibrium and sometimes even more. For such extractants the equipment must be adjusted (e.g., by using multiple mixers in series). A slow reagent is not a no-go factor, but definitely a drawback.

f. Minimum Solubility in the Feed. The problem exists mainly with simple extractants, like alcohols, esters, and ethers. Yet, a volatile solvent having a solubility of even a few percentage points is not necessarily a no-go factor. The dissolved solvent is stripped or distilled from the raffinate. This costs money, but in the end may be cheaper than using more expensive and less efficient alternatives. For example, the IMI extraction processes for manufacturing of phosphoric acid and KNO_3 work profitably with an alcohol and introduce stripping of the solvent from the raffinate as a standard part of the process.

g. Chemical Stability Under Process Conditions. This is a very important, and sometimes overlooked, aspect. The solvent must stay in the system for years. Any decomposed solvent must be replaced, and this is expensive. Moreover, unless the solvent is volatile and is periodically distilled, all the decomposition by-products accumulate in the solvent, causing yet more damage (e.g., by creating surface phenomena).

This is the domain of the chemist. Long-term stability experiments should be done to learn about any possible decomposition. If decomposition is detected, satisfactory steps must be taken to prevent it. Such measures range from lowering the temperature of the process through nitrogen blanketing and addition of antioxidants up to adequate sidestream treatment to remove any accumulated species (see Section VII.D.3.d).

h. Be Commercially Available at Reasonable Price. What is a reasonable price for a solvent? It is a function of the expected solvent losses, the price of the product, the production rate (in terms of tons of product per ton of solvent per year) and the profitability of the process. There is no way to give even an approximate value to this parameter because it depends on the profitability of the entire process, in which the SX may be only one small section.

i. Be Safe and Environmentally Friendly. Again, it should be emphasized that a hazardous solvent is not necessarily a no-go factor. A plant for purification of phosphoric acid by the IMI method, using isopropyl ether as a solvent, has been working successfully for about 30 years, although the ether is known to be highly flammable and explosive. But of course a safe and nontoxic solvent should be preferred, if possible.

2. How to Start the Search?

As claimed before, it is the task of the chemist to choose and compose the right solvent.

> *Remember*
> The solvent is not only the extractant. It contains usually a diluent, and sometimes also a modifier. Their properties and relative amounts are as important as the choice of the right extractant.

Nobody starts developing an SX process from scratch. There is a great deal of available data about recommended solvents for various applications. One of the best single sources is the Handbook of Solvent Extraction, edited by Lo et al. [7]. For hydrometallurgy there is a comprehensive guide to solvents in the book Solvent Extraction by Ashbrook and Ritcey [4]. The full list of all extraction systems used until 1970 provided by Francis [8] is inconvenient to use but may give you some good ideas. A great deal of data appears in the proceedings of various ISEC conferences.

It is always a good idea to contact the major manufacturers of extractants: Cytec, Avecia (Acorga), Cognis (Henkel), Sherex, and so on. These vendors have good professional staff and usually are happy to help a potential customer, providing not only information but also a commercial sample of the recommended extractant for preliminary experiments. And of course there is the personal experience and expertise of the researchers involved in the project.

Usually more than one suitable solvent is found. Sometimes it is the same extractant but with different diluents or modifiers. In other cases we have different extractants to consider. They may be similar compounds, made by different manufacturers, but often two totally different solvents seem suitable, and the better one cannot be chosen from theoretical considerations alone. The comparison between the various solvents may be long and tedious, and yet sometimes they are so similar in bottom-line performance that the final choice is based on nonprofessional considerations like the reliability of the manufacturer or the smell or color of the compound.

3. Concentration of the Extractant

For physical extraction, the solvent consists of one or two simple components, such as alcohols, ethers, esters, aliphatic, or aromatic hydrocarbons, and sometimes a diluent.

For chemical extraction the solvent consists of an extractant that is usually diluted with a diluent, typically a petrochemical solvent, like kerosene. Sometimes it contains also a modifier, usually a high alcohol, which is added to avoid the creation of the third phase (see Section III.D.3) or to improve the properties of the solvent.

The concentration of the extractant may vary typically from as low as 2% up to 70%. At the lower end, the concentration of the extractant is a function of the concentration of the solute in the feed and the O:A phase ratio:

$$C_e = \frac{x_{\text{feed}} 100 a}{(y_{\text{max}} R)} \qquad (3)$$

where C_e is the concentration of the extractant in the solvent (vol%), x_{feed} the maximum concentration of the solute in the feed, y_{max} the maximum concentration of the solute in the extract, R the volumetric phase ratio (O:A), and a the safety factor.

The value of a is usually about 1.1 (lower when the recovery is not critical, and higher if sharp fluctuations may bring x_{feed} above its maximum design value or if higher concentration of the solute in the raffinate creates environmental or safety hazards). A higher concentration of the extractant ensures full extraction, but on the other hand increases the cost of the solvent. At the high end, the concentration of the extractant depends on the physical properties of the extractant (viscosity, density, interfacial tension, solubility in the diluent and the feed, etc.).

For chemical extraction, in the initial experiments check just one concentration of the extractant. If there are no clues in the literature regarding the concentration, use 10% for dilute feeds (<2% of solute) and 30% for a more concentrated one.

III. LABORATORY DEVELOPMENT OF AN SX PROCESS

When the analysis of the process is finished and the potential solvents have been chosen, the laboratory work begins.

A. Limiting Conditions

By *limiting conditions* we refer to an experiment in which the *loading phase* is in large excess, so that after contact with the *receiving phase*, its composition hardly changes. For extraction, the feed (or pregnant liquor, as it is called in hydrometallurgy) is the loading phase and the washed (or barren) solvent is the receiving phase. For stripping/washing and scrubbing/purification the solvent is the loading phase.

The results of the limiting conditions experiments are similar to the maximum concentration in the *receiving phase*, which can be obtained in a multistage plant, under the correct phase ratio. For every battery, the experiments may go both ways. To find the maximum possible loading, purification/scrubbing or product concentration, the *loading phase* must be in large excess. To find the maximum recovery, from either the feed, the extract, or the contaminated solvent, the receiving phase must be in excess.

It goes without saying that a large amount of aqueous solution can be prepared much more easily and accurately than a large amount of loaded solvent. For this reason, the aqueous solution should be used as the loading phase whenever possible. For the washing battery, this procedure may be applied *only* if you are sure that the washing is possible (i.e., that the solute can be reextracted from the solvent by using water).

What phase ratio should be used for a limiting conditions experiment? To comply with the definition, the amount of the loading phase should be 5–10 times higher than expected in the real multistage process. In this early stage of development we usually have no idea what the phase ratio should be (otherwise this entire step would be unnecessary). So, a phase ratio of 10:1 is recommended for extraction and reextraction, and 1:1 for purification/scrubbing.

When the initial experiments are done, we know the approximate value of the distribution coefficient D_{lc}. For further limiting condition experiments the phase ratio should be $10/D_{lc}$. The mass transfer experiments should be carried out, under limiting conditions, for every battery in the process.

B. Experimental Equipment

The equipment used for the experimental work listed in the sections that follow is simple and, if not available in the laboratory, may be purchased at low cost. The experiments may be done either in separating funnels or in an agitated vessel. If a temperature other than ambient is to be considered, a thermostatic bath is needed for both cases.

The *separating funnel* is a stand-alone unit; 250 mL funnels are most convenient, but any funnel between 100 and 1000 mL may be used, according to the tested phase ratio, availability of raw materials, and the size of samples for analysis.

The *agitated vessel* is a graduated cylindrical vessel that houses a mechanical agitator. The agitator is connected to a variable speed motor, and the speed of rotation is measured by a tachometer. The vessel is immersed in a thermostatic bath. Inside the vessel there is a thermometer and some baffles, to ensure uniform mixing.

For screening experiments and for equilibrium isotherms the vessel may be of any shape and size. But to compare the results of kinetic or hydrodynamic experiments with those of other researchers, one should use some a standard piece of equipment (see Section V.A). Experiments in separating funnels are usually much more rapid and easier. For the initial screening of solvents (Section III.D), where accuracy is not essential, separating funnels are recommended.

For preparation of equilibrium isotherms, the separating funnel and agitated vessel methods are equivalent.

For the study of hydrodynamic properties, experiments in separating funnels are less accurate (especially in nonambient temperatures are to be considered) and more difficult to scale up (see Section V.A).

C. Analytical Methods

Analytical methods usually are not developed by process engineers, and the process is beyond the scope of this chapter. Yet a few points must be taken into consideration prior to any laboratory work.

    1. *Are there reliable analytical methods for all the important species in both phases?* Often the concentration can be analyzed only in one phase (in extraction of ions it is usually the aqueous; in extraction of organic molecules, the solvent). This is not convenient, but it is not a no-go factor: if high accuracy is not critical, we can calculate the concentration in the second phase by difference, provided phase volume changes may be neglected. If the residual concentration we are looking for is very low—that is, if most of the material was extracted or stripped, and we are looking for the residue—calculation by difference introduces a very large error and is useless. If the phase, which can be analyzed, is the major one, the calculations may introduce a huge error as well. For this case the solutes should be back-extracted to the phase from which they can be analyzed and sent to analysis. Reextraction and analysis take much time, easily doubling the total amount of work invested in the equilibrium isotherms.

At phase ratios near 1:1, in the preliminary stage, you may analyze one phase only. In the construction of exact equilibrium isotherms for simulation there is no choice but to do the full reextraction and analytical work for both phases. In limiting condition experiments there is no way to calculate the loading, and the reextraction is a must.

    2. *What is the accuracy of the analytical method and is it sufficient?* There are two aspects of the analytical accuracy: the *relative error* and the *detection limit*.

        a. *The relative error* is usually less critical, since even an average cumulative error of 10% (including errors in sampling, weighing, and analysis) enables us to find the optimal solvent and even to carry out a reliable simulation (see Section IV).

b. If the *limit of detection* is above the maximum allowed concentration of the species in the product or the raffinate, a suitable analytical method must be developed prior to any experimental work.

When designing your experimental work, bear in mind that more than 50% of your manpower resources are used for analytical work. This means that the right procedure for sampling is the single most important parameter in your experimental work design.

D. Screening of the Solvents

Initially, a potential solvent should be tested for loading, phase separation, and third-phase formation.

1. Loading

a. Extraction. Take 200 mL of the feed and 20 mL of solvent, and shake vigorously for 5 min (or 300 shakes, if you prefer counting to looking at a stopwatch). Analyze the extracted feed and the extract for all species that are present in the feed and may be important, either as products or as contaminants.

This is a limiting conditions experiment. The final concentration in the solvent is similar to the maximum concentration that can be obtained in a multistage plant.

Now calculate the distribution coefficient at limiting conditions, for every component:

$$D_{lc,i} = \frac{y_{lc,i}}{x_{lc,i}} \tag{4}$$

The selectivity of the solvent is calculated as follows:

$$S_{pi} = D_{lc,p}/D_{lc,i} \tag{5}$$

where i is the index of any component, and p stands for the product. The *minimum* estimated phase ratio in the battery is calculated from the D_{lc} value of the product:

$$R_{min} = \frac{1}{D_{lc,p}} \tag{6}$$

In many processes, especially in hydrometallurgy, the pH can have a significant influence on the extraction. This is the right time to evaluate it. Make your limiting condition experiments over the entire relevant range of pH. Usually three or four experiments will give you an adequate idea of the influence of the pH. Use only the best result for further analysis of the performance of the solvent.

If $D_{lc,p}$ or $S_{lc,i}$ is too low, go to the next solvent.

b.   Stripping.   If the process requires stripping and the results of the extraction seem acceptable, prepare 50 mL of extract, using the procedure just described. (If you are an optimist, and have plenty of solvent, make the limiting condition extraction experiment in a 1000 mL funnel, with 70 mL solvent and 700 mL feed. Use 20 mL of extract for analytical purposes, and the rest for stripping experiments.)

If you assume that *washing* (i.e., stripping with water) should be feasible, put 20 mL of extract and 200 mL of water in a separating funnel and shake for 5 min. After phase separation, send a sample of the aqueous phase to be analyzed.

If the washing is poor, repeat the experiment using hot water.

If you obtain reasonable results, calculate the differential loading. If not, repeat the experiment with a stripping agent.

If stripping is out of question (e.g., in the extraction of salts, acids, or bases), go to the next solvent.

c.   Purification/Scrubbing.   You already know the selectivity of your solvent between the product and various contaminants, under the limiting conditions. In this early stage of screening, it should be enough to decide whether to go on with the solvent or to reject it. Yet, if the purification is of critical importance, you may test it by shaking 50 mL of the extract and 50 mL of the final product or the scrubbing solution in a separating funnel for 5 min. After phase separation, send a sample of the solvent to be analyzed.

Remark

A 1:1 phase ratio is the limiting condition, for scrubbing, since the flow of the scrubbing solution is usually very low.

2.   Phase Separation

As a matter of fact you have done already the phase separation experiment, while testing the loading under limiting conditions.

If the phase separation was good, you do not have to check it further. Even if it was bad, however, a chance remains that the solvent can be used. Under limiting conditions you could have overloaded the solvent—this is a well-known phenomenon in extraction of phosphoric acid and of some organic compounds with high density.

First, find the density of both phases *before loading*. If the difference between them is smaller than 0.05 g/mL, try to increase it by changing the composition of the solvent (e.g., by changing the type or concentration of the diluent). If this does not help (e.g., in the case of a simple one-component solvent like toluene or pentyl alcohol), forget the unsatisfactory solvent and go on to the next one.

If the density difference seems satisfactory, put similar volumes of the solvent and the feed solution in a separating funnel and shake for 5 min. Watch the phase separation. If the phase separation was reasonable, go on. If it was problematic, check the viscosity of both phases. If the feed is viscous, all you can do is apply heat, and this often turns out to be the right solution. In the case of viscous solvent, a low-viscosity diluent and/or viscosity-breaking modifier should be added to the extractant. Heating of the system lowers the viscosity very efficiently: on the average, heating by 10°C decreases the viscosity by 50%.

If the viscosity is low, measure the interfacial tension. If it is low (<2 mN/m), you cannot expect good phase separation. Try to add a modifier or change the diluent. If nothing helps, go to the next extractant. You will come back to the first one only if no alternative is found.

3. Third-Phase Formation

During the study of phase separation, watch carefully to see whether a *third phase* is formed. The phenomenon of three immiscible liquid phases is rather common in extraction from an aqueous feed to a solvent. The organic compound or complex that is created by the interaction between the extractant and the solute may be too polar to dissolve in the diluent or unloaded solvent, but not polar enough to be dissolved in the aqueous phase. In this case, the light solvent phase contains mainly the diluent, with small amount of the extractant. The third phase contains mainly the loaded extractant, with small amount of the diluent, in the case of a composite solvent or the loaded solvent in the case of a simple one. As Ashbrook and Ritcey [4] put it, "the third phase should be avoided like the plague." If a third phase occurs, add some modifier to the solvent. High alcohols (from octanol and above) and phenols (e.g., nonyl phenol) are generally used as modifiers. The concentration of a modifier varies between 5 and 25% of the volume of the phase.

When the solvents have been screened, you analyze the results and try to decide which of the solvents seems best. In this stage it is quite normal for no candidate to be the clear winner. If this is the case, continue to the next stage of experiments with more than one solvent.

IV. PROCESS SIMULATION

A. Liquid–Liquid Phase Equilibrium Data (Equilibrium Isotherms)

The preliminary tests proved which solvents are apparently suitable for your process. Now it is time to quantify these results, to find the best solvent for the application.

Remember that *for a composite solvent, the change of diluent, the addition of different modifier, or the choice of a different extractant concentration defines a new solvent.* In the screening stage, it is enough to carry out the experiments at minimal and maximal expected concentrations of the extractant, and interpolate between them.

At this point there are two possible routes to continue: you may either determine the kinetic and hydraulic data (Section V) before the simulation, or make the equilibrium isotherms and simulation first. The physical experiments are rather rapid, and any solvents that can be ruled out owing to slow kinetics/diffusion/phase disengagement need not be used for the simulations.

The order recommended in this chapter is based on the assumption that if the experiments of Section V can rule out a solvent, it probably should have been eliminated earlier by the screening experiments (Section III) because of either unsatisfactory loading due to bad mass transfer (caused by either kinetics or diffusion) or problematic phase disengagement. The preferred solvent is chosen by using simulation (see Section IV.B), and for this we need to know the distribution coefficient D for every component that is to be extracted.

For a common problem (e.g., extraction of phenol from an aqueous solution or copper from a sulfuric acid leach into some standard extractants), there are equilibrium data in the literature. Extractant manufacturers will supply not only the data, but even a primary simulation, giving the approximate phase ratio, number of theoretical stages, and yield. In the case of simple physical extraction, where a thermodynamic model can be applied, one should refer to the data banks of commercial simulators or try to calculate the distribution coefficient D using one of the thermodynamic models described in detail in the literature (see Ref. 3 or 9).

In many cases, however, the available data do not help or are not sufficiently good. If our system cannot be described accurately by a liquid–liquid equilibrium model on the one hand, and no available data describe it well enough on the other, there is no choice but to generate the experimental equilibrium data and derive from them the distribution coefficient D

The experiments are simple, but time-consuming. The equipment is simple, too: separating funnels or mixer vessels, and if temperature is to be considered, a thermostatic bath as well. Most of the time is spent on the analytical work.

THE EQUAL SPACE NET. The most accurate—and expensive—way to generate equilibrium data is the *equal space net.* In this method, first we determine the number of parameters to be considered and their extreme values. The range of each parameter is divided into a few sections, at constant distance. The number of the sections depends on the purpose of the experiments, the importance of the parameter, and the available resources. For each value of the parameter, the entire range of experiments for all the other parameters must be done.

In this method, the number of experiments N is large:

$$N = \prod_{i=1}^{m} n_i \tag{7}$$

where N is the total number of experiments, n the number of experiments for each parameter, and m the number of parameters.

This method is the most accurate and sound from the statistical point of view because it gives the same weight to every part of the multidimensional space.

If we have just one extractable solute in the feed, the composition of which is otherwise constant, the pH is known or irrelevant, and the temperature of the process is constant, then even the most accurate equilibrium isotherms (e.g., 15 points) can be done rapidly. Even if we have to consider the temperature *or* the pH, and make the experiments at 3 or 4 pH/temperature values and 10 concentrations, this procedure is reasonable and no alternatives are recommended.

But if the concentrations of two or three components at various pH values and temperatures are to be considered, the equal space method can be hardly considered effective any more. For these cases, a statistically based experimental design should be used. In these methods, a relatively small number of experiments are carried out. Analysis of the results enables us to identify the important effects and interactions and to derive the correlations for the distribution coefficient. There exist some good software programs for experimental design (e.g., Design-Ease, Design-Expert, StatGraphics).

1. Experimental Methods

There are two ways for generation of the equilibrium data: the limiting condition method and the variable phase ratio method.

a. The Limiting Condition Method. The principle of the limiting condition method is described in Section III.A. In this method, the feed solutions, at the desired compositions, are prepared in the laboratory. For convenience, always do the experiments by loading a solvent phase, *not vice versa*. The phase ratio should be 5–20 times higher than D_{lc} (the value of which you have already found).

The limiting conditions method is recommended whenever possible. It assures that the equilibrium points are exactly where you wish, whether you are working in the equal space net method or using a statistically based algorithm. It cannot be used, however, when the feed is a product of an earlier stage in the process and contains various contaminants. These might affect the phase equilibrium and cannot be reproduced reliably in a synthetic solution (or, the reproduction may take too many resources). This restriction may apply to leach liquors in hydrometallurgy, distillate cuts in petrochemistry, or reaction products in either the organic or inorganic chemical industry.

If it is known that no more than two components are significantly extracted, the equal space net can yet be used: a semisynthetic feed solution is prepared by enriching the *raffinate* of the examined process with the solute components, up to the desired concentration.

If a real raffinate is not available, it can be obtained by extraction of the real feed under limiting conditions. But this requires a large amount of solvent: for every limiting condition experiment, you need enough solvent to enable a duplicate analysis to be made. Its volume depends on your analytical procedure, but on the average takes about 50 mL solution per point, that is, about 500 mL of feed (assuming a 1:10 phase ratio). For 10 points—the minimum for a one-solute isothermal case—you need 5 L of feed solution. For 50 points— two solutes or one solute at various pH and/or temperature values—you need about 25 L of the feed; that is, you must prepare 25 L of raffinate, using about 250 L of solvent for this pilot-scale operation. If this is not feasible, the limiting condition method cannot be used.

Before you start working, analyze two or three feed solutions on a random basis, to be sure that the solutions were prepared correctly. If the results are wrong, find the reason and prepare the solutions once more. If correct, do the experiments. Send for analysis only solutions at equilibrium. The composition of the feed solutions is now immaterial; there is no point in analyzing it further. Always remember that there are two time-consuming steps: preparation of the solutions and their analysis. The experiments themselves are rapid.

b. The Variable Phase Ratio Method. This method, recommended by Ashbrook and Ritcey [4], is much simpler and more rapid. The feed (either the real solution or an artificial one) is equilibrated with the washed solvent at various phase ratios, typically between 1:10 and 10:1, or, to be more precise, between $1/10 D_{lc}$ and $10/D_{lc}$. Usually 7–15 experiments are made. We start at the lowest solvent-to-feed ratio and nearly double it for every subsequent experiment. A reasonable set of experiments for $D_{lc} \sim 1$ is 1:10, 1:5, 1:3, 1:2, 1:1, 2:1, 3:1, 5:1, 10:1.

If the equilibrium data for very low concentration are important or if D_{lc} is well below 1, then instead of working with experimentally impractical phase ratios, like 1:100, the raffinate from earlier experiments can be used as a feed for the low concentration end.

If the influence of temperature and/or pH is important, a set of experiments should be done *for each temperature and pH to be considered.*

Remark

For a process in which the pH is not to be controlled (i.e., a process that changes along the operating line), the method is adequate. If the pH is to be constant throughout the process, it must be corrected in every sample. In this case the limiting conditions method may prove to be more rapid and more convenient.

Such a wide phase ratio range gives us the D value for every possible concentration in the process. It is far more rapid than using the limiting conditions, since we do not have to prepare the feed solution, a time-consuming job that invites experimental errors. Two drawbacks are associated with the faster process, however:

1.   Since the points are not set at equal distances, part of the space is covered more densely than the rest of it. This can be easily corrected: after receiving the results, add the missing experiments, choosing the phase ratio so that finally you get an approximately equal space net.

2.   All the points are on the operating line. In many cases that is not a problem. If your goal is to find the best solvent for extraction of a single species from a well-defined feed, then generation and correlation of the data along the operating line is the right thing, and any possible phenomena beyond this line are of no practical concern. However, if you are to find the influence of the change in the concentration of *various components* on the extraction of a solute or on the selectivity of the solvent, then collection of data along an operating line is useless. Moreover, it is harmful, because the simultaneous changes in the concentrations may cause coupling of effects, and so mask the real interactions in the system. For such a case, the variable phase ratio method cannot be used.

Note

1.   This step is the most time-consuming of all your preliminary work. If you are dealing with a multicomponent extraction and still have a few solvents to choose from, you may start with a preliminary database: check two temperatures at most and use the variable phase ratio method —for screening of solvents, it is accurate enough. Once you have decided upon your solvent, a full comprehensive database should be prepared to enable reliable optimization of the flow sheet.

2.   Many experiments may be needed to find the optimal pH value. So, the influence of pH should always be checked, as a stand-alone parameter, for the maximal and minimal concentrations and temperatures only. Once the optimum pH has been found, the entire comprehensive database should be prepared at this pH. If fluctuations in the pH are expected, two or three additional values may be incorporated in the database, to predict the influence of deviations in pH on plant performance. Note that doing this will double or triple the cost of the preparation of the database, unless you turn to a statistical method.

For a single solute the total amount of work in preparation of the database is small, so there is no point in preparing a simplified one for the screening stage.

2. Experimental Procedure

The experimental procedure is identical for both methods.

a. Experiments at Ambient Temperatures. Place in a separating funnel the calculated amounts of both phases, and shake manually or using a shaker for 2–10 min. Alternatively, the solutions are placed in a *closed* agitated vessel and stirred for 2–10 min. The duration of the experiment is set according to the kinetics and diffusion rate of the process. If you have no previous information about it, make a few kinetic experiments (see Section V.A), or allow 10 min for every experiment.

Allow 5–15 min for phase separation (according to the phase separation ratio, which was found during the preliminary experiments).

If the concentration of the solute in the solvent is known to be lower than in the feed ($D < 1$), centrifugation of the resulting extract is needed to get rid of the aqueous entrainment. For $D > 1$, if the phase separation was carried out carefully, and 10–20% of the extract adjacent to the aqueous phase was drained, then centrifugation is not necessary (although it is always considered to be good experimental practice).

For every experiment you should also determine the densities of both phases. You need these data to establish the correlation for specific gravity.

Remark

If the pH has to be kept constant, the use of a mixer vessel, where the pH can be monitored and corrected continuously, is recommended. The control of the pH in a separating funnel is possible, but inconvenient and more time-consuming.

b. Experiments at Various Temperatures. The most accurate way is to carry out all the experiments in a mixer vessel with a thermostatic bath, set to the desired temperature. The containers with both solutions should be kept in the same bath. The procedure is the same as the preceding one. For separating funnels, the experiments may be done in a thermostatic bath, using a dipped shaker. Alternatively, both solutions should be heated in a thermostatic bath at a temperature 2–3°C higher than the desired one, placed in a separating funnel, shaken at ambient temperature, and then put back in the thermostatic bath. After a few minutes, the funnel is shaken again, outside the bath, and returned to the bath for phase separation. This procedure is repeated two or three times. It is reliable as long as the process temperature is no more than 30°C above or below the ambient temperature.

Remarks

1. At high temperatures, partial evaporation of the diluent or of dissolved gases may occur, especially if NH_3 or HCl is present. Such evaporation may cause the stopper of the funnel to pop out. Remember to release the pressure after every few shakes. If a high concentration of NH_3 is present, the experiment should be done in a vented mixer vessel only.

2. To avoid evaporation of volatile components, never do the experiments in an open mixer vessel.

B.   Building a Correlation for *D* and ρ

Now you have enough data to simulate your plant and to set the optimal flow sheet for each solvent.

The first stage in simulation is to use a computer to build the correlations of the distribution coefficient for every component, and of the density ρ (sometimes given as *S.G.* for specific gravity) of both phases. If a simple correlation is expected (e.g., *D* = const or *D* = *a* + *bx*), this can be done immediately using Excel's Draw function. However, since the functions tend to be nonlinear, a nonlinear regression may be needed.

1.   Units

If you are using a simulator, your correlations must be in units (usually moles per liter, mole fraction, etc.) that are consistent with the simulator's algorithm.

Note

Most simulators enable you to input the data as weight percent or grams per liter, but they carry out their calculations in molar units (mol/L). In such a case your correlations must be in molar units too. If you are using a graphical method, you may prepare your equilibrium graphs in any consistent set of units.

2.   The Distribution Coefficient *D*

It is impossible to predict the equation for *D* for a general case. It is a function of the concentrations of all the solutes, the pH, and the temperature. If enough data are available, we may incorporate into it also the concentration of the extractant and/or modifier.

In the simplest case, *D* may be a constant at a constant temperature. For example, the *D* value for extraction of sulfuric acid, at a given temperature, by Cyanex 923 (a mixture of four alkyl amines), was found by Rickelton [10] to be nearly a constant at constant temperature over the entire range of concentrations of the acid. On the other hand, the *D* curve for extraction of the same sulfuric acid by TEHA, with octanol as modifier, takes on an S shape, which requires a complicated function to describe properly. The correlation, developed for this process by Gottlibsen et al. [5], has four parameters for the isothermal case:

$$y = a_1 + \frac{a_2}{(1 + x/a_3)^{a_4}}$$

To express y (or D) also as a function of T and the concentration of the modifier, the parameters a_3 and a_4 were expanded, and the final equation has nine parameters:

$$a_3 = a_5 T + (a_6 + a_7 T)(a_8 \cdot Oct + Oct^2)$$

$$a_4 = a_9 + a10 \cdot Oct + a_{11} Oct^2$$

where Oct is the concentration of the modifier (vol%) and y and x are the concentrations (mol/kg) of H_2SO_4 in the solvent and the aqueous phase, respectively.

A model to describe the use of a composite solvent to extract four major ions (Ca, Mg, Cl, and Br) and water from Dead Sea brine was developed by Kogan [11]. The implicit function, based on it, also uses nine parameters, but it takes 70 (!) lines of Fortran code to express. The methodology for preparation of the correlations requires a good understanding of regression techniques and is beyond the scope of this chapter. Yet, the following points should be carried in mind:

a.   It is most important that the experimental data cover the entire range of concentrations that are feasible for the process, including a reasonable safety factor. *There is nothing more misleading than extrapolation of experimental data.*

b.   Correlations based on a physical (or chemical) model are usually more accurate than arbitrary functions.

c.   The maximum average error for a correlation should not exceed 10%; otherwise, the accuracy of the simulation may not be sufficient for the design of the final plant.

d.   If a few data points show strong deviations from the recommended function, while the others show a reasonable fit, recheck them. There may be an experimental or analytical error. Rechecking is often the best way to find wrong experimental points.

3. The Density

Correlation for the density ρ of any stream is usually a simple function of the concentration of the solute, x. Its most detailed shape is:

$$\rho = (\rho_0 + \Sigma a_i x_i + \Sigma b_i x_i^2)[1 - \alpha(T - T_0)] \tag{8}$$

where α is the isobaric thermal expansion coefficient and ρ_0 is the density of the *fully depleted feed or fresh solvent* at the reference temperature T_0 (usually 25°C, but any other temperature may be selected as well).

Notes

    1.   The correlation is valid for any consistent set of units both for concentration and temperature.

    2.   For an aqueous feed, which contains nonextractable species, ρ_0 is the density of the *raffinate* at limiting conditions. There is no need to include in the equation the concentration of the solutes that are not extracted, unless the extraction of the desired solutes causes this value to tend to vary significantly.

    3.   For most practical purposes, the quadratic term may be omitted. It is needed only if a very high change in the density of a phase (>0.3 g/cm³) occurs.

The value of α can be found in the literature for many solvents. For most organic compounds it is about 1×10^{-3}/°C, that is, a drop of 0.1% in the density for heating by 1°C. For water this value is much lower—about 0.3×10^{-3}/°C at 30°C. For brines it goes up to 0.5×10^{-3}/°C. This means that heating by 30°C decreases the density of an average solvent by 3%, of water by 1%, and of concentrated brine by 1.5%. This behavior explains why heating may improve phase separation when the solvent is the light phase. A good source for these data is the CRC Handbook of Chemistry and Physics [12]. If you cannot find the coefficient for a specific solvent, try to interpolate from the values of related compounds in the literature rather than developing a coefficient experimentally. Even if you have the technical ability to measure a derived coefficient with high accuracy (and it is not easy), the additional accuracy of your simulation will be negligible.

C. Process Simulation

The simulation is used to design the flow sheet (i.e., to determine the number of stages in each battery as a function of the phase ratio), and it predicts the influences of temperature and variations in feed composition. The results of the simulation enable us to rule out inferior solvents and to optimize the flow sheet for the remaining ones.

    From the moment you have it, the simulator is going to follow you along all the remainder of the development process. Once you have your distribution coefficient correlation prepared, the use of the simulator is rapid and accurate. Never hesitate to use it.

If you cannot get a computerized simulator, you can use the McCabe–Thiele graphical method for a single solute. This method is not very accurate and also not very convenient, but it is better than nothing. For efficient multicomponent extraction, the computerized simulator is almost a must.

1. Steady State Process Simulators

The standard steady state simulator calculates the flow rates and compositions of output streams for a given plant configuration and operating conditions: number of stages and location, flow rate and composition of all the input streams, process temperature, and pH. In good simulators the expected entrainment and stage efficiency for each stage may be added. Using parametric runs, the simulation enables us to design the flow sheet, that is, for given solvent and recovery, to determine the number of stages in each battery as a function of the phase ratio, and to predict the influence of pH, temperature, and variations in feed composition. For a given solvent, it points out the optimal pH and temperature. If several solvents are considered, the parametric runs help us to pick the best one.

The steady state simulators calculate the mass and energy balances either stagewise or for continuous contactors. The stagewise approach that calculates the plant stage by stage is most suitable for a mixer-settler plant. It may be applied successfully also to columns, assuming that we can estimate the height of a transfer unit. In this case the number of stages stands for the number of transfer units (NTU). These programs are available in most commercial process simulators (ASPEN, HISYS, PRO/II, CHEMCAD, etc.). Still, these modules are usually inferior to the modules for distillation on the same simulators. They do not enable us to assign a temperature and entrainment to every stage, and the insertion of user-supplied thermodynamic correlations (which in distillation is quite rare, but in extraction is a must) is not straightforward. For this reason, many companies that deal with extraction—research institutes like IMI (TAMI), equipment and engineering companies like Bateman and Koch, and solvent manufacturers like Avecia—have developed their own simulators.

The continuous simulators were developed for columns, where the change of concentrations is continuous and not by discrete steps. Usually a specific program for each type of column is needed. Their biggest advantage is the ability to calculate the true height of the column, not as a multiple of transfer units. Their disadvantage is that some of the parameters required for the calculations may be difficult to establish. One of the best simulators of this type was developed by Pratt for perforated plate pulsed columns [13].

The biggest investment in the simulation (after acquiring the simulator and learning how to use it) is to prepare the equilibrium isotherms. After this has been done, optimization of the flow sheet and testing of any configuration are rapid and easy. A trained engineer can readily perform 100 parametric runs per day and optimize the most complicated flow sheet *for a given solvent* in a few days. Choosing the optimal solvents for a multicomponent extraction, and designing the optimal flow sheet for the chosen solvent(s) should take a few weeks—once all the equilibrium data have become available (see an example in Ref. 14).

The results of such an optimization are more accurate than any experimental procedure and usually need far less manpower. Sometimes the results of the simulation are quite unexpected. Once you have them, they seem perfectly logical; but without a simulation you might never have thought about them, at least not before a pilot plant was run.

Still, the simulation results tell us nothing about *five important factors*: kinetics, flux (i.e., the rate of phase separation), crud formation, accumulation phenomena, and solvent deterioration. These factors are determined later, for the optimal solvent(s) found by the simulation (see Section V).

2. Graphical Solution (McCabe–Thiele Method)

The McCabe–Thiele method enables us to find the number of stages for given recovery of a *single solute* in a single battery, at constant pH and temperature, using a valid graph of the distribution at equilibrium. If data for different solvents, temperatures, and pH values are available, their performance can be compared. Different phase ratios may be tested too, by changing the slope of the operating line. The amount of work needed for parametric runs is much higher than in the computerized simulation, and the accuracy is lower. This procedure may provide an adequate solution for simple cases but is not enough to optimize a plant.

3. Preliminary Simulation for Screening of Solvents

For the extraction of a single solute, the preliminary simulation stage is simple because you have to deal with only two batteries: extraction and stripping/ washing. This work may be completed in a few hours.

For extraction of two or more components, the procedure is more complicated because purification of the product often is involved, and various flow sheets may be considered. At this stage choose a standard configuration, and use the product as reflux. The correlations at this stage are based upon a simplified database, and therefore any simulation beyond the operating line may be misleading.

   a. For every solvent, find the number of stages needed to get the desired recovery in the extraction and, if applicable, also in the stripping and purification, at various phase ratios, through parametric runs.

b. Compare the results. The analysis is rather complicated (see Section VI.A), but it usually enables you to rule out most of the remaining options.

c. Make an initial guess and choose:

    i. *Phase ratio*: twice the minimum amount of the *receiving phase* found in limiting condition experiments: $R = 2/D_{lc}$ for extraction and $R = 0.5/D_{lc}$ for reextraction and purification.

    ii. *Number of stages*: 5 if $D_{lc} < 2$; otherwise, 3.

d. Find the number of stages needed to get the desired recovery and/or purity at this phase ratio.

e. Change the phase ratio and repeat step d.

       It is difficult to draw general guidelines for the recommended range of the parameters to be tested. Use your intuition, and if in doubt carry out a few extra runs—the runs are rapid and easy. Remember that for the current purpose you do not have to arrive at the optimum. You always can—and should—consult the simulator in later stages of the project.

    Note

    A phase ratio O:A < 1:7 is impractical (see Section VII.B.1.e). If you find that less solvent may be used, decrease the concentration of the extractant (if applicable). If the solvent is to be washed or stripped, use the 1:7 ratio anyway. Only if it is to be distilled decrease the phase ratio as indicated by the simulation.

f. Run the entire plant, using the local optimum for each battery from step e.

    Remark

    For a single solute this procedure may also be done using the graphical method.

4. Computerized Simulation for Optimization of the Flow Sheet

The procedure for simulation is similar to that for preliminary screening, but now the correlations are based on a comprehensive database. This is the time to test any nonstandard configurations, to get the optimal flow sheet. But remember a basic rule: ***Never put a reflux over a single battery.***

    For optimization you may need the results of the kinetic and hydrodynamic tests (see Section V), but if a few solvents seem feasible at this stage, be aware that probably you do not have enough data to choose the best one.

    For real optimization you must know the cost of the units in the industrial plant. This cost depends on the preliminary design of the industrial plant, which in turn is based on the flux, the type of dispersion, and other parameters that can be obtained only after piloting (e.g., entrainment of the solvent in the aqueous phase, auxiliary operations).

At this stage you may find, from the simulation, the flow rate of every solvent needed to achieve the required production rate under all process constraints and the number of stages needed to achieve this. Of course we prefer a solvent that performs at minimum flow rate and in a minimum of stages. However, sometimes such a solvent may have severe drawbacks (e.g., high price, high toxicity, low flux, increased solvent losses due to higher solubility in the aqueous phase, lower stability). You may not have enough data to make the optimization. Use your best judgment and consult your client. But if in doubt, go on with more than one solvent.

For every solvent, try to find the optimum flow sheet of the plant: number of stages, phase ratio, setting of reflux streams, and so on. This is not trivial, either. You do not yet know the cost of an extra stage (either a mixer-settler or increased height of the column) or an increased diameter of the units. So it is difficult to decide whether it would be better to decrease the phase ratio and increase the number of stages or vice versa.

Usually, if the phase ratio O:A is below 1:3, we try to save stages and increase the phase ratio. The increase in the total flow will be rather small. On the other hand, for O:A above 3:1, try to decrease the phase ratio. The savings in total flow rate and solvent inventory may be substantial. As usual, *use your intuition.* If an additional stage saves 20% of total flows, add it—but if only 2%, forget it.

Do not, however, worry too much: this important point is not critical at this stage. Choose a reasonable design for your pilot plant, and get the maximum performance out of it (see Section VIII). Once the pilot is over, you will have, as a fringe benefit, validation of your simulator and all the necessary data for design. Now the full optimization, using all the data from preliminary design, may be done. Usually this is done together with potential equipment manufacturers, who will fill in the missing figures, and so reveal the optimum design.

Probably this will not be your responsibility any more. A project engineer will take over at this stage, keeping you as a consultant.

V. PHYSICAL, HYDRODYNAMIC, AND KINETIC DATA

By now you have ruled out most of your options. You have typically one or two solvents to consider, and maybe have still to determine the concentration of the extractant and the choice of the diluent or the modifier. But you have a great deal of relevant data. You watched the phase separation in your equilibrium experiments and probably have some feeling for how the setup works at various concentrations. You also know the density of both phases, over the entire range of concentrations. Furthermore, during the equilibrium experiments you necessarily learned something about the kinetics.

Now we are going to generate the physical, hydrodynamic, and kinetic data systematically. If you have not decided yet about the final composition of your solvent, these experiments may help you to determine which extractant to use and at what concentration, and which diluent and/or modifier to add.

But even if the composition of the solvent is set, carry out all the experiments anyway. They are relatively rapid and simple, and will help you to design your pilot plant better. These data are also essential if you want to consult either the solvent manufacturer or a potential equipment supplier. With such data, you can compare the current process with studies you have made in the past or will encounter in the future.

A. Experimental Equipment

There are two basic ways of doing the experiment. As we saw earlier, in one we use an agitated vessel; in the other, a separatory funnel.

Hydrodynamically, experiments in a separatory funnel resemble more closely the behavior in a column, while an agitated vessel simulates a mixer-settler. The experiments in an agitated vessel are more accurate and reproducible, since all the relevant parameters may be fixed and measured accurately. They enable us to work with both types of dispersion, which in a separating funnel is between difficult and impossible. Hence, the experiments are often made in an agitated vessel, even if the process is considered for columns.

The comparison of the phase separation times for various solvents is meaningful only if full hydraulic similarity is maintained. This similarity requires that the following parameters be kept constant:

Geometry of the vessel (shape, diameter, total height of the liquid, setting of baffles)
Geometry of the impeller
Location of the impeller in the vessel
Speed of rotation of the agitator
Type of dispersion (O/W or W/O)
Phase ratio
Duration of mixing
Temperature

Because the pH may influence both phase separation and mass transfer, remember to control the pH during your experiments. It is quite easy to keep this similarity within a given laboratory. All we have to do is to use a standard type of vessel or beaker and the same impeller, and to keep the rest of the parameters constant. To be on the safe side, prepare a few identical impellers and keep them dedicated for this type of experiment. But the resulting data cannot be compared with those from any experiment that was carried out at a different vessel. This makes interlaboratory comparisons practically impossible.

Manufacturers of extractants are aware of this problem and issue detailed instructions concerning the vessel in which the experiments that are to be reported to them should be done. Cognis even offers its customers a standard test vessel and an impeller, free of charge.

The one-liter vessel described here is based on recommendation of Acorga (Product leaflet 310-E) and on IMI (TAMI) experimental procedure. It is a jacketed, cylindrical, flat-bottomed glass flask, 10 cm in internal diameter and at least 14 cm deep. It is equipped with a bottom outlet with a tap, so that liquids can be run off from the bottom. The outlet pipe and valve must be wide enough to drain the content of the one-liter flask to a graduated cylinder in 5–10 s. The vessel has a ground glass flange to take a suitable lid with inlet ports to accommodate a stirrer, thermometer, and sample addition tube. It is further equipped with a removable stainless steel or Teflon baffle comprising four vertical plates 10 mm wide and equidistant from each other. The vessel is water-jacketed, water being pumped from a thermostatic bath. A vertical centimeter scale is inscribed on the outside of the jacket.

On the other hand, many researchers in hydrometallurgy use a cubic vessel. This is more similar to the rectangular mixer in the mining industry, easy to prepare (usually from transparent plastic), and because of its right angles, it does not need baffles. The standard test vessel of Cognis, for the LIX™ reagents, is a one-liter square PVC box, $8.9 \times 8.9 \times 15.4 \ cm^3$. It is not jacketed and has no bottom outlet.

The results from both types are equally good, but there is no way to translate results from one system to the other.

The impeller is based on both Cognis and Acorga recommendations. It should be made of stainless steel, polypropylene, or Teflon. Its diameter is 5 cm and thickness 3 mm, with six blades regularly spaced beneath a circular disk, and two spoiler blades fixed to the upper surface. The blades should be 3 mm high. The impeller is held in position by a Teflon gland in the lid of the vessel. It is driven at the required speed (600–2200 rpm) by a variable speed motor.

The location of the impeller depends on the phase ratio. It should be mounted so that the top of the disk is 10 mm below the surface level of the organic phase for O/W dispersion and 20 mm above it for W/O dispersion (assuming that the organic phase is the lighter one).

B. Physical Data

The following physical data are required *prior* to your experiments.

1. Determine the density and viscosity of the feed, raffinate, product, strip solution, and purification/scrubbing liquor. The composition of the aqueous streams is usually independent of the chosen solvent.

2. *For every solvent,* determine the density and viscosity of the extract and washed solvent, and if applicable, also of the purified extract.

The viscosity should be measured at two or three relevant temperatures. For the density, you should have already all data, set as a correlation. If not, measurement at the expected process temperature is sufficient (see Section IV.A).

The difference in density should be calculated between all interacting pairs: feed–extract, washed solvent–raffinate, washed solvent–strip solution (or wash water), and purified extract–product.

3. Measure the interfacial tension between the interacting pairs of solutions at process temperature. If the process will be carried out below 50°C, this may be determined at ambient temperature just as well.

Remark

For composite solvents, steps 2 and 3 should be repeated for *various diluents, modifiers, and different concentrations of the extractant.* Usually it is enough to interpolate between all the measured properties at the minimal and maximal expected concentrations of the extractant and the modifier.

4. Compare the data among themselves and relative to other processes that you have encountered.

There is no exact way to tell the hydrodynamic behavior from the physical data. Still you get some key information (see Section III.D). You are looking for high interfacial tension and difference in density, and for low viscosity. Analysis of the results may help you to rule out a few of the combinations that you had in mind—for example, a high concentration of the extractant, which causes a too high viscosity or a too small $\Delta\rho$, or a diluent that causes either a small $\Delta\rho$ or low interfacial tension.

The lowest difference in density generally occurs between the following:

a. The strip solution (or wash water) and the loaded solvent (either extract or purified extract).

b. The extract and the raffinate.

These solutions usually are not in direct contact, so the condition of $\Delta\rho$ > 0.05 does not apply to them. Still, if the density of the extract is higher than that of the strip solution or the raffinate, this might be problematic, especially if the use of columns is considered. The following pairs of solutions are in direct contact and may have low differences of density:

a. The wash water (sometimes also the strip solution) and the washed solvent (at the dilute end of the washing battery)
b. The washed solvent and the raffinate (at the dilute end of the extraction battery)
c. The feed and the extract (at the concentrated end of the extraction battery)

For these solution, the condition of $\Delta\rho > 0.05$ applies directly.

5. Measure the solubility of the solvent in both the raffinate and the aqueous product (if applicable). For most commercial extractants and diluents the solubility in water is supplied by the manufacturer or is published in handbooks. You may assume that at a given temperature and pH, the solubility in any aqueous phase is *smaller* than in water. The solubility of most solvents is far below their entrainment in the aqueous phase. There are some commercial solvents, however, with significant solubility in water (e.g., alcohols such as butanol or iso-pentyl alcohol). High solubility of the solvent means that solvent recovery treatment is required.

Example 5
An organic acid is to be extracted from aqueous brine. The brine contains 3% acid, and its density before and after the extraction is about 1.07 g/cm³. The proposed extractant is toluene, and its density was found to be 0.84 + 0.010y g/cm³ as a function of the acid concentration y that is the weight percent of the acid in the solvent. The distribution coefficient of the acid, at limiting conditions, is 10. This means that the limiting condition phase O:W is 1:10, but at this phase ratio we get phase inversion—the solvent becomes heavier than the aqueous phase. What phase ratio is recommended for the process?

From the extraction correlation, we see $y_{max} = D_{lc}x_{feed} = 10 \times 3 = 30\%$ acid. However, the maximum allowed density of the solvent is $1.07 - 0.05 = 1.02$; that is, the density may rise (above that of pure toluene) by 0.18 g/cm³ at most. This corresponds to a loading of 18%. The recommended O:W phase ratio under which the concentration of the solute in the solvent does not exceed 18%, is 1:6 ($30/18 \times 10$) or less. At a higher phase ratio the phases will not disengage, or phase inversion will occur.

C. Hydrodynamic Data

The hydrodynamic data, which can easily be generated in the laboratory, are the rates of phase separation under various process conditions. There are two basic ways of performing the experiment: by using an agitated vessel and by using a separating funnel (see Section III.B).

1. Agitated Vessel

The vessel is filled with both solutions, at the desired phase ratio. The system is kept in a thermostatic bath, with very slow mixing, until it reaches the desired temperature. After a few minutes, for total phase disengagement, the mixer is started at the set speed and run for 5 min. Then the mixing is stopped, and the time necessary for the phases to separate is measured. The comparison of the phase separation rates for various solvents, as a function of temperature, type of dispersion, and speed of the mixer may be meaningful only if full hydraulic similarity is kept (see Section V.A). After the mixing is over, the vessel should be drained through the bottom tap into a measuring cylinder. This procedure has two main advantages:

1. In all the experiments the impeller is in exactly the same location, so no errors due to its mislocation may arise.
2. The volumes in a graduated cylinder can be measured far more accurately than in the mixing vessel.

A more serious problem is how to set the impeller so that the desired dispersion is obtained. The initial experiments should be carried out *at a phase ratio of 1:1*. Locating the impeller as indicated in Section V.A usually, but not always, will enable you to get the desired dispersion. If you get the same dispersion at both levels of the impeller, this means that this is the "natural" dispersion, and inverting the phases might be difficult. This is important information for the design stage. But if the "inverted" phase dispersion is to be investigated anyway, the phase that is supposed to be continuous should have a volume at least twice as large as that of the other one. The level of the mixer should be changed accordingly. In later stages, it is possible to check the phase separation also at the real phase ratio of the process.

2. Separating Funnel

The desired amount of both phases is placed in a separating funnel, which is shaken, either manually or by a shaker, for 5 min or 200 shakes. After the shaking, the time of phase separation is measured.

Usually the shaking is done manually because it is difficult to find a mechanical shaker having an operation sufficiently vigorous for this type of experiment. This procedure is sufficient for reliable screening of various solvents for a process, *as long as the same person does the shaking*. If ambient temperature is to be considered, the manual option is much more rapid, and its accuracy is adequate. For work at elevated temperature, both solutions should be heated to 5–10°C above the desired temperature, poured rapidly to the separating funnel, shaken as needed, and set in a thermostatic bath to separate. Again, for comparative experiments this is accurate enough.

Note

1. In a separating funnel, with a 1:1 phase ratio, you always obtain the natural dispersion. If "inverted" phase dispersion is to be investigated, the phase that is supposed to be continuous should be at least twice as large as the other one.

2. Both in an agitated vessel and in a separating funnel, always use fresh solvent and never recycle solutions to the next experiment. Tiny droplets, which remain in the separated continuous phase, may influence significantly the duration of phase separation.

D. Kinetics and Mass Transfer

The kinetics and mass transfer experiments are also semiquantitative, and their main purpose is to give us some basic idea about the required residence time in the system and the intensity of mixing. They should be done for every extractant, and when in doubt, as well as for various modifiers and diluents with different chemical characteristics (e.g., an aromatic vs an aliphatic diluent). There is no need to do them for similar diluents, and definitely not for various concentrations of the extractant in a diluent.

Since the influence of temperature on kinetics and diffusion is very significant, the experiments should always be carried out at the process temperature. If this temperature is not constant or not known, two or three temperatures should be tested.

If applicable, the behavior at both dispersions should be checked. While the kinetics is not affected by the type of dispersion, the diffusion may vary significantly.

1. Experimental Procedure

The procedure is similar to the phase separation experiments and can be carried out either in a mixer vessel or in a separating funnel, the considerations being similar to those in the preceding section. Since several samples are taken from the vessel during the experiment, the individual volume of the solution should be large enough to enable this sampling without disturbing the experiment.

As a rule, we will sample and analyze only one phase—the one in which the final concentration of the solute is expected to be lower; hence the changes in the loading are more significant and can more readily be detected. For example, in tracking of extraction from an aqueous phase to the solvent, we always sample the aqueous phase because the concentration of the solute there is expected to be lower.

The experiments should be carried out at a 1:1 phase ratio, except in the following cases.

1.   The phase ratio in the process is very high, so that working at a 1:1 ratio brings us to an unrealistic range of concentrations. For example, for a purification or scrubbing battery, where the expected O:A phase ratio is 30:1, working at 1:1 with the real aqueous solution will rapidly cause total purification, and no real conclusions may be drawn. In this case, you should carry out the experiment at an intermediate phase ratio (e.g., 10:1) and sample only the large phase, in this case the solvent.

2.   Checking the mass transfer in both dispersions. In a separating funnel, only the natural dispersion is obtained at a 1:1 phase ratio. Even in a mixer vessel, setting of the impeller in the desired phase does not guarantee the desired dispersion. If the "inverted" phase dispersion is to be investigated, the phase that is supposed to be continuous should be at least twice as large as the other one.

For the 1:1 phase ratio, the experiment may be carried out in the standard one-liter vessel, as described earlier, or in a 500–1000 mL separating funnel. Starting with 150–300 mL of each phase, we can draw 4–8 samples of 10–20 mL each (depending on the analytical procedure) and complete the experiment with an adequate amount of both liquids.

a.   Mixer Vessel.   The height of the impeller is set so that it is dipped in the phase that we want to be continuous, and its speed is set to the desired value before the liquids are entered (but not started). Both solutions are introduced at the desired temperature, so that interaction between them is minimal. the mixer is operated for 15 s and stopped. After phase separation we use a pipette to draw carefully a sample of the tested phase. The mixer is restarted for another 15 s, and then stopped, and a sample is taken after phase separation. This procedure is repeated for 1, 2, 3, 5, and 10 min of mixing. There is no need to draw samples from the second phase unless we want to verify our analytical results. The second phase should be tested only from the solutions in equilibrium, since it may be used for the preparation of the equilibrium isotherm or for verification (if the isotherm is already known).

Now the solutions are discharged, and the experiment is repeated with fresh solutions, under exactly the same conditions but at a different speed of agitation. The experiment should be done at at least three speeds.

The following procedure is recommended for setting the speed of the mixer:

1.   Fill the vessel with the solutions at the set phase ratio.

2. Start slowly to mix, and increase the speed until you get full dispersion. This is the lowest speed you will use.
3. Do your experiment at this speed.
4. Increase the speed by 50% and repeat the experiment.
5. Double the initial speed and repeat the experiment.

Remember

The mixing energy increases with the third power of the speed. By doubling the speed you increased the mixing intensity eightfold. This is sufficient for this experiment. Yet, if the mass transfer at the maximal speed was much higher than at the lower one, you should consider increasing further the speed of the agitator

b. Separating Funnel. Fill the funnel carefully with both liquids at the desired temperature. Shake *gently* 10 times, put the funnel in an isothermal bath at the desired temperature, and allow the phases to separate. If you are sampling the aqueous phase, drain it from the bottom of the funnel. The solvent may be drawn by a pipette from the top of the separating funnel. Repeat the procedure with a further 10, 20, 40, 70, and 150 shakes (totaling 300 shakes, which is equivalent to 5–10 min of mixing). Now drain the separating funnel, fill it with fresh solutions, and repeat the entire experiment with *vigorous* shakes. Do this once more, with medium shakes.

Remark

The results of the experiments in a separating funnel depend on the technician. The entire set of experiments must be done by the same person, and on the same day if possible. The results cannot be compared with results from any other laboratory. If there is available an automatic shaker, with variable intensity of shaking and resulting reasonably strong mixing, the procedure can be done automatically for various periods of time and at different shaking intensities, in a manner similar to the manual one. Such results can be compared with similar experiments from other locations, if the latter are done on the same type of shaker and under similar conditions.

Only the natural dispersion can be tested.

E. Solvent Stability

1. Reasons for Decomposition

The most common reasons for decomposition of the solvent are as follows:

Oxidation
Reaction with contaminants
Self-decomposition to smaller molecules

To test possible decomposition, keep a sample of the solvent, mixed together with the real feed solution, at the process temperature and pH, over an extended period. For a full test, the experiment should be run for 3 months, with the aqueous phase changed every day. If you are short of time, an accelerated test at an elevated temperature may be run. Consult the manufacturer and/or an organic chemist colleague regarding the cost of such a test.

If you expect oxidation, carry out two experiments in parallel, in one bubble air into the mixer vessel and in the second bubble nitrogen, and compare the results.

2. Analysis of Results

The results are at best semiquantitative, but they enable us to analyze the process under the following conditions:

a. If in all the experiments equilibrium is reached, this means that both the kinetics and diffusion are rapid and easy, and the solvent is strongly recommended from this standpoint.

b. If the loading increases with increasing duration of mixing/shaking but is hardly affected by the intensity of mixing, the limiting factor is the kinetics of the process. Further kinetic experiments are recommended, to determine the residence time that is needed for completion of the reaction up to the desired degree. If the process is relatively slow (required residence time >5 min, which is rare but not impossible), this solvent is obviously inferior to more rapid ones. But this is by no means a no-go factor.

c. If the loading increases with intensity of mixing but does not improve at longer residence times (at least not for a high rate of mixing/shaking), we have a case of slow diffusion and resistance to mass transfer, but rapid kinetics. This is a disadvantage, but it may be overcome by the use of suitable equipment.

The following example may help to clarify the suggested approach.

Example 6: Extraction of "Hatulat": A Novel Aromatic Compound
Hatulat, a new aromatic homolog of phenol, is synthesized in a batch reaction and obtained as a 5% solution in the aqueous phase at the bottom of the reactor. Its boiling point is 200° C. The task of the extraction team is to extract it, so that its concentration in the raffinate is below 10 ppm (an environmental constraint). The extract will be distilled, to recover the Hatulat and to remove any contaminants. The maximum allowed concentration of the solvent in the distilled phenol was 500 ppm (marketing constraint). The entrainment of solvent in the raffinate is not considered a problem for all the contemplated solvents. The annual production rate of Hatulat is estimated at 2000 tons. Since Hatulat is analyzed in both phases by means of HPLC, only small samples are needed. The limit of detection is 5 ppm.

Step 1: General Considerations

1. The problem is well defined no data are missing.
2. The extraction must be extremely efficient—recovery of 99.98% (!!!) is required. You need an efficient solvent and probably also many stages.
3. For the small production rate, columns are the natural choice.
4. Since the solvent must be distilled, a simple physical extractant, as volatile as possible, should be used.

Step 2: Choice of Solvent

The compound is new, so no data about it can be found in the literature, but knowing that it is a homolog of phenol, you look for suitable extractants for phenols. Extraction of phenol is well known in the literature. In a rapid search we find that three solvents might be suitable: methyl isobutyl ketone (MIBK), butyl acetate (BuAc), and toluene. The values of D for phenol between both MIBK and BuAc and water are reported to be above 30, compared with 3 for toluene. The distillation team said that toluene is much easier to distill, but if necessary this can be done with the other two solvents, using a longer column, higher reflux ratio (i.e., higher steam consumption) and a more complicated procedure.

At this stage there are three solvents to be tested. The data for phenol may serve as guidelines, but you cannot rely on them and must generate your own.

Step 3: Limiting Condition Feasibility Test

a. Loading. In this case the limiting condition should be tested in both directions:

a. To check the maximum loading, O:W = 1:50 (i.e., 20 mL of solvent and about 1000 mL of feed).

Note

D_{lc} for phenol is between 3 and 30, so it probably is above 1 for Hatulat. This is why a really high phase ratio is chosen, to fulfill the condition of $1:10D_{lc}$.

b. To see if the concentration of 10 ppm in the raffinate can be approached, the experiment must be carried out twice: extraction at O:W = 10 and repeated extraction of the raffinate at O:W = 10.

Table 1 gives the concentration of Hatulat in various streams.

Analysis of Results

Toluene is significantly inferior, but does this rule it out? No, for two reasons:

a. Since the phase inversion prevents us from taking full advantage of the high loading of MIBK and BuAc, their relative advantage is probably smaller.

Table 1 Concentration of Hatulat in Various Streams

Stream	Concentration (wt%) of Hatulat using		
	Toluene	MIBK[a]	BuAc[a]
Feed	5%	5%	5%
First raffinate (O:W = 10)	1000 ppm	250 ppm	125 ppm
First extract (O:W = 10)	5000 ppm	5000 ppm	5000 ppm
Second raffinate (O:W = 10)	20 ppm	<5 ppm	<5 ppm
Second extract (O:W = 10)	100 ppm	25 ppm	12 ppm
Extract (O:W = 50)	15%	40%	50%
D_{lc}	3	8	10
D at low concentration	5	20	40
$R_{min} = 1/D_{lc}$	1:3	1:8	1:10

[a] For MIBK and BuAc the loading caused phase inversion.

   b.   Toluene is known to distill much better. We will have to distill larger amounts of toluene than of the other solvents, but if the distillation is much easier, this may be worthwhile. And a capacity of 15% solute is quite decent.

b.   Phase Separation.   In our standard test (1:1 phase ratio), the phase separation times were as follows: toluene, 30 s; MIBK, 40 s; BuAc, 65 s. Here the trend is opposite to the loading. All three solvents separate reasonably well, but toluene is superior. Therefore, at this stage we cannot rule out any solvent.

Step 4:   Simulation

This is an easy case. We can use the variable phase ratio method to prepare rapidly three databases at ambient temperature, make correlations, and compare results.

The correlations were done in weight percent (our simulator allows this). The correlations obtained for D_{lc} were as follows: toluene, $D_{lc} = 5 - 0.4x$; MIBK, $D_{lc} = 20 - 2.4x$; BuAc, $D_{lc} = 40 - 6x$, where x is the concentration of Hatulat in the aqueous phase.

The density of the aqueous phase was found to be constant over the entire range: 1.1 g/cm^3. The density of the solvent phase, for all solvents, was $\rho = \rho_0 + 0.01y$, where y is the concentration of Hatulat in the solvent phase (wt%) and ρ is in g/cm^3. This means that the density of all the solvents increases by 0.01 g/cm^3 for every 1% of the solute. The solvent density ρ_0 (g/cm^3) is 0.87 for toluene, 0.80 for MIBK, and 0.88 for BuAc.

The results of the simulation, for a plant that meets the constraint of 10 ppm in the raffinate are given in Table 2.

Table 2 Simulation Results for 10 ppm in the Raffinate

Solvent	Number of stages	Phase ratio O:W	Total flow (% of feed flow)	Hatulat in the solvent (%)	Density (g/cm^3)
Toluene	17	1:2.5	140	14	1.01
	13	1:2	150	12	0.99
	10	1:1.8	155	10	0.97
MIBK	7	1:5	120	25	1.05
	5	1:3	133	17	0.97
BuAc	5	1:7	115	30	1.17
	4	1:3	133	17	1.04

Remark

Total flow, as percentage of the feed flow, is a measure of the possible saving by increasing the phase ratio. As long as the total flow is below 150%, the potential for saving is small.

Analysis of Results

The difference in the total flow rates is small. Also some of the advantage of MIBK and BuAc, the higher D is lost, since limits on the density prevent us from loading them above 17% for BuAc and 25% for MIBK. So, there is no advantage to BuAc over MIBK, and only a small one over toluene. For toluene we need more than double the number of stages, but otherwise it does not seem too bad.

Step 5: Kinetic and Hydrodynamic Tests

The phase separation test was done in the preliminary experiments (step 3). The mass transfer test is done in separating funnels, at a phase ratio of 1:1. Table 3 gives the concentration of Hatulat in the raffinate.

Table 3 Concentration (wt%) of Hatulat in the Raffinate

Solvent	Number of weak shakes			Number of strong shakes		
	10	20	50	10	20	50
Toluene	1.1	1.0	1.0	1.05	1.0	1.0
MIBK	0.45	0.4	0.4	0.4	0.4	0.4
BuAc	1.5	1.3	1.0	0.25	0.2	0.2

The experiments show similar rapid kinetics for all three solvents: 10 shakes is practically enough to reach equilibrium. For toluene and MIBK there are no problems of diffusion—weak and strong shakes yielded same results—but in BuAc weak shakes failed to reach equilibrium.

Step 6: Conclusions

First, discard the BuAc, although its D_{ic} is the best. Its high distribution coefficient is of no practical value—its maximum loading is lower than that of MIBK, owing to the higher density, and it has inferior mass transfer and phase separation.

There are not enough data to choose the better of the other two solvents because you do not know the real problems of distillation of each. Try to work out with the distillation people which solvent is better at the bottom line. Issue your report and let the client decide. He may rule out one of the solvents as a result of some specific consideration of his site. For example, he may choose toluene because he is short of steam and has only a simple distillation column, which he does not wish to upgrade. Or he may choose MIBK because he has plenty of steam and a big distillation column, which works only part time, and can be used for distillation of the MIBK at no extra cost, whereas the bigger column for extraction with toluene would cost extra money. Otherwise, the pilot plant of the entire process should be run with both solvents before the final choice is made.

Remark

The "Hatulat" in this example is a purely imaginative compound, and its properties were chosen to make the demonstration clear. Still, all the calculations based on the "experimental results," including the simulation (which was done using the IMI simulator) were carried out in a rigorous way.

VI. SUMMARY OF THE LABORATORY STAGE

The experimental work in the laboratory is over. All the data were generated and summarized, and the simulation was carried out.

The simulator was used (see Section IV), to find the recommended flow sheet. The residence time per stage was obtained from the kinetic experiments (see Section V). For mixer-settlers, slow kinetics means a bigger mixer, or two or three mixers in series. For columns, this may reduce the maximum possible flow, to increase the residence time τ. For centrifugal contactors it is simply a no-go factor. Slow kinetics is an important, although too often neglected, piece of information, that must be emphasized clearly in your report.

This is the time to analyze the work, and to give the client the intermediate results. Usually at this stage there are not enough data to design an industrial plant. However, the existing data are sufficient to reject the process, to recommend bench-scale experiments, or to go directly to a pilot plant.

If the results in Section V were conclusive, or if there are process considerations suggesting that you do so, you may recommend testing only one preferred dispersion, and erect the equipment accordingly. Otherwise both dispersions should be tested.

A. Process Rejection

The SX process should be rejected at this stage in three cases:

1. The process cannot meet one or more of the three basic demands: recovery, product quality (including purity, concentration, and appearance), and environmental constraints.

At this stage, it is up to the client to change the process definition, and so make it feasible. The client may, of course, go to another expert and get a second opinion, *but you should never recommend going to a pilot plant for a process that at the laboratory stage and in simulation could not meet the client's requirements.*

2. The only feasible solvent is too expensive (whether because of its own high price, high solvent losses due to instability, low loading, expensive recovery process, etc.) or creates an extreme safety hazard.

3. The SX process is too expensive owing to the complex flow sheet: many batteries, many stages, expensive auxiliary unit operation, and so on. This factor should be considered seriously in this stage if alternative unit operations that may carry out the desired separation seem more favorable.

It is always frustrating for researchers to have to reject a process they had tried to develop. Keep in mind that it is neither a failure nor a calamity to find out that a process is not feasible. However, failing to recognize this at an early stage, and wasting further resources, *is* a professional failure.

B. Recommendations for Equipment

At this stage you have a recommended flow sheet for every solvent, based on the simulation. You should also be able to determine which type of equipment is recommended for the industrial plant. This is essential, since the pilot plant must be run with the type of equipment that is designated for the industrial one.

Sometimes it is difficult to decide, at this stage, what equipment seems optimal. As a matter of fact it may be difficult even after the piloting is over. If two different types of equipment seem to be attractive in a similar way, you should recommend testing them both in the pilot. This is expensive—but cheaper than missing the more efficient equipment.

The recommendations should be part of your intermediate report. It is up to the client (i.e., the one who is paying for this and eventually will use the plant) to approve or reject your recommendations, based on site considerations and company policy. "Company policy" often means a conservative approach and reluctance to innovate, but nevertheless this concern is legitimate because it is the client's money.

1. General Guidelines

All types of SX equipment may be divided into three main classes: mixer-settlers, columns, and centrifugal contactors. They represent different principles, and the experimental procedures for piloting them are quite different. Deciding whether to use columns or mixer-settlers, and of what type, is difficult. The centrifugal contactors are usually used only if everything else fails, but if you can use mixer-settlers or columns efficiently, do it.

The choice of SX equipment is complex, and beyond the scope of this text. Several authors have recommended various criteria for the choice of optimal equipment. A fundamental analysis of technological feasibility of extraction processes was made by Blumberg [1]. A comprehensive study by Stevens and Pratt [15] divides the available equipment into 22 subclasses and points out a wide range of criteria to choose from them. The most recent book on liquid–liquid extraction equipment was published in 1994 by Godfrey and Longsdail [16]. Following are a few general guidelines.

1. If three or more theoretical stages are needed, and the flow rates are below 200 m^2/h, columns are the recommended solution.
2. For one or two theoretical stages, mixer-settlers are usually the right choice. For small flow rates, consider batch extraction.
3. In mixer-settlers the major phase is usually the continuous one. To invert phases, the minor phase must be internally recycled to make it the larger one in the mixer. In columns, the major phase may be dispersed in the minor one. In a Karr column this can be done up to a phase ratio of 10:1. In a pulsed column there is no limit to the phase ratio, but above 20:1 the back-mixing makes this inefficient. Furthermore, in columns there is no internal recycle.
4. Although a single mixer-settler can deal with higher flow rates than a column, this does not mean that columns should be ruled out for extreme flow rates. Often erection of a few columns in parallel may be altogether more economical than one huge mixer-settler. In a most prominent case, WMC use in their uranium extraction plant in Olympic Dam, South Australia, *10 pulsed columns in parallel*, to deal

with a flow rate above 2000 m^3/h. The company found it more efficient than erecting a few trains of mixer-settlers.

2. Batch or Continuous Processing?

Most extraction processes are continuous, and only these are considered in this text. Still, this is the place to mention the existence of a group of industrial batch extraction processes. In the fine chemicals industry, the extraction of the product from the aqueous mother liquor or the extraction of organic traces from the aqueous raffinate is often done in a batch process, if the number of extraction stages is up to three. The piloting of such processes, usually done as a part of the organic process development, is not dealt with here.

C. Recommendations for Piloting

If the SX process seems feasible, usually a pilot plant is recommended, to enable a reliable scale-up of the data and erection of a plant.

There are three main goals for the pilot plant:

1. Verification of the recommended flow sheet—piloting of the process
2. Performance of the equipment (flux, mass transfer, entrainment, preferred dispersion, etc.) and piloting of the equipment
3. Detection and treatment of accumulation phenomena (crud formation, deterioration of the solvent, foaming, deposition of solids, etc.)

The piloting equipment may be divided to four categories:

Bench-scale pilot plants
Equipment pilots
Full-scale pilot plants
Demonstration plants

The flow sheet can be verified by using a *bench-scale unit*. The performance of the equipment is checked in a dedicated pilot plant. The behavior of the process as a whole, including the quality of the product, yield, production rate, and detection of accumulation phenomena may be spotted in a full-scale pilot plant. For a full verification of the process, including materials of construction, control scheme, detection and treatment of all accumulation phenomena, and solvent deterioration, a demonstration plant is needed.

It is rare to run all four stages. For mixer-settlers and centrifugal contactors, usually three stages are carried out, with *either* a pilot plant *or* a demonstration plant being erected.

For columns two steps may be sufficient: the bench-scale and *either* a pilot *or* a demonstration plant. For the extraction of fine chemicals, usually only the bench-scale pilot plant is run.

1. Bench-Scale Unit

The bench scale is the smallest unit that enables the verification over long periods of time of the entire flow sheet *at steady state*, using the solvent in a closed loop. Its results either verify the results of the simulation or modify them. The bench scale also enables us to spot serious accumulation phenomena and, if real solutions are used, the creation of crud.

a. Mixer-Settlers. For mixer-settlers, the bench-scale unit is usually made from small rectangular plastic boxes of 1–5 L capacity. Each box consists of a mixer and a settler. The boxes are fed by metering pumps, and the flow within the batteries is maintained by pump-mixers (i.e., the mixers act also as pumps). These units may be purchased from numerous producers, or they can be easily constructed by any research institute with a good workshop. The units are so compact that a simulation of a 40-stage plant can be run on a 10 m² floor (bench) area. A typical total flow rate through a bench-scale mixer-settler plant is 1–10 L/h. The small flow rate is very convenient because it decreases the logistic problem of feed supply and waste disposal.

The mixer-settler bench-scale plant may give some information about the expected flow rates, but it is not designed to find the real flux in the plant. It should be run at low flux, to ensure low entrainment and full mass transfer. A successful bench-scale plant should always be followed by an equipment pilot plant, in which the hydrodynamics are studied.

Additional drawbacks of the bench scale are as follows.

1. Since the flows are controlled by metering pumps and verified by manual measurement of the flow, control is difficult. Work at this scale is tedious and for low flow rates may be inaccurate.
2. Sampling of intermediate streams tends to disturb the steady state and may influence the results.
3. Heating or cooling is problematic, owing to large losses to the surrounding, unless the units were built with a heating/cooling mantle and adequate insulation.
4. Application of internal recycle is done by using the metering pumps, and for a long plant this is complicated. Often only the natural dispersion, occurring when the minor phase is dispersed in the major one, is tested.
5. Solvent losses are far higher than in a bigger plant, owing to sampling, evaporation, and spillage. This makes accumulation phenomena more difficult to detect.

To obtain reliable results from a bench-scale plant, a few weeks of continuous operation without process problems is recommended; see Section VII.A.

Note

Even when columns are considered for the process, a mixer-settler bench scale is sometimes run in the preliminary stage to verify and optimize the recommended flow sheet and to detect accumulation phenomena. It is recommended especially for a complex process (≥ 3 batteries) and if the simulation was poor.

b. Columns. The term "bench-scale columns" may be applied to the smallest possible columns. Their diameter varies between 15 and 40 mm, depending on the type. They are used in the fist stage of piloting, when columns are considered to be applied to the industrial plant. Usually only one column is used, and it simulates a different battery every time.

Sometimes a 15 mm column is sufficient for scaling up to industrial size. This is the case with the Karr column (although usually a 25 mm column is used). Such a column should not be treated as a bench-scale but as a full-scale pilot plant. The bench-scale columns are utilized when the minimum diameter that enables reliable scaling up involves flow rates above 50 L/h. For example, the recommended diameter for reliable scale-up of the BPC—a disk-and-doughnut pulse column—is 100 mm, with flows of 200–400 L/h. A smaller column, with a diameter of 40 mm, is used as a bench scale.

The height of a bench-scale column is usually limited by the height available in the laboratory, the head of the feed pumps, and other technical constraints.

It is not essential to obtain the required recovery and product quality in the bench-scale piloting unless it is an explicit demand of the client. Since in the laboratory stage and in the simulation you were able to meet the process constraints (otherwise the project should have been rejected without a pilot), and final verification will be done in the pilot plant, extrapolation at this stage being allowed and rather safe.

Compared with mixer-settlers, bench-scale columns have two advantages:

1. Even the smallest column enables fair estimation of the flux. For example, the 40 mm pulsed column predicts the flux with an accuracy of 20–30%, close enough to allow you to choose the optimal solvent and even to estimate roughly the size and cost of the industrial plant. The running of the pilot-scale column afterward is a short and easy task.
2. The flux in any column is far higher in a column, typically 15–50 $m^3/m^2/h$, than in the mixer-settler bench scale, 1–4 $m^3/m^2/h$. As a result, the concentration profile in the column approaches the steady state within an hour, compared with a whole day for one battery of mixer-settlers. A one-day run of 8 h supplies data equivalent to 10–

15 shifts in mixer-settlers. And there is no need to run the column in shifts to obtain representative results.

But columns have their disadvantages, too:

1. Detection of accumulation phenomena is impractical.
2. After every stop the concentration profile has to be reestablished. (In mixer-settlers the profile of concentrations remains constant when the flows are stopped. This allows the entire plant to be restarted with the existing profile.)

The procedure for running bench-scale columns and pilot-scale columns is exactly the same. For the columns that may be scaled up from the bench scale, no distinction exists.

The bench-scale columns are recommended in the following cases.

1. A preliminary study of a new process, for which no mixer-settler bench scale was run.
2. The feed is a product of a pilot plant operation. This applies either to the extraction of an organic product from an aqueous mother liquor or to the recovery of organics from the aqueous waste stream. In both cases the amount of the stream is small, and the smallest possible column is desired. For such an operation columns are always the right choice (unless centrifugal extractors should be applied). Because the final column is small anyway, no further optimization is justified, and *the results may be applied directly to the design of the production plant*, without any further piloting.
3. An artificial solution that simulates a stream of a hydrometallurgical operation is used. Again, the feed is expensive (it must be produced artificially), and as it does not contain all the foreseeable contaminants, the study of accumulation phenomena and crud formation is not possible anyway. For this case, once the experiments in the bench-scale column are finished successfully and the recommended solvent is chosen, a full-scale pilot plant or demonstration plant must be run, using the real solutions, to learn about the influence of various contaminants on the flux, entrainment, and crud formation.

Note

If the columns are to be applied to an existing plant, either as an additional separation or to replace mixer settlers, the bench scale may be skipped and a pilot-scale column erected and run directly on site.

c. Centrifugal Extractors. It is possible to run a small centrifugal extractor, with flow rates of 100 mL/h and below. Whether such equipment can be used directly to scale up to the final plant, or an intermediate stage of piloting is needed, depends on the manufacturer of the equipment. It is important to remember that unlike the case of a mixer-settler or a column, one cannot "improvise" a centrifugal extractor.

2. Equipment Pilot Plant

Dedicated piloting of the equipment is needed for mixer-settlers only. For columns and centrifugal contactors it is an integral part of the bench-scale and/or the process pilot.

The equipment pilot is designed to find the optimal phase continuity and predicts the flux, the mass transfer, and the entrainment. There is no need to build a full plant to obtain these data, and usually one unit is sufficient. If standard equipment is considered, its size should be according to the manufacturer's recommendations. For nonstandard design or a problematic application (i.e., foaming, heavy crud, emulsification, slow kinetics, internal recycle, inversed phases, etc.), the size of the unit should be above 10% of the final one. For new models of the mixer or the settler, a full-scale prototype is recommended.

The unit should be tested at conditions simulating the most difficult ones in the process for each battery. Usually this applies to the last stage in the extraction (maximal viscosity and small $\Delta\rho$), the first stage of washing/stripping (minimal $\Delta\rho$), and anywhere in the purification (extreme phase ratios).

The unit should be obtained from the manufacturer from whom you expect to purchase the full-scale equipment. If you have your own pilot unit, this may be used, if it has the recommended mixer and settler for the future plant. Otherwise, the data from the pilot may provide only an idea about the hydrodynamic properties of the system relative to other processes that you encountered. Data generated with one type of equipment cannot be scaled up to another one.

3. Process Pilot Plant

In the development of extraction processes the pilot plant stage is a must. It serves to verify the recommended process and equipment. The piloting must be done in equipment similar to the proposed industrial one and large enough to enable reliable scaling up. It should consist of all the auxiliary units that are expected: solvent stripping/distillation, solvent treatment and makeup, crud removal, treatment of the feed and/or the raffinate, and so on.

a. Size.   The size of the pilot plant is determined by scaling-up procedures that are available for the specific equipment. Usually this should be the smallest size that enables reliable scale-up and design.

For columns the diameter is set by the manufacturers, and for many types the appropriate dimensions may be found in the literature. For all the common types it does not exceed 100 mm, *regardless of the final size of the plant.* The combined flow rates in such a column are up to 500 L/h. For some types of column the bench-scale size is sufficient—a Karr column can be scaled up from a 15 mm unit, with flow rates of 10 L/h. In petrochemistry and hydrometallurgy industrial plants, the flow rates are well above 100 m³/h. This means that the flow in the pilot plant is far below 1% of that in the industrial plant. Still, this may also be the size of a demonstration plant, enabling the detection of all accumulation phenomena, including solvent decomposition and regeneration, crud removal, and recovery of entrained solvent from the aqueous waste. This is analogous to distillation columns with structured packing, where results from a 50 mm column may be scaled up to any known diameter.

The pilot column should be high enough to ensure the desired recovery. Its height is set to the required number of theoretical stages (NTS), as predicted by the simulation, multiplied by the height of an equivalent theoretical stage (HETS) that was found in the bench-scale experiments. If the process demands were met in the bench-scale experiments, the pilot can be scaled up directly from them. If it is not feasible to supply the required NTS in a single column, two or more columns in series may be employed.

For mixer-settlers the minimum size is usually larger. Unless a smaller unit is recommended by the potential manufacturer of the equipment, the pilot should hold 3–5% of the final plant flows.

For centrifugal extractors, follow the manufacturer's instructions.

b.   Location.   The pilot may be located either in the R&D center or in the site of the future plant. The choice of location is usually administrative. If the location of the R&D center (whether it is the research department of the client or an external consultant) is more than an hour's drive from the plant, running the pilot at the client's location is cheaper, quicker, and more efficient, provided the logistical problems are not prohibitive. If the experiments must be done close to the real site (e.g., to have the real raw material or because of logistic/ environmental considerations), then there is no choice but to erect the pilot plant there.

4.   Demonstration Plant

The demonstration plant must be erected at the final site, and it must use the raw materials of the future plant. For mixer-settlers and centrifugal extractors, its size should be about 10% of the final plant, to enable reliable scaling up. For columns, the size of the pilot scale is adequate. Still, clients often request the use in a demonstration plant of equipment larger than the minimum recommendation of the manufacturer. The higher flow rates are needed to supply larger amounts of product, to be used either for market development or as a feed to larger equipment downstream. An oversize demonstration plant has an additional advantage: the large amount of the product covers most of the operating costs and enables a prolonged run without additional budget allocation. The same one or two operators per shift are needed for running a plant with flow rates of 100 L/h or 100 m³/h.

When is a demonstration plant erected? There are five typical situations.

a. In a new SX-based process and/or when new raw material is used, if there is some uncertainty about the quality of the product and the possible accumulation phenomena. In such cases the plant must be able to supply large quantities of product, for market evaluation.

b. If there are known problems in the process, the solution of which can be proved only in a long run (e.g., crud formation, solvent decomposition and poisoning). The plant should be large enough to feel the full impact of the problems and their solutions, but usually may be smaller than in the former case.

c. If a new technology is being applied to a working SX process (e.g., installation of a settler with partitions instead of a plain one or the replacement of mixer-settlers by columns). In this case the size of the plant should be according to the manufacturer's recommendations. Such a plant usually works in parallel to the existing one, and the results are compared routinely.

d. If the SX section is a noncritical part of a new process or new technology that is to be tested, as a whole, in a demonstration plant. The size of the demonstration plant is set according to the constraints of the critical parts of the process. As long as this size is above the minimum required by the SX equipment manufacturer, the run provides you with free-of-charge (from the SX point of view) verification of the process. On the other hand, the SX team must be able to operate efficiently its section under varying conditions that may be dictated to them.

e. If the client insists on it, without any real technological justification. Maybe he has to convince his investors that it works, or just wants to cover himself in case of a potential failure of the full-scale plant. Try to convince the client to build the plant on the smallest applicable scale, and run it. You really have no other choice besides quitting your job. Probably you will learn something new from this run too, and you may add an obvious success to your CV.

In all five cases, the demonstration plant can be used *instead* of the full-scale pilot plant.

How long should a demonstration plant run? At least three months to verify new equipment for an existing process and one year for a new process. If problems arise, the running-in may be longer, since the plant should be run for 6 months without any new process problem. The longest running-in known to the author was a DSB demonstration plant for extraction of calcium bromide from the Dead Sea. Six years were needed to overcome all the accumulation phenomena and conclude the project successfully. Since the value of the product covered most of the operating costs of the running-in, the company could afford the long run.

In a demonstration plant there are no excuses, and the facility must meet *all* the process demands. If any part fails, it should be corrected and reverified, since this is what the plant was built for.

D. Design of the Pilot Plant

The design of the pilot plant is of crucial importance. It practically determines the flow sheet of the industrial plant. The running-in may change some secondary assumption—for example, the number of stages or phase ratios—although these are not expected to change significantly (unless the simulation was really poor and no bench scale was run).

If you recommend going on with the process, your interim report should include a basic design of the pilot plant. No mechanical details are required, but the recommended flow rates, type and size of the equipment, number of batteries, number of stages for mixer settlers, and NTS and/or height for columns and all auxiliary units must be included. For a full-scale pilot or demonstration plant, always include an agitated tank for sidestream treatment of the solvent (even if you do not know yet what treatment, if any, will be needed).

If both you and the client lack sufficient background in the design of the equipment, you should at this stage contact the prospective suppliers of the equipment and ask them to build or rent you the pilot plant equipment. This step is necessary because most manufacturers insist on running a pilot with their equipment and under their supervision anyway, to provide a process guarantee for the future plant. Moreover, consulting the manufacturers at this stage may help you to correct some conceptual errors, or to choose from a few possible types of equipment that do not require any experimental work. If possible, try to contact more than one supplier for every type of equipment. This will enable you to choose the best, and as a fringe benefit you may learn something useful.

There are a few important points that may be overlooked by the manufacturers.

1.   The bench-scale and the pilot plant should be *transparent*. If possible, make them from glass, otherwise from any suitable polymer. Solvent extraction

is a visual unit operation, and there is no substitute for direct observation. This is critical in columns (without it there is no real way to run an efficient pilot plant—see Section VII), but it is very important also in mixer-settlers and centrifugal extractors. Moreover, the experiments should be videotaped, for documentation and future analysis [17]. If the equipment cannot be transparent, there must at least be long sight glasses.

2. It is important to make the internal components from the same materials that you want to use for the industrial column. For O/W dispersion the internal components should be hydrophilic (e.g., from metal). For W/O they should be hydrophobic (e.g., from a polymer). If both dispersions are to be considered, try to order two sets of internal components, and use the right ones for every dispersion. (Note that this is not always possible, since process considerations may leave you with one material only.)

3. The equipment should have plenty of sampling points. There must be a convenient way to sample both inlets and outlets, with minimal dead time. In a column, three to seven additional sampling points should be available along the active part of the column. In a mixer-settler, it should be possible to sample the dispersion between the mixer and the settler and both outlets from the settler. In an equipment pilot, a few sampling point along the settler should be available.

4. Do not save on sensors. The pilot plant generates data, and for this sensors are needed. In addition to standard measurements, like flow rates at all inlets, temperature, and pH (where applicable), it is strongly recommended that you measure the flow rates at all outlets, even if no change in the relative volume of the flows is expected (see Section VII.B.3.c on why this is so important). In a column, four flow meters at most are required. In mixer-settlers it is important to measure also all the flow rates of the advance and recycle flow from every settler. This may be quite expensive, but it is necessary. And if possible, add some automatic control. Be sure that the flow rates are accurate. Manpower in a pilot plant is usually limited, so let it concentrate on sampling and follow-up, instead of measuring flow rates with a bucket.

5. For a column, be sure that fine control of the discharge flow rate is possible and that the flow does not exceed 120% of the expected maximum. This is important especially if the light phase is considered to be the dispersed phase (O/W dispersion). This may be obtained by using two valves in series: one to restrict the flow to the allowed maximum and the second to regulate it. The second valve may be an automatic on–off valve or a manual needle valve. The first one should be manual and must be calibrated before the experiment starts. Alternatively, two valves in parallel may be used: a larger one that is set by feed-forward only, and transfers 80–90% of the discharge, and a smaller one that is used for feedback control. Additionally, a large valve for emergency draining is recommended.

Do not expect the supplier of the equipment to do the design for you. Do it yourself; it may be critical while the column is running. This is a critical stage. If you are not sure, consult your colleagues or even an independent consultant. Do not let anybody rush you. A mistake at this step may be very costly later.

VII. PILOTING

After the client has approved your report and allowed you to go on to a pilot plant, four further steps are possible:

1. Run a bench scale of the entire plant.
2. For mixer-settlers and centrifugal contactors, test the performance of one unit in a dedicated equipment pilot. For columns this step is included in steps 1 and 3, hence is not needed as a separate phase.
3. Run a complete pilot plant, with flow rates high enough to test the performance of the equipment (usually in the range of a few hundred liters per hour), for verification of the former results.
4. Erect a demonstration plant on site and run it for a prolonged time, using the real solutions. The demonstration plant usually includes not only the extraction plant, with its auxiliary units, but also other parts of the process to be implemented.

It is rare to run all four stages. For mixer-settlers and centrifugal contactors usually three stages are carried out, with *either* a pilot plant *or* a demonstration plant being erected. For columns, two steps are sufficient: the bench scale and either a pilot or a demonstration plant.

For a mixed plant that includes both columns and mixer-settlers, the piloting has four stages:

1. A bench-scale mixer-settler pilot of the entire plant
2. A bench-scale column pilot for the sections that are designed to be run in columns
3. An equipment pilot plant for the mixer-settlers
4. A full-scale pilot or demonstration plant for the entire plant

Piloting of centrifugal extractors is done according to manufacturers' instructions and is not dealt with in this text.

A. Running-In of Bench-Scale Mixer-Settlers

The running-in is long. The bench scale may be operated in shifts, especially if a long plant is simulated, but may be shut down over weekends. The length of the run depends on the number of stages, the flux, and the budget. The more stages and the lower flux, the longer the run, since the residence time is longer. An average run takes a month at least. The logistics, starting with manpower and ending with raw materials, waste disposal, and analytical support, are not trivial.

1. Design of the Layout and the Experiment

The layout of the bench-scale pilot plant should follow the optimal configuration that according to the simulation fulfills the process demands: number of batteries and stages, location of input and output streams, and the phase ratio in each battery. To be on the safe side, use a few more stages than predicted by the simulator. In this way you get the required concentrations and see the real phenomena at the end of the batteries, even if the simulator was too optimistic or the stages are not really ideal. This is important because most unexpected problems tend to happen at the most concentrated or diluted points (i.e., at the ends of the battery). Try not to change the number of stages and the phase ratio during the run unless you cannot meet process demands.

If more than one solvent is to be tested, the layout should be adjusted for each solvent.

Dispersion: A bench scale is not the right equipment to find the recommended dispersion, and it is usually run at the natural one. Still, if there is a good reason to prefer one dispersion over the other at this stage, the impellers of the plant should be set so that they are dipped in the phase that should be continuous.

What Flux to Use? This is the most difficult question. The flux should be low enough to ensure full equilibrium in the mixer and low entrainment from the settlers. Still, the duration of the run is determined by the total of residence times in the bench-scale equipment rather than by chronological time. Increasing the flux by 30% may save a week of running-in.

Residence Time: The residence time in each mixer is

$$\tau_m = \frac{V_m}{F_S + F_A} \tag{9}$$

where τ_m is the residence time in the mixer (min), V_m is the volume of the mixer (L), and F_S and F_A are the flow rates of the solvent and the aqueous phase, respectively (L/min).

The value of τ_m must be larger than the minimal residence time, established in the kinetic experiments, and usually $\tau_m > 1$ min. *Note that τ_m is equal for both phases.*

For the settler, we write

$$\tau_{S_j} = \frac{V_{S_j}}{F_{S_j}} \tag{10}$$

where τ_{S_j} is the residence time of phase j in the settler (min), V_{S_j} is the volume of the phase in the settler (L), and F_{S_j} is its flow rate (L/min).

To estimate the total residence time in the plant, the value of τ_m that is much smaller than that of τ_S is neglected, and the resulting equation is

$$\tau_i = \sum_{i=1}^{n} \frac{V_{S_j}}{F_{S_j}} \tag{11}$$

where τ_j is the residence time of phase j in the plant (min), n the number of cells, V_{S_j} the volume of the phase in the settler (L), and F_{S_j} its flow rate (L/min). Usually V_{S_j} is assumed to be half the total volume of the settler. A reasonable value of τ in a settler is $10-15$ min for the more rapidly moving phase. This corresponds to a flux below 1 $m^3/m^2/h$. To obtain representative results from a bench scale, it should be run for $50 \times \tau$ of the solvent or $10 \times \tau$ of the slowest phase, whichever is longer. As a rule of thumb, 500 h is considered a reasonable time for running a bench-scale pilot, but the residence time consideration should be the dominant one.

Example 7
A process is to be tested in a bench scale, using 2.5 L cells. The simulation suggests that it should be carried out in three batteries: six stages for extraction, eight for purification, and six for stripping (a total of 20 stages). The effective volume is 500 mL in the mixer and 2000 mL in the settler. The phase ratio O:W is 1:2 in the extraction, 1:1 in stripping, and 20:1 in the purification. How long to run the pilot?

Solution
The highest flow rate is that of the aqueous phase in the extraction, so this is the most rapid phase. To get a τ of 15 min in a one-liter volume (half of the settler), the flow rate should be 4 L/h or 67 mL/min. From the phase ratio, the solvent and the strip solution flow rates are 2 L/h (33 mL/min), and the purification flow rate is 100 mL/h. This is obviously the slowest one. From Eq. (11) τ is about 10 h per stage, or 80 h for the entire battery.

The residence time of the solvent, in the entire plant, is $\tau = 20 \times 0.5 = 10$ h, where 20 is the number of stages and 0.5 comes from 1 L volume and 2 L/h flow rate.

To fulfill the condition of 10 τ, the bench scale must be run for at least 800 h (i.e., about 34 round-the-clock days) because of the purification battery. The solvent completes in this time 80 cycles. Obviously, the constraint of 10 τ of the slowest phase is dominant in this case.

Note
1. If there were no purification battery at all, τ would be 6 h only (with the solvent being the slow phase), and 50 τ could have been achieved in 300 hours or 13 days.
2. The numbers just given are only an indication. The true flow rates are tested during the running of the plant. They may be double the values or half of them. This is why the τ_m may be neglected in Eq. (11).
3. If the interface in the purification battery is lowered to 25% of the height, the residence time will drop by 50%.

2. Operational Details

a. Start-Up. During the start-up the concentration profile is being built up. In a long plant that includes purification, it may take more than a week.

Fill up the cells. For a new process, fresh solvent is used. The extraction battery is filled with the feed, the stripping/washing with strip solution/water, and the purification/scrubbing with the purifying/scrub solution, respectively. This procedure ensures the quickest buildup of the profile. If pure product is not available and cannot be easily prepared, then:

i. The purification battery should be filled with either the feed solution or with water (whichever acts better, according to the simulator).
ii. The plant is operated at *total reflux* (i.e., the entire product is recycled to the purification) until the concentration profile has been built up.

The pumps of the solvent and the feed are started first. Only after the solvent that enters the stripping/washing battery has been loaded should the strip solution/wash water be allowed to advance. If the scrubbing/purification was filled with the right solution, the washing battery yields the correct product from the very first moment, and the feed solution of the scrubbing/purification should be prepared according to the simulation. Otherwise, the entire product leaving the washing battery should enter the scrubbing/purification, until its composition is correct.

The follow-up is done according to the analysis of the compositions of the outlets of all batteries.

b. Continuous Run. During the continuous run the plant should be kept at a steady state, without changes of the parameters, unless changes seem necessary to obtain the correct product or to analyze unanticipated problems. The main purposes of this run are to get concentration profiles and to learn about rapid accumulation phenomena.

This is a long and boring step, but indispensable for verification of the process.

c. **Sampling**. Taking of profiles, especially in a long plant (>15 stages), is problematic. On the one hand, it has to be done as rapidly as possible, to represent the steady state. On the other hand, sampling may take the plant out of steady state. A simple calculation shows this: the average size of the samples is 50 mL, to enable pretreatment and duplicate analysis. For 20 stages, this means a liter of each phase. This may be similar to or even higher than the hourly feed rate of some streams to this battery, or 5–10% of the solvent inventory. Such sampling will take the plant out of the steady state and harm any study of accumulation phenomena.

There are two ways to do the sampling: the better one is to shut down the plant and then sample. The profile is exact, but a few hours will elapse before the plant has returned to steady state. This is always the way to take the last profile. The other way is to sample one cell every few minutes, to enable the plant to return to steady state. This is more difficult and less accurate, and is recommended for short plants only. Usually a profile is taken once or twice a week, and no more than two or three profiles for each run. Samples of the outlet streams from all the batteries are taken routinely, for control purposes, as in any plant.

d. **Shutdown**. When the experiment is finished, take a full profile. At this time large samples—100 mL at least—may be taken. Repeat the mass transfer and phase separation tests (see Section V), using the recycled solvent from the relevant stages. Compare the results with those of a fresh solvent, to find possible deterioration. Drain all the remaining solvent into a container.

Check the cells for deposits, and if any are found, analyze them. Check to determine whether they are just suspended solids or have crystallized on the walls. See whether there is any sign of incrustation on the impeller. Whereas sedimentation of fine solids is a part of the life in many extraction processes, crystallization is a very severe problem.

Analyze the composition of the solvent itself, to be sure that there was no degradation or decomposition.

3. Analysis of the Results

The results of the run tell you how accurate the simulation was. If the simulator was exact, then the results of the bench scale should be equal to those of the simulation (if the cells have 100% efficiency). If the results from the bench scale are equal to the simulation, *better* or *slightly* worse, then probably both the experiment and the simulation are reliable (unless both were really wrong and in the same direction). If they are far inferior, maybe the stages were not acting well—either entrainment or incomplete mass transfer is to be blamed. Check the profile against the simulation. If the results of some of the stages are similar to the simulation and a few are inferior, check those stages. Otherwise, maybe the simulation was too optimistic. If this is the case, correct the operating conditions: the simplest correction is to decrease the flow rate of the feed (i.e., increase the phase ratio) and/or to increase the rate at the stripping/scrubbing inlets. A more complicated (and less recommended) solution is to add stages.

If there is an obvious deterioration in the results with time, try to find its source. Consult your chemists about what to do. *Until this problem is solved, you do not have a valid process.* The recommended solution often cannot be tested on a bench scale. It is possible to go to a full-scale pilot to test it, provided this seems feasible. Other than crystallization in the cells, this is the worst single problem that may be detected.

B. Running-In of Columns

The columns can be divided into three main types: gravitational or static, agitated, and reciprocating and pulsed. The analysis that follows (Section VII.B.1) applies to columns of all types.

1. *Gravitational or Static Columns*   In gravitational columns no external energy is supplied to cause mixing of the phases. The flow through the column is due to the difference in density only (i.e., gravitational flow).

    For every given system (composition of input solutions, phase ratio, and temperature) the maximal flux and the mass transfer characteristics for this column are constant.

2. *Agitated Columns*   Some type of mechanical rotating agitator causes the mixing of the phases in agitated columns. The maximum flux and the mass transfer (i.e., the HETS) are functions of the speed of rotation.

3. *Reciprocating and Pulsed Columns*   In both types, the content of the column is moved up and down, by a mechanical agitator or by pulsation. These columns have two mechanical parameters that influence the flux and the mass transfer: the amplitude of the step and its frequency.

1. Basic Concepts

The understanding of the basic terms that apply to the performance of columns is essential to the following analysis.

a. Settler. A column has two settling sections, below and above the active part of the column. Because their task is to separate phases, they are named settlers, exactly as in the mixer-settler units. The settlers are always an integral part of the column.

One of the settlers is called *static* and is full with the continuous phase. Its task is to capture the droplets of the dispersed phase, which are entrained in the continuous one. In the other settler, called *active*, the droplets of the dispersed phase coalesce and separate from the continuous one. For W/O dispersions the lower settler is active, while for O/W dispersions it is the upper one. Since their tasks are different, the size and design of both settlers in industrial columns may be different. In experimental systems they are usually similar, since the experimental column should be able to perform in both dispersions.

b. Holdup. The holdup is defined as the amount of the dispersed phase in the dispersion. It is dimensionless, and usually is given as *volume percent or volume fraction*

The desired holdup varies for one process to another. For an average process, a holdup above 8% is considered "good"; that is, the dispersion is sufficient to achieve good mass transfer and reasonable residence time of the dispersed phase. Holdup below 4% is considered low, and usually indicates poor mass transfer. On the other end, a holdup above 35% is a warning that phase inversion may occur, and we would be working with the wrong type of dispersion. *A holdup above 50% is a sure indicator of phase inversion.*

c. Slip Velocity. Slip velocity is defined as the relative velocity with which the dispersed phase flows through the column relative to the continuous one:

$$v_s = \frac{U_d}{e\phi} + \frac{U_c}{e(1 - \phi)} \tag{12}$$

where v_s is the slip velocity (m/s), U_d the velocity of the dispersed phase (m/s), U_c the velocity of the continuous phase (m/s), e the fractional void volume of the column, and ϕ the holdup of the dispersed phase (volume fraction).

This quantity was defined by Gayler and Pratt [15], who also developed a theoretical expression that enables the prediction of the maximum velocity and holdup from experimental data. The equation, developed for a perforated plate pulsed column, recently was found to fit a disk-and-doughnut column, as well [18].

d. Flux. Flux in a column is defined as the combined flow rates of both phases, F_c and F_d, per column cross-sectional area, A. It is usually measured in cubic meters per meters squared per hour:

$$J = \frac{F_c + F_d}{A} \tag{13}$$

Maximal flux, J_{max}, is obtained when one is working near the edge of flooding and is a function of the phase ratio.

The flux in a column is usually in the range of 15–50 m³/m²/h. In some processes extreme values may be obtained: the author has observed flux above 70 m³/m²/h in both reciprocating and pulsed columns.

e. Type of Dispersion. The dispersion that is created when two mutually insoluble liquids are shaken in a separating funnel at a phase ratio of 1:1 is called "normal" and is more stable than the other one. If the flows of both phases are not equal, then the major phase tends to be the continuous one.

Which phase should be the continuous one? There is no general rule, and the following parameters, which often contradict one the other, should be taken into consideration.

ENTRAINMENT. There is always some entrainment of the dispersed phase in the continuous one. Columns are efficient in creating low entrainment, yet it is difficult to get below 500 ppm. On the other hand, the continuous phase leaves almost no entrainment in the dispersed phase. Therefore, it is recommended to keep as the continuous phase the one having the more harmful entrainment.

Usually, the solvent is more expensive on the one hand and more harmful in the waste on the other, making the W/O (i.e., organic continuous) dispersion the recommended one.

PHASE SEPARATION. Often there is a significant difference between the rate of phase separation of the two dispersions. Slow phase separation lowers the flux, and so the dispersion with rapid separation is preferred.

It is difficult to predict which dispersion separates faster. When there is a large difference in the viscosity of the liquids, usually the more viscous liquid should be the dispersed phase, because the drag force on the droplets is mainly a function of the viscosity of the continuous phase. In some cases, the one type of dispersion causes severe surface phenomena, like accumulation of crud and stable emulsions, whereas these do not happen with the other type.

PHASE RATIO. The natural tendency is to disperse the minor phase in the major one. This arrangement is easier and more stable from the operational point of view. In mixer-settlers this is the only possible dispersion. In most columns a major phase can—within known limits—be dispersed in the minor one and remain stable.

MASS TRANSFER. Again, this is a parameter that is difficult to predict. Usually, as long as we work in the standard range of phase ratios (i.e., up to 1:7), there is no significant influence of the dispersion on the mass transfer, provided the holdup and the residence time of the more rapidly flowing phase are high enough. Above this ratio, it is difficult for the droplets to meet one the other, in order to coalesce and break again. As a result, the rate of mass transfer drops, and a higher HETS results. Dispersion of the major phase in the minor one may improve significantly the mass transfer and decrease HETS.

But it should be kept in mind that the dispersion of a large amount of liquid in an almost stationary phase causes large back-mixing in the continuous phase and may be unstable, and in some types of columns even impossible.

PRICE OF THE SOLVENT.   Working in an aqueous–continuous dispersion decreases the amount of the solvent in the plant because holdup of the solvent in the columns is smaller and because the static settler is always full with the continuous phase (i.e., the aqueous rather than the solvent one). In a large plant, this is always welcome from the safety point of view and may readily make a difference of 100 m³ of solvent. If an expensive solvent ($20–$30/kg) is used, a saving of a couple of million dollars can result.

EASE OF OPERATION.   From the operational point of view, it is more convenient to disperse the heavy phase in the light one. Usually this corresponds to W/O dispersion. The operation is more stable, and the detection of flooding easier.

f.   Flooding.   Flooding occurs when the rate of input of the dispersed phase is higher than its rate of discharge. This happens when the column is operated above its flooding slip velocity (i.e., above the maximal flux).

There are three types of flooding.

i.   Accumulation of the dispersed phase in the settler near its inlet.   From the operational point of view, this is not real flooding, but simply a decreased flow rate of the dispersed phase. The excess of the inlet accumulates in the static settler of the column that is normally full with the continuous phase. It goes without saying that such behavior is a no-go factor. To start with, the column does not operate under the correct phase ratio, since a part of the dispersed phase no longer flows through it. This is bad enough in itself, but if the flooding continues, the dispersed phase will accumulate in the settler section and finally flow through the outlet of the continuous phase. From the operational point of view, this is the worst thing that can happen in a column.

This type of flooding, if detected in its early stages, is easy to solve. Simply decrease the flux and/or the agitation energy. The column will stabilize, and the accumulated phase will disappear within a few minutes. No emergency measures, such as shutting down the inlets, are necessary. Still, it may be a good idea to close the inlet of the dispersed phase until its level in the settler section has dropped.

ii.   A "plug" of continuous liquid layer appears in the column itself.   True flooding occurs when a "plug" of one phase blocks the entire area of the column, preventing any flow through it. The bottom line is similar to the accumulation case: the wrong liquid flows through the outlet. But the plug is much more problematic and can be positively detected only in a transparent column. Otherwise, it is difficult to distinguish between this and the "accumulation" type of flooding, whereas the plug type cannot be treated like the latter. Both inlets and the outlet must be shut down, *at a high agitation energy input*, to allow the phases to separate, and only then may the column be restarted.

　　iii. Phase inversion. Phase inversion fits into the definition of flooding. It means that the column cannot be operated *at the desired dispersion* above the flooding velocity. It is detected readily by measuring the holdup. Increase of the holdup above 50% means phase inversion. The recovery procedure is similar to that for the plug flooding: both inlets and the outlet must be shut down, but at a low agitation energy input, to allow full separation of the phases. The column has to be refilled so that it is full with the desired continuous phase, and only then the operation is restarted.

　　If the stable dispersion has the more rapid settling characteristics, then after the phase inversion the system may work successfully at higher flux. However, often the opposite is true, and phase inversion is followed by accumulation flooding. If the type of dispersion was chosen from process considerations, the inversion is a no-go factor even if it allows a higher flux.

g. Flooding Curve. A flooding curve is the graphical description of the maximum flux against the agitation energy input. It is used for any type of column that uses some input energy to achieve the phase dispersion. For reciprocating and pulsed columns, the energy is expressed as the product of the amplitude and the frequency of the pulse or stroke; for agitated columns, the speed of the rotation serves. Its general shape resembles a *normal distribution function* (see Fig. 3a). To optimize the performance of the column, it is important to understand this shape.

　　What happens in a column without any energy input (e.g., a packed column)? The stream of the phase to be dispersed is winding its way through the continuous phase. During its flow through the internal components of the column, this stream splits into smaller and smaller substreams, until finally it is dispersed into drops of reasonable size. The same happens in every type of column with the energy input shut down. Working at this point yields the *zero-energy flux*. In a few processes this flux is reasonable even for columns that are designed to be energized. In these cases the flooding curve shows mainly the right-hand side of the generalized curve. At zero energy, the flux starts near the maximum (or sometimes even at it) and decreases with increasing energy (see Fig. 3b). In most cases, however, the flux at zero energy is rather low.

　　When agitation or pulsation is started, the movement of the liquids helps to mix both solutions, that is, to disperse one of them into the other. The higher the energy, the sooner a well-developed dispersion is obtained. The maximum flux is obtained when the energy introduced is just enough to disperse the entire phase that is set for dispersal as soon as it enters the column. If the energy is increased further, the droplets become smaller and smaller and the maximum slip velocity decreases, owing to the increase of the inertia (or drag) forces that oppose the gravitational force and slow the flow down.

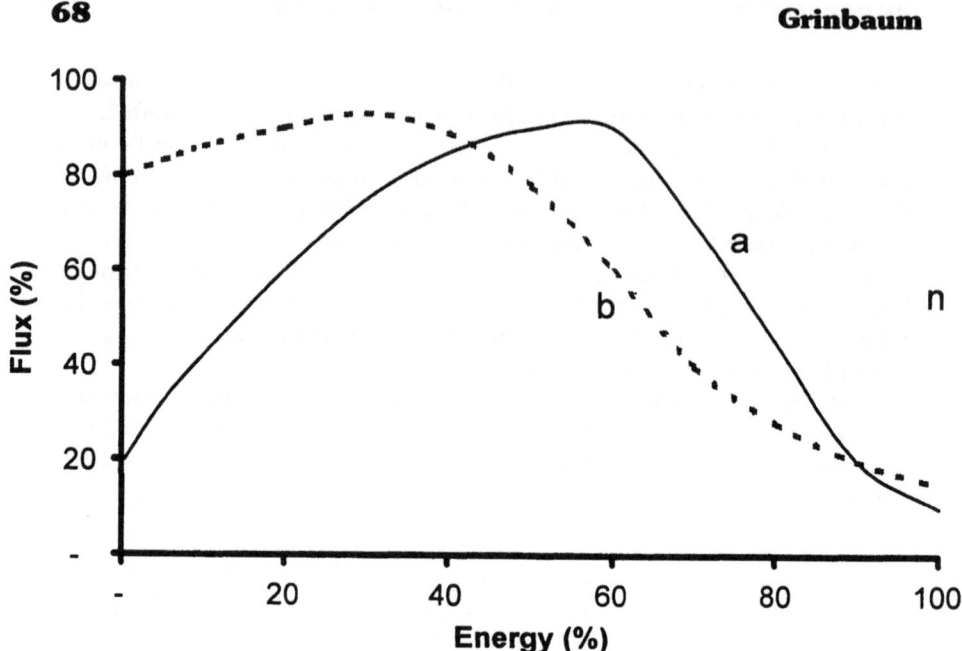

Figure 3 Flooding curves: (a) standard shape and (b) low-energy case.

h. Residence Time. It is important to remember that the residence time is different for each phase in the column. For the dispersed phase, it is equal to:

$$\tau_d = \frac{Ah\phi e}{F_d} \tag{14}$$

and for the continuous one to

$$\tau_c = \frac{Ah(1 - \phi)e}{F_c} \tag{15}$$

where F_d and F_c are the flow rates of the dispersed and continuous phases, respectively, A is the cross-sectional area of the column, h is the total height of the column in which extraction takes place, ϕ is the holdup, and e is the fractional free area in the column. The residence time can also be expressed as a function of the flux J:

$$\tau_d = \frac{(1 + R)h\phi e}{J} \tag{16}$$

$$\tau_c = \frac{(1 + R)h(1 - \phi)e}{JR} \tag{17}$$

where R is the phase ratio, F_c/F_d

2. What Information Can be Generated from an Experimental Column?

The experimental column enables us to evaluate five important factors, which are discussed in the subsections that follow.

a. The Maximum Flux. Running of the column gives the maximum flux, at a given phase ratio, as a function of the energy input (i.e., the *flooding curve*). For a static column the maximum flux is obtained in a single experiment, since no energy input exists. For an agitated column, the curve of flooding vs the speed of rotation is obtained. For reciprocating or pulsed columns, the energy is given as the *product* of the frequency and the amplitude of the stroke. Many researchers assume that this product represents the energy input, regardless of the individual parameters [15,19,20]. Recent (confidential and unpublished) research shows that this assumption is not accurate for pulsed columns and that the best results are obtained when the maximal frequency is used.

Note that the energy input is measured in speed units (cm/s or m/s). This is a convention that allows easy measurement and scale-up. Calculation of the real energy is complex, and for process development and scale-up purposes quite irrelevant.

In the rare case of a flat optimum, when the same maximum flux and HETS are obtained at various energy inputs, an extraction process should be carried out with the minimal required energy to decrease the entrainment and save on energy costs.

b. The Composition of the Outlet Streams and the Recovery. The *recovery* of a battery is defined as the percentage of the solute that was extracted or stripped in the column:

$$\eta = \frac{100(x_{in} - x_{out})}{x_{in}} \tag{18}$$

where x_{in} and x_{out} are the concentrations of the extracted or stripped component in the inlet and outlet streams, in any consistent units, and η is the recovery (%).

c. Height Equivalent of Theoretical Stage (HETS). The number of theoretical stages (NTS) needed to achieve the recovery that was obtained in the experiments is known from the simulation. The average height of theoretical stage (HETS) is:

$$HETS = \frac{h}{NTS} \tag{19}$$

where h, as before, is the total height of the column in which extraction takes place.

d. Holdup Through the Column. The holdups are used to calculate the residence time of each phase and the slip velocity. They also enable better qualitative understanding of the process, mainly in the developmental stage.

Low Holdup (<4%): The column does not get enough energy. If flooding combined with low holdup is observed, then probably we are along the left-hand side of the flooding curve, and increasing the energy may improve the flux. Low holdup combined with a high HETS means that the agitation energy should be increased, to decrease the HETS. In the rare (and lucky) case of low holdup combined with a low HETS, there are no problems of mass transfer, and there is no reason to be concerned about the low holdup.

High Holdup (>35%): The maximum possible energy is applied. Further increase of the energy may cause flooding or phase inversion. If the HETS is still high, the flux should be decreased before the energy is increased.

Extremely High Holdup (>50%): A sign of phase inversion. This is the standard way to detect flooding in industrial columns, before the settler section is flooded with the wrong phase.

The combination of holdup and HETS describes the kinetics and rate of diffusion in the process. Low holdup combined with a low HETS shows rapid kinetics and rapid diffusion. High holdup combined with a high HETS means the opposite—either slow diffusion or a residence time in the column that is too short to approach liquid–liquid equilibrium.

e. Entrainment. The entrainment is measured in both outlet streams. The entrainment of the continuous phase in the dispersed one is usually small and is a direct function of the system properties and energy input. The entrainment of the dispersed phase in the continuous one is, additionally, a function of the size and shape of the settler section of the column. High entrainment means that the settler is underdesigned. Usually experimental columns are operated with rather small settlers, to decrease the total residence time in the column on the one hand, and to spot possible problems of phase disengagement on the other one. High entrainment is definitely not a no-go factor. It just tells the equipment manufacturer that a larger settler is needed for the industrial column. In bench-scale columns the settlers are very small, and the entrainment is not measured nor considered.

3. Experimental Procedures

Two types of experiment can be carried out: *hydraulic tests*, in which only the hydrodynamics are tested, and *mass transfer experiments*, where both the hydrodynamics and the mass transfer are analyzed. In hydraulic tests both solutions are recirculated, whereas for mass transfer, the solutions must flow in an open loop, without circulation.

a. Hydraulic Tests. In a hydraulic test, about 5–30 L (depending on the column height and diameter) of the aqueous solution and of the solvent are circulated through the column in a closed loop. In these experiments no mass transfer occurs, since after a short time both phases are in equilibrium. The hydraulic tests are used for all bench-scale columns with energy input.

They enable initial evaluation of a process, by testing the influence on the flooding curve of temperature, type of dispersion, various modifiers, and other factors. Hydraulic tests are much cheaper than mass transfer experiments, since no chemical analyses are required, only small amounts of solutions are needed, and the experiments are much more rapid. They are reliable only if the net amount of mass transfer is below 10%, so that the difference in physical properties of both phases, before and after the extraction/stripping, is small.

If the experiment is carried out on-site, with real solutions, there is no need to recycle the solutions. The column is operated in open loop, but without waiting for chemical equilibrium and without taking samples for chemical analysis. *In this case the restriction of 10% mass transfer does not apply.*

Hydraulic experiments can be done in a column as short as one meter. Since the dynamic response of the column is proportional to the volume, the experiments become longer as the length of the column increases, and there is no advantage in a longer column for this purpose. Still, the hydraulic tests are so quick (between 15 and 30 min per point), that usually it is not worth the effort to save time by shortening an existing column.

HOW TO DESIGN HYDRAULIC EXPERIMENTS. Always start at the *lowest reasonable flux and energy.* Go on increasing the energy. When the column floods, decrease the energy to the initial one, increase simultaneously both flow rates by 10–20%, and start increasing the energy until flooding takes place. Repeat the procedure until you reach a flux for which the column is flooded for any energy input. This is the *maximum possible flux.*

In the initial step of the work, try to increase the energy when flooding occurs. If you are on the left-hand side of the flooding curve, increased energy will take the column out of the flooding. If this fails, you know that you are on the right-hand side of the flooding column and must not try it again.

Measure the holdup for every flux and energy setting, and measure the entrainment near the maximum flux.

The hydraulic test is relatively quick: an approximate flooding curve, for a given dispersion, may be obtained after 2–3 days of operation.

b. Mass Transfer Experiments. The mass transfer experiments have two purposes:

1. Verification of the maximum flux and holdup that were obtained in the hydraulic tests, while work under true process conditions proceeds over a longer period.
2. Evaluation of the mass transfer in terms of the recovery of the process as a function of the energy input and the residence time (i.e., flux). When the recovery is known, the equilibrium data for the process can be used to calculate the HETS.

Mass transfer experiments require relatively large volumes of solutions and a minimum of 15–30 h of running the column to obtain sound results for a single dispersion. The appropriate volume of both solutions must be available.

HOW TO DESIGN MASS TRANSFER EXPERIMENTS.   The mass transfer experiments start at 60% of the maximum flux that was found in the hydraulic tests. The column is run at this flux with three levels of energy, up to the maximum obtained in the hydraulic test. The flux is increased in several steps up to the maximum, and the column is run for every flux at two or three energies. If the results show that the HETS decreases with increasing the energy, increase the energy input further, while decreasing the flux, according to the flooding curve.

If the HETS remains high for the given type of column, and the kinetics are known to be rapid, consider other means to improve the diffusion, such as changing the type of the dispersion, or heating or changing the concentration of the extractant and/or the phase ratio.

If the best recovery obtained by the column is below process criteria, a longer column is needed.

The experimental simulation of a longer column is easy. For extraction, the column is rerun with fresh solvent and the raffinate from the preceding run is used as a feed. For purification/stripping, the purified/stripped solvent is rerun against fresh purification/strip solution. This procedure gives approximately the equivalent of a column with double length. It can be repeated again and again, so that with the existing column one can simulate a process that requires many stages.

Note

The rigorous experimental procedure requires that the first extraction/stripping/purification be done not with a fresh solution but with the raffinate/extract of the second step. This procedure is less convenient and is not usually applied, since the difference in the results is not large.

Is the second run really necessary? If a reliable simulation is available, it calculates the NTS achieved by the column. The HETS may be calculated from the experimental results by using Eq. (19), and the final height of the column can be calculated by extrapolation, without additional experimental work. The second run is used in three cases: no reliable simulation is available; problematic mass transfer may be expected in the diluted end of the process; or the client is reluctant to extrapolate and wishes to see the real thing coming out of the plant.

The output of the mass transfer experiments is a table of outlet compositions as a function of the flux and energy input. It may be readily translated to a table of HETS vs flux. Typically, two or three sets of results per dispersion should be given to the potential suppliers of the column, from which they calculate the optimum. For more exact results, the profile of concentrations in the column may be used to calculate the HETS in various parts of the column.

Remarks

1. The column should not be used as a tool to determine the kinetics and the diffusion properties of the system. This can be done in a quicker and more reliable way in simple laboratory experiments (see Section V). On the contrary, the laboratory results should help in the analysis of the results from the column.

2. The experiments that are done in the column should be used for validation of the laboratory results. If decreasing the flux, at constant energy input, improves the mass transfer (i.e., decreases the HETS), the kinetics of the system tends to be slow. Contrariwise, if increasing the energy at constant flux improves the mass transfer, it seems that the diffusion in the system is slow.

Usually the results from the column agree with the earlier laboratory results. If not, go back to the laboratory and repeat the experiments. Either they were done wrong the first time, or there is some bizarre aspect of the column that should be analyzed and understood before scaling up to an industrial plant.

c. How to Detect Flooding. Quick and reliable detection of flooding is essential for successful operation of the column, and the technique deserves detailed explanation.

For the sake of clarity, in this section (only!) the case of the dense phase being dispersed in the light one is referred to as W/O dispersion, while dispersion of the light in the heavy phase is named O/W. In the rare case that the solvent is the heavy phase, just reverse the notation. First let us recall how a column is operated and controlled.

A column has two settlers: one of them is *static*, and full with the continuous phase, whereas in the second, called *active*, the droplets of the dispersed phase coalesce and separate from the continuous one. For W/O dispersions the lower settler is active, while for O/W dispersions it is the upper one. The level of the interface in the active settler is always controlled by the rate of withdrawal of the separated heavy phase from the lower settler. The light phase flows from the top by overflow. There is no practical way to do it otherwise.

For *W/O dispersion*, the withdrawal is made from the active settler, this being a direct control action. If it is too slow, the level of the aqueous phase in the lower settler rises, and vice versa. In both cases it cannot cause accumulation of the aqueous phase in the upper settler, and it has nothing to do with flooding.

For *O/W dispersion*, the withdrawal is made from the static settler and affects the level in the upper one through the mass balance. Increasing the rate of withdrawal of the heavy phase from the lower settler lowers its level in the upper one, but may also cause the light phase to accumulate temporarily in the lower settler, even if the column is far from flooding. This makes the detection of flooding more complicated, since the presence of solvent in the lower settler is not an obvious sign of flooding.

ACCUMULATION FLOODING. This type of flooding happens when the flow rate of the dispersed phase in the column is lower than its feed rate. Then the dispersed phase accumulates near its inlet, in the static settler.

For W/O dispersion the detection of accumulation flooding is straightforward. If the aqueous phase accumulates in the upper settler, then flooding has occurred. There can be no other interpretation.

For O/W dispersion the detection of accumulation flooding is similar, but it is not unique. If solvent accumulates in the lower settler, then either flooding occurred or the discharge of the aqueous phase from the bottom is too high. How can we distinguish between these possibilities? There are two ways to find quickly that the discharge rate is too high:

If the flow rate from the bottom is higher than the aqueous feed.
If the upper settler drains and the overflow stops or slows down.

(This is why it is essential to have four flow meters in a column, on both inlets and outlets; see Section VI.)

If the discharge rate from the bottom of the column was too high, decrease it. Then if the solvent level in the lower settler rises above its feed point, there is no flooding. But if the level of the solvent continues to drop, either accumulation flooding or phase inversion has occurred.

If the flow rate of the aqueous outlet is correct, then it is flooding, due either to accumulation, plug formation, or phase inversion. To distinguish between them, look at the holdup in the bottom of the column: if it is above 50%, then phase inversion took place; if below 50%, accumulation or plug fooding has occurred.

On the other hand, too slow discharge of the aqueous phase may conceal, for a few minutes, an accumulation flooding. It should now be clear why operation and control of the column, especially on the laboratory scale, are more difficult for O/W dispersions.

PLUG FLOODING.   When a plug of one phase blocks the entire area of the column, preventing flow through it, "true" flooding occurs. The blockage causes accumulation of the solvent in the lower settler and of the aqueous phase in the upper one, regardless of the type of dispersion. The plug can be positively detected only in a transparent column. Otherwise, it may be detected by elimination only: if it is neither accumulation flooding nor phase inversion, then it is a plug. Plugging may happen anywhere along the column, and you must search its entire length to find it. Do not even try this in an opaque column. The plug is common in some column types (e.g., the Kuhni column) but in others (e.g., pulsed columns), it is quite rare.

PHASE INVERSION.   Phase inversion is detected easily by measuring the holdup. Increase of the holdup above 50% means phase inversion. In a transparent column it can often be detected visually if both phases have different colors. The color of the dispersion in the column is similar to that of the continuous phase. Change of color signals that phase inversion has occurred.

Note

For O/W dispersion, if the level of the solvent drops below its feed point, then *phase inversion* may occur, even if the column is operated below its flooding point. This is especially true when the solvent is the major phase. In every case of phase inversion after the solvent dropped below its feed point, restabilize the column and check it again under the same conditions.

4. Operation of a Column

The operating instructions apply to both bench scale and pilot scale columns.

a. Preparations to Run a New Application.   If the application is to be run with artificial solutions, they must be prepared first. If both phases are colorless, the addition to the feed of a pigment that does not undergo extraction is strongly recommended. For aqueous solutions of acids and salts, methylene blue is the best, since it is never extracted to a solvent, whereas phenolphthaleine is recommended for the basic range. For every case consult with analytical chemists and check the pigment in the laboratory before piloting.

If real solutions are to be used, the solvent is always colored. In the best case the color is light enough to be transparent. Often, however, the solvent is nearly black, making visual observation of flow patterns impossible for a W/O dispersion and difficult for an O/W one. The aqueous phase may be anything from transparent to dark.

For off-site tests, prepare adequate amounts of the solvent and of the aqueous solution, according to the expected scope of the work.

Assign a separating funnel for each sampling point and mark it with the number of this point. You are going to use these funnels for the entire experiment without washing them. It is important to use the same funnel for a given point every time.

b.   Start-Up.   The procedures are similar for columns of all types. It is important to follow exactly the order of operation, to prevent phase inversion or flooding during the start-up.

1.   Fill the column up to the aqueous inlet with the solution that is to be continuous one.

2.   Start the mixing/pulsation, at the desired speed for frequency and amplitude.

3.   Start to feed the solvent and the aqueous solution into the column simultaneously, at a low flux. If you are restarting a column and wish to operate it at a known flux, the initial flows should be 60% of the final ones.

4.   When the active settler is one-quarter full with the dispersed phase, open the aqueous discharge valve and control it to the expected flow rate. Remember that the available discharge flow rate should not exceed 120% of the expected maximum (see Section VI.D).

5.   Increase the flows slowly, until you reach the starting point.

6.   For mass transfer experiments, the column may be considered to be in steady state after 5 residence times τ of the slower phase (see Section VII.B.1). For hydraulic tests, the steady state is very quick, and flooding usually can be detected after 10–15 min. In the early stages of the experiment allow 15 min after a change in energy or flow. In the vicinity of the maximum, or after you have gone through a flooding, allow 30 or 15 min of stable operation, whichever comes *last*.

Sometimes the residence time for one of the phases is so long that waiting 5τ is impractical. This is especially true at low fluxes and under high phase ratios. In such a case refer to the residence time *in the column only*. It saves time, but the composition of the slow phase at the outlet may not any longer be regarded as a steady state composition.

The slow phase is not necessarily the minor one. As long as the phase ratio is below 3, the *continuous phase* is the slow one.

If the slow phase is the dispersed one, we can decrease its τ by keeping the interface near to the top/bottom of its settler. But if the slow phase is continuous, its residence time cannot be decreased.

c.   Continuous Operation.   After reaching steady state, take samples. Increase the energy input. If flooding occurred, decrease the energy to the initial one and increase both flow rates simultaneously by 10–20%. Adjust the opening of the discharge valve. If flooding occurs, decrease the flows to their earlier value and increase the energy. Allow 15 min to stabilize and increase the flow again.

If the column gets out of the flooding conditions, go on increasing the flux and/or energy, until you get to a flux, for which flooding occurs at any energy input.

The preceding flux is the maximum allowable one. The experiment is finished.

d. Sampling. After the column has reached a steady state, samples should be taken from all the sampling points to the separating funnels. When applicable, a sample from the solvent outlet should be taken into the Immhoff cone, to measure the entrainment.

If the experiment is not at ambient temperature, measure the temperature of the top and bottom samples immediately after sampling.

After phase separation, the holdup is measured in all the separating funnels. In mass transfer experiments the separated phases are chemically analyzed, together with all the input and output streams. The entrainment in the Immhoff cone may be measured after at least 4 h. To obtain the best results, close the Immhoff cone, to prevent evaporation and accidental dropping of water into the cone, and leave it for 24 h.

Drain the separating funnels, but never wash them before the end of an experiment.

Remark

The main constraint for taking a large number of samples is the cost and the availability of analytical services. To prevent unnecessary sampling, the following is recommended: wait double the normal time before you take your first profile, and if the plant is not really stable, do not send the samples to analysis at all. Five to ten profiles per battery is sufficient for a good understanding of the process. Do not take a profile after a minor change of a parameter. Just sample the extract and the raffinate, and measure the holdups.

Remember

The column is very sensitive to changes in flow. Never open sharply the manual discharge valve; this action takes the column out of steady state for a long time and may cause phase inversion.

The following example describes a work routine for full analysis of a hypothetical battery.

Example 8

A process was developed for extraction of a product "Ez" by means of a novel proprietary solvent. The simulation found that four theoretical stages are needed for extraction and one for stripping. The concentration of Ez in the feed is 41 g/L. The required recovery of the Ez is 85%. Additional data are available from the simulation and the laboratory experiments: the phase ratio O:W is 2:1, ambient tem-

perature is adequate, equilibrium is achieved in 30 s (i.e., relatively quick kinetics), and both dispersions give similar results regarding the rate of phase separation and mass transfer. The project manager decides to use a BPC disk-and-doughnut pulsed column for extraction and a mixer-settler for stripping. A limited stock of real aqueous feed is available from an existing plant (which uses a different solvent). The solvent is to be prepared by you. Design and carry out a test to check the feasibility of the extraction in the column.

1. Basic analysis

Because this is a preliminary study, and the solvent is limited and expensive, a bench-scale column should be used. The smallest available BPC column is a 40 mm unit. The HETS in this type of column is in the range of 80–150 cm, so a column with 4 m active height may provide four theoretical stages and has a good chance of giving the required recovery in one run. Its volume is about 1.25 L/m, or total of 5 L. The maximum combined flow rate through such a column is 75 L/h (at a flux of 60 $m^3/m^2/h$). To ensure τ above 3 min, 5 L settlers are recommended.

> *Remark*
> In a bench-scale column the entrainment is measured for qualitative impression only, so a 5 L settler seems adequate.

To save time and solutions, as much of the work as possible should be done in closed-loop hydraulic tests. Both dispersions are to be tested.

Table 4 gives you an idea of the flow rates and residence times in the column at various flux values. Always have such a table ready before you start an experiment.

A few important conclusions arise from Table 4.

a. Residence Time and Kinetics. The extraction process is known to be completed in 30 s *per stage*. It means that for a four-stage column, the minimum residence time should be 2 min. Table 4 indicates that for a holdup of 10% this condition is fulfilled in W/O dispersion only up to a flux of 35 $m^3/m^2/h$, and for O/W hardly ever. The conclusion is that a higher holdup is necessary. Remember that τ of the dispersed phase is linear with the holdup. If the holdup rises to 20%, all values double; for W/O τ is more than sufficient for every reasonable flux, and for O/W a flux of 35 $m^3/m^2/h$ becomes feasible. Hence we can formulate two guide lines.

1. The HETS is expected to decrease with strong pulsation not only as a result of improved diffusion (which is known to be good), but mainly because it causes higher holdup and a higher τ of the solvent.
2. If a flux above 50 $m^3/m^2/h$ can be achieved, O/W dispersion probably will be inferior because of kinetic problems.

Table 4 Flux and Residence Time in a 40 mm Column, 4 m High, with 5 L Settlers; Phase Ratio O/W = 2; Average Holdup of 10% Assumed

Flux (m³/m²/h)	Flow rates (L/h)		Residence time (min)							
			Including settlers volume				Excluding settlers volume			
			W/O		O/W		W/O		O/W	
	Solvent	Aqueous	Solvent	Aqueous	Solvent	Aqueous	Solvent	Aqueous	Solvent	Aqueous
20	16.8	8.4	34.1	21.5	10.8	68.2	16.2	3.6	1.8	32
25	20.9	10.5	27.3	17.2	8.6	54.6	13.0	2.9	1.4	26
30	25.2	12.6	22.7	14.3	7.2	45.5	10.8	2.4	1.2	22
35	29.3	14.7	19.5	12.3	6.1	39.0	9.3	2.1	1.0	19
40	33.5	16.7	17.1	10.8	5.4	34.1	8.1	1.8	0.9	16
45	37.6	18.8	15.2	9.6	4.8	30.3	7.2	1.6	0.8	14
50	41.8	20.9	13.6	8.6	4.3	27.3	6.5	1.4	0.7	13
55	46.1	23.0	12.4	7.8	3.9	24.8	5.9	1.3	0.7	12

Still, keep in mind that for an analysis of the results, both dispersions must be tested.

b. Residence Time and Duration of an Experiment. As noted earlier, full steady state may be assumed after 5τ of the slowest phase, including the settler volume. From Table 4 we see that for O/W dispersion this is readily fulfilled for the solvent (less than an hour at all reasonable fluxes) but is an unrealistic condition for the aqueous phase, where 3–4 h at medium flux is needed. For W/O the opposite happens: after 60–90 min the aqueous phase is in steady state, while for the solvent about 2 h is needed for a medium flux. This does not mean that the mass transfer experiments should take hours. After 90 min we have more than 5τ in the column itself, in both dispersions, for any flux above 20 $m^3/m^2/h$. Therefore, the concentration profiles in the column are at a steady state. The only value that is doubtful, is one of the outlet streams. During the tests this does not matter, since we treat the composition in the point nearest to the outlet as the outlet.

Still, after the mass transfer experiments are finished and analyzed, a *verification run*, in which the column is run at the recommended conditions during at least 5τ, should be made.

2. Hydraulic experiments

The frequency of the pulsation was 120 strokes per minute. For technical reasons it was constant.

a. Start-Up. The feed vessels were filled with 20 L of aqueous phase and 20 L of fresh solvent. For W/O dispersion the column was filled with solvent up to the aqueous inlet. For O/W it was filled with the aqueous phase.

The experiments started at a flux of 20 $m^3/m^2/h$, that is, 16.8 L/h solvent and 8.4 L/h aqueous feed (from Table 4). The initial amplitude of the pulsator was 3.6 mm.

b. Operation. The follow-up of the experiment is listed in the form illustrated by Table 5.

The test is over. The maximum total flow rate is 34.8 L/h (i.e., a flux of 27.5 $m^3/m^2/h$).

1. The initial experiments, between 9:00 and 12:00, were very quick. As long as the column has not yet been flooded, 15 minutes are sufficient to determine that no flooding is occurring. After flooding, you have to stabilize the column and only then to restart it.

2. In the example given, at 12:00 the column was almost flooded. Note that the holdups increased sharply in the vicinity of the flooding point. It was no use to increase the flow further. The energy was decreased, and then the flux was increased. It took 45 min to restabilize the column and another 15 min to realize that the system was flooded.

Table 5 Extraction of Ez: Hydraulic Test in Closed Circuit for W/O[a] Dispersion

Time (h:min)	Temp (°C)	Feed (L/h) Aqueous	Feed (L/h) Solvent	Pulse amplitude (mm)	Holdup (%) SP 1	Holdup (%) SP 2	Holdup (%) SP 3	Holdup (%) SP 4	Holdup (%) SP 5	Entr (%)	Flood, Y/N
9:00	35	8.4	16.8	3.6	8	6	5	<4	<4	<0.1	N
9:15	37	8.4	16.8	4.8	—	—	—	—	—	—	N
9:30	41	8.4	16.8	6.0	12	8	6	4	<4	—	N
9:45	43	8.4	16.8	7.5	—	—	—	—	—	—	N
10:00	44	8.4	16.8	9.0	15	10	6	5	<4	—	N
10:15	45	8.4	16.8	11.0	—	—	—	—	—	<0.1	N
10:30	46	8.4	16.8	13.0	—	—	—	—	—	—	Y
11:00	45	10.5	21.0	3.6	13	9	6	<4	<4	—	N
11:15	44	10.5	21.0	5.3	—	—	—	—	—	—	N
11:30	45	10.5	21.0	7.5	14	9	6	5	4	<0.1	N
11:45	44	10.5	21.0	9.0	—	—	—	—	—	—	N
12:00	45	10.5	21.0	11.0	30	18	11	9	5	—	Almost
12:45	45	10.5	21.0	3.6	—	—	—	—	—	—	N
13:00	46	12.6	23.2	3.6	—	—	—	—	—	—	Y
13:30	46	11.6	23.2	4.8	15	9	7	5	<4	—	N
14:00	46	11.6	23.2	6.0	23	17	20	8	6	<0.1	N
14:30	46	11.6	23.2	7.5	28	21	12	9	7	—	N
15:00	45	11.6	23.2	9.0	—	—	—	—	—	—	Y
15:30	45	12.6	25.2	4.8	—	—	—	—	—	—	Y
16:00	45	12.6	25.2	6.0	—	—	—	—	—	—	Y
16:30	45	12.6	25.2	7.5	—	—	—	—	—	—	Y

[a]Key: SP, sampling point; Entr, entrainment; Flood, occurrence of flooding.

3. Increasing the energy enabled stable operation. This time we allowed 30 min for the experiment, to be sure to avoid flooding. Finally, after three successful experiments, we got flooding at 15:00.

4. The amplitude was decreased to its lowest stable value in the preceding flux (i.e., 4.8 mm).

5. After stabilization, the flux was increased again, to 30 $m^3/m^2/h$. This flux was too high, and there was flooding again and again. Stabilization took 10–20 min. Flooding was observed in less than 10 min. So, these experiments took 30 min each.

6. At 16:30, after three attempts to change the energy, we gave up: just in time to go home.

7. Samples for holdup were taken every 30 min, provided there was no flooding. It is useless to sample anything from a flooded column. Note that the holdup increases with increasing the flux and the energy as well. For low holdup "<4%" is written. Holdup below 4% is considered to be very low, and it does not matter whether it is 3 or 1%.

8. The entrainment was below 0.1% (i.e., <1 mL in a 1000 mL separating funnel). This is considered to be a low entrainment.

The next day the O/W dispersion was examined. The procedure was similar and does not need to be included here. The maximum flow rates were 13.8 L/h feed and 27.6 L/h solvent, corresponding to a flux of 32 $m^3/m^2/h$. It is higher than in W/O by almost 20%.

3. Open loop mass transfer experiments

When the open loop mass transfer experiments were started, we knew already that the maximum flux was about 27 $m^3/m^2/h$ for W/O and 32 $m^3/m^2/h$ for O/W. The tanks were filled with 300 L of solvent and 150 L of aqueous solution, to enable about 12 h of work.

The W/O was tested first. To be on the safe side, the experiments were started at 25 $m^3/m^2/h$ (i.e., with 21.0 L/h solvent and 10.5 L/h aqueous feed). The initial amplitude was 3.6 mm. After 2 h of smooth operation, the first profile was taken. The experiment is described in Table 6.

Remark

The holdups, which were similar to those of the day before, are not given here.

Note the point at which flooding "almost" occurred. This means that there was very slow, yet significant, accumulation of the aqueous phase in the upper settler. If this is slow enough to enable continuous work of 2 h, without flooding the settler, it is accepted as a limiting case. This is the best thing that can happen, because it means that we are really on the flooding curve.

Table 7 gives the results for O/W dispersion, and Table 8 lists the analytical results of the best profiles for each dispersion.

Table 6 Extraction of Ez: Open Circuit Mass Transfer Tests for W/O Dispersion

Time	Temp (°C)	Feed (L/h) Solvent	Aqueous	Amplitude (mm)	Flood	Extract (g/L)	Raffinate (g/L)	Remarks
8:30	37	21.0	10.5	6.0	N	—	—	Start
10:45	42	21.0	10.5	6.0	N	16.7	6.8	Profile E-1
12:45	45	21.0	10.5	9.0	N	17.4	5.2	Profile E-2
14:30	45	23.2	11.6	6.0	N	16.4	6.6	Profile E-3
16:10	46	23.2	11.6	7.5	Almost	17.4	5.0	Profile E-4

Table 7 Results for O/W Dispersion

Time	Temp (°C)	Feed (L/h) Solvent	Aqueous	Amplitude (mm)	Flood	Extract (g/L)	Raffinate (g/L)	Remarks
8:30	37	23.2	11.6	6.0	N	—	—	Start
10:45	42	23.2	11.6	6.0	N	17.0	6.5	Profile E-5
12:45	45	25.4	12.7	9.0	N	17.3	4.8	Profile E-6
14:30	45	25.4	12.7	6.0	N	16.0	6.5	Profile E-7
16:10	46	27.6	13.8	7.5	N	17.4	5.3	Profile E-8

Table 8 Analytical Results (g/L) of the Best Profiles for Each Dispersion

Stream	E-4 Aqueous	Solvent	E-6 Aqueous	Solvent
SP 1	12.1	1.2	10.5	1.4
Sp 2	17.3	3.0	19.1	3.8
Sp 3	20.3	4.9	21.4	6.0
SP 4	24.3	6.8	28.4	9.5
SP 5	34.3	12.8	37.2	12.5

Analysis of the results:

1. *Recovery*   This is the most important parameter. In both dispersions the criterion of 85% recovery was met in two profiles. The difference between the results of all four profiles that met the criteria is within the experimental accuracy. The profiles are smooth and show that if the column was taller, higher recovery could have been achieved. Increased energy improves the results.
2. *Flux*   The flux at O/W was 20% higher, and the recovery similar. Unless special considerations rule against it, this is the preferred dispersion.
3. *Working point*   For W/O, the recovery at the maximum flux was the best, so this is the recommended point. For O/W, the recovery in profile 6 is slightly better than in the maximal flux of profile 8. Since the difference is within the experimental accuracy, the higher flow should be recommended.

Final Conclusions

For the verification run, it is recommend to work in O/W, 27.6 L/h solvent and 13.8 L/h aqueous feed, pulsation with amplitude of 7.5 mm and 120 strokes per minute. If the verification run, which should take 8 h at least, confirms the results obtained in profiles 4 and 8, these are the recommended data for the designers. If the results are inferior, make another run at the conditions of profile 6 and compare.

This is the end of the experimental work, and it is time to write the final report. Since the work was done in a bench-scale column, it must be verified in a full-scale pilot, typically in a 100 mm column.

Remark

Ez is a synonym for the extracted component. The data in the tables were taken from real runs, the details of which cannot be disclosed.

5. *Optimization*

If liquid–liquid equilibrium data and a simulator are available, the composition of the outlet streams may be compared with the results of the computerized simulation. This enables us to estimate the average HETS, as a function of the energy input and the flux (i.e., the residence time).

The target of the optimization of an extraction process in a column is the combination of the best mass transfer (i.e., minimal HETS) and the maximal flux. This optimum is sometimes described by the means of the volumetric efficiency η_v (m³/h) [20], which is defined as follows:

$$\eta_v = \frac{J}{\text{HETS}} \tag{20}$$

where J is the total flux (m³/m²/h).

The volumetric efficiency is inversely proportional to the volume of the column that is required to do a given extraction job.

This optimization is done for a given phase ratio and has nothing to do with the optimization of the number of theoretical stages (NTS) to phase that applies to all types of extraction equipment and was dealt with in Section IV.C.4.

In a process with rapid diffusion, working at the maximum flux does not cause a significant increase in the HETS. In this case the optimization is easy, since the maximum possible flux is also the process optimum. Sometimes, however, working at the maximum possible flux yields a poor mass transfer (i.e., a high HETS). The only way to decrease the HETS is to increase the energy input, decreasing the flux. In these cases η_v goes through a maximum. The conditions corresponding to the maximal value of η_v represent the *apparent* optimum performance of the test column.

Note that this optimum is apparent, since the trade-off between the cost of a wider column (determined by the flux) and a higher one (determined by the HETS) is not straightforward. This can be done only by the supplier of the column.

C. Equipment Pilot of Mixer Settlers

The bench-scale results proved experimentally the feasibility of the process. If mixer-settlers are the recommended option for all or a part of the plant, an equipment pilot plant, consisting of one or two mixer-settlers, is needed.

1. General Considerations

There are three strategic points to be determined before designing the pilot.

a. A Pump-Mixer or a Pump and a Mixer. A pump-mixer is more space- and cost-effective. There are two main types: axial pump mix (APM) and turbine pump mix (TPM). They were both developed at IMI and are described elsewhere [21]. Their main disadvantage is that in pump-mixers there is an unwanted linkage between the flux and the mixing intensity, and sometimes the intensity needed to obtain the pumping head is too high for the process, causing high entrainment and lowering the flux in the settler. Short residence times are implied, as well.

An agitator that does not have to act as a pump may be designed to suit any process demand (e.g., short or long residence time, gentle or high shear mixing). Its drawback is the need of pumps to transport both phases to the next stage. In addition to the cost and maintenance problems, the suction of a solvent with an entrained aqueous phase through a centrifugal pump may cause a stable emulsion, and for some processes this is a no-go factor.

b. An Empty or a "Furnished" Settler. A "furnished" settler has partitions that enhance the coalescence. Development of these units at IMI is described elsewhere [21]. The flux in a "furnished" settler is higher by 200–300% than in an empty one. "Furnished" settlers work well with a small amount of floating solids, but they cannot be used when heavy crud is present. This is why most settlers in hydrometallurgy are empty.

c. How Many Mixers to Put in Series. Multiple mixers in series must be used if pump-mixers are used and a residence time exceeding one minute is necessary. The number of mixing boxes is readily calculated from knowledge of the required τ from the kinetics and the available one from the equipment manufacturer.

> *Remark*
> For very large flow rates it may be more cost-effective to use two or three mixers in series instead of a big one. This configuration ensures good mixing with decreased residence time distribution, and excessive problems of shortcuts and dead volumes in huge mixers are avoided. However, this consideration belongs to the design of the industrial plant and is not a part of the process development.

2. Design of the Pilot Plant

The choice of the specific type and size of mixer and settler should be made in consultation with your supplier(s) or based on your experience. Unless this aspect is a part of your expertise, both the mixer and the settler should be designed and supplied by your anticipated equipment manufacturer, who probably will also assist you in piloting and in the analysis of the experiments. Remember that if the mixer or settler do not have full geometrical similarity to the future ones, the experiments are useless.

Location

For an existing process, the recommended location is the site of the plant, to enable the pilot to run with the real solutions. Once the pilot has been erected at the site, it can be relatively easily connected to various batteries, thus giving representative results from all parts of the plant from just one experimental unit. This siting also solves the problem of transport and storage of large amounts of solutions. Otherwise, it has to be erected in the research center.

Transparency

It is most important to build both the mixer and the settler shells from transparent materials, since it is essential to see the dispersion band in the settler and the phase dispersion in the mixer. The size of the pilot plant usually rules out glass. If possible, a polymer like PVC, Perspex or polycarbonate should be used. Even a polymer that can endure only a few weeks may be satisfactory, since the total time the pilot is operated is short. The shaft and the impeller may be made from any suitable material, preferable dark.

Internal Reflux

In mixer-settlers a phase ration $R_{cd} > 6$ or $R_{cd} < 1.5$ usually implies internal reflux. If the process is expected to operate in this range, the experimental unit should enable it.

One or Two Stages?

Usually one mixer-settler is enough for the equipment pilot. A second stage almost doubles the cost of the pilot plant and the space it takes. If the pilot is erected on-site, one stage is *always* sufficient. Still, two stages are recommended under the following conditions:

    1.   *Two batteries are tested simultaneously.* If both extraction and stripping are to be tested in the research center, using two mixer-settlers in series makes the run much quicker and easier by providing both loaded and stripped solvent simultaneously.

    2.   *Pump-mixers are tested.* Although it is possible to carry out all the tests using one stage, it is quicker and more representative to use two stages in a closed loop, permitting you to see the influence of changing the process parameters within the loop. If pump-mixers are tested for a new process, with two or more batteries, the use of two stages is strongly recommended.

Size

The size of the settler is set by the manufacturer. To decrease scale-up problems, the height of the settler in the pilot plant is often equal to that of the industrial settler, between 1.5 and 2 m. The area is usually set to 0.1 m^2 (i.e., yielding a volume of 150–200 L), although various manufacturers may use different sizes.

    The characteristic flux in an empty settler is 3–10 $m^3/m^2/h$ and in a "furnished" one 10–25 $m^3/m^2/h$. This implies that the expected flow rates in the pilot plant are between 300 and 1000 L/h in an empty settler and between 1000 and 2500 L/h in a furnished one. Note that these flow rates are much higher than the flow rates through a pilot column. The average τ in the settler varies between 10 and 30 min in an empty settler and between 5 and 15 min in a furnished one.

    The volume of the mixer is set so that it provides the desired residence time. For a process with an approximate τ of 60 s, the mixer volume should be between 7 and 40 L. Again, it must be large enough to satisfy the minimum requirement of the manufacturer.

The mixer is always cheaper than the settler, and thus it is good engineering procedure to overdesign it, to be sure that the most expensive unit—in our case the settler—is the flow-limiting feature (i.e., the bottleneck).

3. Experimental Procedure

In general, the running of an equipment pilot for mixer-settlers is similar to the hydraulic test for columns. The advance phase ratio O:W is set according to the simulation and the bench-scale results and is not a parameter for these experiments.

If internal recycling is to be implemented, start with the *minimum recycle* that causes the phase ratio in the mixer to be in the $1.5 < R_{cd} < 4$ range. When the working conditions for this recycle have been set, the influence of this ratio on the type of dispersion, mass transfer, and entrainment, at given flux and speed of the mixer, may be analyzed. The main parameters to be determined are given next.

a. Type of Dispersion: O/W or W/O. To satisfy this parameter, all the experiments, listed shortly, must be carried out for both dispersions, to evaluate the best one. Remember that different dispersions may be used in various stages of the battery. This is a practice that should be encouraged.

Example 9
How to design a battery in which the flux and mass transfer were found to be much better with O/W dispersion, but the entrainment of the solvent in the raffinate must be as low as possible.

Solution
All the battery should be run O/W, except for the stage from which the raffinate leaves. This stage should be operated W/O, to decrease solvent entrainment, with a larger mixer and settler, to ensure full mass transfer and reasonable aqueous entrainment.

b. Mass Transfer. The residence time in the mixer and the mixer speed required to obtain full mass transfer should be determined. Mass transfer is a parameter of the mixer only; thus the performance of the settler may be disregarded while mass transfer is being tested. Remember that the shortest possible τ is the one determined in the laboratory kinetic experiments (see Section V.D).

The experiments are rapid and easy, but they require a great deal of chemical analysis: for every flux through the mixer, the composition of both outlet streams must be measured as a function of the speed of the mixer. If the desired continuity cannot be obtained, the phase ratio is too high, or the mass transfer too slow, so apply internal recycling. Since the composition at equilibrium is known from the simulation, it is easy to find the minimum speed for which equilibrium is reached at the given flux.

The output is a table of residence times (i.e., flux in the mixer) vs speed of the mixer and composition of the outlet flows. From this table you obtain the minimum residence time needed to reach equilibrium vs the speed of the mixer.

To save time, start with a simple equal space net:

i.   Choose three speeds of the mixer: n, $1.5n$, and $2n$, where n is the minimum speed required to achieve full dispersion of both phases in the mixer and the required flow rate (for a pump-mixer). Since the mixing energy is proportional to n^3, the energy input in the last experiment is eight times higher than in the first one.

ii.   Choose three residence times: τ_k, $1.5\tau_k$, and $2\tau_k$, where τ_k is the minimum residence time found in the kinetic experiments (see Section V.D), and calculate the corresponding flow rates of both phases.

iii.   Carry out the nine experiments and analyze the compositions. Use the results to design the rest of the experiments (see Example 10, below).

c.   Pumping Capacity and Head.   The pumping capacity and head for a pump-mixer should be determined as a function of the speed of the impeller. These are the simplest and quickest experiments. Up to 10 points are sufficient, with each experiment taking about 15 min.

Note

For a given flux, the speed of the impeller that is required for the pumping may differ from the one needed for complete mass transfer. For design purposes and settler flux experiments, use the higher of these two values.

d.   Entrainment in the Settler.   The entrainment in the settler should be determined as a function of the flux and the speed of rotation of the mixer. Unlike columns, there is no "flooding" in mixer-settlers, and therefore the definition of the maximum flux is not unique. It is determined either by the width of the dispersed band in the settler or by the entrainment of the dispersed phase in the continuous one.

The tests are done by measuring the entrainment at various fluxes and speeds of rotation. Again, an equal space net is recommended. These data will be used for optimization of the flux vs the entrainment. All that needs to be set prior to the experimental work is the maximum value of the dispersed band and the tolerated entrainment. The tests are stopped as soon as one of these values is exceeded.

Note that while the flooding in a column is straightforward, the maximum entrainment and the width of the dispersed band depend on the process and on the designer's criteria.

A reasonable entrainment may vary between 100 ppm and 5%, depending on the damage it causes. The entrainment causes three main problems:

i. *Solvent losses* This is the most sensitive parameter. In many processes, solvent losses above 50 ppm entrainment require special treatment of the aqueous product or the raffinate. The allowed entrainment is usually lower if O/W dispersion is used rather than a W/O one.

ii. *Product contamination* This is usually less critical than solvent losses and may vary from 100–200 ppm in one process to 0.1–1% in another.

iii. *Back-mixing* This is the least sensitive parameter. It depends on the phase ratio, since the back-mixing caused by the entrainment is equivalent to:

$$\alpha = \frac{F_c \varepsilon}{F_d} = R_{cd} \varepsilon \qquad (21)$$

where α is the back-flow, ε is the entrainment (in volume percent or volume fraction), and R_{cd} is the ratio of continuous phase to dispersed phase. For a phase ratio of 1:1, the back-mixing due to 1% entrainment is only 1%, but at $R_{cd} = 20$ it causes back-mixing of 20%.

Note

Solvent losses or product contamination apply only to the stages from which either an aqueous or a solvent stream is leaving the battery. Those stages often have larger settlers or two settlers in series. For all other stages, the entrainment causes back-mixing only. In many processes, especially when $R_{cd} \sim 1$, entrainment as high as 3–5% is acceptable for the inner stages.

Chemical analysis is required only during the mass transfer test. No chemical analysis is needed during the testing of the flux and the entrainment in the settler because the degree of completion of the extraction is irrelevant.

4. Analysis of the Results

a. Flux in Settler vs Entrainment. The optimization of entrainment vs flux is one of the important results of the equipment pilot and deserves some explanation. There is always a trade-off between the entrainment and the flux. Higher flux enables a higher production rate. Higher entrainment means either more stages, to compensate for the back-mixing and product contamination, or a higher flow rate of the purification and strip feeds to the battery (i.e., a lower yield of the process). The bottom line is a higher flux at the cost of higher solvent losses and either more stages or lower yield. A good simulator allows the assignment of entrainment to any single stage [22], hence the optimization of the plant without excessive experimental work.

Remember that different values should be assigned to the entrainment in the middle of the battery and at its end.

b. Flux in Mixer vs Flux in Settler.   Quicker mixing creates finer droplets. This means a more rapid mass transfer (i.e., a smaller mixer) on the one hand, but slower coalescence (i.e., a larger settler) on the other. The optimum is not straightforward; rather, it depends on the relative price of the mixer and the settler. Unless this is a part of your expertise, leave it to the equipment manufacturer. Moreover, it is possible that the mixer/settler volume ratio in your pilot plant is not optimal. This is another value to be determined by the equipment manufacturer, based on the pilot data.

As in columns, prepare the full table of speed of rotation, τ in the mixer, and the maximum flux in the settler that met the entrainment determined in the preceding section. The equipment manufacturer will choose the most efficient alternative for you.

Example 10
The mixer-settler that was recommended in Example 8 (Section VII.B.4) must be tested.

1. Basic analysis
In a single mixer-settler, the incentive to use a pump-mixer is small. But in this case the client explicitly requested the use of a TPM. Since the reaction duration is about 30 s, a single TPM is sufficient. The maximum speed of the impeller is 80 rpm.

Since the real solution is not expected to have an excessive amount of solids, a furnished settler (i.e., with internal components) may be used. The area of the settler, as recommended by the manufacturer, is 0.1 m^2, and its height is 1.8 m. Assuming a flux of 20 $m^3/m^2/h$ (a reasonable range for this type of equipment), the flow rate through the unit will be about 2 m^3/h. The volume of the mixer, to enable τ of 30 s at this flux, is about 20 L.

The phase ratio $R_{O:W}$ is 10. The phase separation characteristics are similar, and so the obvious choice is to use the normal dispersion (i.e, W/O). Because $R > 4$, an internal recycle is installed. A loaded solvent is needed; hence the experiments must be done along with the tests of the column from the preceding section. The column supplies the mixer-settler with loaded solvent and takes the stripped one. Still, the tests must be separate. Typically, the column is run on one day and the mixer-settler on the next. Remember that the output of 8 h work of the column is sufficient for 2 h of the mixer-settler.

2. Hydraulic test of the mixer

In the second stage the flow of both phases through the mixer and the head developed by the TPM are measured as a function of the speed (rpm) of the latter. Note that these experiments are very sensitive to the configuration of your plant. The length and diameter of the pipes and the pressure drop through the flow meters and the valves may influence your results. All you can do is note exactly what you have on every pipe, and the TPM manufacturer will calculate from this information.

No internal recycling is applied in this stage. The phase ratio is kept at 10:1 by regulation of the aqueous flow through a valve. The results of the experiments are listed in Table 9.

Remark

Because a flow rate above 2000 L/h causes a residence time below the kinetic constraint, the maximum reading on the solvent flow meter was 2500 L/h. At a flux of 60 $m^3/m^2/h$, the τ was below the lower limit. There was no point in testing a higher speed of the mixer, and it was obvious that the pumping capacity is not the limiting factor.

3. Mass transfer for the TPM

The extract contains 17.1 g/L Ez and the stripping solution contains no Ez. The D value for this range is 0.0091; that is, the expected equilibrium composition is 165 g/L in the product and 1.5 g/L in the washed solvent. The tests were carried out at two residence times:

    a.   *High residence time*: τ = 60 s, F_s = 1100 L/h, F_a = 110 L/h; TPM speed = 46–70 rpm

Table 9   Results of Hydraulic Tests

Speed of mixer (rpm)	Feed rate (L/h)		Residence time (s)
	Solvent	Aqueous	
20	0	0	—
30	200	20	330
35	500	50	132
40	1000	100	66
45	1500	150	44
50	2000	200	33
55	2500	250	26
60	<2500	300	<24

b. *Minimum residence time*: $\tau = 30$ s, $F_s = 2200$ L/h, $F_a = 220$ L/h; TPM speed = 58–80 rpm

The results of the experiments, listed in Table 10, are not unexpected. The system is known to have good diffusion characteristics, and the mixing is quite intensive. No doubt the equilibrium could have been reached with slower mixing. This is one of the disadvantages of a pump-mixer. The results also prove that as far as mass transfer is concerned, there is no need of internal reflux.

4. Entrainment vs flux and speed of rotation

Table 11 gives the entrainment results.

Analysis of Results

1. Settler

The settler can maintain a flux of 24 $m^3/m^2/h$ with reasonable entrainment as long as the speed of the mixer is low. Entrainment of 0.4% (i.e., 4% of the product going back to the extraction) may be acceptable to this process. The simulator shows us that to keep the required recovery of 85%, the solvent flow rate must be increased by 3%, which is acceptable. If the speed of the mixer may be further lowered, the entrainment would decrease.

2. Mixer

a. Pumping performance is satisfactory: the TPM can pump far more than we need.
b. Mass transfer is excellent. Even at the minimum speed, needed to get τ of both 30 and 60 s, 100% mass transfer is obtained.

However, the combination is problematic.

Table 10 Results of Mass Transfer Tests

Speed of mixer (rpm)	Ez in product (g/L)		Mass transfer (%)	
	$\tau = 30$ s	$\tau = 60$ s	$\tau = 30$ s	$\tau = 60$ s
41	—	164	—	100
48	—	167	—	100
55	—	163	—	100
60	160	166	100	100
65	168	165	100	100
70	164	166	100	100
75	166	—	100	—
80	163	—	100	—

Table 11   Results of Entrainment Tests

Speed of mixer (rpm)	Flow rate (L/h)		Flux (m³/m²/h)	Entrainment (%)
	Solvent	Aqueous		
40	1000	100	11	<0.1
60	1000	100	11	0.2
80	1000	100	11	0.8
45	1650	165	18	0.1
65	1650	165	18	0.3
80	1650	165	18	1
55	2150	215	24	0.4
68	2150	215	24	1
80	2150	215	24	3

If a mixer with τ of 30 s is to be used, its speed is 50 rpm at least. In the pilot plant it probably should be higher, to compensate for longer pipes, extra valves, and so on. And with time the solvent may deteriorate, causing higher entrainment. So some safety factor should be taken. It means that either the settler operates at flux of 24 m³/m²/h but with a larger mixer ($\tau = 40-60$ s), which allows a lower speed of agitation, or a mixer with a residence time of 30 s is used, but the design flux should be lowered to about 20 m³/m²/h. Usually larger mixers are used, since they tend to be cheaper on one hand, and should be able to compensate for a probable deterioration in the mass transfer in the future.

Since the pump-mixer caused excessive entrainment, it is natural to reconsider the use of a regular mixer and pumps. How to do this in the quickest way? Just add pumps and use your TPM at a low speed. The results obtained are listed in Table 12.

Now the trade-off is clear. For slow mixing, the entrainment is low even at a flux of 24 m³/m²/h (0.2% at most). For a residence time of 60 s, full mass transfer is obtained at any reasonable speed of mixing. For $\tau = 40$ s, full mass transfer is obtained at 40 rpm, with resulting low entrainment of 0.1%. To get full mass transfer at $\tau = 30$ s, however, the mixing must be quite vigorous, about 50 rpm, resulting in high entrainment. Since you need mixing of about 40 rpm, it is better to stay with the TPM. Slower mixing harmed the mass transfer, and a mixer with a residence time of 30 s is too small. A larger one, with $\tau = 40-60$ s, is needed. Consult the manufacturer. But if the difference in price is not too high (and it should not be), make τ of 60 s and be sure that it works. Now you can design your settler for a flux of 24 m³/m²/h without hesitation.

Table 12 Results at Low Mixing Speed

Speed of mixer (rpm)	Flow rate (L/h)		Flux (m³/m²/h)	Residence time (s)	Entrainment (%)	Ez in product (g/L)	Mass transfer (%)
	Solvent	Aqueous					
20	1000	100	11	60	<0.1	153	92
30	1000	100	11	60	<0.1	167	100
40	1000	100	11	60	<0.1	163	100
20	1650	165	18	40	<0.1	139	85
30	1650	165	18	40	<0.1	152	92
40	1650	165	18	40	0.1	166	100
20	2150	215	24	30	0.1	124	75
30	2150	215	24	30	0.15	140	85
40	2150	215	24	30	0.2	149	90

D. Pilot Plant of Mixer-Settlers

A full-scale pilot plant of mixer-settlers is run when both the bench scale and the equipment pilot are finished. Its purpose is the verification of earlier results, finding the actual maximum flux and mass transfer and observation and treatment of rapid accumulation phenomena.

If plans for the full-scale plant include both columns and mixer-settlers, a pilot of the recommended configuration should be run, after the bench-scale test has been run both for the mixer-settlers and for the columns, and the equipment pilot for the mixer-settlers was finished. The piloting should follow simultaneously the procedure listed next and the procedure for piloting of columns (see Section VII.B).

1. Determination of Mass Transfer

Mass transfer can never be measured directly, but it may be evaluated in a simple experiment. The outlet from the mixer is sampled into two separating funnels. The dispersion from the first one is separated and analyzed. The second is shaken manually for 3 min, and the phases are separated and analyzed. The difference in concentration between the samples gives a clear picture of the mass transfer, using either stage efficiency or the mass transfer factor (MTF) [22].

a. Stage Efficiency. Murphree defined stage efficiency for distillation as the ratio of the actual concentration change within a stage to the change that would have occurred if equilibrium had been reached. Later this was adapted to solvent extraction:

$$E_M = \frac{y_{n+1} - y_n}{y_{n+1} - y_n^*} \tag{22}$$

where the numbering of the stages, n, follows the rule that $y_{n+1} > y_n$.

As pointed out by Lloyd [23] and others, the application of Murphree's efficiency to solvent extraction is problematic, especially if the concentration profile is flat or passes through a maximum.

b. Mass Transfer Factor. The mass transfer factor is defined as the ratio of the actual concentration of the solute in the solvent and its equilibrium value:

$$\text{MTF} = y/y^* \tag{23}$$

where y is the concentration of the solute leaving the mixer in the solvent and y^* is the respective concentration after the shaking.

Note that the MTF tells us nothing about the *actual amount* of mass transfer that took place. The MTF value can be readily applied to simulation:

$$y = K \times \text{MTF} \times x \tag{24}$$

2. Design of the Pilot Plant

The design of the mixers and the settlers is done according to the results of the equipment pilot. The units usually are smaller than in the equipment pilot, but they should be big enough to enable reliable scale-up of its results.

The pilot plant should have the full configuration of the future plant, including auxiliary units (e.g., for solvent regeneration). It is important to use the same sensors and control scheme as proposed for the industrial plant. The pilot serves to test them too. It is recommended, although not mandatory, to use the type of automatic control that is proposed for the industrial plant. This enables direct transfer of the control scheme, without wasting time and money on adaptation to a new control system.

Never save on sensors in the pilot plant. Try to measure every parameter, even if its importance is not obvious. Measure the temperature in every settler. Extraction is very sensitive to temperature, and it is necessary to know the temperature in every unit (deviations up to 10°C within a battery are common in a pilot plant). If the pH is important, install two electrodes in every settler, and be sure they are not in a spot where the solvent could contact them. Measure also all the flow rates—not only those of inlets and outlets, but also those of most of the interim flows. If an internal recycle is applied, it should be measured too. The mixers should have variable speed drives, with an output to a display.

It is not essential for the units to be transparent, although for the settler this is helpful. If the settlers are opaque, a sight glass along the entire length of the settler, with multiple connections to the settler, is required. A sight glass that is a transparent strip on the settler itself is rather problematic and is not recommended, especially if crud is expected.

a. Location. Since mixer-settlers are usually used for high flow rates (for small plants, columns are always recommended), and a pilot plant is on the order of 1% of the final plant, the flow rates in a pilot plant range from hundreds to thousands of liters per hour. Erection of such a plant in a research center creates logistic problems of feed, product, and waste management. For an existing process, the recommended location is the site of the plant, to enable operation with fresh and real solutions, without any problems of transport and storage of very large amounts of solutions.

b. Duration. This depends on the complexity of the process, its novelty (a new solvent for a known process requires shorter piloting than a totally new application), and the budget. If a demonstration plant is planned later, one month of running-in is sufficient. Otherwise 2–4 months is recommended, with at least one month of smooth operation, without the discovery of any new accumulation phenomena.

Remark
Erection of a demonstration plant after a full-scale pilot is not recommended and is quite rate.

3. Experimental Procedure

The main goal of the pilot plant is to prove that everything done so far was correct. It is, therefore, difficult to derive an exact experimental program. Usually the plant is started at conditions about 10% below the optimum found in the former stages, and after stabilization and analysis of its behavior we start pushing it up. Theoretically, if the earlier stages were perfect, the plant should stabilize on the apparent optimum and stay there. But this never happens, and that is why pilot plants are needed.

The four recommended stages of the piloting are described next.

a. Stabilization and Initial Analysis.   The plant is run a little below the apparent optimum. All the data are measured and analyzed: inlet, outlet, and intermediate flows, internal recycle, temperature, speed of mixers, entrainment, concentrations of both phases leaving every settler, and mass transfer. After a few days of steady operation, all the process values that relate to this steady state are evaluated. They serve as a basis for comparison for the future.

b. Adjustment of Mixers and Internal Recycle.   The plant is started at the same speed of rotation for all mixers and standard internal recycle. The initial results may show that for some stages this speed is too high, causing excessive entrainment, while for others it is too low, causing low mass transfer. It is quite normal to set every mixer at a different speed. The trade-off between entrainment and mass transfer was evaluated in the preceding step (see Section VII.C). The effect of the internal recycle on both mass transfer and entrainment may be tested readily by changing it and analyzing the outlet. (As a matter of fact, it is easier to test the internal recycle in the pilot plant than in the equipment pilot.)

c. Maximum Flux.   After the speed of mixers and the internal recycle have been set, it is time to start and increase the flux. This is done by increasing simultaneously the inlet flows at the same phase ratio, by steps of 5%. Check the entrainment after 2 h. If it exceeds the allowed level in any settler, analyze the mass transfer in that settler. It it is high, decrease the speed of its mixer and measure again. If everything seems steady, measure the entrainment again after 12 h. If the entrainment is within process limits, the flux may be increased further. If not, *this is the maximum permissible flux.*

d. Accumulation Phenomena and Solvent Regeneration. It is impossible to predict what accumulation phenomena will occur. A few may be foreseen from the stability tests (Section V.E) or from the bench-scale experiments (see Section VII.A). The others just come unexpectedly—this is why a pilot plant is needed. Hopefully the running-in can last long enough to discover most of the accumulation phenomena and treat them. If not, the problems will appear, *bigger*, in the industrial plant.

Often the procedure for solvent regeneration is known before the pilot is started and is included as a part of it. This procedure may include the following steps.

    i. *Distillation* For volatile solvents. It is applied to the washed/stripped solvent.

    ii. *Chemical regeneration* When some ions create a stable compound with the solvent and poison it, stripping with the right reagent regenerates the solvent. The treatment is usually done on a sidestream, and the resulting aqueous stream is either disposed of or recycled to the process. An optimization schemes should be developed as a means of minimizing the cost of the regeneration without losing on solvent performance. In processes with a nonvolatile solvent in a closed loop (e.g., hydrometallurgical processes), such regeneration is a common procedure.

    iii. *Filtration* To remove solid particles that accumulate in the settler and/or float with the solvent. This is done on a sidestream, usually by means of a decanting centrifuge. Later the filter cake is treated chemically to release the solvent from it, both for economical and environmental reasons. This procedure should also be part of the pilot plant.

E. Demonstration Plant

A demonstration plant is erected for mixer-settlers, columns, centrifugal extractors, and any combination of them, to simulate the future plant. The data needed for erection and running of the demonstration plant are similar to those that are needed for the full-scale pilot.

The running-in is divided into two periods: piloting and verification. In the first period, the experimental procedure is similar to the running of a pilot plant, both for columns and for mixer-settlers (see earlier sections). The main difference is that the demonstration plant is not transparent, and the sensors must serve as the eyes of the operators, as in the industrial plant. The length of this run is similar to that of running-in of a pilot plant.

The verification run is much longer. During this run few experiments are done, and the main goals are to keep the plant working at steady state, to observe the accumulation phenomena, and to get true values of the production rate and the yield. The shift supervisors at this stage are usually plant operators, not skilled technicians, and the plant is run 7 days a week.

VIII. PROCESS BOOK (RECOMMENDATIONS AND SCALING UP)

All the experimental work is finished. It is time now to analyze the results and write the process book. Often unexpected problems have been encountered, and solved by heating/cooling or addition of reagents, stages, or auxiliary units. These measures solved the problems, but increased the costs.

The most important question now is: *Is the recommendation to erect the industrial plant still valid?*

To answer this crucial question, the following factors should be tested.

1. The product quality, the yield, and the production rate of the plant
2. The cost of the raw materials per ton of product, including energy for heating/cooling, solvent losses, and solvent treatment
3. The layout of the industrial plant, including all auxiliary units

Compare all three factors with the estimate on which the recommendation to go to piloting was based.

A. Project Rejection

The project should be rejected in the following cases:

1. *The product quality is unsatisfactory.* In rare cases the pilot or demonstration plant fails to make a final product having the desired quality, even with an increased number of stages and a higher purification stream or O:W ratio. Usually an unacceptable result is due to contaminants not present in the artificial solutions in the laboratory and bench-scale stage that either lower the solvent loading/selectivity or are coextracted.

2. *The cost of the raw materials is too high.* Most unexpected costs are related to the solvent: high losses due to entrainment and decomposition, expensive stabilizers, high cost of recovery of entrained solvent from the aqueous outlet streams, and so on. Additional costs may be due to the demand of excessive heating on cold days or cooling on hot ones and higher flow rates of purification/scrubbing solutions.

3. *The expected cost of the industrial plant is far above the initial estimate.* This may be caused by three main factors:

a. The flux is below expectation (usually due to surface phenomena in the solvent and/or floating crud).
b. Increased NTS and size of the equipment are required to compensate for lower solvent loading and/or selectivity or higher entrainment that were discovered during the piloting.
c. Auxiliary units for solvent/crud/entrained-solvent treatment had to be added to overcome problems detected during the piloting.

Most of the problems are linked directly to the solvent, and this is natural. In *solvent* extraction the *solvent* is the heart of the process, and it determines its feasibility and profitability.

The other potential source of problems is sedimentation of solids.

If the product quality is too low or the variable costs are too high, you may and should reject the project in your final report. In the case of higher fixed costs, this is not up to you. State clearly the increased costs and their cause, make a note about "decreased profitability," and send your report to the client. You do not have enough data to reject the process for reasons of fixed costs. In every case, make recommendations on how to overcome the obstacle.

In the case of decomposition or poisoning of the solvent, ask your chemists how to stabilize or regenerate it. If nothing helps, consider an alternative solvent.

If the phase separation deteriorates and the entrainment increases, try more efficient solvent treatment, and in parallel try some modifiers or heating. If nothing helps, consider switching to equipment that can cope better with difficult phase separation (e.g., columns instead of mixer-settlers). Do not give up—leave this decision to your client.

Sometimes the failure is your fault because problems could have been identified and dealt with in earlier stages of the project. More often you are not to blame, but in any case it is much better to discover the cause of failure now than after erection of the industrial plant. *Never help the results to fit your expectations.* A researcher must be optimistic, but also realistic.

B. Process and Plant Description

If the original recommendation for utilization of the process is still valid, all the data that were mentioned in Section II must be supplied as a part of the process package.

1. Solvent Composition and Properties

In this stage there may remain only one recommended solvent. The following must be specified:

a. The chemical and commercial names of all the components of the solvent: extractant(s), diluent(s), modifier(s), stabilizer (if applicable), and their Material Safety Data Sheets (MSDS)
b. The concentration of each component, including the allowed deviation and the harm caused by larger deviations
c. The viscosities at two temperatures, and the interfacial tension with the feed and product
d. An analytical procedure to check the composition (including the concentration of known contaminants and products of decomposition), and an applicable test of how to check the performance of the solvent (loading, stripping, and phase separation)
e. Data about solvent stability and any special precautions to be taken (e.g., nitrogen blanketing)

2. Temperature and pH Range for Each Battery

The temperature and pH range should be specified for each battery, as well as scenarios to be expected in the event of larger deviations.

3. Composition of All Aqueous Streams

Even if the composition of some of the streams is not up to you, it is essential to state the range for which the process is valid. If a deviation in feed concentration occurs in the future, the client must know what to change to get the proper product from the plant. This is also your insurance policy: you developed a process for a given feed, and if its composition changes beyond the defined limits, a drop in the recovery or production rate is not your responsibility. In some processes there exists a lower limit, below which the process is no longer feasible. This should be clearly stated, especially if the process is designed for hydrometallurgy, where changes in the concentration of the feed are frequent and high. For example, the process book of calcium bromide extraction states that the plant should be shut down below certain combinations of feed composition and temperature. "A documented bug is a feature"—but as long as it is not documented, it remains a bug.

4. Process Characteristics

Process characteristics include product(s) concentration and purity, overall yield of the process, concentration of the product(s) in the raffinate, and entrainment of solvent in all outlet aqueous streams.

Note
Although the concentration of the product in the raffinate determines the yield of the process and vice versa, sometimes both values are required.

This is important mainly when the recovery is very high and there is a specific constraint—either environmental or psychological—on the raffinate. For example, in extraction of uranium, in some plants the environmental constraint limits the allowed concentration of uranium in the raffinate to 2 ppm, regardless of its concentration in the feed. In another case, a manufacturer demanded that the concentration of the product in the raffinate be below 10 g/L, down from 700 g/L in the feed. There was no real reason for this demand beside the fact that it is easier to sense that something is wrong when the raffinate rises from 10 g/L to 12 g/L than when the recovery is down from 98.6% to 98.3%.

5. Plant Layout

Plant layout concerns include the following:

a. The number of batteries and the NTS in each battery.
b. The location and flow rate of all streams in the plant, including the advance phase ratio (i.e., the phase ratio of the inlets) in every battery.
c. The exact recommended configuration of the plant, that is, the type of equipment recommended in any battery. Remember that it is quite normal to use several types of solvent extraction equipment for various batteries and even within a battery.

6. Solvent Extraction Equipment

The description of the solvent extraction equipment should give the following items.

a. Detailed description of every type of equipment used in the pilot, including manufacturer's name, number of units, dimensions (height, width, and length or diameter, effective volume), energy input (speed of rotation/vibration/pulsation), and internal recycle (if applicable).
b. The flux, internal recycle (if applicable), dispersion, and W/O entrainment in every unit, at the optimum, and if available—also as a function of the energy input, temperature, and so on.
c. Description of the performance of each unit, modifications made during the piloting, and recommended changes for the industrial plant (if any).

7. Auxiliary Equipment

Equipment should be recommended for the following:

a. Solvent regeneration and stabilization
b. Crud/solids removal

c.   Removal of entrained solvent from aqueous effluents
d.   Evaporation/distillation units (if applicable).
e.   Pumps, filters, and so on

For every unit, describe the exact type, size, flow rates, material of construction, and so on.

8.   Measurement and Control

Measurement and control considerations include the following:

a.   A description of the control strategy of the piloting: open and closed control loops, manual and automatic operations, and so on, as well as recommendations for the industrial plant.
b.   Full description of the sensors and final control elements, including performance of each unit, modifications made during the piloting, and recommended changes for the industrial plant (if any).

Remark
The drawings and specifications made to erect the pilot should be used at this stage.

9.   Scaling Up

Scaling up is the most difficult part, but most of it is not your task. The actual scale-up of the solvent extraction units will be made by the manufacturers, whereas the design and sizing of pumps, piping, storage tanks, and so on belongs to the design team. All you may do is recommend the diameter (or width and length) of the columns or settlers. For most types of equipment this is quite easy, since the flux in every unit is known and does not change in the scale-up. Yet in some columns (e.g., the Scheibel column), the flux is a function of the diameter. You may also have to check the mechanical flow sheet (P&I) before it is sent to the client.

C.   Description of the Experimental Work

All the experimental work that was carried out during all the stages of the development must be summarized in your report. This should not be a part of the process book, but should be included in separate reports for each stage. If these have yet not been prepared, because you were busy with the piloting, now is the last occasion to write them, before the project is closed and you are transferred to another task.

It is most important to document all the work that was done, *including all the unsuccessful experiments* that convinced you to change the solvent composition, equipment, or operational procedures. It is not enough to document only the latest success. Otherwise, in a few years' time somebody may repeat the work associated with all those ideas that you tested and rightly abandoned.

Remember

What is not written was as good as never done, and a negative result is as important as a good one.

REFERENCES

1. R Blumberg. Liquid–Liquid Extraction. London: Academic Press, 1988, pp 1–73.
2. G Ritcey. Development of industrial solvent extraction processes. In: J Rydberg, C Musikas, and G Choppin, eds. Principles and Practices of Solvent Extraction. New York: Dekker, 1992.
3. Y Marcus, AS Kertes. Ion Exchange and Solvent Extraction of Metal Complexes. London. Wiley-Interscience, 1969, pp 425–858.
4. AW Ashbrook, G Ritcey. Solvent Extraction, Part I. Amsterdam: Elsevier, 1984.
5. K Gottlibsen, B Grinbaum, D Chen, G Stevens. Hydrometallurgy 56: 293–307, 2000.
6. ES Perez de Ortiz. In JD Thornton, Ed. Science & Practice of Liquid–Liquid Extraction. New York: Clarendon Press, 1992, pp 157–205.
7. TC Lo, MH Baird, C Hanson, eds. Handbook of Solvent Extraction. New York: Wiley, 1983.
8. AW Francis. Handbook of Components in Solvent Extraction. New York: Gordon & Breach, 1972.
9. S Walas. Phase Equilibrium in Chemical Engineering. London: Butterworths, 1985.
10. WA Rickelton. In: DH Longsdail and MJ Slater, eds. Solvent Extraction in the Process Industries. London: Elsevier, 1993, pp 731–736.
11. L Kogan, Ph.D. thesis. Casali Institute, The Hebrew University, Jerusalem, 1994.
12. CRC Handbook of Chemistry and Physics 6-14, 79th ed. Boca Raton, FL: CRC Press, 1998/1999.
13. HRC Pratt. Solvent Extr Ion Exch 1(4): 669, 1983.
14. B Grinbaum. In: DH Longsdail and MJ Slater, eds. Solvent Extraction in the Process Industries. London: Elsevier, 1993, pp 715–722.
15. G Stevens, HRC Pratt. In: JD Thornton, ed. Science & Practice of Liquid–Liquid Extraction. New York: Clarendon Press, 1992, pp 492–582.
16. JC Godfrey, DH Longsdail. Liquid–Liquid Extraction Equipment. New York: Wiley, 1994.
17. B Grinbaum. In: M Valiente, ed. Proceedings of ISEC '99 (in press).
18. G Stevens, A Ichia. Hydrometallurgy (in press).
19. DH Longsdail, JA Jenkins, KS Robinson. In: T Sekine, ed. Solvent Extraction 1990. Amsterdam: Elsevier, 1992, pp 1345–1350.

20. AE Karr, RE Cusack. In: T Sekine, ed. Solvent Extraction 1990. Amsterdam: Elsevier, 1992, pp 1333–1338.
21. E Barnea. In: TC Lo, MH Baird, and C Hanson, eds. Handbook of Solvent Extraction. New York: Wiley, 1983.
22. B Grinbaum, SA Angel. In: T Sekine, ed. Solvent Extraction 1990. Amsterdam: Elsevier, 1992, pp 1235–1240.
23. P Lloyd. Principles of industrial solvent extraction. In: J Rydberg, C Musikas and G Choppin, eds. Principles and Practices of Solvent Extraction. New York: Dekker, 1992, pp 237–266.

2

Design of Pulsed Extraction Columns

Alfons Vogelpohl

Technical University of Clausthal, Clausthal-Zellerfeld, Germany

Hartmut Haverland

Gebrüder Lödige Maschinenbau GmbH, Paderborn, Germany

I. INTRODUCTION AND DESIGN CONCEPT

Pulsed sieve-plate extraction columns were introduced by Van Dijck [1] in 1935 and developed in the form of a reciprocating plate column and a liquid pulsed sieve-plate column. Liquid pulsed sieve-plate extractors were first applied to the solvent extraction of radioactive materials in the 1940s. Now packings also are being used as internals, and ordered packings have become increasingly attractive in recent years.

Numerous studies have proved that pulsed extractors are characterized by high throughput and high separation efficiency. These favorable properties and the simple construction of the equipment have resulted in wide application in the chemical and metallurgical industry.

The performance of pulsed extraction columns strongly depends on the behavior of the dispersed phase, which is governed by the geometry of the column and the internals, the type of agitation, and its intensity, as well as the physical properties of the liquid–liquid system involved. In view of the many decisive parameters, we developed and verified a new method for the design and simulation of pulsed extraction columns, based on single-drop experiments and on drop population modeling. Since the real liquid–liquid system to be used in the technical column may be applied in the single-drop experiments, the design inherently takes into consideration the presence of impurities or

107

surface active agents, which may have a dramatic effect on the drop size distribution and thus on fluid dynamics as well as mass transfer.

To facilitate the single-drop experiments, several pieces of so-called standard apparatus were developed which use minimum amounts of test substances and allow for the measurement of the characteristic rising velocity, the breakage, the coalescence, and the mass transfer of defined drops under defined conditions. The results from these experiments then serve—in combination with other information—as the input data for a population balance model. Solving the population balance model allows to describe the influence of the operating parameters on fluid dynamics and mass transfer under either steady state or dynamic conditions.

The validity of this design and simulation method has been proved by comparison of simulated drop size distributions, holdup profiles, flooding curves, and concentration profiles with measured data from the literature and our own experimental data in pilot columns of up to 225 mm in diameter.

II. POPULATION BALANCE MODEL

The conventional design of extraction columns is based on an averaged drop size; in most cases, the so-called Sauter mean diameter is used. Because of the strong influence of drop size on almost every phenomenon in extraction, it is evident that a model that takes the drop size distribution into account should provide a superior description of the performance of an extraction column. In this work the population model proposed by Casamatta and Vogelpohl [2] was used. The model is based on the definition of a function $P(h, d)$ representing the volume distribution of drop sizes with the diameter d at the height h in the column. Integrating this function over the range of drop sizes gives the holdup of the dispersed phase at the height h:

$$\varepsilon(h) = \int P(h, d)\partial d \tag{1}$$

The hydrodynamics of the continuous phase is described by the dispersion model, resulting in the following equation for the flow rate of the continuous phase at height h:

$$F_c(h) = A[1 - \varepsilon(h)]u_c(h) - AE_c(h)\frac{\delta[1 - \varepsilon(h)]}{\delta h} \tag{2}$$

The differential flow rate of the drops of class d at height h may be expressed as follows:

$$\delta F_{\rm d}(h, d) = AP(h, d)\delta u_{\rm d}(h, d) - AE_{\rm d}(h, d)\frac{\delta P(h, d)}{\delta h}\delta d \tag{3}$$

The unsteady drop population balance at height h leads to the basic differential equation of the model:

$$\frac{\delta P(h, d)}{\delta t} = -\frac{\delta[P(h, d)u_{\rm d}(h, d)]}{\delta h} + \frac{\delta}{\delta h}\left[E_{\rm d}\frac{\delta P(h, d)}{\delta h}\right] + R(h, d) \tag{4}$$

The terms of the right-hand side of this equation represent, respectively, the convective, diffusive, and productive contributions to the accumulation term. The productive term $R(h, d)$ takes into account the breakage and coalescence of drops that modify the drop size distribution $P(h, d)$. For the boundary conditions and further details, see Ref. 3. Integration of the balance equation requires in particular information on drop velocity, breakage, and the coalescence of drops. Simulation of mass transfer in extraction columns is based on equations similar to the Eqs. (1)–(4) and requires additional information on the mass transfer coefficient as a function of the drop diameter and the driving force.

III. EXPERIMENTAL STUDIES

A. Standard Apparatus and Test Systems

All the standard apparatus consists in principle of the following (Fig. 1):

> Up to four calibrated glass tubes with an inner diameter of 72 mm and a length of 300 mm each (i.e., a maximum total length of 1200 mm). One or two of the glass tubes serve as the test section (TS), representing a characteristic compartment of the type of column to be used in practice.
> A reciprocating pump (P) to produce a pulsation of the liquid in the test section.
> A Hamilton pump (H) to produce drops of a defined diameter.
> An injection pump (IP) to transport the drop into the test section.
> The HYDROMESS measuring device (DSM) for the online measurement of drop size or drop size distribution including a recirculation pump (RP) [4].
> A high speed video camera (HVC) for studying specific details.
> A PC (PC) for data processing and control purposes.
> A reservoir for the organic phase (ROP).

Depending on the kind of study, the test section may be equipped with a plate, a packing, or some specific internals (e.g., to measure the mass transfer coeffi-

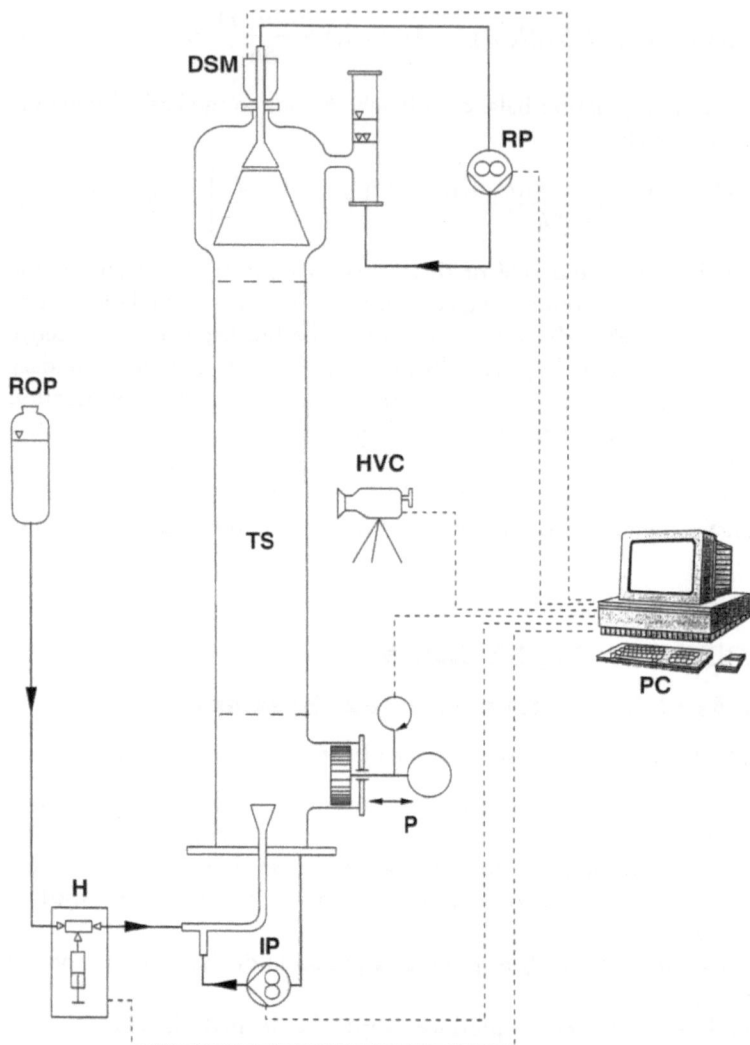

Figure 1 Basic layout of standard apparatus: ROP, reservoir of the dispersed phase; H, Hamilton pump; IP, injection pump; P, pulsator; TS, test section; HVC, high speed video camera; PC, personal computer; DSM, drop size measurement device; RP, recirculation pump.

cient). Single drops of a specified diameter are produced by the Hamilton pump and transported into the column via a gear pump, both pumps being triggered by the PC. If necessary (e.g., if the drops are passing through the holes of a plate in the test section), the introduction of a drop into the column is also recorded with respect to the position of the pulsating piston.

To be able to compare own results with data from the literature, most of the experiments were carried out with test systems as recommended by the European Federation of Chemical Engineering (EFCE) [5], namely,

Toluene/water (high interfacial tension)
n-Butyl acetate/water (intermediate interfacial tension)
n-Butanol/water (low interfacial tension).

In each case the organic phase was distributed in the aqueous phase. Acetone and succinic acid served as the components transferred between the phases in the mass transfer experiments.

B. Experiments

The parameters changed in the different studies were as follows:

The test system
The kind of internals
The drop size
The pulsation intensity (energy input) of the pulsating device as the product of the frequency and the amplitude of the pulsation

1. Single-Drop Velocity

A basic parameter in modeling the fluid dynamics in extractors is the velocity of the rising drop as a function of the drop size, since this velocity is used most often to calculate the characteristic slip velocity. For a pulsating flow and/or internals in the column, this velocity may differ significantly from the velocity of the freely rising drop as given, for example, by the well-known correlation of Hu and Kintner [6]. Therefore, a standard apparatus was developed as described next.

a. **Apparatus and Measuring Technique.** The apparatus used to determine the velocity of single drops as a function of the test system, the pulsation intensity, and the drop size is shown in Fig. 2. In addition to the basic layout as already discussed, the standard apparatus is equipped with a mirror system M, which—in combination with a digital clock triggered by the PC—allows one to record the time at which a drop passes two markings in the column. The lower marking is positioned about 350 mm above the orifice where the

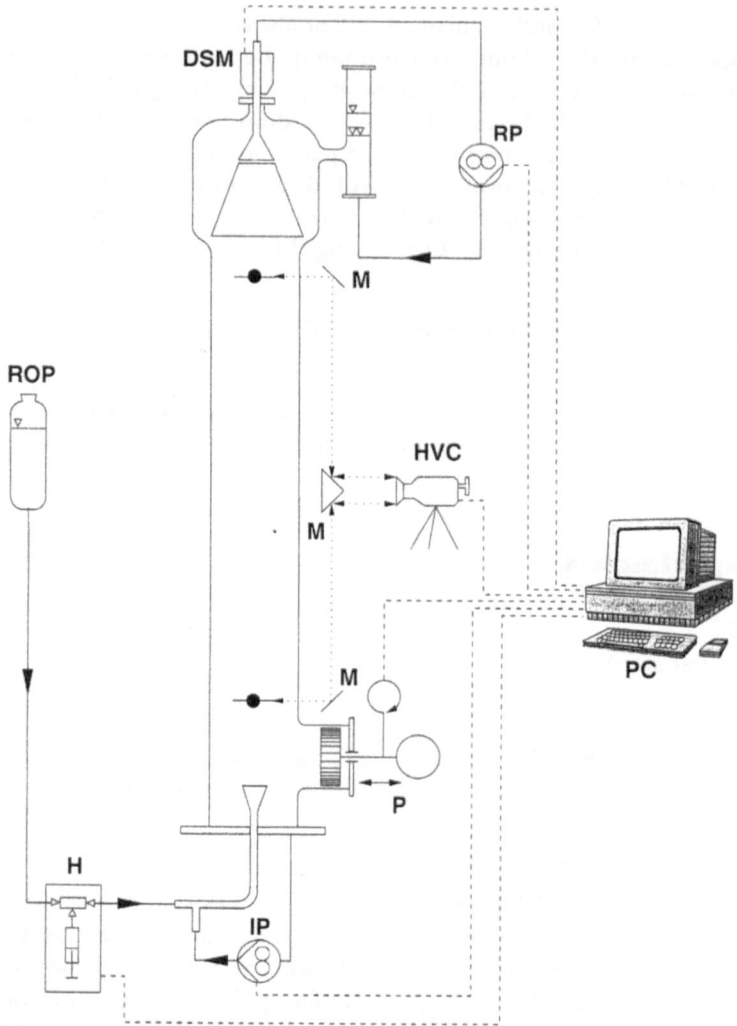

Figure 2 Standard apparatus for measuring the drop velocity: M, mirror system; other abbreviations as in legend for Fig. 1.

drop is produced by the Hamilton pump, to ensure that the rising velocity is no longer affected by the drop formation. The upper marking is located 700 ± 1 mm above the lower marking. From this distance and the trigger time, the average rising velocity is determined with an accuracy of better than 1%. The drop sizes and the amount of drops required to obtain statistically relevant results are set via the PC, which triggers the Hamilton pump and the related equipment accordingly. An online drop measurement device at the top of the apparatus serves to control the drop size. If it is necessary to determine the effect of an internal, like a plate or a packing, on the rising velocity, such internal is positioned between the lower and upper marks.

The apparatus and the measuring technique are discussed in detail by Haverland [7] and Leu [8].

b. Experimental Results and Correlations. As shown in Fig. 3 for the toluene(d)/water system, the general dependency of the droplet velocity rising with the drop size is observed and is in good agreement with, for example, the correlations of Hu and Kintner [6], Ishii and Zuber [9], or Klee and Treybal [10] and is hardly affected by pulsation, as shown in Fig. 4 [11]. For the *n*-butyl acetate(d)/water system, however, commonly used correlations completely fail to describe the detailed dependency, even for freely rising drops, as shown in Fig. 5. If the drops are rising through an ordered packing, the drop velocity is strongly reduced as a result of the influence of friction and impacts within the packing. In particular, the motion of larger drops is hindered and, in comparison to the velocity of freely rising drops, the maximum of the velocity is

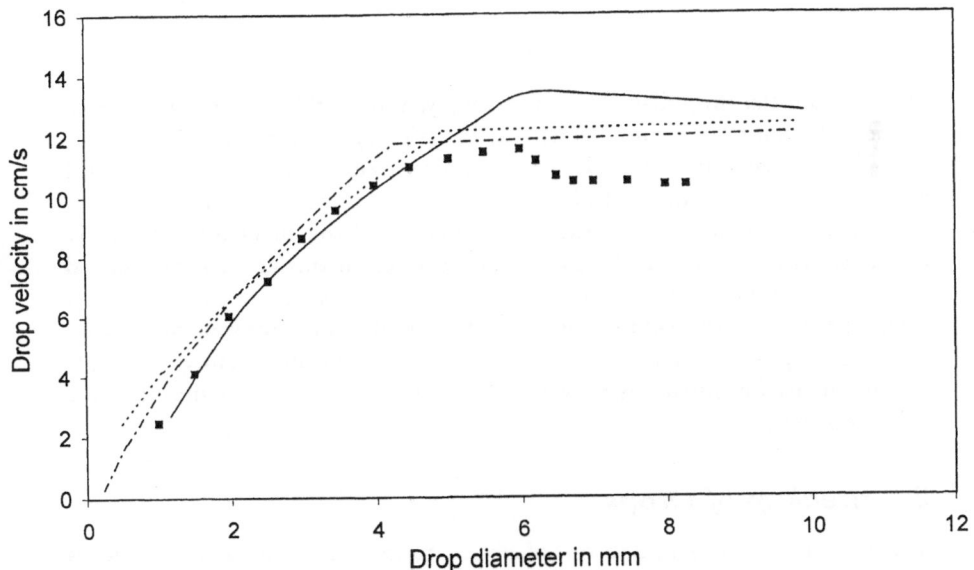

Figure 3 Single-drop velocity for the toluene(d)/water system: squares represent measured data point. Results of Hu and Kinter [6], solid curve; Ishii and Zuber [9], dashed curve; and Klee and Treybal [10], dot–dash curve.

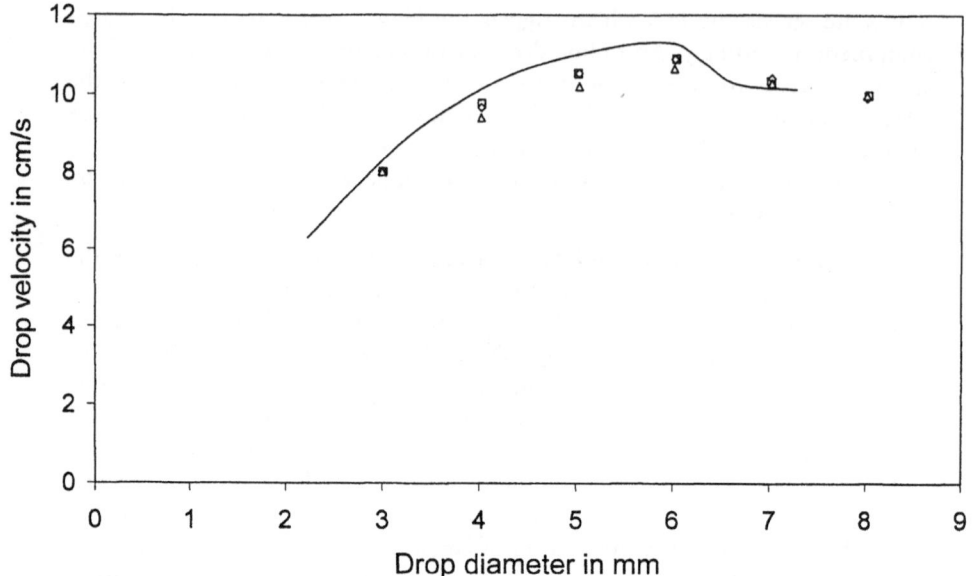

Figure 4 Effect of pulsation on single-drop velocity in the toluene(d)/water system. Solid curve represents no pulsation. Pulsation intensity (mm/s): □, 20; ◇, 28; △, 36.

shifted to smaller drop diameters. Surprisingly, also for this system no influence of the pulsation on the drop velocity in the packings was found [11,12].

For modeling the effect of packing on drop velocity, Leu [8,13] developed a concept based on a detailed physical description of drop movement within a packing as a function of the packing geometry and the characteristic velocity of a freely rising drop. All the parameters required in this concept are derived from experimental single-drop data. The work of Leu seems to indicate that it should be possible to describe the effect of internals on the movement of drops by using single-drop data only and thus confirms the importance of simple measurements for situations in which correlations presented in the literature fail completely.

2. Breakup of Drops

The breakup of drops due to turbulence or to internals (e.g., a sieve plate) is an important phenomenon in extraction columns because of its effect on drop size distribution and thus both on fluid dynamics and mass transfer. A sufficiently accurate prediction of this phenomenon, based on first principles, has not been achieved and, owing to the large variety of internals and to the above-

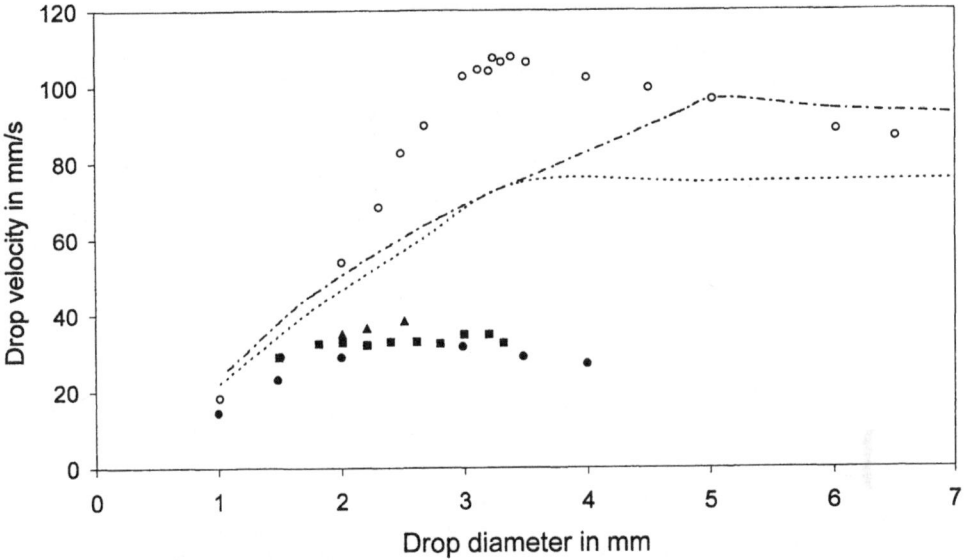

Figure 5 Single-drop velocity for packing with Montz-Pak B1-50 of an *n*-butyl acetate(d)/water system versus no packing (O). Results of Hu and Kintner [6], dot–dash curve, and Klee and Treybal [10]. Pulsation intensity (mm/s): ●, 0; ■, 10; ▲, 20.

mentioned effect of impurities, surface active ingredients, and so on, is most unlikely to be developed.

For these reasons a standard apparatus has been designed which allows one to measure the breakup of single drops under conditions to be expected in a technical column.

a. Apparatus and Measurement Technique. The standard apparatus for the measurement of the breakup of single drops as shown in Fig. 6 has the same layout in principle as the other apparatus and may be equipped with a plate or a packing within the test section.

b. Experimental Results and Correlations. For characterizing the breakup behavior of drops passing a plate or a packing, a special diagram developed by Haverland [7] is used (Fig. 7), where the drop size is plotted as a function of the pulsation intensity. In general, three main regions are found: no breakup (region A) and complete breakup (region B), and an intermediate "stochastic breakup" region (region AB). In region A the interfacial forces stabilizing the drop cannot be overcome by the forces caused by the impact with the plate or packing structure and, thus, no drop breakage occurs. In region B the resulting forces always cause drop breakup. While the drop behavior in the main regions

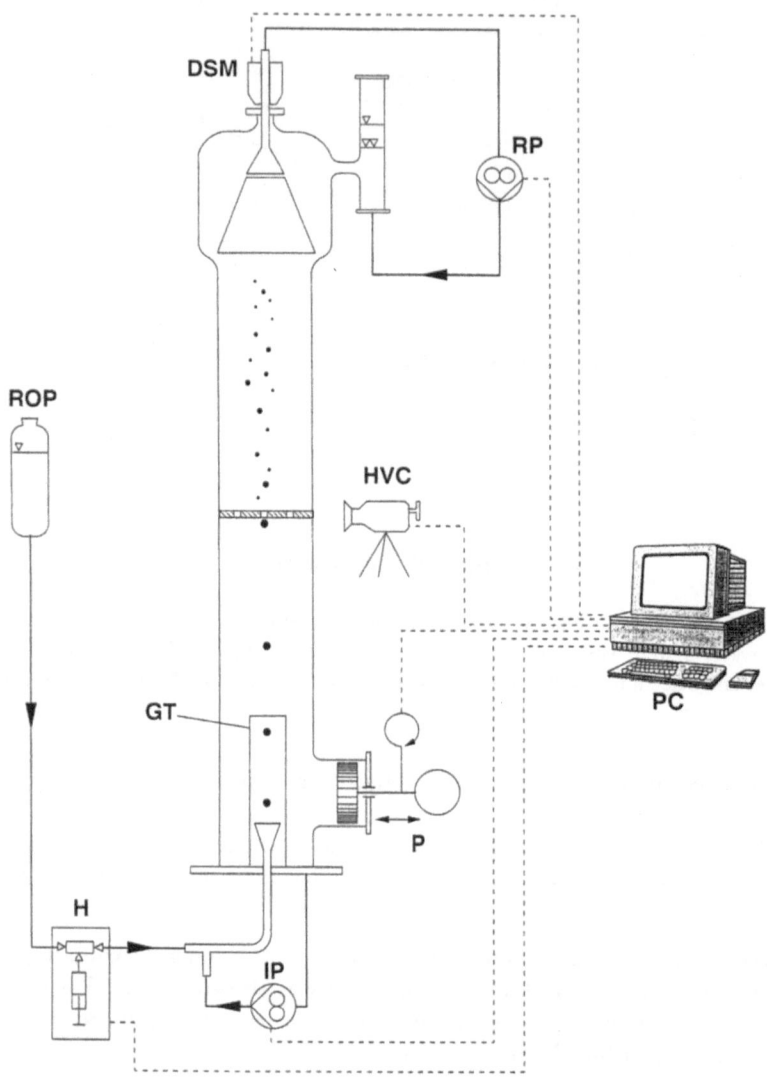

Figure 6 Standard apparatus for measuring the breakup of single drops: GT, guide tube; other abbreviations as in legend for Fig. 1.

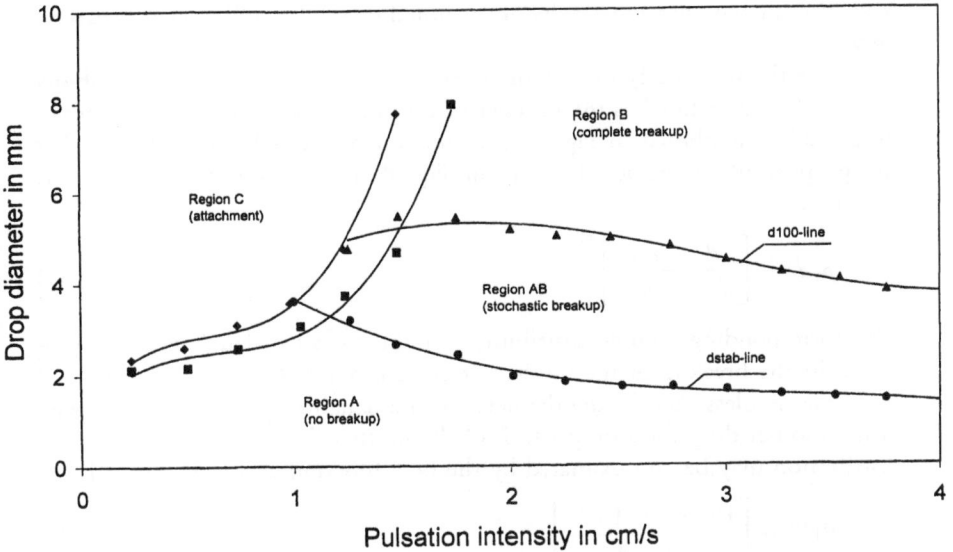

Figure 7   Breakage of single drops passing a sieve plate in a toluene(d)/water system: hole diameter, 2 mm; hole spacing, 4 mm; free area, 22.7%.

A and B is exactly determined, the stochastic behavior in the intermediate section AB is described by a breakup probability $0 < p < 1$. In addition to regions A, B, and AB, there exists a region C in which the drops attach to the plate. The area between the two left-hand curves in Fig. 7 indicates intermediate sections, though of less technical importance, between regions A and C, AB and C, and B and C, respectively. In these regions—as in AB—the performance of the drops follows stochastic laws and can be described only on a statistical basis.

The single-drop experiments supply the basic parameters required for the modeling of drop breakage. These parameters are the maximum stable drop diameter and the minimum diameter of total breakup characterizing the region AB in dependence on the pulsation intensity as well as the breakup probability and the daughter drop size distribution resulting from drop breakage. Drop breakage is also affected by mass transfer [14]. In systems with a negative gradient of the interfacial tension $d\sigma/dx$, drop stability decreases as a result of changes in system properties due to the presence of the transferred component as well as interfacial disturbances caused by the mass transfer. Therefore, the characteristic drop diameter is lowered under mass transfer conditions.

Figure 7 shows typical results for the toluene-(d)/water system and a sieve plate with holes of 2 mm in diameter and a free area of 22.7%. Experiments have also shown that the results are a function of the pulsation intensity af only.

For the technically important region of $1 < af < 4$ cm/s, the probability of a breakup p, defined as the number of drops broken up over the number of drops added, is plotted in Fig. 8 as a function of the pulsation intensity. For the purpose of modeling, the data in Fig. 8 may be approximated by the function:

$$p(d) = \left[\frac{d - d_{stab}}{d_{100} - d_{stab}} \right]^{m} \tag{5}$$

The corresponding volume distribution of the so-called daughter drops produced in the breakup of the mother drops is given in Fig. 9 as a function of the dimensionless ratio of the diameter of the daughter drops to the diameter of the mother drop for a drop size Γ of the mother drop of 5 and 7 mm. The distribution may be approximated by the beta function as a function of Γ:

$$B(\Gamma) = \left[\frac{\Gamma^{q-1}(1 - \Gamma)^{r-1}}{b(q, r)} \right] \tag{6}$$

where the parameters p and q are determined from the average size and the variance of the dimensionless drop size distribution [3,7].

Similar results were obtained for stainless steel ordered packings as shown, for example, in Figs. 9–11 and described in detail by Miller et al. [15]. It is interesting to note in Fig. 9 that the region of attachment C is missing for this specific packing. In addition, the d_{100} line is shifted to larger drop diameters for two packing elements with a height of 100 mm each and turned at an angle of 90° relative to one another. This fact also shows up in Fig. 11 with a higher probability of drop breakage for the packing made up of two elements. The volume density distribution of the daughter drops produced is given in Fig. 12 for mother drop diameters of 4.0, 5.0, and 6.0 mm. The probability of breakage and the drop size distribution of the daughter drops may again be approximated by the same correlations discussed earlier. This indicates that breakage occurs by the same mechanisms for both plates and packings and that it should be possible, therefore, to develop generalized diagrams like the one in Fig. 13, in which the average diameter of the drop size distribution and the normalized standard deviation are correlated for three different packings [8].

3. Coalescence of Drops

The coalescence of drops may be described by two parameters: coalescence time and coalescence probability. Coalescence time is the time interval between the collision of two drops and coalescence, and coalescence probability is the number of successful drop coalescences out of the number of drop collisions. Several

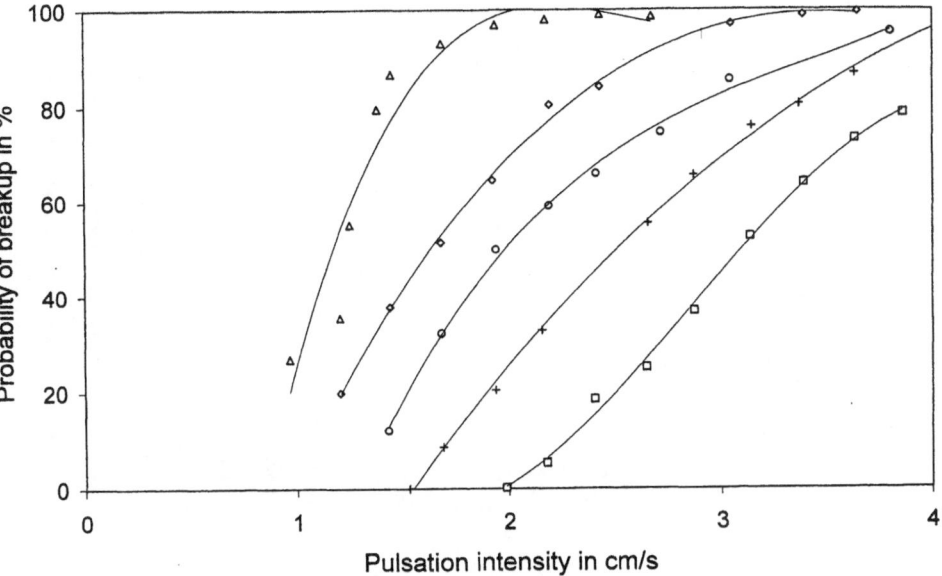

Figure 8 Probability of breakage for single drops passing a sieve plate in a toluene(d)/water system: hole diameter, 2 mm; hole spacing, 4 mm; free area, 22.7%. Mother drop diameter (mm): □, 2; +, 2.5; ○, 3; ◇, 4; △, 5.

models have been developed to predict interdrop coalescence, but because of the high sensitivity of coalescence to fluid and interfacial dynamics, physical properties, and mass transfer, the results are only qualitative [3,15,16]. Therefore, a number of experimental setups have been implemented to study the coalescence of drops, but the results from these experiments, too, are thus far quite unsatisfactory [15–17].

For the measurement of coalescence time, the problem is often reduced to the collision of one drop with a plane interface [18,19]. The disturbance caused by the drop on the interface due to collision is detected, and its duration is defined as coalescence time. The measurement of coalescence probability is investigated in pilot-scale extraction columns (e.g., by addition of a tracer to a fraction of drops entering the column) [20,21]. The drop size and tracer distributions are measured at the entrance and the exit of the column. From these data, the coalescence parameters can be calculated with the knowledge of single-drop breakup parameters. This method, however, uses a tracer that can influence the interface activities concerning the coalescence behavior, and the pilot-scale experiments are costly.

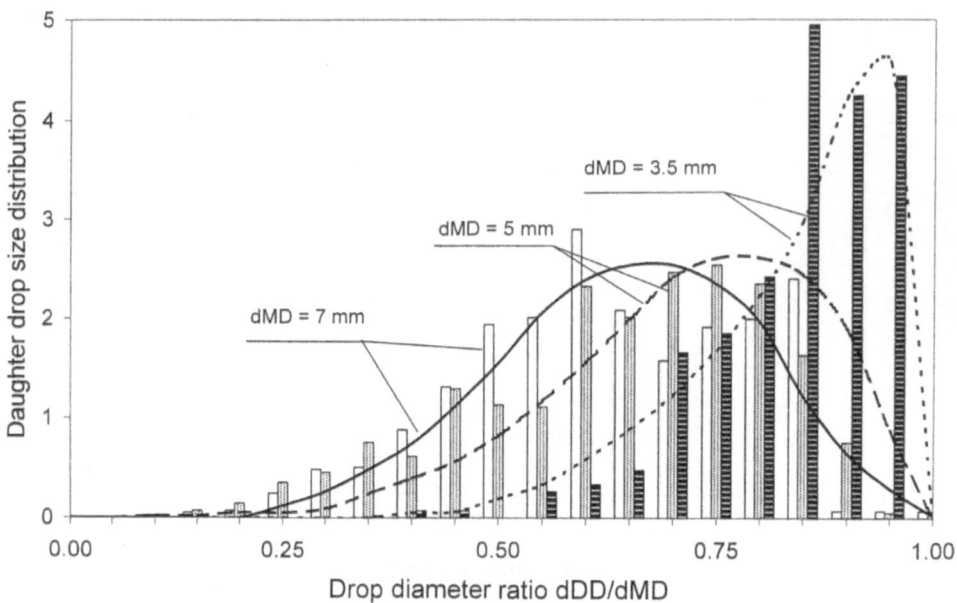

Figure 9 Daughter drop size distribution produced by breakage at a sieve plate in a toluene(d)/water system: dDD, diameter of daughter drop; dMD, diameter of mother drop.

a. Apparatus and Measuring Technique. The standard apparatus used in our measurements is shown in Fig. 14. The core of the apparatus is a cone, where a swarm of drops with the same diameter d_s and number n is held in place by a countercurrent flow of the continuous phase established by the recirculation pump (RP). The drops are produced with the Hamilton pump. A single drop with the diameter d_E rises into and through the drop swarm because of its higher rising velocity u_E ($d_E > d_s$). The size of the single drop after contact with the swarm d_C is measured with a photoelectric suction probe. The drop sizes before and after contact are compared. If $d_C \neq d_E$, coalescence has taken place, if $d_C = d_E$, no coalescence has taken place. Observation of 50 single drops gives a coalescence probability $k_C = f(d_E, d_s, n)$ dependent on the sizes of the single drop and the swarm drops and the number of drops in the swarm.

b. Experimental Results. Experiments have been carried out with the *n*-butyl acetate(d)/water system. The number of drops in the swarm, which is

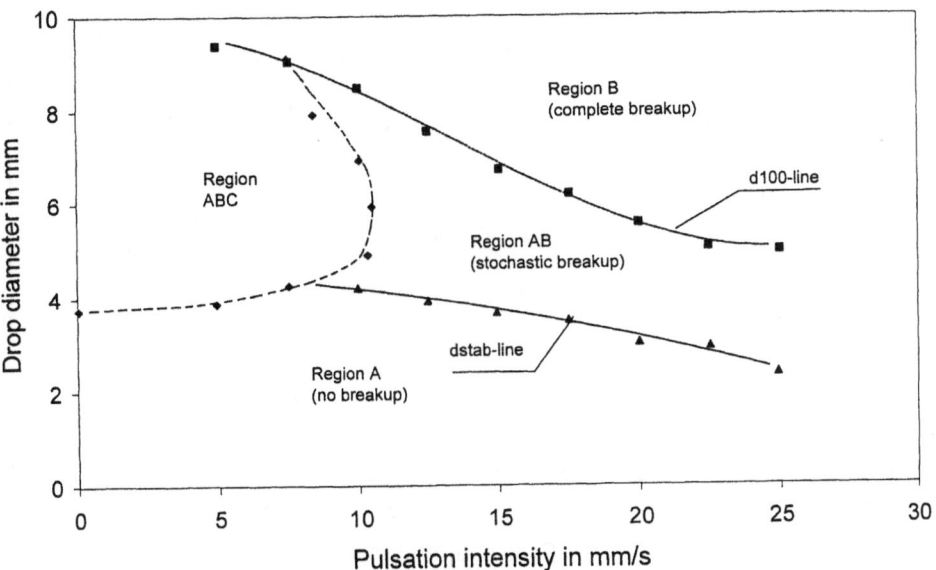

Figure 10 Breakage of single drops passing an ordered packing in a toluene(d)/ water system (Montz-Pak B1).

limited by coalescence of the swarm drops among one another, was varied from 20 up to 40. Figure 15 shows the coalescence probability dependent on the number of drops in the swarm for a single drop diameter of d_E = 3.5 mm and a swarm drop diameter of d_S = 3.0 mm.

In general, one expects increasing coalescence probability with increasing number of drops in the swarm. This is confirmed for 20 and 30 drops. At higher numbers, the drops start to influence one another in the swarm. The swarm extends over a larger part of the cone, so the drop density becomes lower as the number of drops increases. An equivalent holdup of the dispersed phase can be calculated by measuring the volume of the swarm of drops and the number of drops. Figure 16 shows the coalescence probability dependent on the holdup for a single drop of diameter d_E = 3.5 mm. The expected trend of increasing coalescence probability with increasing holdup is confirmed.

The number of data is still too small to permit us to present correlations for the probability of coalescence $k_C = f(d_E, d_S, n)$ and implement them into the drop population model. But the first experiments show that the standard apparatus is suitable for the study of the coalescence of single drops within a swarm of drops [22].

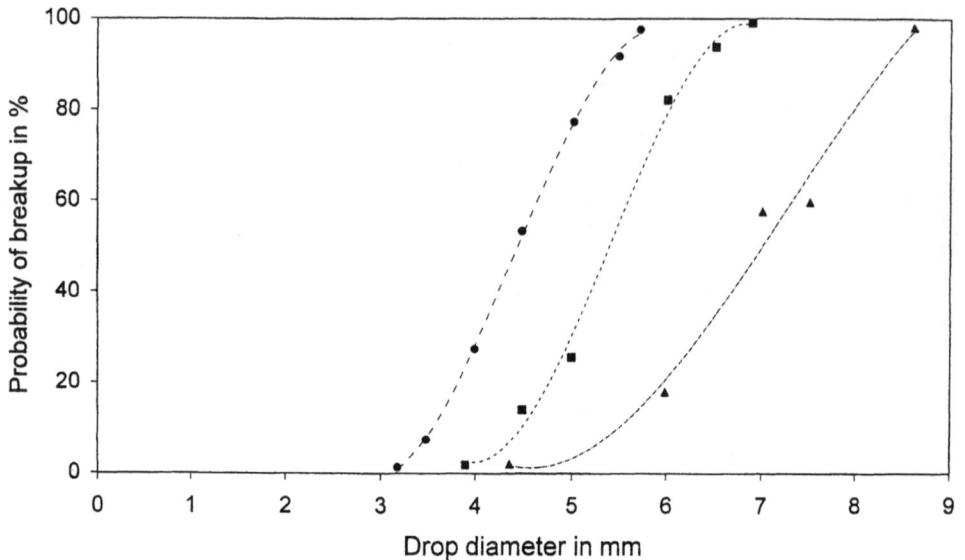

Figure 11 Probability of breakage for single drops passing an ordered packing toluene(d)/water system (Montz-Pak B1). Pulsation intensity (mm/s): ●, 10; ■, 15; ▲, 20.

4. Mass Transfer of a Single Drop

According to the Maxwell–Stefan equation, the flux of each component in a multicomponent system depends in a complex way on the concentration of all components and their diffusivities. In modeling mass transfer in extraction columns, therefore, the Maxwell–Stefan equation is normally used in the simplified form of an overall mass transfer coefficient. Since prediction of a mass transfer coefficient is still uncertain, an experimental determination is necessary if reliable data are desired.

a. Standard Apparatus and Measuring Technique.   The scheme of the standard apparatus is given in Fig. 17. Since the amount of substance transferred to or from a single drop is rather low, a chain of drops is produced at the bottom of the apparatus instead of single drops. For this purpose a special two-phase jet (TPJ) has been developed, which allows investigators to produce a sequence of very uniform drops [22]. The chain of drops leaving the two-phase jet is protected from the turbulence created by the pulsator by a guide tube (GT). Above the guide tube, the diameter of the drops may be measured by the drop size measuring device (DSM). In the absence of a drop size measuring

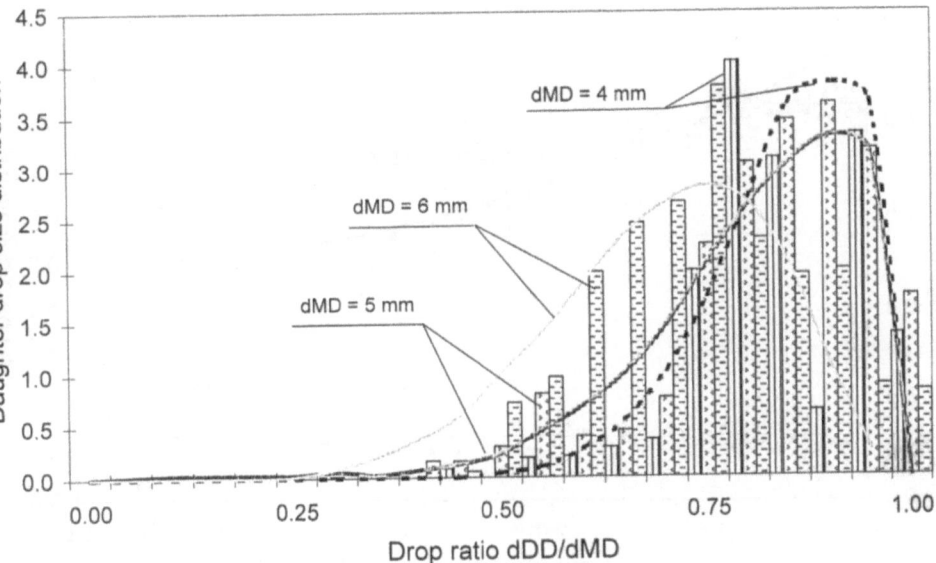

Figure 12 Daughter drop size distribution produced by drop breakage in an ordered packing, (toluene(d)/water) system: dDD, diameter of daughter drop; dMD, diameter of mother drop.

device, the drops rise freely through the test section, which may be equipped with either a sieve plate or a packing. Before and after the test section, the drops and the continuous phase may be sampled by using the sampling ports (SDP and SCP, respectively) to determine concentration. Taking certain corrections due to the sampling procedure into account, the average mass transfer coefficient for single drops is then determined from the defining equation [23,24]:

$$\beta_{od} = \frac{d}{6\Delta t} \ln \frac{y^* - y_1}{y^* - y_2} \tag{7}$$

b. Experimental Results. The average mass transfer coefficient was determined for freely rising drops as well as in the presence of a sieve plate or a packing dependent on the drop size and the pulsation intensity. In all cases the mass transfer coefficient increased with increasing drop diameter, whereas a pulsation may or may not have an influence, depending on the system. The lower mass transfer coefficients observed in the presence of a packing are due to the reduced drop velocity. Figures 18–20 show typical results for the test section without internals or equipped with a sieve plate or a packing. As would

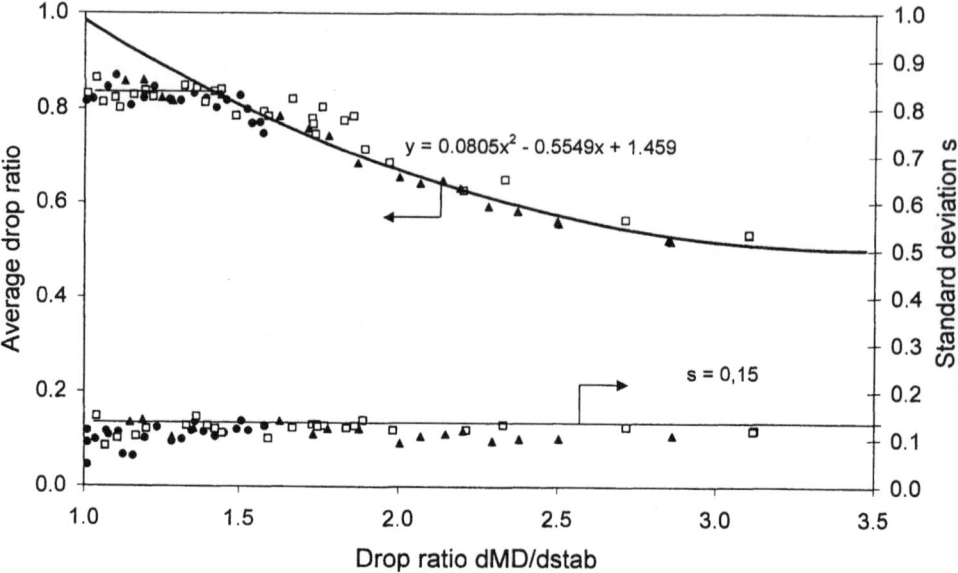

Figure 13 Average diameter and standard deviation of the drop size distribution produced by breakage within an ordered packing, toluene(d)/water system: dMD, diameter of mother drop; d_{stab}, stable drop diameter; y = ordinate, x = abscissa.

be expected, a mass transfer direction from the dispersed phase to the continuous phase results in a lower mass transfer coefficient, as shown in Fig. 18 for the toluene(d)/acetone/water system. The change in the characteristic behavior at a drop size of about 4 mm is caused by the fluid dynamics within the drop passing from a circulating to an oscillating regime. An increase in the pulsation intensity does improve the mass transfer for toluene(d)/acetone/water but has no effect for the *n*-butyl acetate(d)/acetone/water system, as shown in Figs. 19 and 20, respectively. For a sieve plate, however, the mass transfer coefficient increases for both systems by about 20% independent of the pulsation intensity. For additional results, see Refs. 11 and 25.

IV. SIMULATION OF PULSED EXTRACTION COLUMNS

The modeling of the fluid dynamics and mass transfer behavior of an extraction column reported here is based on a simplified stagewise drop population balance

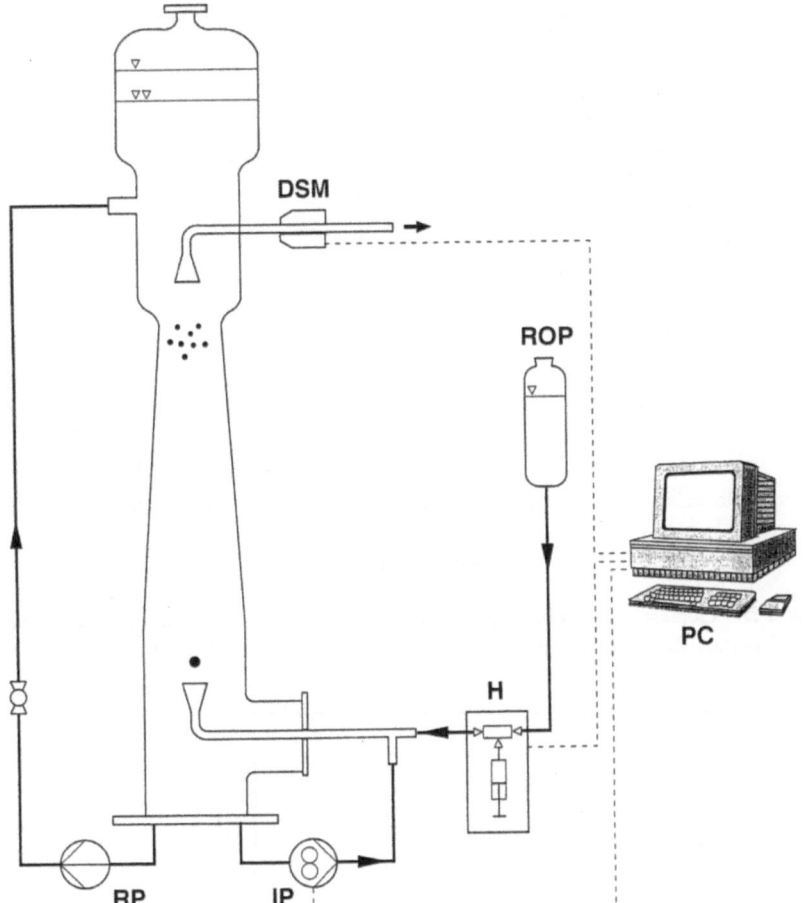

Figure 14   Standard apparatus for measuring drop–drop coalescence; abbreviations as in legend for Fig. 1.

model proposed by Haverland [7] and applied to sieve plates [7,25,26] and to packings [8,11]. Figure 21 summarizes the strategy of this simulation model.

If the influence of coalescence is negligible, the volume drop size distribution $q_n(d)$ above a sieve plate or a packing element n may be calculated from the distribution below this element $q_{n-1}(d)$, using breakup probabilities $p(d)$ and resulting daughter drop size distributions $q_{DD}(d)$ determined in single-drop experiments:

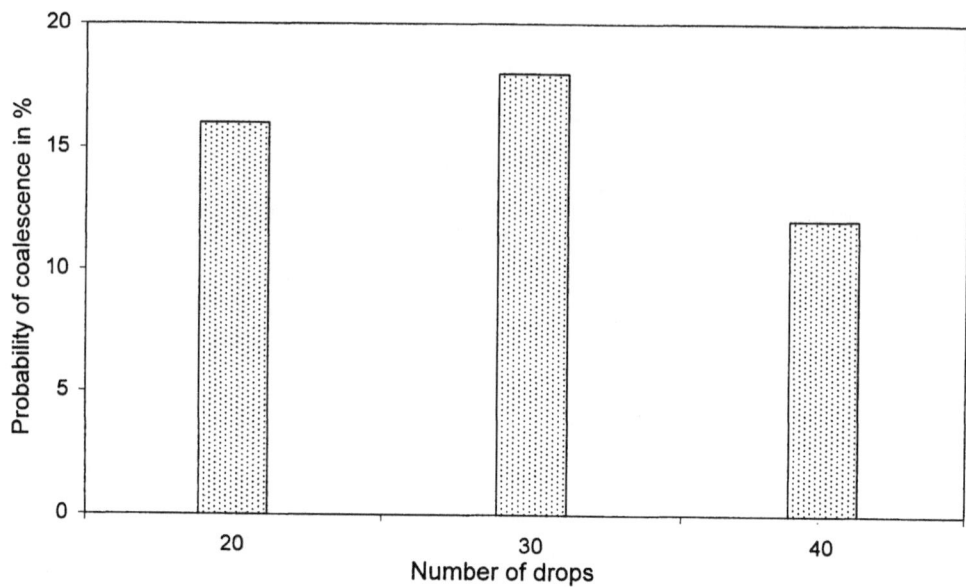

Figure 15 Coalescence probability versus the number of drops in the swarm for an *n*-butyl acetate(d)/water system with diameter of single drop of 3.5 mm.

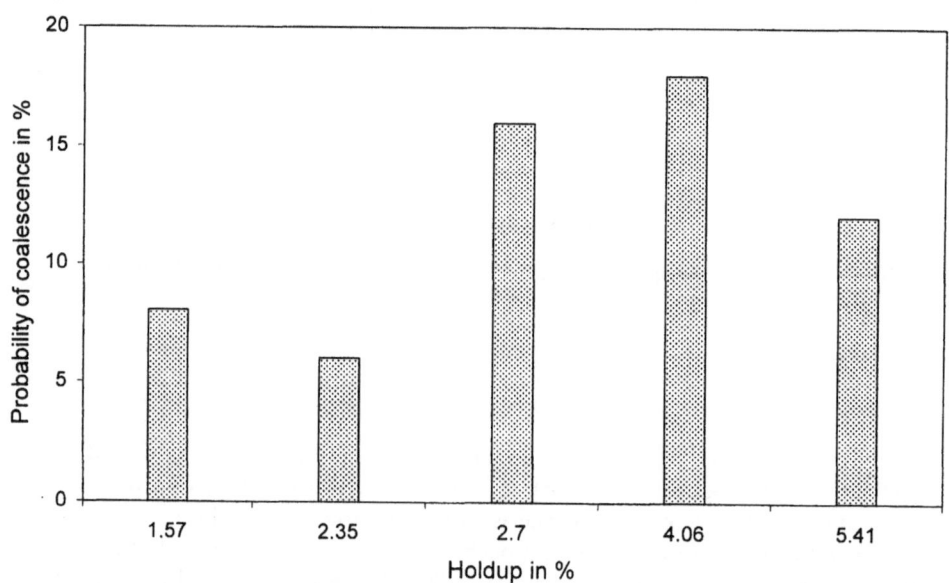

Figure 16 Coalescence probability versus holdup for an *n*-butyl acetate(d)/water system: diameter of single drop, 3.5 mm; swarm drop diameter, 3.0 mm.

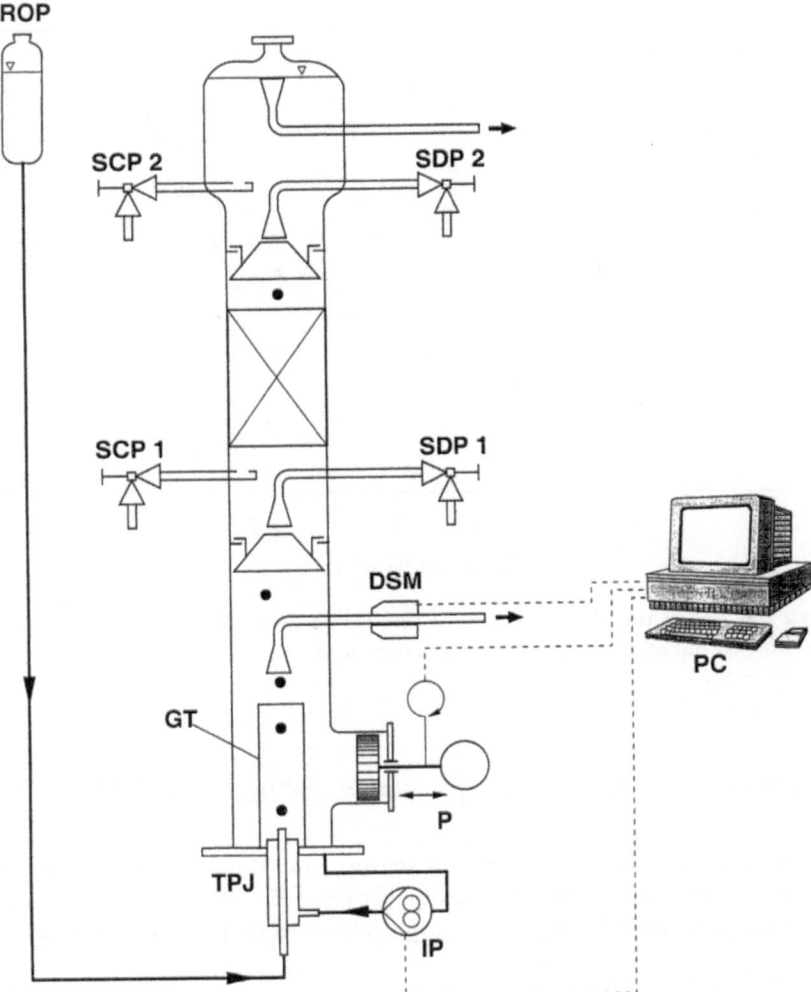

Figure 17 Standard apparatus for measuring the mass transfer of single drops: GT, guide tube; TPJ, two-phase jet; SCP, sampling port continuous phase; SDP, sampling port dispersed phase; other abbreviations as in legend for Fig. 1.

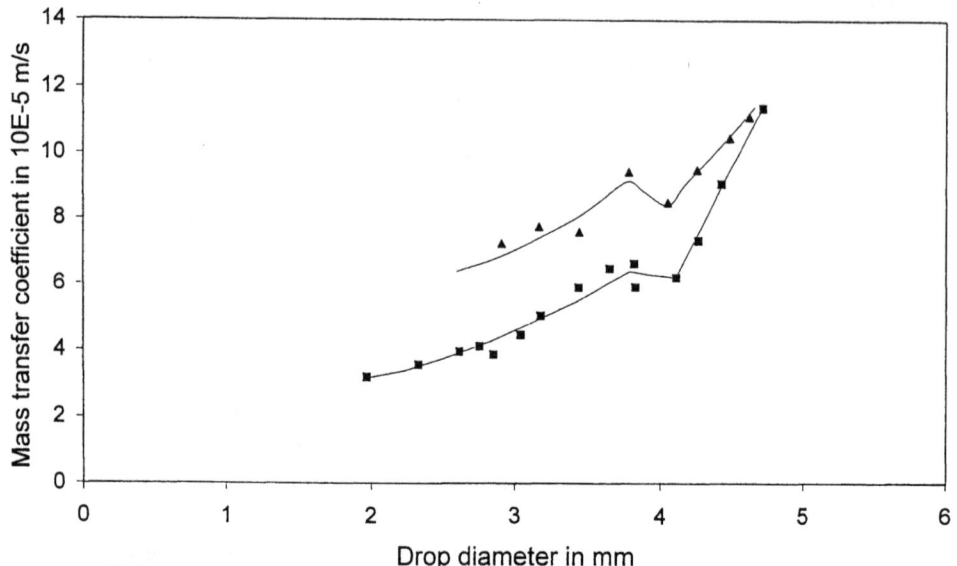

Figure 18 Mass transfer coefficient for single drops in a toluene(d)/acetone/water system with no pulsation. Mass transfer direction: ▲, d → c; ■, c → d.

$$q_n(d) = q_{n-1} - p(d)q_{n-1}(d) + \int_d^{d_{max}} p(d^*)q_{n-1}(d^*)q_{DD}(d^*, d)\delta DD^* \qquad (8)$$

The drop size profile may be determined starting from a measured or calculated inlet distribution, using the relationship between slip velocity, holdup, and phase velocity, as given by Eq. (9) for a polydispersed drop size distribution:

$$\frac{u_c}{1-\varepsilon} + \frac{u_d}{\varepsilon} = \sum u_{slip,i}q(d_i)\Delta d \qquad (9)$$

and the slip velocity between the drop class i (class width Δd) and the continuous phase:

$$\frac{u_{d,i}}{\varepsilon_i} + \frac{u_c}{1-\varepsilon} = u_{slip,i} = u_{c,i}(1+\varepsilon)^m \qquad (10)$$

with

$$u_{d,i} = u_d q(d_i)\Delta d \qquad (11)$$

and m an appropriate correlation factor.

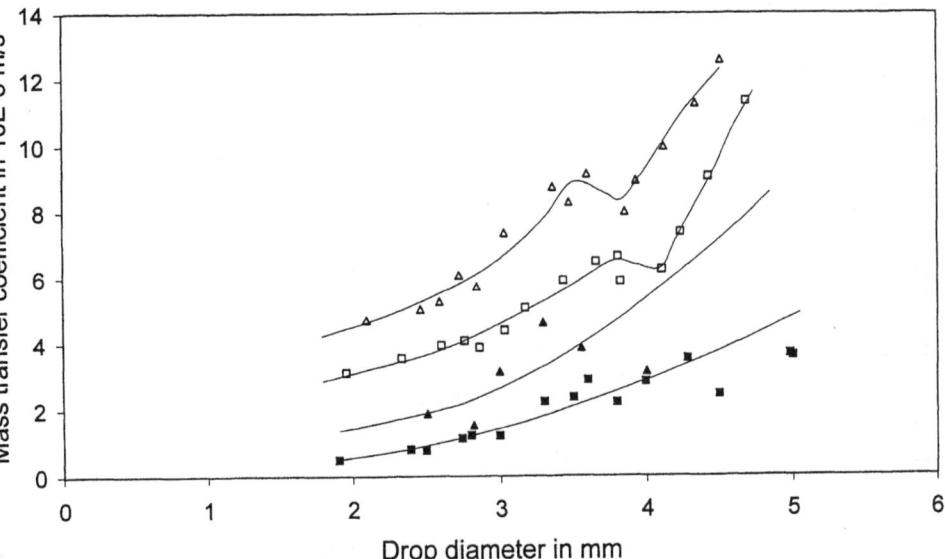

Figure 19 Mass transfer coefficient for single drops in a toluene(d)/acetone/water system, mass transfer direction c → d. Pulsation intensity for no packing (mm/s): △, 0; □, 10; concentration in the continuous phase, 15 mass%. Pulsation intensity for Montz-Pak B1-500 packing (mm/s): ▲, 0; ■, 10; concentration in the continuous phase, 5 mass%.

The cross-sectional holdup Φ and the holdup of each class Φ_i may be computed by iteration with the drop swarm velocity derived from the characteristic single-drop velocity for the sieve plate or in the packing. The restricted mobility in the packing and a drop motion along an exactly defined direction given by the packing geometry cause neighboring drops to rise in groups that remain in a stable configuration over a certain period. In contrast to known results from extraction columns of other types, the swarm velocity in structured packings may be enhanced because the drag coefficient of a group of moving drops is reduced in comparison to the drag coefficient of a single drop. The number of drops rising in a group increases with increasing holdup, therefore the influence of the swarm on the drop velocity is correlated by the factor $(1 + \Phi)^m$.

The calculation of mass transfer performance is based on mass balances for a differential column cross section with the volume dV, using computed drop size distributions and holdup profiles as shown earlier. The mass transfer rate from the continuous phase into the dispersed phase is considered for each class of drop sizes, with the mass transfer coefficient determined in the single-drop experiments:

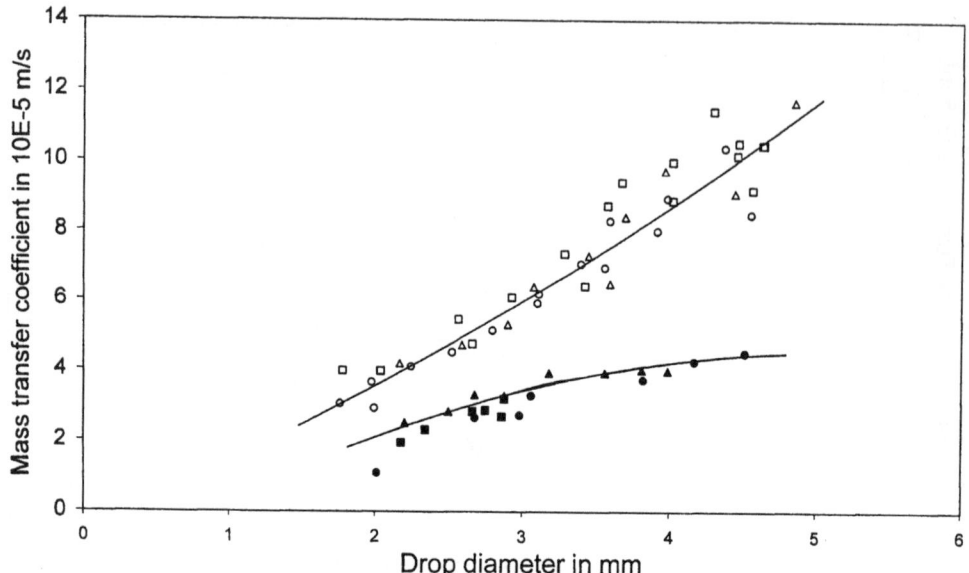

Figure 20 Mass transfer coefficient for single drops in an *n*-butyl acetate(d)/acetone/water system, mass transfer direction c → d; concentration in the continuous phase, 15 mass%. Pulsation intensity (mm/s) for no packing: ○, 0; △, 5; □, 10; for packing with Montz-Pak B1-500; ●, 0; ▲, 5; ■, 10.

$$N = \sum N_{\mathrm{d},i} = \sum \beta_{\mathrm{od},i} \frac{6\varepsilon_i}{d_i} \rho_{\mathrm{d}} [y(x) - y] \Delta V \tag{12}$$

The one-dimensional dispersion model is used to describe back-mixing in the continuous phase, with an axial dispersion coefficient determined in pilot-scale columns [11,27–30]. Further details of the calculation of concentration profiles are presented by Hoting et al. [31].

Because of the lack of reliable coalescence data, coalescence has not been taken into account in our calculations except for some modeling with the objective of studying the sensitivity of the drop size distribution with respect to coalescence, as reported elsewhere [2,22,26,30].

V. COMPARISON OF EXPERIMENTAL AND SIMULATED RESULTS

For testing the validity of the concept of designing extraction columns based on single-drop experiments, numerous fluid dynamic and mass transfer studies

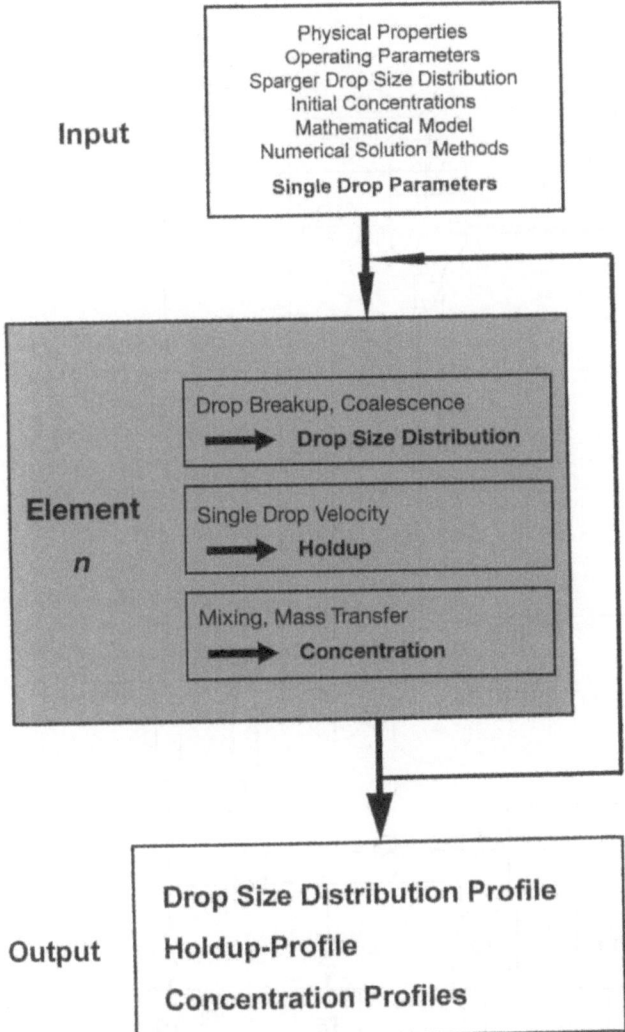

Figure 21   Simulation strategy.

have been carried out in two pulsed columns with a nominal diameter of 80, 100, and 225 mm and an active height of up to 4.3 m.

The investigations included measurements of the drop size distribution, dispersed phase holdup, maximum throughput, and mixing in both phases. The following parameters were varied:

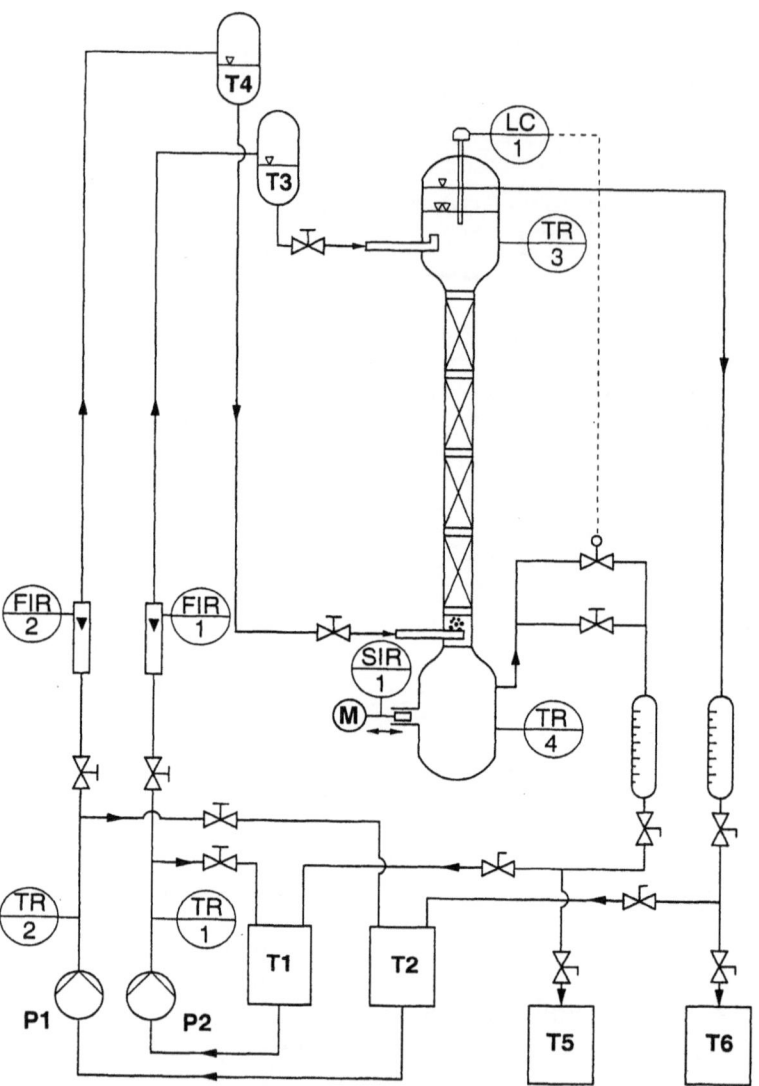

Figure 22 Schematic layout of the pilot-scale columns: M, pulsator; P, pumps; T1, T3, T5, holding tanks, continuous phase; T2, T4, T6, holding tanks, dispersed phase; FIR, flow indicator recorder; LC, level controller; SIR frequency indicator recorder; TR, temperature recorder.

Liquid–liquid system (EFCE test systems)
Sieve-plate geometry or type and height of packing
Cumulative load (60 and 90% of maximum throughput)
Pulsation frequency (from 0.0 to 2.5 Hz)

Most of the experiments were performed at a phase ratio of unity, 44 sieve plates with a spacing of 100 mm, or a packing height of 3 or 4 m and a pulsation amplitude of 8 mm. The test systems proposed by the EFCE (i.e., toluene/water and n-butyl acetate/water) were employed as liquid–liquid systems for the investigations. The lighter organic phase was always dispersed in the aqueous phase. The mass transfer experiments were carried out with toluene(d)/acetone/water and n-butyl acetate(d)/acetone/water, and the acetone was always transferred from the continuous to the dispersed phase. All details regarding the design of the column and the data acquisition, the physical properties of the test systems, the operating parameters, and the geometry of the

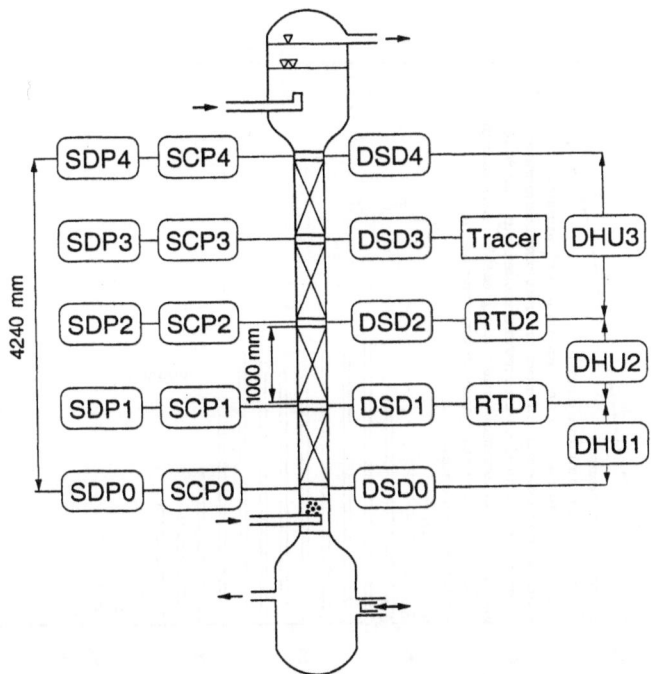

Figure 23 Points of measurement for the pilot-scale columns: SDP, concentration dispersed phase; SCP, concentration continuous phase; DSD, drop size distribution; RTD, residence time distribution; Tracer, tracer injection point; DHU, differential pressure.

internals are given by Lorenz [25] for the sieve-plate columns and by Coulal-oglou and Tavlarides [17] for the packing columns. The principal layout of the pilot-scale plant is given in Fig. 22. The plant is equipped with analytical instruments (Fig. 23) that permit measurement of the concentration of the dispersed (SDP) and the continuous (SCP) phases at four positions. The drop size distribution (DSD) is measured above the sparger and also at the same four positions. An injection of a tracer allows the determination of the residence time distribution (RTD) and thus the dispersion coefficient. Pressure sensors finally permit us to obtain average holdups (DHU) over the column sections.

A.   Results from a Sieve-Plate Column

Figure 24 compares measured and calculated drop size distributions at different column heights. The calculated data are based on the distribution produced by the sparger at the bottom of the column and measured single-drop data. The measured averaged drop size in the form of the so-called Sauter diameter and

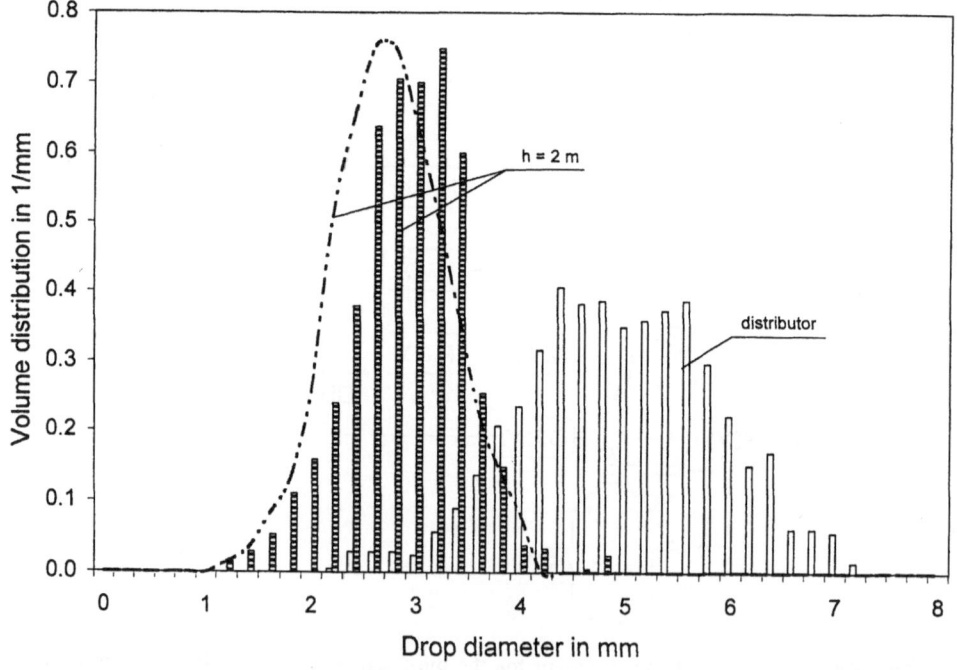

Figure 24   Comparison of measured and simulated volume drop size distributions of a sieve-plate column for a toluene(d)/water system. Hole diameter, 4 mm; free area, 39%; pulsation intensity, 20 mm/s; h, height of the column.

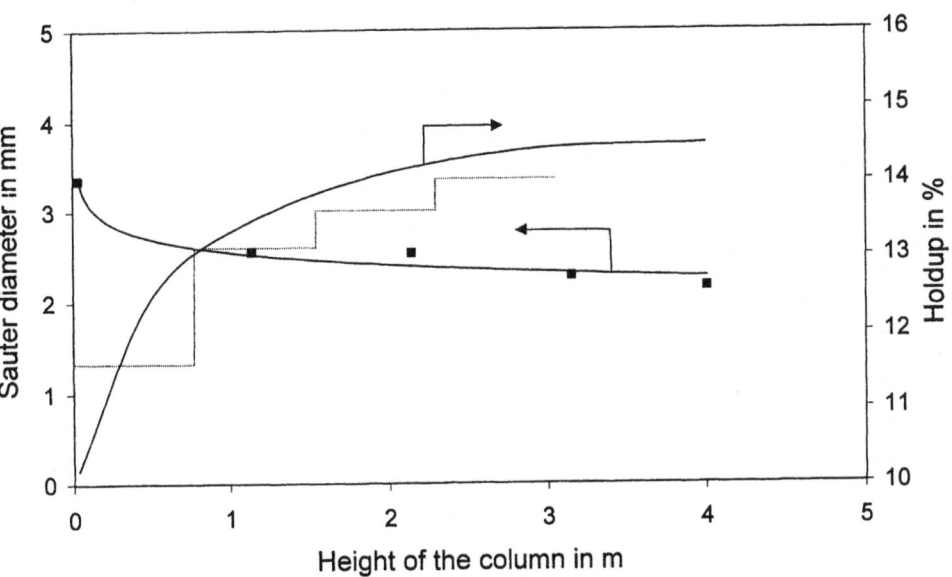

Figure 25 Comparison of measured and simulated profiles of the Sauter diameter and the local holdup for a sieve-plate column in an *n*-butyl acetate(d)/water system. Hole diameter, 4 mm; free area, 39%; pulsation intensity, 15 mm/s; cumulative load of continuous and dispersed phases, 22 m^3/m^2 h. Square data points, measured Sauter diameter; dashed curve, measured average holdup; solid curve, results of the simulation.

the local holdup as a function of the column height are compared with results of a simulation in Fig. 25. In both cases the measured and the calculated data are in good agreement. Figure 26 compares measured and calculated concentration profiles with a very satisfying result [25,26].

B. Results from a Packing Column

A comparison of measured and calculated drop size distribution for an ordered packing is given in Fig. 27. However, the *n*-butyl acetate(d)/water system shows a tendency for coalescence, and since coalescence of drops is not included in the simulation, the agreement is not as good as with the toluene(d)/water system. A satisfactory agreement again is found in Fig. 28, comparing holdup and Sauter diameter data. Figure 29 compares measured and simulated concentration profiles. The agreement is not as good as in Fig. 26, perhaps because of the small differences in acetone concentration of the two phases [31].

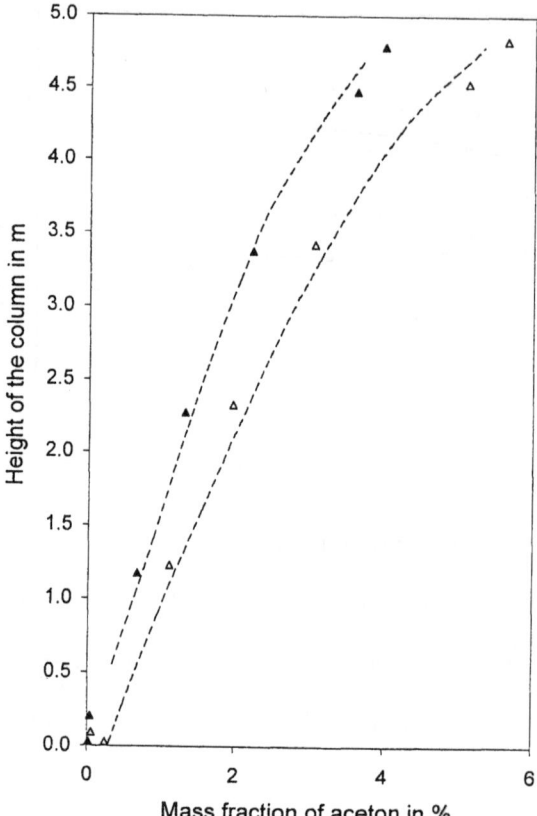

Figure 26 Comparison of measured and simulated concentration profiles of a sieve-plate column in a toluene(d)/acetone/water system: ▲, measured concentration organic phase; △, measured concentration aqueous phase; dashed curves, results of the simulation.

VI. SUMMARY AND OUTLOOK

The comparison of simulated drop size, holdup, and concentration profiles with experimental results shows that single-drop parameters obtained in simple experiments in laboratory-scale standard apparatus may be used successfully to predict fluid dynamics and the mass transfer in pulsed sieve-plate and packed extraction columns, at least for systems with high to moderate interfacial tension.

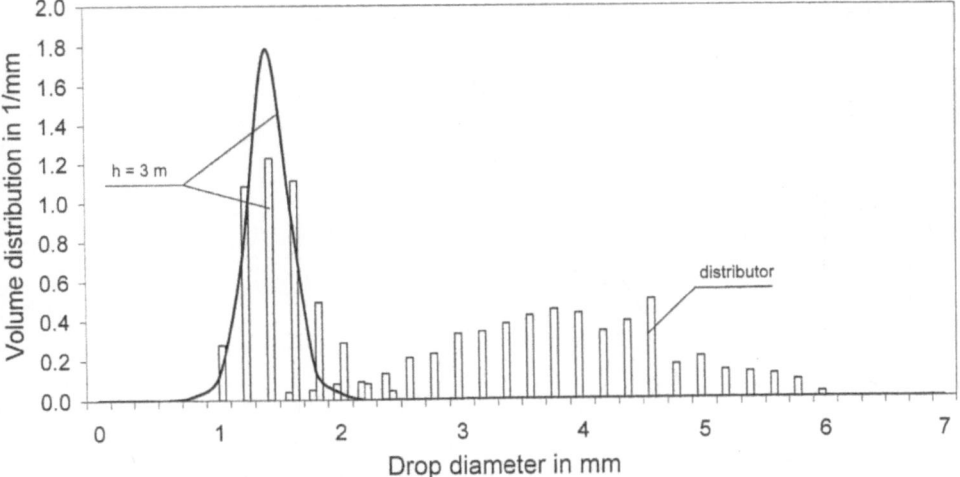

Figure 27 Comparison of measured (bars) and simulated volume drop size distributions for an *n*-butyl acetate(d)/water packed column (Montz-Pak B1-350) system: pulsation intensity, 20 mm/s; *h*, height of the column; phase ratio, 1/5; cumulative load, 34.3 m³/m² h.

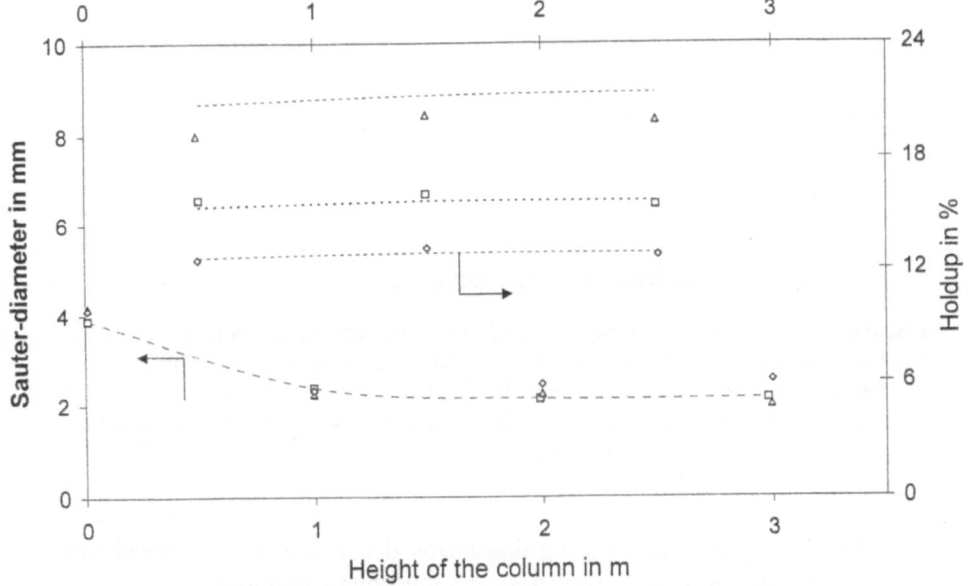

Figure 28 Comparison of measured and simulated profiles of the Sauter diameter and the local holdup for an *n*-butyl acetate(d)/water packed column (Montz-Pak B1-350) system: pulsation intensity, 10 mm/s; phase ratio, 1/1; cumulative load (m³/m² h), △, 25.7; □, 34.2; ◇, 42.4.

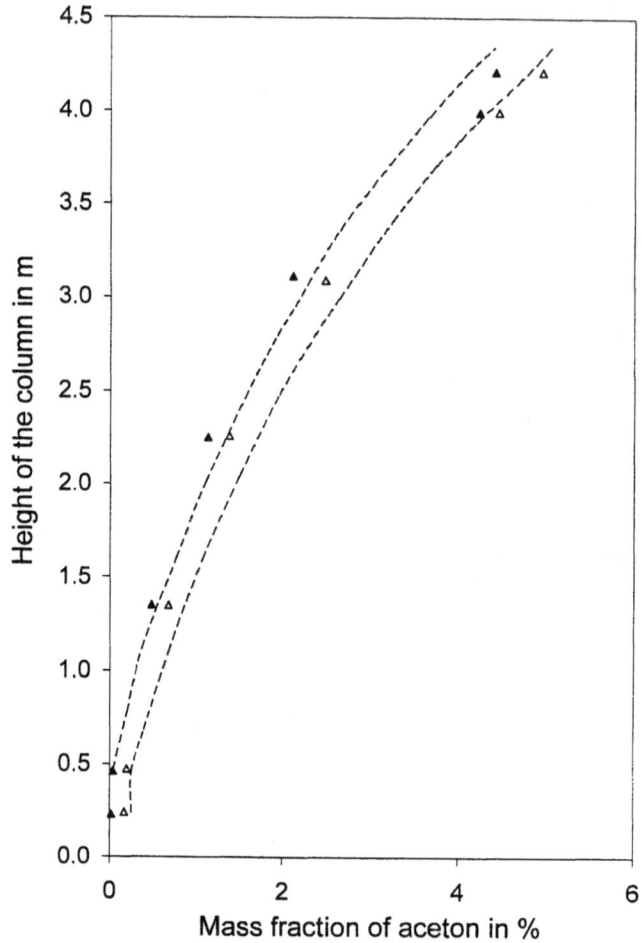

Figure 29 Comparison of measured and simulated concentration profiles of an *n*-butyl acetate(d)/water packed column (Montz-Pak B1-350) system: pulsation intensity, 15 mm/s; phase ratio, 1.2/1; cumulative load, 33 m³/m² h. Solid triangles, measured concentration organic phase; open triangles, measured concentration aqueous phase; dashed curves, results of the simulation.

The main advantage of using single-drop data obtained in so-called laboratory standard apparatus is seen from the following conditions:

1. The drop size is the most important single parameter with respect to both fluid dynamics and mass transfer.
2. The single-drop data are obtained under controlled conditions.

3. The standard apparatus may be equipped with the internals to be used in the technical column.
4. The drop size may be determined by using the fluids to be used in the technical column.
5. The use of standard apparatus permits comparison of experimental data with data obtained by different researchers.
6. A large database may be accumulated as a basis for the development of physically sound correlations.

The concept and the data presented here are the result of ideas and experimental work carried out in more than 20 years of cooperative research, and we hope that these materials will be useful to everybody interested in the fascinating field of liquid–liquid extraction.

NOTATION

Symbol	Meaning	Unit
A	cross-sectional area	m^2
a	amplitude of pulsator	m
B	beta function	—
b	denominator of Eq. (6)	—
d	diameter of a drop	m
d	differential operator	—
E	dispersion coefficient	m^2/s
F	flow rate	m^3/s
f	frequency of pulsator	s^{-1}
h	height	m
k	probability of coalescence	—
N	mass transfer rate	mol/s
n	number of drops	—
P	drop size distribution	—
p	probability of drop breakup	—
p	parameter of the beta function	—
q	parameter of the beta function	—
R	production term in Eq. (4)	—
S	standard deviation	—
t	time	s
u	velocity	m/s
V	volume	m^3
x	molar fraction in the continuous phase	—
y	molar fraction in the dispersed phase	—

Symbol	Meaning	Unit
Greek letters		
β	mass transfer coefficient	m/s
Γ	ratio of size of daughter drop over size of mother drop	—
Δ	increment	—
δ	partial differential operator	—
ε	holdup of the dispersed phase	—
Φ	holdup of the dispersed phase	—
ρ	molar density	mol/m^3
σ	interfacial tension	N/m
Indices		
C	coalescence	
c	continuous phase	
DD	daughter drop	
MD	mother drop	
d	dispersed phase	
E	single drop	
i	variable, class	
m	exponent of a correlation	
max	maximum	
n	at position n	
od	overall (disperse phase defined)	
s	swarm	
slip	slip	
stab	stable drop diameter (see Fig. 6)	
1	at position 1	
2	at position 2	
100	unstable drop diameter (see Fig. 6)	
*	at equilibrium	

ACKNOWLEDGMENTS

It is very satisfying to have been part of this research, and we thank everybody involved. Our sincere thanks go to those directly concerned: A. Aufderheide, C. Casamatta, C. Gourdon, B. Hoting, T. Leu, M. Lorenz, S. Mohanty, D. Niebuhr, M. Qi, and J. Schröter.

Finally we acknowledge the continuous financial support of the DFG (German Research Foundation) and the AIF (Association of Industrial Research Institutions). Without their financial help, this research would not have been possible.

REFERENCES

1. WJD Van Dijck. US Patent 2,011,186, 1935.
2. G Casamatta, A Vogelpohl. Ger Chem Eng 56:96–103, 1985.
3. C Gourdon, G Casamatta. In: JC Godfrey and MJ Slater, eds. Liquid–Liquid Extraction Equipment. Chichester: Wiley, 1994, pp 137–226.
4. B Genenger, B Lohrengel. Chem Eng Process 31:87–96, 1992.
5. T Misek, R Berger, J Schröter. Standard Test Systems for Liquid Extraction, 2nd ed. London: Institution of Chemical Engineers, 1985.
6. S Hu, RC Kintner. AIChE J 1:42–48, 1955.
7. H Haverland. Untersuchungen zur Tropfendispergierung in flüssigkeits-pulsierten Siebboden-Extraktionskolonnen. PhD dissertation, Technische Universität Clausthal, Clausthal, Germany, 1988.
8. JT Leu. Beitrag zur Fluiddynamik von Extraktionskolonnen mit geordneten Pakungen. Düsseldorf: VDI Verlag, 1995.
9. M Ishii, N Zuber. AIChE J 25:843–855, 1979.
10. J Klee, E Treybal. AIChE J 2:444–447, 1956.
11. B Hoting. Untersuchungen zur Fluiddynamik und Stoffübertragung in Extraktionskolonnen mit strukturierten Packungen. Düsseldorf: VDI Verlag, 1966.
12. B Hoting, A Vogelpohl. Prediction of fluid dynamics and mass transfer in solvent extraction columns with structured packings based on single drop experiments. Proceedings of ISEC '96, Melbourne, 1996, pp 231–236.
13. JT Leu, A Vogelpohl. Influence of the packing geometry on the fluid dynamics of solvent extraction columns with structured packings. Proceedings of ISEC '93, York, 1993, pp 73–80.
14. AK Chesters. Trans Inst Chem Eng 69: Part A, 239–270, 1991.
15. RS Miller, JL Ralph, RL Curl, GD Towell. AIChE J 9:196–207, 1963.
16. LL Tavlarides, M Stamatoudis. Adv Chem Eng 2:199–273, 1981.
17. CA Coulaloglou, LL Tavlarides. AIChE J 32:1289–1297, 1977.
18. C Gourdon, G Casamatta. Chem Eng Sci 46:1595–1608, 1991.
19. N Aderangi, DT Wasan. Chem Eng Commun 132:207–222, 1995.
20. JA Hamilton, HRC Pratt. AIChE J 30:442–450, 1984.
21. C Weiss, V Köhler, R Marr. AIChE J 43:1153–1162, 1997.
22. A Eckstein, A Vogelpohl. Chem Ing Tech 71:480–483, 1999.
23. M Qi. Untersuchungen zum Stoffaustausch am Einzeltropfen in flüssigkeitspulsierten Siebboden-Extraktionskolonnen. PhD dissertation, Technische Universität Clausthal, Clausthal, Germany, 1992.
24. M Qi, H Haverland, A Vogelpohl. Chem Ing Tech 72:203–214, 2000.
25. M Lorenz. Untersuchungen zum fluiddynamischen Verhalten von pulsierten Siebboden-Extraktionskolonnen. PhD dissertation, Technische Universität Clausthal, Clausthal, Germany, 1990.
26. S Mohanty, A Vogelpohl. Chem Eng Process 36:385–395, 1997.
27. D Niebuhr. Untersuchungen zur Fluiddynamik in pulsierten Siebbodenextraktionskolonnen. PhD dissertation, Technische Universität Clausthal, Clausthal, Germany, 1982.

28. D Niebuhr, A Vogelpohl. Ger Chem Eng 4:264–268, 1980.
29. E Aufderheide. Hydrodynamische Untersuchungen an pulsierten Siebbodenextraktionskolonnen. PhD dissertation, Technische Universität Clausthal, Clausthal, Germany, 1985.
30. E Aufderheide, A Vogelpohl. Ger Chem Eng 8:388–394, 1985.
31. B Hoting, M Qi, A Vogelpohl. Determination of single drop mass transfer coefficients in a laboratory-scale standard apparatus. Proceedings of ISEC '93, York, pp 453–460.

3

Purification of Nickel by Solvent Extraction

Kathryn C. Sole

Anglo American Research Laboratories (Pty) Ltd, Johannesburg, South Africa

Peter M. Cole

Hydrometallurgy Consultant, Bryanston, South Africa

I. INTRODUCTION

Although nickel is more abundant in the earth's crust than the combined reserves of copper, lead, and zinc, until recently there were few deposits of nickel that could be economically processed [1]. Since the mid-1990s, however, there has been a tremendous surge of interest in nickel metallurgy, owing to high international nickel prices and forecasts of future market deficits [2,3].

Nickel extractive metallurgy was for most of the twentieth century based primarily on pyrometallurgical technology. Nickel sulfide concentrates were smelted to give a copper–nickel–cobalt matte. This was then further refined using either pyro- or electrometallurgical techniques. However, recent advances in hydrometallurgical process technology and engineering—in particular, the development of flow sheets that combine pressure acid leaching with solvent extraction (SX) and electrowinning (EW)—are today allowing formerly intractable or uneconomic reserves of both sulfide and oxide ores to be economically exploited.

SX has been widely used for several decades in the recovery of copper from low grade materials. Its integration as a purification and separation step into other base metal flow sheets has, however, been much slower. From an initially hesitant acceptance, SX is today also regarded as the preferred unit operation for separating cobalt from nickel, owing in large part to the commercial availability and proven performance of highly selective, chemically stable reagents for this separation.

With the exception of cobalt extraction, there has been far less development with respect to the inclusion of SX in nickel flow sheets. Furthermore, it is only in the past few years that SX has enabled significant advances to be made in the direct processing of nickel leach liquors. Three new dry laterite projects commissioned in Western Australia use pressure acid leach (PAL) technology to maximize cobalt and nickel recovery. Several other hydrometallurgical process flow sheets are under development for the recovery of nickel from both wet and dry laterites, as well as from low grade sulfides and flotation concentrates. While the process routes have significant differences, they all employ SX to produce high grade nickel products, either for the removal of cobalt from the nickel-rich leach liquor or for the purification of the nickel solution up- or downstream of the cobalt removal step [4,5].

This chapter briefly discusses some of the traditional flow sheets, emphasizing the diversity of process routes employed. Recently commercialized flow sheets are then examined in detail, including the separation chemistry and the integration of the SX operation with the up- and downstream processes. Finally, some promising research and pilot plant developments in the hydrometallurgical processing of nickel are presented.

II. MINERALOGY OF NICKEL ORES AND ITS IMPLICATIONS FOR THE CHOICE OF PROCESS ROUTES

Two main sources of workable nickel exist [1]. These are sulfide and laterite ores. The sulfide ores have traditionally been the major contributor to annual nickel production, but with the diminishing reserves of sulfides (55 million tons) [6], the laterites (reserves of 144 million tons) are beginning to account for a larger proportion of refined nickel. In 1995 laterites comprised about 30% of world nickel production [7]; this figure is predicted to rise steadily to over 50% by 2010 [8].

A. Sulfides

Sulfide ores contain typically between 0.5 and 3.0% Ni and can be upgraded by flotation. The deposits are sometimes very rich in the platinum group metals (PGMs) (Pt, Pd, Rh, Ir, Ru, Os), and nickel may be produced merely as a by-product of the processing of the PGMs. Copper and cobalt are also important by-products from these ores. The main deposits occur in Canada, South Africa, Japan, Alaska, and the Commonwealth of Independent States [1].

The inherent energy content of sulfides has traditionally enabled the concentrates to be processed by smelting [9], producing a nickel-rich matte (also containing the PGMs and copper) that is treated further by pyro- or hydrometallurgical methods to obtain purified metal products [9–14]. The past decade has seen a change in focus within the base metal industry away from pyrometallurgy toward "cleaner" hydrometallurgical processes that produce high grade metal products with environmentally acceptable waste products. This trend has been supported by the superior cobalt recovery potential of hydrometallurgical processes. To date, the Sherritt ammonia process is the only completely hydrometallurgical processing route that has been commercialized for nickel sulfides. Several alternative process flow sheets at advanced stages of feasibility evaluation are discussed in Section VII.A.

B. Laterites

Laterites are low grade nickel deposits (1–3% Ni) produced under particular environmental conditions by the chemical weathering of surface rocks. Two chemical subtypes of laterite exist. Limonitic laterites (typically 1.4% Ni, 0.15% Co, and about 40% Fe) are found closer to the surface and contain lower concentrations of silica and magnesium. The deeper saprolitic (silicate) ores have a lower concentration of iron (~15%) and contain about 2.4% Ni and 0.05% Co. The most widespread laterite deposits, the so-called wet laterites, are those formed in high temperature, high humidity tropical rain forest climates, such as Brazil, Cuba, the Dominican Republic, Colombia, New Caledonia, Central and West Africa, Indonesia, and the Philippines. Less common are the dry laterites found in the hot, semiarid areas of Western Australia, Greece, and Albania.

Because of their differing chemical and mineralogical characteristics, the two types of laterite ore require quite different processing routes. There are three traditional methods for the processing of nickel laterites: smelting for saprolitic ores, and ammonia or acid leaching for limonitic ores.

Processing of saprolites has dominated historically because of the higher nickel content of these ores and their amenability to established pyrometallurgical processing. As recently as 1996, about 80% of nickel production from laterites originated from the pyrometallurgical processing of wet saprolitic ores [15]. The relative proportions of the various elements in these ores make them ideally suited to the production of ferronickel (which contains about 20% nickel). Alternatively, a nickel sulfide matte is produced. A constraint on the smelting of saprolitic ores is the necessity of maintaining a 0.66:1 ratio of Mg to Si in the slag, to minimize the problems of excessive corrosion of the refractories and high slag viscosity. The upgrading of the smelter product is similar to that of the sulfide ores.

As the nickel-rich saprolitic ore reserves decrease because of exploitation, the incentive to process the lower grade limonitic ores increases [16]. Smelting limonitic ore to produce ferronickel is impractical because of the low Ni/Fe ratio of the ore. The higher cobalt content of limonitic ores also makes them more suited to hydrometallurgical processing, inasmuch as pyrometallurgical routes do not usually produce a cobalt by-product. Most laterite ores cannot be efficiently preconcentrated (as is the case with sulfide deposits) because the nickel and cobalt atoms are uniformly dispersed through the mineral lattices, and so only limited physical upgrading is possible. This means that any hydrometallurgical approach to the processing of these ores is required to treat large volumes of relatively dilute leach liquors. Limonitic ores have traditionally been processed by using the Caron process (a reduction roast followed by an ammonia/ammonium carbonate leach) or by pressure leaching in sulfuric acid. These processes are discussed in more detail in Section V.

III. CHEMISTRY OF NICKEL SEPARATIONS BY SOLVENT EXTRACTION

Cobalt and copper are the base metals most commonly occurring with nickel in solutions produced from the solubilization of nickel-containing materials. Other elements from which nickel may have to be separated include iron, zinc, calcium, magnesium, and manganese. Because copper and cobalt occur adjacent to nickel on the periodic table, their chemistries are very similar. In the past, separations of these elements have been based on selective precipitation or oxidation reactions and on differences in the coordination chemistry of various complexes [17]. Today, the availability of specialty ion-exchange resins or solvent extractants has simplified these separations considerably.

As discussed by Warshawsky [18], the chemistry of nickel does not lend itself to the development of a nickel-selective extractant, particularly in acid media. The rate of exchange of most ligands with nickel by displacing the inner hydration sheath of water molecules is well known to be kinetically far slower than complexation with other base metal cations. Nickel SX is, however, ideally suited to integration with nickel EW where the final product of choice is high-purity nickel cathode. Nickel EW from sulfate media typically transfers about 30 g/L Ni and requires the advance electrolyte (loaded strip liquor) to have a pH of 3.5–4 [19]. Since the reduction of 30 g/L Ni will generate approximately 50 g/L H_2SO_4, the acid concentration of the return electrolyte will be very suitable for the stripping of nickel from the loaded organic phase.

Except for ammonia-based systems, approaches to purifying nickel leach liquors have traditionally concentrated on the removal of the other elements from the nickel liquors. Only recently have other technologies based on the direct extraction of nickel itself been considered to be commercially viable.

A. Chloride Media

Several commercial SX processes for the purification of nickel from copper and cobalt are based on hydrochloric acid systems. In chloride solution, nickel is purified by taking advantage of the fact that cobalt and copper form anionic complexes in chloride media (MCl_4^{2-}), while those of nickel are neutral ($NiCl_2$). The anionic complexes can be extracted from the leach liquor by using either a tertiary or quaternary amine extractant.

The extraction of cobalt and copper (M) by a tertiary amine can be represented as follows:

$$R_3N(org) + H^+(aq) \rightarrow R_3NH^+(org)$$

$$2R_3NH^+(org) + MCl_4^{2-}(aq) \rightarrow [(R_3NH^+)_2MCl_4^{2-}](org),$$

where R denotes an alkyl group or hydrogen.

The order of extraction of various base metal chloride complexes by a tertiary amine as a function of chloride concentration is shown in Fig. 1 [20]. Most divalent base metals are readily separated from nickel; however, their subsequent purification may still be required. Iron(III), for example, extracted by the amine as the $FeCl_4^-$ species, is typically separated prior to the amine SX circuit by hydrolytic precipitation or by SX using tri-n-butyl phosphate (TBP).

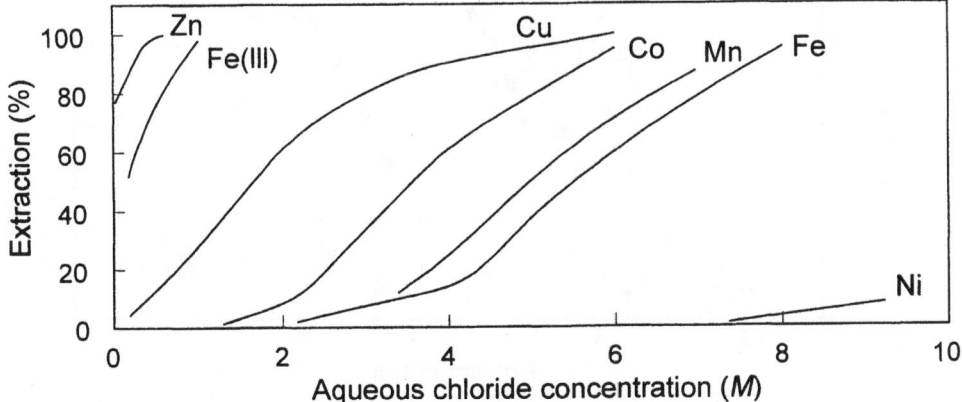

Figure 1   Extraction of some divalent base metals and Fe(III) from chloride solution by Alamine 336 hydrochloride. (From Ref. 20.)

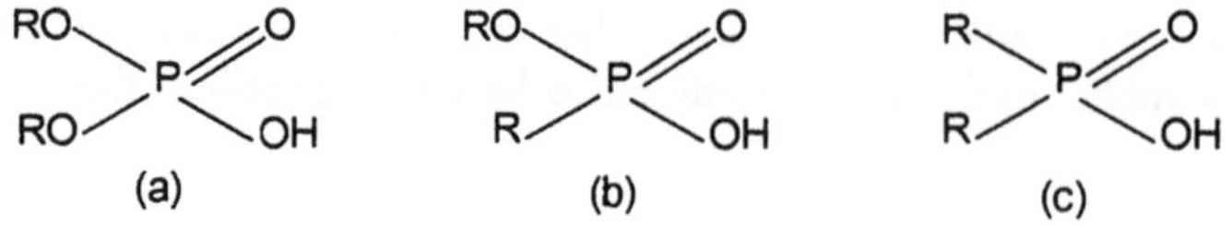

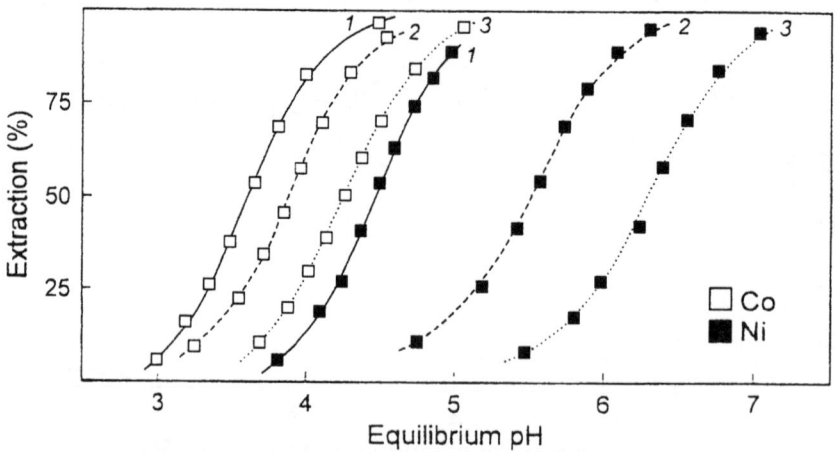

(a)   (b)   (c)

Figure 2 Structures of the organophosphorus extractants employed for the solvent extraction separation of cobalt and nickel. (a) Dialkylphosphoric acid. (b) Alkyl alkyl-phosphonic acid. (c) Dialkylphosphinic acid.

B. Sulfate Media

The SX separation of cobalt from nickel in acidic sulfate media has been practiced commercially for many years, and various innovative flow sheets have been developed [21]. Organophosphorus acids and carboxylic acids are typically employed.

1. Organophosphorus Acids

In sulfate media, cobalt is separated from nickel by using the dialkylphosphorus acid family of extractants (Fig. 2). The basicity of these reagents increases with decreasing distance of the alkyl chain from the central phosphorus atom. The consequent influence on the separation of nickel and cobalt is clearly demonstrated in Fig. 3, which shows the pH dependence of the extraction by three organophosphorus acid analogs [22].

Figure 3 pH dependence of the extraction of nickel (■) and cobalt (□) by 1, di(2-ethylhexyl)phosphoric acid (pK_a = 3.51); 2, 2-ethylhexyl 2-ethylhexylphosphonic acid (pK_a = 4.77); and 3, di(2-ethylhexyl)phosphinic acid (pK_a = 6.03). (From Ref. 22.)

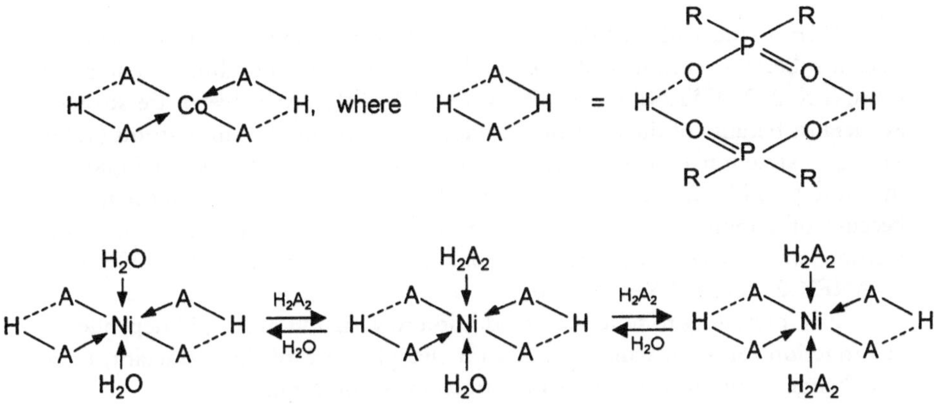

Figure 4 Comparison of tetrahedral cobalt complexes and octahedral nickel complexes formed with a dialkylphosphinic acid. The dimeric hydrogen-bonded structure, typical of this family of extractants, is also shown. (From Ref. 24.)

In commercial reagents, the change from the phosphoric acid structure [exemplified by di(2-ethylhexyl)phosphoric acid (D2EHPA)] to the phosphonic acids (such as Daihachi's PC-88A or Ionquest 801, supplied by Rhodia) to the phosphinic acids (Cytec's CYANEX* 272 and PIA-8 from Daihachi) results in a corresponding improvement of the Co:Ni separation factors from 14 to 280 to 7000 [23].

If the extractant is represented as the dimeric hydrogen-bonded species (Fig. 4), the extraction reactions for cobalt and nickel can be written as follows [24,25]:

$$Co^{2+}(aq) + 2H_2A_2(org) \rightarrow CoA_2 \cdot H_2A_2(org) + 2H^+(aq)$$

$$Ni^{2+}(aq) + (x + 1)H_2A_2(org) + (4 - x)H_2O$$
$$\rightarrow NiA_2 \cdot (H_2A_2)_x \cdot (H_2O)_{4-x}(org) + 2H^+(aq)$$

where $x = 2$–4, depending on the extractant concentration. The extraction is selective for cobalt over nickel because the formation of the tetrahedrally coordinated cobalt species is more favorable in the organic phase than the octahedrally coordinated nickel species (Fig. 4). Furthermore, the cobalt complex is hydrophobic and so readily transferred to the organic phase, while the nickel complex can contain one or two coordinated water molecules in its inner sphere and is therefore more hydrophilic [24,26].

---

*CYANEX® is a registered trademark of Cytec, Inc.

During the early 1990s, sulfur-substituted organophosphorus acid extractants became commercially available. The mono- and dithio analogs of CYANEX 272 (CYANEX 302 and CYANEX 301, respectively) are stronger extractants because of the electron-withdrawing effect of the sulfur atoms (Table 1). The extraction of base metal cations from higher acidity leach liquors is therefore possible [27,28]. CYANEX 302 has been withdrawn from the market because of stability problems. Figure 5 compares the pH dependence of extraction of selected base metal cations in sulfate solution by D2EHPA, CYANEX 272, and CYANEX 301.

It is interesting to note that in addition to variation in the pH dependence of extraction for a particular cation with changing pK_a of the extractant, there is a change in the relative order of extraction of the cations:

D2EHPA: Fe(III) > Zn > Ca > Mn > Cu > Mg > Co >> Ni

CYANEX 272:   Fe(III) > Zn > Cu > Mn > Co > Mg > Ca > Ni

CYANEX 302:   Cu > Fe(III) > Zn > Co > Mn > Ni >>> Mg ~ Ca

CYANEX 301:   Cu > Zn > Fe(III) > Co ~ Ni >> Mn >>> Mg ~ Ca

This can provide some flexibility in reagent choice to achieve the desired selectivity for particular applications.

Although some older plants still use D2EHPA or PC-88A for the separation of cobalt from nickel (see Section IV), today the preferred reagent is CYANEX 272. Despite a considerable cost premium, this reagent is used to produce about 40% of the world's cobalt [23]. Not only is the Co:Ni separation factor for CYANEX 272 substantially higher, but the phosphinic acid is more

Table 1   Structure and Properties of Sulfur-Substituted Organophosphinic Acid Extractants

Extractant	Structure[a]	pK_a
CYANEX 272	R—P(=O)(OH)—R	6.37
CYANEX 302	R—P(=S)(OH)—R	5.63
CYANEX 301	R—P(=S)(SH)—R	2.61

[a]R is the 2,4,4-trimethylpentyl alkyl chain.

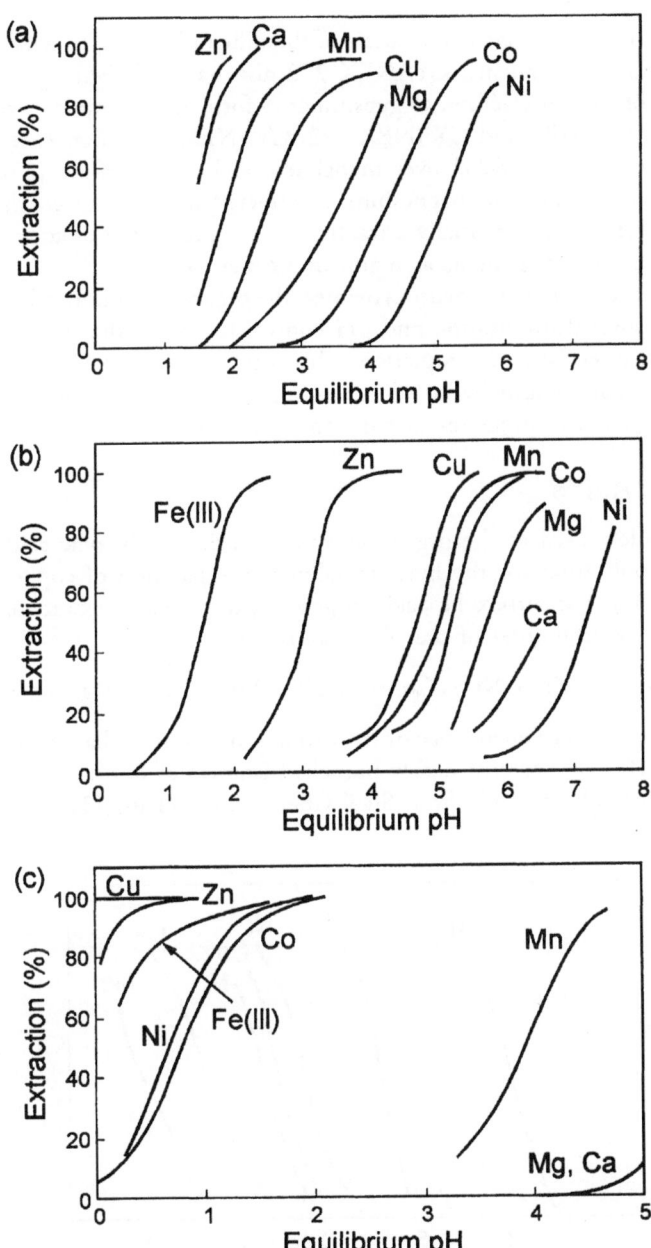

Figure 5 Illustrative pH dependence of the extraction of cations by (a) D2EHPA, (b) CYANEX 272, and (c) CYANEX 301. (Adapted from Ref. 28 and unpublished data.)

stable with respect to the oxidative degradation of the diluent by Co(III), which can be a problem in these systems. CYANEX 272 also has the advantage of being selective for cobalt over calcium, so gypsum crud formation is minimized.

In contrast to D2EHPA and CYANEX 272, CYANEX 301 offers excellent selectivity for cobalt and nickel over manganese at low pH values [29]. Furthermore, neither calcium nor magnesium is extracted under these conditions, making CYANEX 301 potentially attractive for the processing of laterite leach liquors that have these elements as major impurities [30].

In contrast to the amines and hydroxyoximes, the organophosphorus (and carboxylic) acids require neutralization and pH control to ensure that the extraction reactions go essentially to completion. This is typically achieved by the addition of ammonia or sodium hydroxide solution, both of which represent significant operating costs to processes that use these reagents.

2. Carboxylic Acids

Although not as widely used as the organophosphorus acids, carboxylic acids have found some application for the hydrometallurgical separation of copper, nickel, and cobalt [31]. For carboxylic acids in general, the order of extraction of metal cations follows the order of their hydrolysis constants:

$$Fe(III) > Al > Cu > Zn > Ni \sim Co \sim Fe(II) > Mn > Ca \sim Mg$$

Figure 6 shows the pH dependence of the extraction of various base metal cations by Versatic 10, a mixture of highly branched isomers of C_{10} monocarboxylic acids commercially available from Shell Chemical Company [31]. The

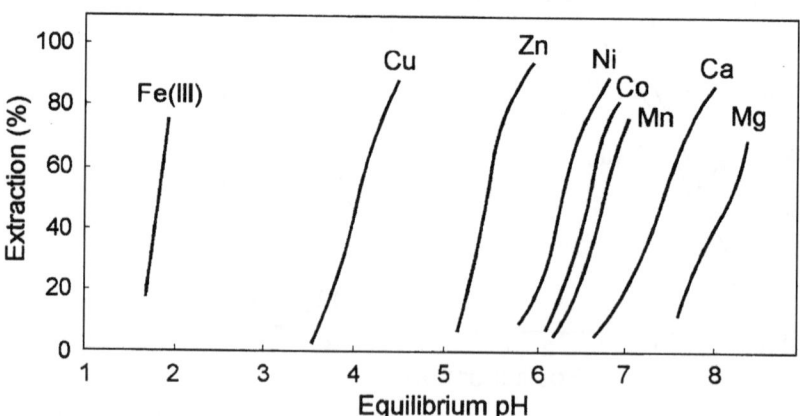

Figure 6   pH-dependence of the extraction of some metal cations by 0.5 M Versatic 10 in xylene. (Adapted from Ref. 31.)

extraction reactions and complexes formed are similar to those of the organo-phosphorus acids, although the carboxylic acids generally do not remain di-merized when complexed.

Nickel is extracted in the pH range 6–7.5, depending on the experimental conditions. The extraction reaction is [32]:

$$Ni^{2+}(aq) + \left(\frac{m}{2}\right) H_2A_2(org) \rightarrow NiA_2 \cdot (HA)_{m-2} \ (org) + 2H^+(aq)$$

where $m = 2$ for high solvent loading and $m = 4$ for low loadings. A similar extraction reaction can be written for cobalt. Although Versatic 10 is selective for nickel over cobalt, there is insufficient difference in the pH dependence of extraction to enable reagents of this type to be used to separate nickel from cobalt. The selectivity for nickel and cobalt over calcium and magnesium is, however, exploited in some flow sheets.

C. Ammoniacal Media

Nickel is extracted from ammoniacal systems by using oximes that have a struc-ture similar to that shown in Fig. 7. Nickel exists in solution as either the tetra-, penta-, or hexaamine complex, depending on the concentration of free ammonia. The extraction reactions for nickel can be written as follows [19]:

$$Ni(NH_3)_4^{2+}(aq) + 2HA(org) \rightarrow NiA_2(org) + 2NH_3(aq) + 2NH_4^+(aq)$$

or

$$Ni(NH_3)_6^{2+}(aq) + 2HA(org) \rightarrow NiA_2(org) + 4NH_3(aq) + 2NH_4^+(aq)$$

where HA represents the oxime extractant.

The kinetics of nickel stripping from oximes are slow relative to the rate of extraction [19]. As is evident from these reactions, the equilibrium position and extraction of nickel depends on the concentration of free ammonia (Fig. 8) [33]. Stripping can therefore be achieved by using a solution of strong

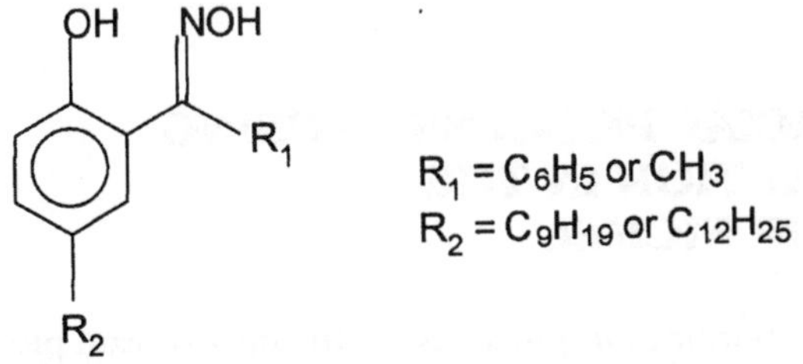

$$R_1 = C_6H_5 \text{ or } CH_3$$
$$R_2 = C_9H_{19} \text{ or } C_{12}H_{25}$$

Figure 7 General structure of oximes used for the extraction of nickel from am-monia solution.

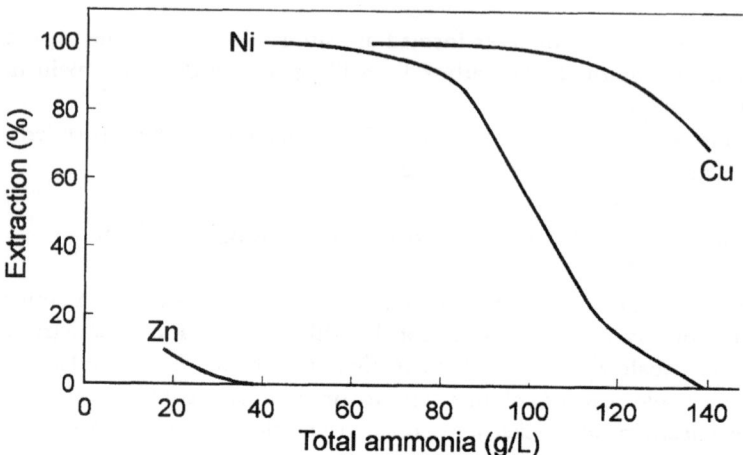

Figure 8   Extraction of nickel, copper, and zinc from ammonia solution by LIX 84. (Reprinted from Ref. 33, The Chemistry of Metals Recovery Using LIX Reagents, 1997 ed, by permission of Cognis Corporation.)

ammonia and ammonium sulfate or ammonium carbonate to reverse the equilibrium. Alternatively, sulfuric acid may be used for stripping:

$$NiA_2(org) + 2H^+(aq) + SO_4^{2-}(aq) \rightarrow 2HA(org) + Ni^{2+}(aq) + SO_4^{2-}(aq)$$

      Although not shown in Fig. 8, Co(II) is preferentially extracted over nickel via similar extraction reactions. Once loaded onto the organic phase, Co(II) is oxidized to Co(III), which is not readily stripped. To prevent the extraction of cobalt in the first instance, it is oxidized to Co(III) ahead of the SX circuit, since trivalent cobalt is not extracted under these conditions. The ammonium cation is also complexed by the oxime, which can lead to ammonia losses in the system if operating conditions are not adequately controlled.

IV. EARLY COMMERCIAL PROCESSES USING SOLVENT EXTRACTION IN THE PURIFICATION OF NICKEL

SX was originally used in nickel purification processes for the downstream processing of partially purified solutions in refinery flow sheets, rather than for upgrading liquors derived from the leaching of whole ore.

A. Amine Extractants

One of the earliest applications of SX in nickel refining was in the matte leach process developed by Falconbridge, Canada, and installed in the late 1960s at Kristiansand, Norway [34,35]. The process comprised an acidic leach of the matte that solubilized most of the iron along with the nickel, cobalt, and copper. Iron(III) was removed first by SX using TBP, with extraction of the $H^+FeCl_4^-$ ion pair by a solvation reaction. The anionic complexes of copper and cobalt were then extracted from the nickel chloride liquor by using trioctylamine (TOA). Nickel was crystallized as the chloride salt. This plant operated from 1968 until 1983 [36]. Thereafter a new plant was built, based on a much improved chloride leach of the matte feed [37]. The copper–nickel–cobalt matte is now leached under oxidative conditions ($Cl_2/CuCl_2$), producing a liquor containing approximately 200 g/L Ni and 60 g/L Cu. Copper is removed from solution by cementation with matte. The iron is removed by chlorine oxidation and hydrolysis with nickel carbonate. Cobalt is then extracted with TOA and cobalt metal electrowon from the chloride strip liquor. The raffinate is a purified $NiCl_2$ solution from which nickel is recovered by electrowinning.

Cobalt extraction with TOA has also been applied commercially by Union Minière in Belgium in the refining of alloy scrap. In this flow sheet, zinc (as the anionic complexes $ZnCl_3^-$ and $ZnCl_4^{2-}$) is removed from the leach liquor ahead of the cobalt–nickel separation using an amine [38]. Eramet's le Havre–Sandouville refinery in France treats metallurgical waste using a chlorine leach, followed by selective amine extraction of cobalt [39].

B. Oxime Extractants

The first plant to use SX to extract nickel directly was that operated by SEC Corporation in El Paso, Texas [40,41]. Started in 1971, it produced 0.5 ton/day nickel cathode, along with 2 tons/day copper. The feed liquor was the ammonia releach solution from the precipitation of copper sulfate in the nearby Phelps Dodge copper refinery. This contained ~80 g/L Cu, 23–40 g/L Ni, and small amounts of other base metals. Following copper removal by SX, the residual solution was purified by nickel SX with LIX 64N,* a hydroxyoxime extractant. The circuit comprised three extraction, one wash, and two strip stages. This process separated nickel not only from cobalt but also from zinc and copper (see Fig. 8).

---

*LIX® is a registered trademark of Cognis Corporation.

C. Organophosphorus Acid Extractants

Since the first application of D2EHPA to remove cobalt from nickel solution at Eldorado Nuclear in Canada in the early 1960s, the organophosphorus extractants have found widespread application for this separation [42]. The Pyrites Company in the United States used D2EHPA to separate cobalt from nickel in leach liquor derived from the pressure acid leaching of scrap.

Anglo Platinum's base metal refinery in Rustenburg, South Africa, also uses a D2EHPA circuit to remove cobalt from a nickel liquor [43]. The process employed at this site was designed in the 1970s and relies largely on precipitation as the purification technique for the nickel stream. Following matte leaching, copper, iron, and other impurities in the nickel liquor are removed in a series of precipitation reactions. The final element to be removed is cobalt, precipitated as the cobalt(III) hydroxide in a process developed by Outokumpu (see Ref. 17):

$$Ni(OH)_3 + CoSO_4 \rightarrow Co(OH)_3 + NiSO_4$$

Nickel(III) hydroxide is produced electrolytically. The chemistry takes advantage of the difference in stability of the hydroxides of cobalt and nickel at pH 5.6–5.7. The purified nickel solution is electrowon to produce saleable nickel cathodes. The precipitated cobalt hydroxide cake contains about 40% nickel, as well as other coprecipitated impurities, such as magnesium, calcium, and iron. Following redissolution of the cake, the leach liquor is purified by precipitation of most impurities, followed by extraction of the cobalt with D2EHPA. The nickel value is recovered from the SX raffinate and the loaded strip liquor is crystallized to produce high grade $CoSO_4$ crystals.

The original SX circuit is still used. This comprises seven extraction stages, six scrubbing stages, and three stripping stages. Two final stages are used for the removal of trace impurities from the stripped organic phase and regeneration of the extractant. The feed to the SX circuit typically contains 15–18 g/L Co and 5–8 g/L Ni at pH 4.5–4.8. Extraction is carried out at 40–45°C to enhance the separation between cobalt and nickel. Any magnesium in the feed solution is coextracted with cobalt and eventually reports to the cobalt product (see Fig. 5a). Coextracted nickel is scrubbed from the loaded organic phase at pH 6 by means of a bleed of the cobalt sulfate loaded strip liquor. Cobalt is stripped with 10% sulfuric acid, regenerating the extractant to its acidic form. The stripped organic phase is then contacted with 20% sulfuric acid to remove coextracted magnesium, manganese, and calcium, before being regenerated to the sodium form. This D2EHPA circuit upgrades the SX feed solution from a Co/Ni ratio of 2:1 to 20,000:1 in the loaded strip liquor. Cobalt recovery across the entire circuit exceeds 96%, and across the SX circuit is better than 98%.

In a process installed in 1975, the Nippon Mining Company used a series of SX steps to treat liquor derived from the pressure acid leaching of mixed cobalt nickel sulfide [44]. Zinc was first extracted by using D2EHPA at pH 2–3, followed by cobalt extraction with the more selective alkyl alkylphosphonic acid extractant PC-88A. After adjustment of the pH to 8.5, nickel was extracted by means of LIX 64N. Nickel was recovered from the sulfate strip liquor by electrowinning.

Since the early 1980s, the Chinese have operated a nickel refinery at Jinchuan where zinc is removed with P204 and cobalt is then selectively extracted with P507, leaving a purified nickel solution. P204 and P507 are both phosphonic acid extractants [42].

D. Carboxylic Acid Extractants

An innovative process was implemented in 1975 by Sumitomo Metal Mining Company at its refinery in Niihama, Japan, for the purification of a nickel stream derived from the pressure acid leaching of mixed sulfide precipitates (Fig. 9) [45–47].

A bulk separation of cobalt and nickel is carried out by means of the highly branched carboxylic acid extractant, Versatic 10. The metals are stripped into hydrochloric acid. Cobalt and nickel are then separated by means of TOA and recovered by electrowinning. Zinc is extracted into TOA (see Section III.A) and is prevented from building up in the organic phase by NaOH neutralization of an organic phase bleed.

V. SECOND-GENERATION PROCESSES

During the period from about 1980 to 1995, several novel processes using SX technology were commercialized for the refining of nickel. The first application of SX in the treatment of leach liquor derived from the leaching of whole ore was implemented at Queensland Nickel in Australia. Other processes employed innovative adaptations of the older technology, using modern reagents and revisited flow sheet configurations.

A. Queensland Nickel's Ammoniacal Solvent Extraction Process

1. Original Caron Process

In the Caron process [15], milled laterite ore is heated to 700°C in a reducing atmosphere to convert the nickel and cobalt minerals to the metallic form. This

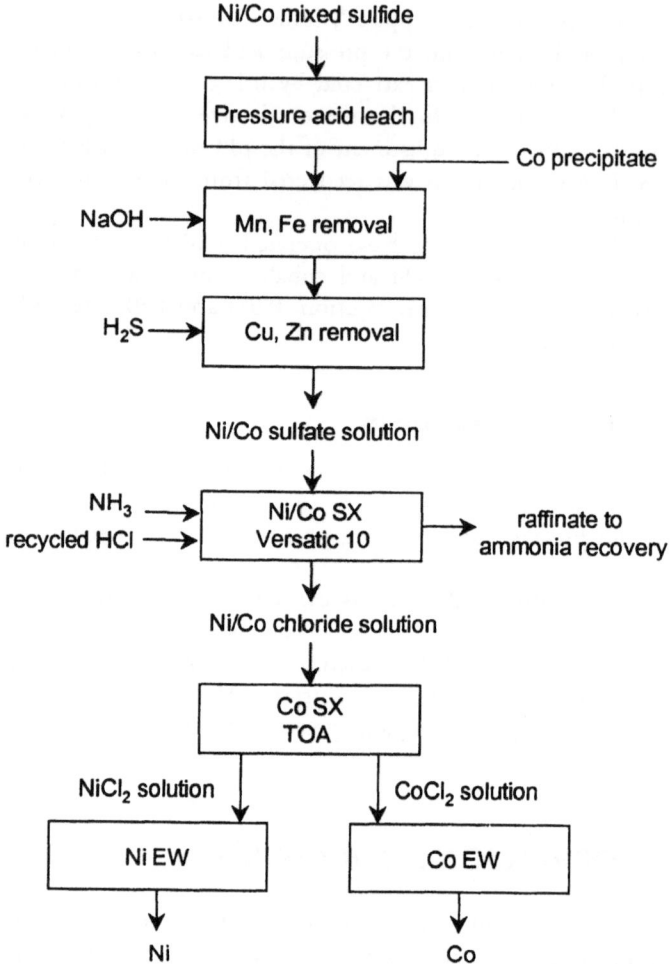

Figure 9 The Sumitomo Metal Mining Company nickel purification flow sheet. (From Ref. 47.)

allows their fairly selective dissolution in ammonia solution under ambient temperature and pressure conditions. Ammonia is removed by boiling and nickel precipitates from solution as the carbonate. Calcination of the carbonate at 1200°C gives a nickel oxide product.

The Caron process was first implemented commercially at Nicaro, Cuba, in 1943. Two other Cuban plants, Punta Gorda (1974) and Las Camariacas (1986), were built using the Caron process for the processing of lateritic material [48,49]. Today, Billiton's Tocatins plant in Brazil uses ammonia leach technology, as does its Queensland Nickel Industries (QNI) Yabulu refinery in Townsville, Australia. The higher magnesia contents of the ores treated by these plants were important criteria in the selection of the ammonia leaching route.

None of the plants using the original Caron process ever operated very profitably. It is only during the past decade, with modifications introduced to this process by QNI, that this process has come to be regarded as a viable technical and economic alternative for the processing of limonitic laterites.

2. Caron Process Modified by Using Ammoniacal SX

The process at QNI today bears little resemblance to that installed in 1974. The original flow sheet featured a reductive roast of the laterite ore followed by leaching in ammonia, removal of the cobalt as a crude cobalt nickel sulfide by precipitation, crystallization of nickel carbonate, and production of a nickel oxide sinter. This impure product was suitable only for making stainless steel and was sold at a significant discount on the prevailing nickel metal price [50].

The current flow sheet (Fig. 10) can be considered to be a modification of the original Caron process, with an ammonia leach circuit for the recovery of nickel [51–53]. The in-house development and installation of an ammoniacal solvent extraction (ASX) process for the purification of the nickel leach liquor has enabled the plant to upgrade its product to class 1 nickel metal, for a significant increase in revenues.

Nickel is extracted from the ammoniacal leach liquor by using the ketoxime-based extractant LIX 84I [19,54]. Nickel is selectively extracted from an aqueous feed containing 9 g/L Ni and 0.3 g/L Co. The aqueous flow rate is 330 m^3/h. The organic phase comprises 30 vol% LIX 84I in the aliphatic diluent Escaid 110. By means of three extraction stages and an advance organic-to-aqueous volumetric flow rate ratio (O:A) of 1.5, the nickel concentration is reduced to less than 0.01 g/L in the raffinate. The loaded organic phase is stripped in three stages with 270 g/L ammonia solution at an O:A of 15 to produce a strip liquor containing 75–80 g/L Ni. Stripping efficiency is about 65%. Cobalt is precipitated from the raffinate as the sulfide.

One of the main disadvantages of the ASX flow sheet is that it requires cobalt in the feed solution to be oxidized from Co(II) to Co(III) prior to the nickel SX. This is because divalent cobalt loads onto the organic phase, but does not strip (Section III.C). Cobalt oxidation is carried out by means of hydrogen peroxide. Because some residual Co(II) is always present in the SX feed solution, it is necessary to remove cobalt from a bleed of the organic phase by a reductive strip using iron [53].

A further disadvantage of this process is that Co(III) causes oxidation of the oxime extractant to a ketone, with a consequent loss of extractant capacity.

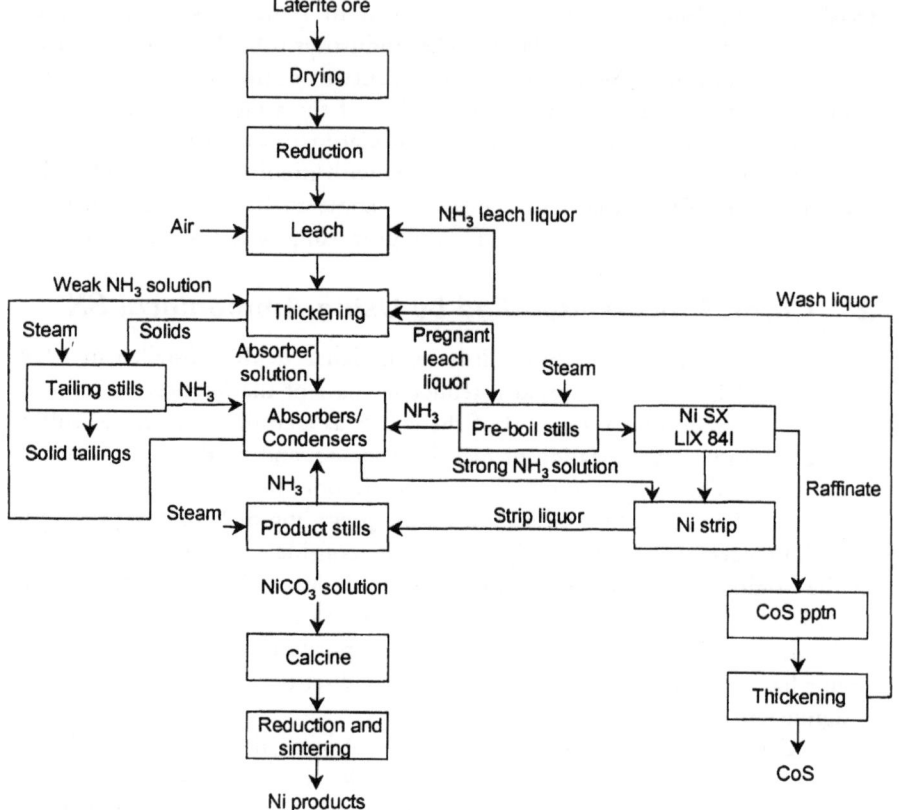

Figure 10 Flow sheet of the Queensland Nickel Industries' process. (From Ref. 53.)

Under high ammonia conditions, it is also possible to convert the oxime extractant to an imine [19]. A reoximation process (typically using hydroxylamine) is therefore also required in the flow sheet to avoid prohibitive reagent replacement costs.

Ammonia transfer to the organic phase, while not affecting the chemistry of the process, also represents an operating cost. Recovery of ammonia from the ammonium sulfate organic wash stage requires a lime boil. An experimental reagent, LIX XI-84-IT, has been shown to minimize ammonia transfer and additionally provide better selectivity for nickel over zinc and magnesium [54].

The impurity levels in the nickel products before and after the introduction of solvent extraction are compared in Table 2 and are testimony to the success of the new installation.

Table 2 Impurity Levels in Nickel Products Before and After the SX Installation at QNI

| Element | Concentration (%) | |
	Pre-SX (NiO sinter)	Post-SX (Class 1 nickel)
Ni	85–90	>99.5%
Co	0.2	0.02
Fe	0.2	0.02
Mn	0.2	0.03
Si	0.5	0.03
Cu	0.005	0.005
Zn	0.1	0.005
S	0.03	0.005
C	0.1	<0.01
As	0.002	<0.0005
Pb	0.002	<0.0005

Source: Ref. 50.

A refinery upgrade using a sequence of SX steps has recently been commissioned at the Yabulu site to produce a purified cobalt product from the sulfide intermediate precipitated from the ASX raffinate [55]. SX with CYANEX 272 is first employed to remove iron and zinc impurities (four extraction, a single scrub, and three stripping stages). This is followed by extraction of cobalt and residual nickel by means of D2EHPA (three extraction, two scrubbing, and four stripping stages) with stripping into an ammonia/ammonium carbonate solution. After oxidation of the cobalt to Co(III), the traces of nickel (and manganese and copper) are removed from the strip liquor by means of LIX 84I. In a step analogous to the ASX process employed in the main plant, an upgraded cobalt solution is produced, from which the final cobalt product is precipitated.

B. Tocatins Process

Billiton's Tocatins plant, commissioned in Brazil in 1984, also employs an ammonia leach to treat laterite ore, although Outokumpu technology is used to modify this downstream process [56]. Cobalt is coprecipitated with nickel as the basic carbonate from the ammonia leach solution during steam stripping. This intermediate product is transported to a refinery site for further processing. There it is redissolved in acidic spent nickel electrolyte, trace copper is removed by cementation with zinc, and then the soluble zinc is removed by SX. Initially the cobalt was precipitated by means of nickel(III) hydroxide, but since 1986 the cobalt has been separated from nickel by SX with CYANEX 272 (one of the earliest applications of this reagent). Both metals are recovered from solution by electrowinning.

C. Process Developments by Outokumpu

The HIKO process, developed by Outokumpu, was commissioned at the Kokkola plant in Finland in 1991 [57]. The plant is currently owned and operated by the OMG group. The process was originally developed for the treatment of high magnesium nickel concentrate from the Hitura Mine but is currently operated on a mixture of feeds.

The feed material is leached under pressure in sulfuric acid. Limestone is added to precipitate iron prior to solid–liquid separation, and copper is removed by sulfide precipitation. Several SX steps are then used to produce a purified nickel stream. D2EHPA modified with TBP is used to remove calcium and zinc, and these impurities are stripped into hydrochloric acid. Magnesium is then extracted in a second SX step. Originally Ionquest 801 (30 vol%) was used, in part because of the relatively low organic phase viscosity exhibited by this extractant under high magnesium loading conditions. It is believed that the lower magnesium levels now being encountered have led to its recent replacement by CYANEX 272. Cobalt and manganese in solution are coextracted with the magnesium and all are stripped into hydrochloric acid, from which cobalt is recovered as a salt. Before being used in the extraction circuit, the stripped solvent is preloaded with ammonium ions in a separate stage. After pH adjustment with ammonium hydroxide, the nickel in the raffinate is extracted, probably by using an oxime, then stripped into sulfuric acid from which a purified nickel salt is produced. The cobalt-catalyzed oxidation of the diluent is minimized by the use of custom-designed "vertical smooth flow" (VSF) mixers, which ensure minimum air and solvent contact.

As a world leader in the development of environmentally friendly smelting technology, Outokumpu further refined its flash-smelting process in the early 1990s to eliminate the converting step [58,59]. Its DON (direct Outokumpu nickel) process, implemented for the expansions at the Harjavalta refinery in Finland, produces a flash-smelting matte (5% iron and 71% nickel) and an electric furnace matte (34% iron and 53% nickel). These are leached separately under conditions that remove most of the copper as copper sulfide and the iron as hematite. The downstream processing of the nickel-rich liquor is carried out by SX. Cobalt is extracted by using CYANEX 272, yielding a purified cobalt electrolyte, while the nickel remains in the SX raffinate. A portion of the nickel is recovered as cathode by electrowinning, while the remainder is recovered as briquettes by pressure hydrogen reduction. Originally cobalt briquettes were also produced at this site by hydrogen reduction, however today the cobalt is further refined at the Kokkola refinery. The Harjavalta nickel refinery is now owned by OMG, although the smelter is still owned by Outokumpu.

A sulfate-based leaching process is also used for treating high grade smelter matte in the new Jinchuan plant in China [14]. This is almost identical to the Outokumpu process, except that it has only two stages of atmospheric leaching. Cobalt recovery from the nickel-rich leach liquor is carried out by SX using CYANEX 272. Further details of the operation have not been published.

VI. MODERN FLOW SHEETS FOR THE HYDROMETALLURGICAL PROCESSING OF NICKEL

A number of new nickel sulfide deposits are under advanced development. These include Voisey's Bay (Canada) and several sites in Australia and the Far East. In most of the proposed flow sheets for these projects, smelting and pyrometallurgical processing have remained the process option of choice. However, there are three new nickel laterite projects for which hydrometallurgical processing routes were selected. Common to all the modern hydrometallurgical flow sheets is the inclusion of pressure acid leaching for the solubilization of cobalt and nickel, followed by solution purification using SX. The application of SX to the downstream processing in these flow sheets can be essentially divided into separations in ammoniacal and acidic sulfate media.

The three nickel projects in Western Australia (Bulong, Cawse, and Murrin-Murrin) were fast-tracked through the feasibility and testing phases, then commissioned during 1998 and 1999. Table 3 compares the typical compositions of the laterite material being treated by these plants [60].

A. Pressure Acid Leaching

The first PAL plant was built at Moa Bay in Cuba in 1959, using technology developed by Freeport Sulfur. Because this company was the world's major supplier of sulfuric acid at the time, there was an incentive to develop a process that would use this reagent. Acid is the largest operating cost component of PAL processes, and this approach is economical only for ores that contain relatively low levels of magnesium (an acid-consuming constituent) [15]. The low magnesia content of this Cuban laterite ($\sim$0.8%) made it attractive for acid leaching.

The Moa Bay PAL, carried out at 250°C and 400 kPa in vertical autoclaves, dissolves 90% of the nickel and cobalt, while most of the dissolved iron and aluminum reprecipitate as hematite and jarosite under these conditions

Table 3 Comparison of the Compositions of Three Western
Australian Laterite Ores Processed by Pressure Acid Leaching

Component	Concentration (%)		
	Bulong	Cawse	Murrin-Murrin
Ni	1.11	1.0	1.24
Co	0.08	0.07	0.09
Fe	21.8	18.0	21.7
SiO_2	41.9	41.5	42.1
Al	2.75	1.71	2.51
Mg	4.62	1.58	4.02
Mn	0.36	0.17	0.4
Cr	0.6	0.92	0.88
Ca	0.03	0.03	0.53
Moisture	35	10	30

Source: Ref. 60.

[15,49,61]. These precipitates cause bad scaling in the autoclaves, pipes, and
heat exchangers, and the plant suffered from considerable downtime to remove
the scaling. Nickel and cobalt are coprecipitated with hydrogen sulfide to produce a mixed sulfide complex that requires subsequent refining, currently carried out at Sherritt's Fort Saskatchewan site in Canada. Although this plant still
operates, no further PAL plants were built for the treatment of nickel laterites
for almost 40 years.

Advances in engineering technology and in materials of construction, and
a more comprehensive understanding of the chemistry of autoclave reactions,
revived interest in PAL flow sheets for nickel during the mid-1990s. Autoclaves
have also become routinely employed in gold metallurgy for the processing of
refractory ores, giving greater confidence in their design and operation [62].

Although laterites can be leached in sulfuric acid at ambient temperatures
and pressures, high recoveries of nickel and cobalt require almost complete
dissolution of the iron minerals with which they are associated. Leaching at
higher temperatures (150–250°C) causes the iron to precipitate as hematite or
jarosite. This not only minimizes the iron reporting to the leach liquor for
downstream processing, but also considerably reduces the acid consumption of
the process. It is for this reason that PAL has become the dissolution option
of choice in many modern laterite and sulfide projects. The significant capital
costs of the autoclaves are justified by the reduction in size and complexity of
the downstream iron removal circuit and associated operating costs.

While several recent reviews have discussed the advantages and process
considerations of the three Western Australian PAL plants from the point of

view of the leaching operations [4,61,63,64], relatively sparse details have appeared in the literature regarding the full-scale operation of the downstream SX operations [65]. This chapter provides a more comprehensive summary of the flow sheets of these processes, with particular attention to the SX circuits.

B. Comparison of the Downstream Processing of Ammonia and Sulfuric Acid Leach Liquors

Table 4 illustrates the compositions of typical leach liquors produced via ammonia and sulfuric acid leaching of laterites [50]. The ammonia leaching process is far more selective than acid leaching, so the latter requires a more complex downstream processing flow sheet. In particular, the relatively high levels of soluble silica in PAL liquors can pose long-term problems for SX circuits because of the tendency of silica to promote the formation of crud. The nickel concentration is also lower in PAL liquors, so higher volumetric flow rates are required to achieve equivalent production levels.

In contrast, acid leaching has the advantage of much higher metal recoveries than similar ammonia leaching processes. Using Cuban examples, Moa Bay recovers more than 90% of nickel and cobalt, compared with 73% nickel recovery and no cobalt recovery at Nicaro. Furthermore, PAL energy consump-

Table 4   Typical Compositions of Laterite Leach Liquors Arising from Different Leaching Routes

	Concentration (mg/L)	
Component	Ammonia leach (QNI Yabulu)	Pressure acid leach (Moa Bay)
Ni	12,000	4500
Co	600	450
SiO_2	<20	2000
Mg	100	2000
Mn	50	1200
Al	<1	2500
Fe	10	300
Cr	<1	400
Zn	25	200
Cu	5	100

Source: Ref. 50.

tion is approximately one third of the Caron process [15]. Ammonia leaching is also limited to the processing of high iron limonites.

Advantages cited for the ammonia leach process include [14] the following: there is no net reagent consumption (in principle); the leaching reactions are selective for nickel and cobalt; and corrosion problems in ammoniacal systems are minimal. A further advantage of the ammonia leach process over the sulfuric acid route lies in the effluent treatment and environmental impacts [15]. Ammonia is recovered from effluent streams by boiling, allowing the reuse of nearly all the ammonia and carbon dioxide required for the metal extraction. The effluent streams therefore have little environmental impact when discharged. In PAL processes, it is not possible to regenerate sulfuric acid for reuse, and the soluble and insoluble sulfate products require disposal. The sulfuric acid effluents are also far more corrosive than those of the ammonia process, incurring higher costs for materials of construction and ongoing maintenance.

C. Cawse

The Cawse project [66–68], owned by Centaur Mining, is one of three Western Australian dry laterite projects commissioned in 1998–1999. This surface deposit contains 0.7% Ni and 0.04% Co, with reserves in excess of 200 million tons. The novel aspect of this flow sheet is the combination of sulfuric acid leaching under pressure with precipitation of an intermediate product, which is then subjected to ammoniacal downstream processing and recovery of the nickel by ASX as nickel cathode (Fig. 11).

The Cawse limonite ore is unique among the Australian laterite projects in that it is able to be upgraded by 30–40% by screening. Following PAL and neutralization of the free acid [which includes iron removal and solid–liquid separation by countercurrent decantation (CCD)], nickel and cobalt are recovered from solution by precipitation as the hydroxides. Although sulfide precipitation is more selective, redissolution of the hydroxide cake is easier. To ensure minimal magnesium and manganese reporting downstream, 90% of the nickel and cobalt are precipitated by using freshly slurried magnesium oxide. The remaining nickel and cobalt are precipitated with lime, and this precipitate (containing 15% of the manganese in solution) is recycled and releached in the PAL discharge slurry. The hydroxide precipitate is releached in ammonia. Ammonia leaching is highly selective, rejecting iron, manganese, and magnesium, although copper and zinc do report to the leach liquor with nickel and cobalt.

Cobalt is oxidized to Co(III), which is not extracted in the SX plant. Prior to SX, the leach liquor is partially steam-stripped to reduce the concentration of free ammonia and ammonium carbonate, allowing for recycle of ammonia and carbon dioxide. A typical composition of the clarified feed liquor to the SX circuit is given in Table 5.

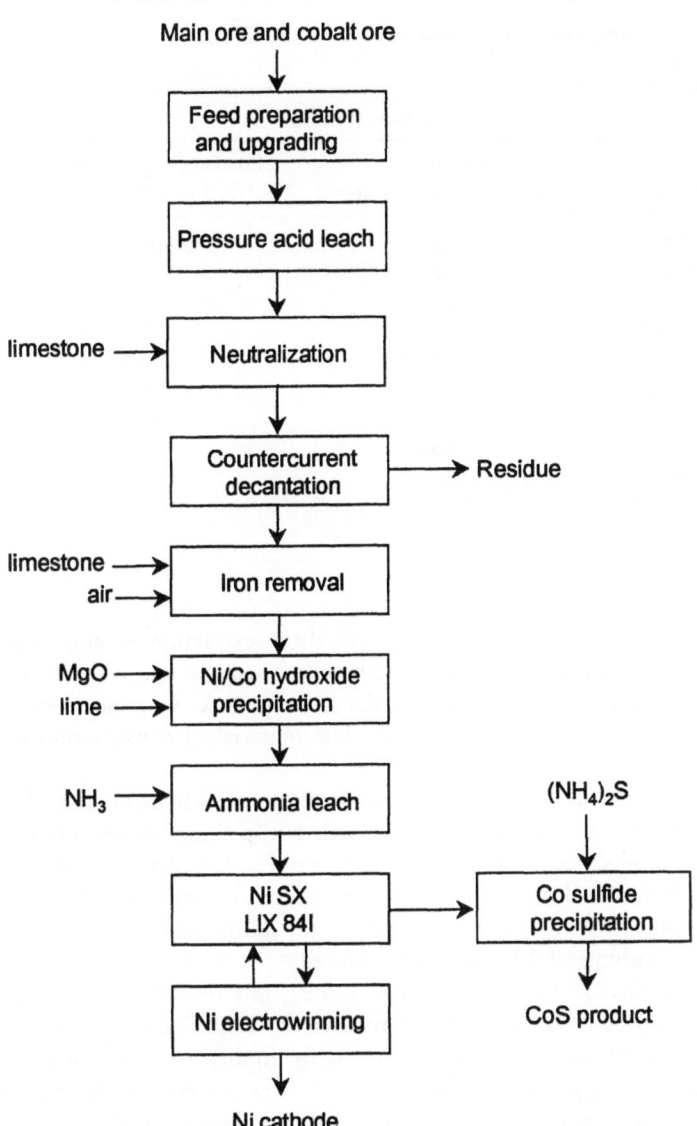

Figure 11   Simplified Cawse process flow sheet. (From Ref. 67.)

Table 5 Typical Compositions of Various Streams in the Cawse Ammoniacal SX Circuit

	Concentration (mg/L)		
Element	Feed liquor	Raffinate (feed to Co precipitation)	Loaded strip liquor (feed to Ni EW)
Ni	11,500	2	100,000
Co	2800	2800	2
Mg	17	15	5
Mn	<0.1	<0.1	2
Fe	<1	<1	<0.1
Ca	7.7	7	7
Cr	1	1	<0.1
Cu	5	<0.1	<0.1
Zn	50	50	3

Source: Ref. 67.

The ASX circuit configuration comprises three extraction stages, two scrub stages, four strip stages, and a single organic wash stage to remove sulfuric acid [66]. The mixers are conventional circular mixing tanks with horizontal reverse flow settlers mounted at ground level. The material of construction is fiberglass.

LIX 84I (30 vol% in Escaid 110) is used as the nickel extractant. As shown from the analyses of the raffinate and loaded strip liquor compositions in Table 5, nickel is selectively and quantitatively extracted. Cobalt(III) amine is rejected to the raffinate, along with most of the other impurities in solution. Coextracted ammonia is removed by scrubbing with dilute sulfuric acid. Organic phase nickel loadings of 13.6 g/L are achieved.

In contrast to the QNI circuit (Section V.A.2), this flow sheet employs sulfuric acid for stripping to facilitate nickel recovery by EW. The acid concentration and flow rate of the return electrolyte (strip liquor) are controlled to produce a loaded strip liquor with a pH of 3–3.5. The aqueous nickel concentration across the stripping circuit is increased by an average of 25 g/L.

In operating practice, it has been found that the SX circuit needs to run all mixers with aqueous phase continuity: organic phase continuity leads to very poor phase disengagement and high entrainment losses. Despite the inclusion of a final wash stage and aftersettler, extractant replacement represents a significant operating cost. In addition to physical recovery of organic phase lost by entrainment, a bleed stream is treated to maintain its loading capacity. Copper and cobalt buildups on the organic phase need to be chemically removed.

Copper is removed by a simple acid strip with 200 g/L H_2SO_4, while cobalt requires a reductive strip [66]. As in the QNI circuit, the oxime is prone to degradative oxidation to the ketone. Initially the organic phase capacity was maintained merely by the addition of fresh extractant. This leads, however, to a steadily increasing viscosity of the organic phase, which exacerbates physical losses. The Cawse circuit now includes a reoximation process [68].

At present, the cobalt-containing raffinate is treated with ammonium sulfide to give a sulfide precipitate. In collaboration with Union Minière, a cobalt refinery on site is planned for the future.

Cawse was the lowest cost nickel producer in the world in 1999 [69], with a nickel cash cost of this process at \$0.11/kg after cobalt credits [5]. During 2000, the plant reached its design capacity of 8500 tons/year nickel and 1900 tons/year cobalt.

D. Murrin-Murrin

Anaconda's Murrin-Murrin, located 250 km north of Kalgoorlie in Western Australia, is another recently commissioned dry laterite project. This surface deposit has reserves of 220 million tons, with an average grade of 1.1% nickel and 0.08% cobalt [4,5]. It is the largest of the Western Australia nickel projects using PAL technology, with a first-phase production capacity of 45,000 tons/year nickel and 3000 tons/year cobalt [70]. If future proposed expansion to bring the second phase of Murrin-Murrin (115,000 tons/year Ni and 9000 tons/year Co) and Mount Margaret and Three Rivers (currently under feasibility studies) into production is fulfilled, Anaconda will become the world's leading nickel producer.

The process flow sheet adopted by Murrin-Murrin is shown in Fig. 12 [6]. The laterite material is leached with sulfuric acid in a six-compartment autoclave, and locally mined calcrete is used to neutralize the leach liquor to pH 2.5. Nickel and cobalt are then precipitated as sulfides, using hydrogen sulfide gas at elevated pressure (105 kPa H_2S pressure) and temperature (95°C). Sulfide precipitation provides for good rejection of the impurity elements. The sulfide product is dissolved in a pressure oxidation leach step where pure oxygen is added. The resultant nickel sulfate leach liquor is purified by SX using CYANEX 272. Zinc is removed first at low pH, followed by a second SX step in which cobalt is extracted at higher pH. Cobalt and nickel metal powders are recovered by hydrogen reduction, also yielding ammonium sulfate as a valuable fertilizer by-product.

The feed to the SX circuit has high nickel (~100 g/L) and cobalt (~8 g/L) tenors. The zinc SX circuit comprises two extraction and three strip stages. In the extraction circuit, pH control is by the addition of 7% (m/m) aqueous

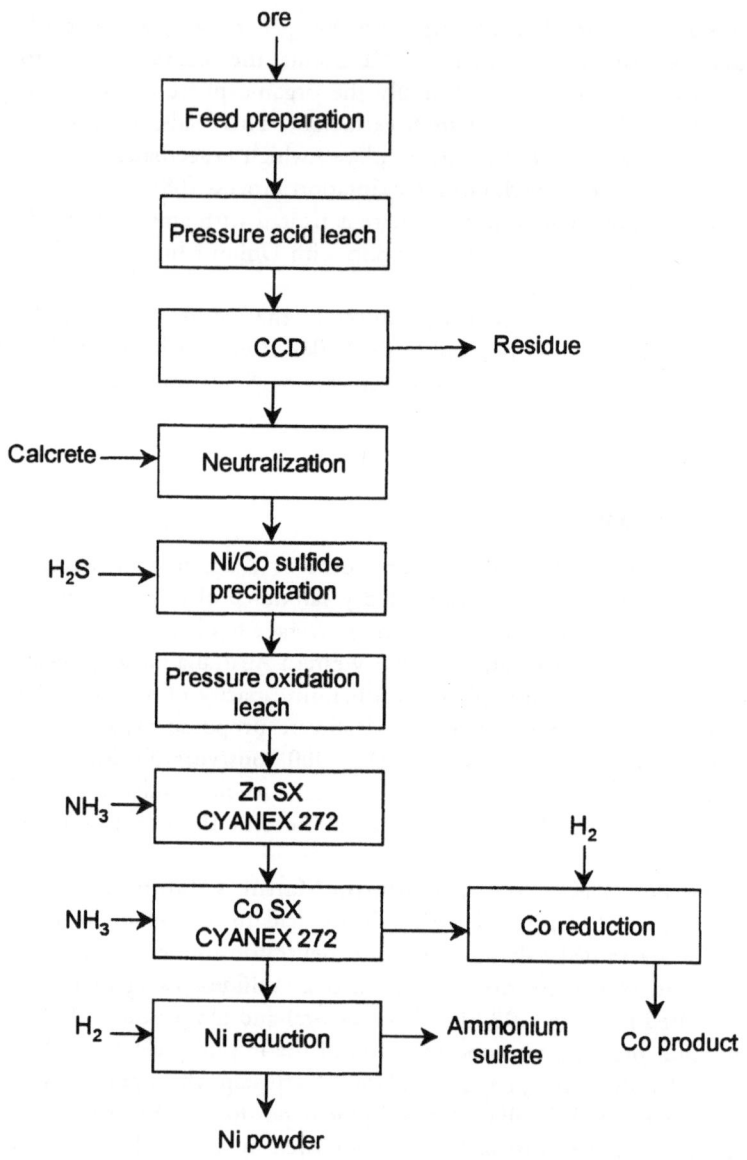

Figure 12 Murrin-Murrin process flow sheet. (From Ref. 6.)

ammonia. The zinc raffinate is passed through a Jameson cell for solvent recovery and then fed to the cobalt SX circuit.

The cobalt circuit has four extraction, two scrubbing, and three strip stages. In the initial design, the stripped organic phase was preloaded with ammonia prior to entering the extraction circuit, to obviate the need for individual pH control in each extraction stage. Coextracted nickel is removed from the loaded organic phase by scrubbing with diluted cobalt sulfate loaded strip liquor, reducing the Co/Ni ratio on the scrubbed organic phase to better than 1000:1. Stripping is effected with sulfuric acid, generating a purified cobalt sulfate solution with a concentration of 85 g/L Co. The loaded strip liquor and raffinate are passed through Jameson cells prior to the recovery of cobalt and nickel, respectively.

The mixer-settlers are relatively small compared with those of Cawse and Bulong. They are constructed of fiber-reinforced plastic (FRP) [4]. All the extraction stages and the cobalt stripping cells have an auxiliary mixer in addition to the pumping mixer chamber: the cobalt scrub cells and zinc stripping cells have single mixing compartments. Murrin-Murrin employs sophisticated on-line analysis of its control streams using an inductively coupled plasma/optical emission spectroscopy (ICP/OES) detection system supplied by Outokumpu. This has enabled extremely stable control of the circuit to be maintained.

One severe problem experienced with the high nickel tenor of the SX feed solution has been the unanticipated crystallization from solution of the nickel ammonium double sulfate [$NiSO_4(NH_4)_2SO_4 \cdot 6H_2O$] under certain operating conditions. At one stage, this necessitated reducing the nickel concentration of this solution and operating at elevated temperature ($\sim 75°C$) to maintain solubility of the salt. Throughput was seriously compromised, and the plant had reached only 50% of design production in mid-2000. This problem was eliminated by preloading the stripped organic phase with nickel ahead of the extraction circuit. The extractant is partially converted from the ammonia form to the nickel form by contact with the nickel sulfate raffinate at pH conditions under which nickel is loaded. Full production capacity by the second half of 2001 was anticipated [71].

The original Murrin-Murrin flow sheet was designed to treat a Ni/Co ratio of 15:1 in the feed material, with capability to treat materials with Ni/Co of 12:1 as necessary. The imminent expansion of operations will also allow hydroxide feed materials to be treated, and the Ni/Co ratio may drop as low as 5:1. A second leach and SX circuit were recently added to the plant to be able to treat these materials. Few details are available concerning this second circuit, although it is believed to use 20 vol% CYANEX 272 for the removal of zinc and cobalt.

E. Bulong

The Bulong nickel operation [72–75], presently owned by Preston Resources Limited, is located close to Kalgoorlie. The plant was commissioned in 1998 and was close to reaching its design capacity of 9000 tons/year Ni and 720 tons/year Co during 2000. A flow sheet of the downstream purification process is given in Fig. 13.

A slurry of the nickel cobalt laterite ore is fed to a six-chamber, titanium-clad, mechanically agitated autoclave. Dissolution conditions of 4500 kPa, 250°C, and a residence time of 75 min are employed. Following flash-cooling

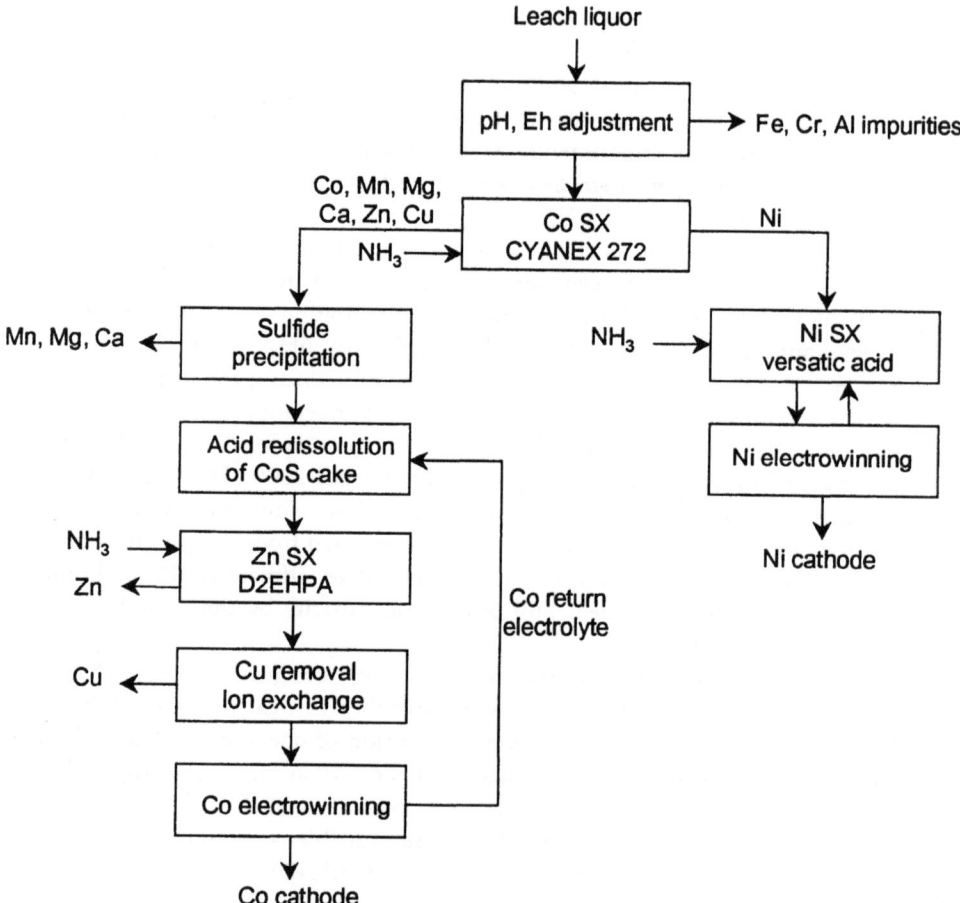

Figure 13 Bulong nickel/cobalt purification scheme.

and thickening, the leach liquor is processed directly for the recovery of the valuable metals. The leach liquor is first treated with limestone slurry for removal of iron, aluminum, and chromium by precipitation, achieving the target levels of 1, 1, and 0.1 mg/L, respectively. Cobalt, zinc, and manganese are removed by extraction with CYANEX 272. Nickel in the raffinate is upgraded and purified by SX with a carboxylic acid. The first SX circuit removes cobalt, zinc, and manganese from the iron-free solution. Because the leach liquor is relatively dilute (~3.5 g/L Ni), large mixer-settlers are required. Bateman mixer-settler units are employed, with a two-stage mixing configuration. The settlers are HDPE-lined concrete [4]. The organic phase comprises 15 vol% CYANEX 272 and 5 vol% tri-n-butyl phosphate (to avoid third-phase formation) in the partially aromatic diluent Shellsol 2046. Extraction is carried out at 50°C to maximize cobalt selectivity over nickel [24]. The original circuit design comprised five extraction, one scrub, and three strip stages. The pH of the extraction stages is maintained between 5.5 and 6, with pH control effected by the addition of gaseous ammonia into the organic interstage pipes. The O:A advance flow rates are typically maintained at 0.45 in the extraction circuit, 4.4 in the scrubbing circuit, and 0.7 in the strip circuit. Except for the final extraction and strip stages, the mixers are operated in aqueous-continuous mode.

Saline bore water is available on site at significantly lower cost than potable water, so this circuit is somewhat unusual in that saline water is used for both the scrub and wash stages. The presence of high concentrations of NaCl in the feed liquor is found to have no detrimental effect on the separation of cobalt and nickel in either the extraction or scrubbing circuits. The scrub liquor comprises a mixture of wastewater (1–2 g/L chloride) and strip liquor. The strip liquor is a mixture of plant water and recycled strip liquor. It contains minimal chlorides (1–2 g/L, carried over by entrainment from the scrub circuit). It is necessary to ensure that the chloride content is controlled adequately, since zinc stripping is compromised at high chloride levels [32].

Table 6 shows illustrative compositions of the various process streams of the cobalt SX circuit [32]. Extractions of cobalt, manganese, and zinc are excellent, with minimum coextraction of nickel. Although less than 10% of the magnesium in the feed was coextracted, magnesium accounts for more than 60% of the organic loading.

The raffinate from the cobalt SX circuit is passed through a series of Jameson cells, followed by adsorption first on anthracite and then using activated carbon filters to remove any entrained organic phase. The liquor is then refined by nickel SX to produce a high grade nickel solution suitable for the production of nickel cathode.

The nickel SX circuit comprises five extraction, two scrub, and three strip stages, with a single additional stage for solvent recovery. The organic phase comprises 9 vol% neodecanoic acid (supplied by ExxonMobil) in Shellsol 2046,

Table 6 Composition of Process Streams in the Bulong Cobalt SX Circuit

Element	Concentration (mg/L)			
	Feed	Raffinate	Loaded organic	Loaded strip liquor
Co	280	5.3	530	7200
Ni	3500	2800	10	120
Ca	500	500	5.8	180
Mn	990	0.4	2300	32,000
Zn	29	<0.2	66	950
Mg	15,000	14,000	3100	37,000
Co/Ni ratio	0.08	—	53	1440

Source: Updated from Ref. 32.

although the circuit originally used Versatic 10. These competitive monocarboxylic acid extractants have very similar formulations. Close to quantitative nickel extraction is achieved with the pH controlled in a staggered profile from 7.0 to 6.2 by the direct injection of gaseous ammonia. Coextracted calcium and magnesium are crowded off the loaded organic phase by scrubbing with a bleed stream of the loaded strip liquor and eventually report to the raffinate. Calcium tends to load and to be progressively displaced by nickel as the organic phase proceeds through the extraction circuit. Return nickel electrolyte is employed as the strip liquor. The acid concentration of the stripping circuit is adjusted by the addition of H_2SO_4 to the first strip stage. This enables nickel stripping efficiencies in excess of 99% to be attained. The nickel-containing strip liquor from this process has a purity exceeding 99%. Nickel cathode with a purity better than 99.5% is produced.

One of the disadvantages of carboxylic acids is their high aqueous solubility and consequent reagent losses to the raffinate, particularly at pH values above 6 [31,38,75,76]. It is therefore necessary to include a solvent recovery stage in the nickel SX circuit. The stripped organic phase is contacted with the raffinate under mildly acid conditions (pH 3), and this serves to minimize organic losses [28]. The loaded strip liquor is also passed through a dual-media filter to remove and recover entrained organics, and through two carbon columns for the recovery of dissolved organic phase prior to nickel EW.

The Bulong flow sheet also includes a third SX operation for the removal of zinc in the cobalt circuit. Cobalt is precipitated from the cobalt-loaded strip liquor as the sulfide to reject manganese and some magnesium. The cobalt sulfide cake is reslurried with demineralized water and fed to an autoclave operated at 120°C and 1100 kPa. The cobalt sulfate from the autoclave is neutralized by the addition of peroxide. Zinc is removed from this solution by SX with 15 vol% D2EHPA in Shellsol 2046. The circuit comprises three extraction, one scrub, and two strip stages. Zinc is stripped from the loaded organic phase by sulfuric acid. The cobalt-containing raffinate is then passed to an ion-exchange (IX) circuit in which residual copper is removed from the liquor by means of Purolite's S950, a diaminophosphonic acid cation exchanger [77].

One of the problems experienced with the operation of the Bulong circuit has been cross-contamination of the organic phases of the cobalt and nickel circuits. This has arisen from organics in return wash waters and from carryover from the cobalt to the nickel circuit. Because of the differing relative selectivities of the two reagents, small quantities of cross-contamination can markedly affect the performance of the circuits. In particular, elevated CYANEX 272 levels in the versatic acid circuit caused excessive calcium extraction, which then reported to the nickel stripping circuit and caused widespread precipitation of gypsum [78]. This caused substantial production losses owing to the need to empty the mixer-settlers to remove the solids.

As an interim measure to deal with this problem, the pH profile across the nickel extraction circuit is maximized at pH 7.0 in stage E2, thereby controlling the gypsum precipitation to a single stage, which facilitates cleaning and minimizes downtime. In addition, the final extraction stage of the cobalt circuit has been converted to a washing stage to reduce the levels of entrained organic materials reporting to the raffinate (and hence to the nickel circuit) by contact with fresh diluent. The extractant is then recovered from the diluent by saponification at pH 9, and can be reused. This temporary measure has resulted in higher cobalt levels in the raffinate, but carryover of CYANEX 272 is greatly reduced and the downtime due to gypsum removal has been reduced to one day in 8 weeks. A separate washing stage is planned in the near future. In late 2000 the plant reached 95% of the design production.

VII. CURRENT DEVELOPMENTS AND PILOT PLANT TRIALS

From the myriad of flow sheets currently used for the processing of nickel materials, it is obvious that there is no ideal process yet. Several interesting alternative process routes are under development for both laterite and sulfide ores. The main aims of the novel processes are to achieve lower capital and operating costs, higher metal recoveries, and environmentally acceptable waste and by-products.

A. Nickel Sulfides

Hydrometallurgical routes have been commercially established for zinc and copper sulfide concentrates, such as at the Hudson Bay Mining Company, Canada, and the recent Gunpowder Expansion Project in Australia. There are also significant research developments in the processing of nickel sulfide concentrates and nickel mattes, which can be broadly categorized into four possible process options:

Bacterial leaching (35–75°C), such as Billiton's BioNIC process
Low temperature pressure oxidation (90–110°C), such as the Activox process owned by Western Minerals Technology in Australia
Moderate temperature pressure leaching (135–165°C), including the processes developed by Dynatec and Cominco Engineering Services (CESL) in Canada
Total pressure oxidation (200–220°C)

All these process options require downstream processing involving one or more SX unit operations.

1. BioNIC Process

The BioNIC process was developed as an economically competitive alternative for the extraction of nickel from sulfide ores and concentrates [79,80]. Billiton has been responsible for the development of the bioleaching step for the dissolution of the metals, collaborating with Mintek (South Africa) in the development of a downstream recovery process. The conceptual flow sheet includes solubilization of the base metals by bacterial leaching and iron removal by precipitation followed by nickel recovery by precipitation, ion exchange, or SX [79,81]. This process is claimed to offer savings of up to 60% on the capital cost and 20% on operating costs compared with those of smelting followed by refining [79].

After extensive demonstration plant testing, during which LME grade nickel cathode was produced, the use of ion exchange as a process option was eliminated because of the prohibitive cost of the resin [81]. Mini-plant [82] and integrated pilot plant testing [83,84] have resulted in the proposal of the flow sheet given in Fig. 14 for the treatment of low cobalt feeds. For high cobalt feeds, the cobalt removal will be ahead of the nickel removal step.

During the pilot plant trials, an integrated flow sheet consisting of all the steps indicated in Fig. 14 was tested [83]. For an aqueous feed containing 8 g/L Ni and 0.13 g/L Co and an organic phase comprising neodecanoic acid as the extractant, a nickel SX plant consisting of four extraction, three scrubbing, and three stripping stages was found to be effective in providing close to 99% nickel and cobalt extraction efficiency with good rejection of calcium and mag-

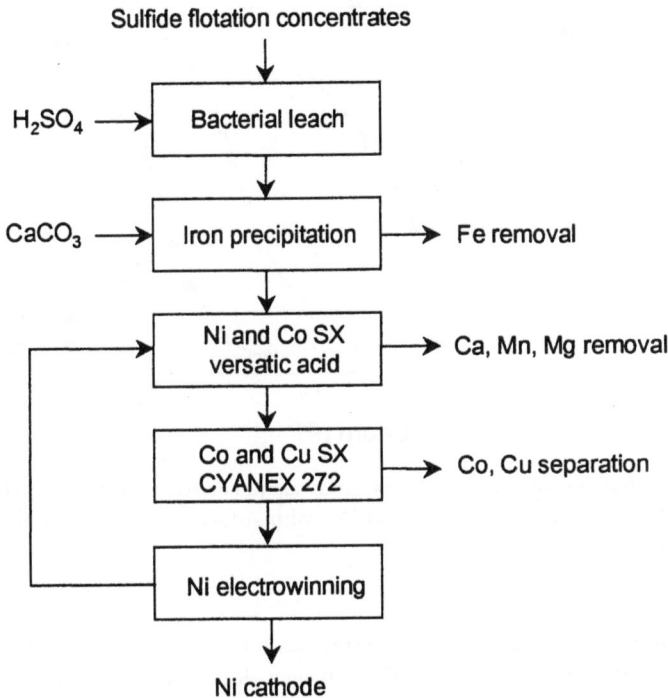

Sulfide flotation concentrates

H_2SO_4 → Bacterial leach

$CaCO_3$ → Iron precipitation → Fe removal

Ni and Co SX versatic acid → Ca, Mn, Mg removal

Co and Cu SX CYANEX 272 → Co, Cu separation

Ni electrowinning

Ni cathode

Figure 14 Proposed flow sheet for the BioNIC process. (From Ref. 83.)

nesium. Stripping was achieved using return nickel anolyte to provide a 3.75 times upgrading of nickel. The precipitation of gypsum, cited as a major problem in the Bulong process (Section VI.E), was overcome by the use of an appropriate pH profile in the extraction circuit and by ensuring that enough capacity was available on the organic phase. During the 8-week trial, no gypsum precipitation was evident. Cobalt was removed from the loaded strip liquor (91 g/L Ni and 0.7 g/L Co) by using CYANEX 272 in a circuit consisting of five extraction, three scrubbing, and three stripping stages. This provided the advance electrolyte to an EW cell, where nickel cathode satisfying the ASTM specification was consistently produced.

Related laboratory and mini-plant test work at Mintek have shown that the use of nitrogen donor compounds in synergistic combination with Versatic 10 can overcome some of the disadvantages of carboxylic acids and improve their selectivity for nickel [82,85,86]. Figure 15 shows the pH dependence of extraction of nickel and calcium by Versatic 10 and by Versatic 10 in combination with a nitrogen donor synergist. The separation is improved markedly in the latter case. In addition, nickel extraction can be carried out at much

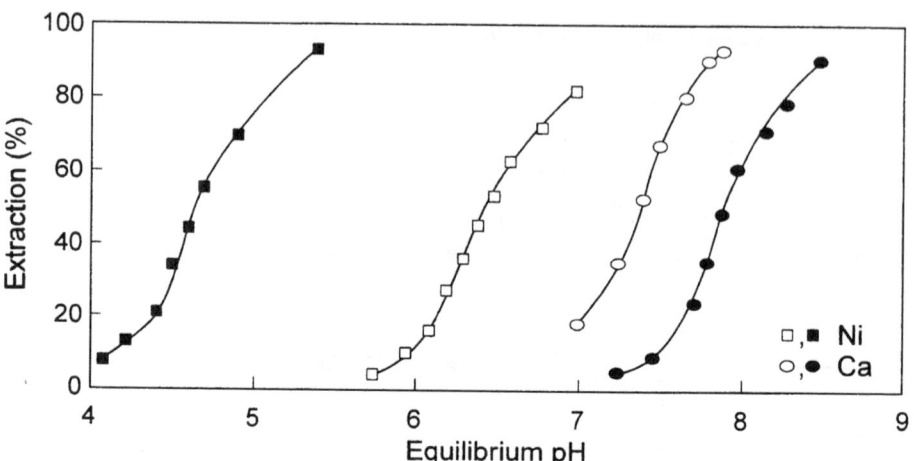

Figure 15 The extraction of nickel (squares) and calcium (circles) by Versatic 10 (open symbols) and by Versatic 10 in combination with 4-(5-nonyl)pyridine (solid symbols). (From Ref. 87.)

lower pH, reducing solubility losses of the extractant and minimizing the tendency to form the hydroxide precipitate [77]. Oxygen donor synergists exhibit similar effects [88].

2. Yakabindie Nickel Process

Yakabindie is one of the largest undeveloped low grade nickel sulfide deposits in the world. A hydrometallurgical process for treating a sulfide concentrate produced from this Australian ore has been tested on a pilot plant scale by Western Minerals Technology and Lakefield Oretest (Australia). The key to the so-called Yakabindie nickel process is the use of Activox leaching technology [89]. Briefly, this involves ultrafine grinding of the flotation concentrate (5–10 μm), followed by a low temperature (100°C) oxidative leach. Although the leach is carried out at elevated pressure (1000 kPa), this is substantially lower than conventional pressure leaching systems, which typically operate near 3000 kPa. The materials of construction, safety considerations, and engineering difficulties all benefit from the use of the less severe leaching conditions. A further distinct advantage of this process is that both cobalt and nickel recovery from the leaching step are claimed to be greater than 95%. CYANEX 272 is used to separate the cobalt and nickel in solution, and then the metals are recovered by reduction [90,91].

An economic evaluation of the project with a metals plant producing 30,000 tons/year nickel metals from concentrate on site at Yakabindie has indicated an operating cost of U.S.$0.66/kg, comparing favorably with equivalent costs for a smelting process route.

3. CESL Process

Cominco Engineering Services Ltd. (CESL) developed a process to treat nickel sulfides to recover the valuable metals as LME grade products [92]. The process is flexible enough to treat the majority of nickel–copper–cobalt concentrates containing a wide range of other impurity elements. Specific aims of the process are as follows:

> Recovery of all the metals to exceed 96% if present in economic quantities
> Sulfur oxidation to sulfate to be minimized to typically <30%
> Use of established technology to minimize risk
> Use of no expensive reagents, limiting the choice to acid, limestone, lime, and oxygen
> Safely disposable by-products or residues
> Minimized energy consumption
> Capital and operating costs 50% of costs for the matte smelting and refining process

Several 5 kg/h integrated pilot plant campaigns treating a pentlandite concentrate (15% Ni, 0.5% Co, 5% Mg, minor Cu) were conducted to produce a workable flow sheet satisfying these process aims [92]. The current flow sheet is illustrated in Figure 16.

Pressure oxidative leaching (17% solids, 150°C, 1400 kPa, 60 min retention time) was conducted on ground concentrate in a sulfate medium containing 12 g/L Cl and 5 g/L Cu as catalysts. Close to 96% recoveries were obtained for nickel and cobalt, and 26% extent of sulfur oxidation to sulfate was recorded. After washing and separation in a CCD circuit, the leach solution (30 g/L Ni, 1 g/L Co, 5 g/L Cu, 3 g/L Fe, 0.4 g/L Zn, 5 g/L Mg) was subjected to a two-stage SX extraction for copper. Copper was reduced to less than 100 mg/L by neutralization of the first-stage raffinate to pH 2 using limestone. Copper strip liquor was recycled to the pressure leach but would normally be sent to electrowinning. Residual Cu, Fe, and Zn were then precipitated to levels below 1 mg/L from the raffinate in a two-stage countercurrent process using slaked lime and recycled magnesium hydroxide; the resulting exiting solution had a pH of 5–6. For higher zinc values, the next step would employ Zn SX.

Nickel and cobalt were then precipitated, along with some of the magnesium, as a hydroxide/gypsum mixture in a multistage countercurrent operation using slaked lime at a pH of 7–8. The mother liquor was limed to pH 9.5 to produce magnesium oxide for reagent recycle and a chloride-containing barren solution for recycle to the pressure leach. The washed hydroxide was

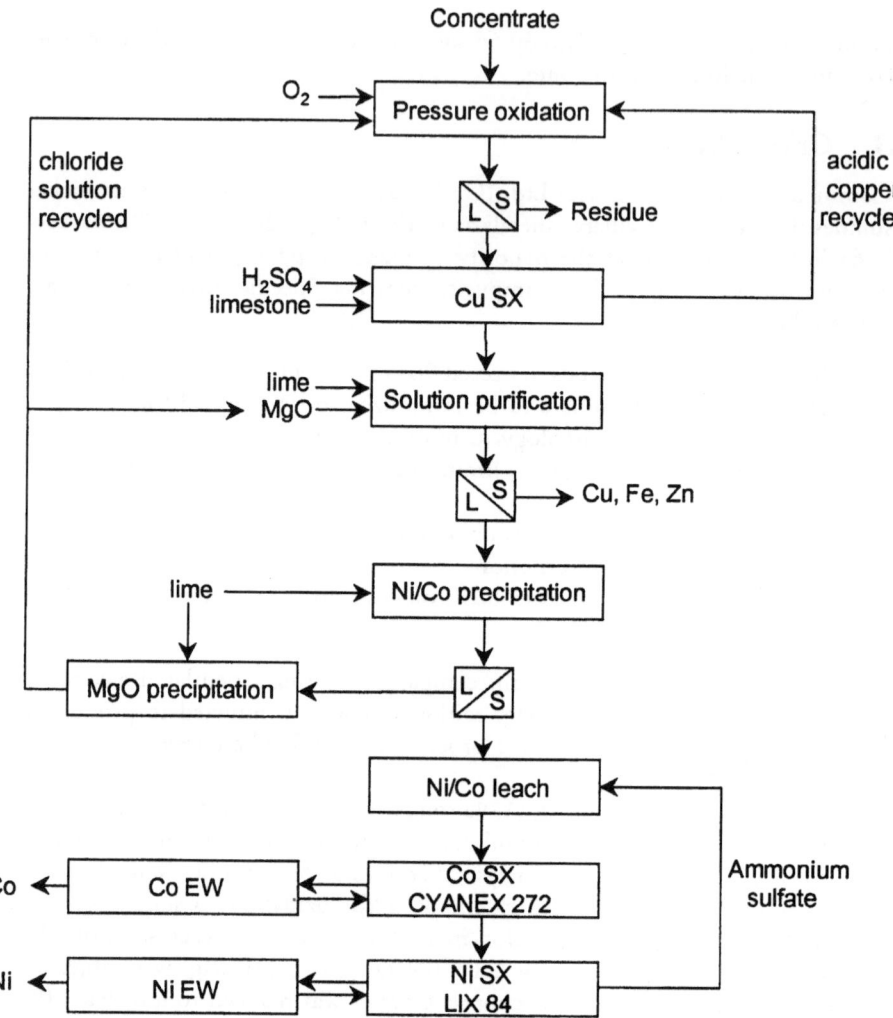

Figure 16   Integrated CESL flow sheet for nickel concentrates. (From Ref. 92.)

then dissolved (40°C and pH 7.5) by using 200 g/L $(NH_4)_2SO_4$ solution to generate a 10 g/L Ni liquor.

About 90% of the cobalt was then extracted by means of CYANEX 272 and stripped with return electrolyte. Limited cobalt extraction was maintained to minimize magnesium and nickel coextraction. The remaining magnesium, cobalt, and traces of nickel were then extracted by CYANEX 272 in a further

SX circuit. The HCl spent strip liquor was recycled to the hydroxide dissolution step. Nickel was then extracted by using LIX 84 and stripped into return electrolyte for electrowinning. The raffinate, combined with various bleeds, was evaporated to produce reconstituted ammonium sulfate solution for recycle.

Various cost studies of this process demonstrate capital and operating costs to be approximately half the costs of the matte smelting and refining route.

4. NorthMet Project

The NorthMet deposit in Minnesota is a large low grade copper and nickel sulfide deposit containing appreciable PGM values [91]. Since 1991 PolyMet Mining Corporation has evaluated this deposit extensively, demonstrating a total resource of 1450 Mton at a grade of 0.43% Cu, 0.11% Ni, 0.12 g/ton Pt, and 0.44 g/ton Pd. Upgrading by flotation was found to be favorable for all the metals.

A downstream processing route has been identified that will ensure simultaneous high base metal and PGM recoveries [93]. The flow sheet features high temperature (220°C) pressure leaching with the addition of a suitable lixiviant. The PGMs are then concentrated from the autoclave liquor by sulfide precipitation. Copper and nickel are recovered in sequential SX and electrowinning steps with appropriate intermediate neutralization/purification steps.

Pilot plant evaluation of the downstream processes was carried out by Lakefield Research (Canada) during 2000 in preparation for the prefeasibility study.

5. Nkomati Project

The Nkomati project, a joint venture between Anglo American Corporation and Anglovaal Minerals to develop a nickel sulfide complex in Mpumalanga, South Africa, has indicated that the preferred downstream flow sheet will be similar to that of Bulong. Following Activox leaching, copper SX-EW, and iron removal, the leach liquor is purified by cobalt SX with CYANEX 272, then nickel SX with versatic acid and nickel EW. Advanced feasibility testing for this project is under way.

Another African nickel sulfide project, Tati Nickel in Botswana, is also evaluating the use of a hydrometallurgical flow sheet, although details remain proprietary at time of writing.

B. Nickel Laterites

The choice of pressure acid leaching over ammonia leaching for many of the new laterite projects is mainly because of the large resources of limonitic ores.

PAL has the advantages of high metal recoveries, lower energy requirements, and favorable operating costs [94].

1. Goro Process

The Goro process, developed by INCO for the treatment of a nickel laterite deposit in New Caledonia, is currently being evaluated in a 2-year, on-site demonstration plant treating 12 tons/day of dry ore. The novel process flow sheet involves the coextraction of nickel and cobalt by SX with CYANEX 301, followed by their separation in chloride medium using an amine extractant. The simplified flow sheet for this process is illustrated in Fig. 17 [95].

The lateritic ore is acid leached at 270°C under pressure. Following solid/liquid separation, the leach liquor is partially neutralized to precipitate Al, Cr(III), Si, Cu, and Fe. Trace quantities of residual copper are removed by IX using a commercially available copper-selective chelating resin with iminodiacetate functionality. Nickel, cobalt, and zinc are quantitatively and selectively extracted from calcium, magnesium, and manganese by CYANEX 301 (Fig. 5c). Extraction is carried out in three countercurrent stages operated at 55°C. Stripping is carried out in four stages using ~6 M HCl at 60°C. The metal concentrations are upgraded by a factor of 20.

Zinc is removed from the loaded strip liquor with a conventional strong- or weak-base ion-exchange resin. The final nickel separation from cobalt in chloride medium is achieved by means of the well-known amine SX process (Section III.A) with Alamine 336, a tertiary amine extractant. Nickel is recovered as NiO by pyrohydrolysis, regenerating the HCl for recycle to the CYANEX 301 strip circuit, while cobalt is currently recovered as the carbonate salt.

As shown in Section III.B.1, the CYANEX 301 extractant is unusual in that it offers good rejection of manganese, magnesium, and calcium over cobalt and nickel under strongly acidic conditions. CYANEX 301 is, however, known to be very susceptible to irreversible loading of copper, which causes degradation of the reagent to a disulfide species [96]. It is extremely important to reduce residual copper as far as possible prior to the first SX circuit. When a conventional cation-exchange resin is used, copper reportedly is removed to levels of less than 0.04 mg/L. Even trace amounts of Cu(II) in solution, however, will catalyze the oxidation of the extractant, leading to a progressive loss of capacity. INCO and Cytec have patented a process for the regeneration of the disulfide form of the extractant [97]. Early indications from the demonstration plant performance are that this method can be used to control extractant losses to within acceptable levels.

2. Syerston

The Syerston ore is unique among the Australian laterite ores [98]. Particular geological history has left an ore low in magnesium, aluminum, and silicate

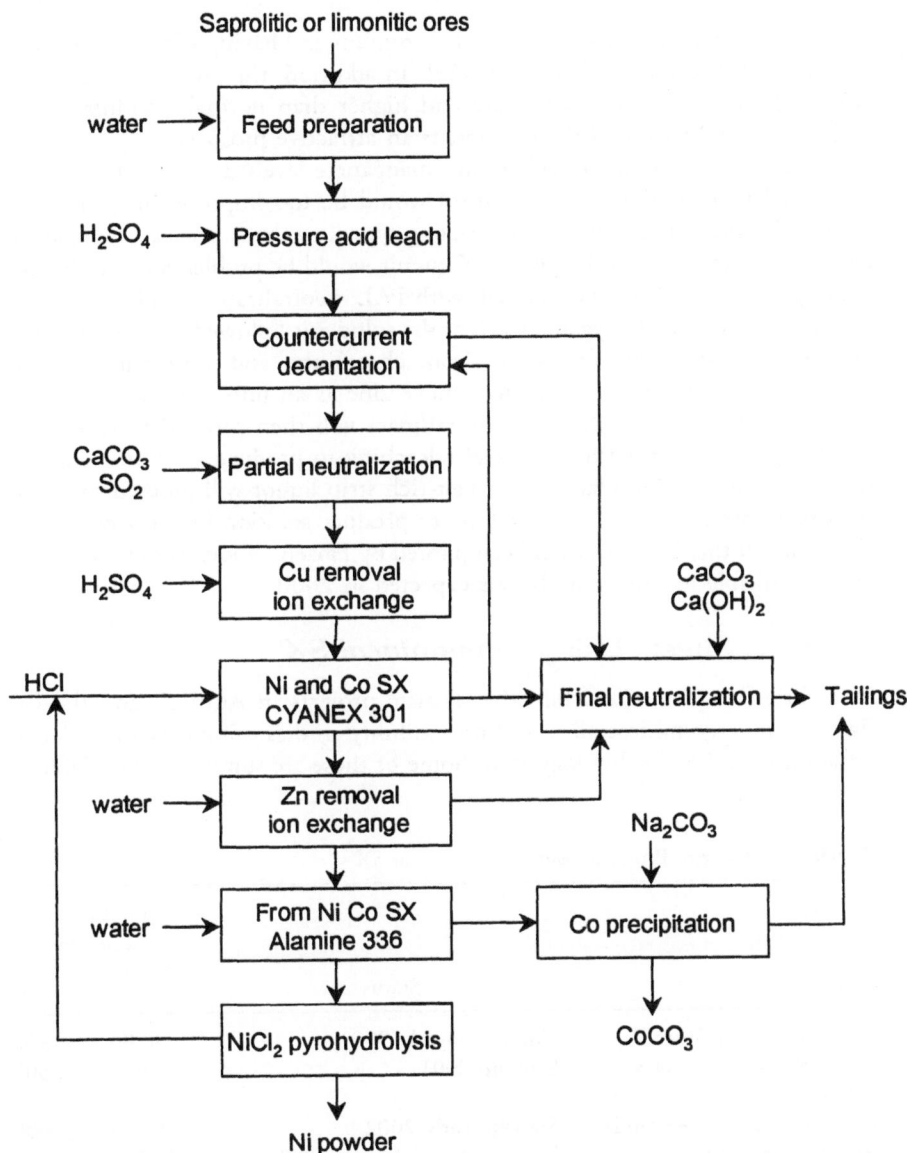

Figure 17 INCO's Goro process for the recovery of nickel from laterites. (From Ref. 95.)

minerals but higher in manganese and chromium and having a lower than usual ratio of nickel to cobalt (5.6 vs 10) [61]. In addition, the presence of a PGM-containing mineral alloy in the ore and higher than normal scandium levels will make the recovery of these elements an attractive prospect.

Because of the higher cobalt and manganese levels in the leach solution, sequential precipitation by pH control would be too imprecise to justify the use of a Cawse-style circuit. With a Bulong-type flow sheet, excessive manganese and chromium entering the direct SX circuit would be problematic. A Murrin-Murrin-type flow sheet is envisaged, with PAL, neutralization, and sulfide precipitation. A standard oxygen releach of the sulfides is followed by neutralization with magnesium oxide to remove iron, chromium, and aluminum. This is followed by sequential SX steps to remove zinc as an impurity and to recover cobalt by using CYANEX 272. The raffinate will then proceed to nickel hydroxide precipitation with MgO and releaching to produce a solution suitable for electrowinning. Similarly, the cobalt-rich strip liquor will proceed to cobalt hydroxide precipitation and releaching to produce solution for electrowinning.

Part of this flow sheet has been piloted by Hazen Laboratories (Colorado), and a bankable feasibility study was expected in 2000.

3. Flow Sheets Using Ammoniacal SX

In addition to Cawse, several other laterite projects in Australia and the Far East are undergoing feasibility and prefeasibility studies using flow sheets based on ammoniacal SX technology [54]. Some of these are summarized in Table 7.

Table 7 Laterite Projects Using Ammoniacal SX

		Production (tons/year)	
Project	Status	Ni	Co
Cawse (Western Australia)	Commissioned 1999	8500	1900
Ramu River (Papua New Guinea)	Start-up 2001	30,000	2500
Marlborough (Queensland)	Start-up early 2002	25,000	2000
Ravensthorpe (Western Australia)	Mining started March 2000	25,000	1900
Weda Bay (Indonesia)	Plant start-up in 2002, feasibility to be completed by early 2002, first phase starting in 2004	30,000	1350

Source: Ref. 54.

a. Ramu River. Piloting for the Ramu River project in Papua New Guinea has been undertaken by Lakefield Research in Canada [99,100]. A nickel–cobalt solution is produced by pressure acid leaching of the laterite material at 250°C. This project employs SX for the recovery of nickel from the ammonia/ammonium sulfate releach solution of the hydroxide produced from lime precipitation of the PAL liquor. Similar to other projects, the SX circuit comprised three extraction stages, two scrubbing stages (using dilute sodium sulfate to reduce the coextracted ammonia from transferring to the strip circuit), and five strip stages. The nickel concentration was reduced from 12 g/L to less than 0.01 g/L in the raffinate by using 30 vol% LIX 84I in an aliphatic diluent. The nickel SX circuit operated at 50°C. Extraction was carried out at an O:A phase ratio of 1.1:1, while the strip circuit operated at O:A = 2.2:1. The scrub ratio was 22:1, and the pH was controlled between 5.5 and 6.5 to ensure efficient removal of ammonia. A loaded strip liquor suitable for the electrowinning of nickel was produced, and the nickel cathode assayed over 99.97% purity.

In the pilot flow sheet, cobalt was removed as the sulfide by precipitation with ammonium sulfide. The loaded strip liquor was further processed through two ion-exchange circuits for purification prior to electrowinning: the Dowex resin XU43578 to remove traces of stripped copper, followed by Bayer's Lewatit OC 1026 to scavenge trace amounts of zinc.

b. Ravensthorpe. The Ravensthorpe nickel project will treat laterite ore from the south coast of Western Australia [101]. This laterite is unusual in that it is able to be upgraded by screening and classification processes to separate out the fine high grade fraction to produce an upgraded autoclave feed. The downstream process has not yet been defined.

c. Weda Bay and Marlborough. The Weda Bay project (Indonesia) to treat lateritic ore [102,103] is evaluating the circuit adopted for Cawse and under evaluation for Ramu River. An ASX route is also under consideration for the Marlborough nickel project in Queensland [104].

4. Equipment

The first three laterite projects commissioned in Western Australia employed mixer-settler contacting units for the SX operations. Important reasons for this choice of well-known equipment were the inherent financial and technical risks associated with the use of PAL technology in these flow sheets, and the desire to minimize perceived risk in other areas. Many of the new projects are now considering the use of pulsed columns, rather than mixer-settlers, as the SX contacting equipment [105–107]. There are many advantages associated with columns for use in these applications, and it will be interesting to see whether column technology will finally make its debut in the base metal industry when these plants come on line early in the new century.

C. Upgrading of Base Metal Refineries

1. Anglo Platinum Base Metal Refinery

Anglo Platinum is the world's largest producer of PGMs. The sulfide ore from the Rustenburg area in South Africa is rich in both PGMs and base metals; it is smelted to give a PGM-containing matte. The nonmagnetic component of the matte contains nickel and copper, which, together with the liquor produced in the leaching of the PGM-rich magnetic component of the matte to dissolve residual base metals, is the feed material to the base metal refinery (RBMR). The current refining process is described in Section IV.C.

Anglo Platinum has recently announced a major expansion, aimed at increasing its annual PGM production capacity to 25 million ounces. Associated with this, RBMR will increase annual nickel production from 21,000 tons to 40,000 tons. While initial expansion capacity is being generated by debottlenecking of the existing plant, various options for the longer-term modernization of the nickel processing flow sheet are under consideration.

The use of solvent extraction with CYANEX 272 to replace the nickel(III) hydroxide circuit is being evaluated. The novel aspect of this approach lies more in its engineering than the chemistry of the process. A pilot plant trial carried out on site in mid-2000 used a Bateman pulsed column rather than a traditional mixer-settler configuration to contact the two phases. Similar to practice in the existing D2EHPA circuit, CYANEX 272 was partially neutralized to the sodium salt prior to extraction to avoid the need for pH control in the column itself.

CYANEX 272 can considerably reduce the number of extraction and scrubbing stages required because of the significantly improved Co:Ni separation factor. Phase separation losses are claimed to be lower when a column configuration is used. Furthermore, because the cobalt–nickel separation efficiency is temperature dependent and SX is typically carried out at elevated temperatures, diluent losses by evaporation are significant in mixer-settlers, even when the cells are covered. The use of columns will also reduce operating costs associated with such losses.

2. Impala Platinum Base Metal Refinery

Impala Platinum's base metal refinery in South Africa uses a process developed by Sherritt to produce 17,000 tons/year nickel, 6000 tons/year copper, and 140 tons/year cobalt as by-products from a PGM-containing matte [12]. The matte is pressure-leached with return copper electrolyte to solubilize the nickel, cobalt, and iron. The leach residue is leached further with sulfuric acid to produce copper sulfate for copper EW, and the residue is sent for PGM recovery. The nickel solution is treated with matte to cement copper and then iron is precipitated. After conversion to the nickel amine by means of return ammonium sulfate solution and anhydrous ammonia, most of the nickel is precipitated by hydrogen reduction. The remaining cobalt and nickel are precipitated as a mixed double salt. This is leached under oxidizing conditions in ammonia and, after various steps of purification, cobalt is precipitated by hydrogen reduction.

A refinery expansion is planned to accommodate a mixed nickel cobalt sulfide concentrate that will be produced from the Philnico laterite deposit in the Philippines. Metal production will be substantially increased to 60,000 tons/ year Ni and 4300 tons/year Co [108]. The proposed new refinery will become an independent company, to be known as SANico. The flow sheet is shown in Fig. 18.

The existing leach circuit will remain to accommodate the PGM-containing matte, with a second leach train introduced to process the mixed sulfide. Copper will be removed from the second leach solution by cementation with NaHS. Iron and manganese are removed by oxidation with an air/SO_2 mixture. The liquor will then be combined with the liquor from matte leaching and the zinc extracted with CYANEX 272. About 1000 tons/year of a zinc by-product will be produced. The raffinate, containing about 7 g/L Co and 100 g/L Ni, will be subjected to SX with CYANEX 272 (15 vol% in an aliphatic diluent) to selectively remove cobalt. Anhydrous ammonia will be used for pH adjustment. The cobalt-free raffinate is subjected to solution adjustment with return nickel reduction end solution and anhydrous ammonia to produce the nickel diammine, and then reduced by hydrogen to produce nickel powder. The cobalt-rich strip liquor (80 g/L Co) will in turn undergo solution adjustment to form the diammine complex, trace amounts of copper will be removed by IX, and cobalt precipitated by hydrogen reduction. Ammonium sulfate will be recovered from a portion of the reduction end solutions after metal stripping.

The proposed SX steps were successfully tested and proven in pilot plant trials carried out by Mintek at the refinery during 2000. The expansion project is planned to begin construction in the third quarter of 2001 and to be completed by 2003.

The major advantages of the new process are as follows:

The overall recovery of cobalt across the refinery is improved from 80% to 98%.

The new circuit is sufficiently flexible to handle feedstocks containing impurities such as zinc that formerly could not be treated.

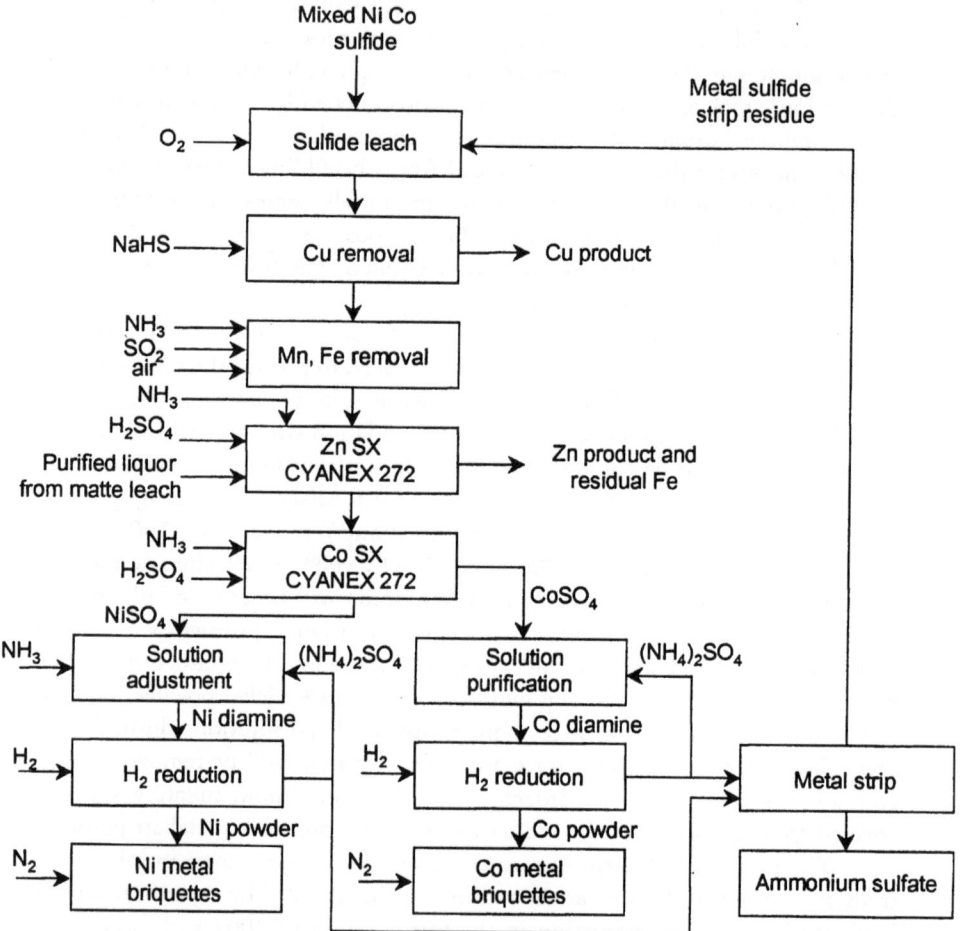

Figure 18 Proposed new flow sheet for the SANico base metal refinery. (Courtesy of Impala Platinum.)

The new circuit will operate using only minimally more staff than are currently employed, enabling expected unit costs to be in the lower quartile for nickel.

The current separation of nickel from cobalt by means of the Sherritt process requires the difficult control of the molar ratio of the diammine formation, whereas the cobalt SX step will give far superior separations more easily.

VIII. CONCLUSIONS

For most of the last century, nickel was processed primarily by smelting and associated pyrometallurgical techniques. The few plants that employed hydrometallurgical processing relied heavily on precipitation methods to achieve separations between nickel, copper, and cobalt. The flow sheets involved numerous recycles and repurification steps to achieve adequate product purities (often at the expense of high operating costs or poor single-pass product recoveries).

Today, advances in engineering technology and materials of construction, coupled with developments in specialized SX reagents and innovative process chemistry, have led to a myriad of process options for the recovery of nickel. Three Australian plants using pressure acid leaching technology have been commissioned, and several other hydrometallurgical flow sheets for the treatment of both laterite and low grade sulfide materials are under advanced feasibility studies and pilot plant development.

Despite the many teething problems of the new plants, these commissioning experiences have led to a far better understanding and acceptance of the capabilities and reliability of SX technology by the base metal industry. The PAL technology employed in these plants was a largely unknown and relatively high-risk approach in nickel processing at the time of construction. Much of the effort in developing these flow sheets was therefore focused on the leaching circuit, with Bulong being the only plant to employ innovations in its downstream purification circuit. Now that many of the materials, engineering, and operational issues associated with the autoclave leaching have been clarified and PAL technology has been proven for this application, the new projects under development are concentrating far more on optimizing the downstream processing routes and allowing a more harmonious integration of the leaching and purification circuits. Considerable variation exists in the approaches taken to solution purification, but all the hydrometallurgical flow sheets employ at least one SX circuit. While some of these approaches remain to be proved in long-term operation, there is no doubt that the adoption of this technology for large, capital-intensive projects augurs well for the future of solvent extraction in base metal hydrometallurgy.

ACKNOWLEDGMENTS

Appreciation is extended to Angus Feather (Mintek) and Jeremy Mann (AARL) for their useful comments concerning the manuscript. We are also grateful to the staff of various refineries who generously provided details and insights concerning their operations. This review is published by permission of Anglo American Research Laboratories (Pty) Ltd.

REFERENCES

1. B Terry, AJ Monhemius, AR Burkin. In: AR Burkin, ed. Extractive Metallurgy of Nickel, Critical Reports on Applied Chemistry, vol 17, London: Society of Chemical Industry, 1987, pp 1–6.
2. D Eldridge. MBM 334(3):69–71, 1996.
3. D Russell. Chamber Mines J (6):33–39, 1994.
4. R Mayze. In: ALTA 1999 Nickel Cobalt Pressure Leaching and Hydrometallurgy Forum. Melbourne: ALTA Metallurgical Services, 1999, 19 pp.
5. J van Os. MBM 337(1):32–34, 1999.
6. R Sridhar. In: Process Metallurgy of Nickel and Cobalt Extraction, CIM Professional Development Course. Montreal: Canadian Institute of Mining, Metallurgy and Petroleum, 1997, 73 pp.
7. WK Brinsden. Australas Inst Min Metall Bull (3):29–32, 1995.
8. G Bacon, A Dalvi, MO Parker. In: ALTA 2000 Nickel Cobalt-6. Melbourne: ALTA Metallurgical Services, 2000, 16 pp.
9. G Bolton, R Berezowsky. The hydrometallurgical production of nickel. Presented at ASPEGN-CIM Joint Conference, St. Johns, Newfoundland, Nov 5–6, 1999.
10. B Terry. In: AR Burkin, ed. Extractive Metallurgy of Nickel, Critical Reports on Applied Chemistry, vol 17. London: Society of Chemical Industry, 1987, pp 7–50.
11. AR Burkin. In: AR Burkin, ed. Extractive Metallurgy of Nickel, Critical Reports on Applied Chemistry, vol 17. London: Society of Chemical Industry, 1987, pp 98–146.
12. DGE Kerfoot, RM Berezowsky. Hydrometallurgical processes for the recovery of nickel and cobalt from nickel mattes, Presented at International Conference on Lateritic Ore Acid Leaching Technology, Moa Bay, Cuba, Nov 11–15, 1991.
13. D Muir, E Ho. In: EJ Grimsey and I Neuss, eds. Nickel '96—Mineral to Market. Victoria: Australasian Institute of Mining and Metallurgy, 1996, pp 291–297.
14. T Anthony, DS Flett. Miner Ind Int (1):26–42, 1997.
15. JG Reid. In: EJ Grimsey and I Neuss, eds. Nickel '96—Mineral to Market. Victoria: Australasian Institute of Mining and Metallurgy, 1996, pp 11–16.
16. RM Berezowsky. Miner Ind Int (1):48–55, 1997.
17. DS Flett. In: AR Burkin, ed. Extractive Metallurgy of Nickel, Critical Reports

on Applied Chemistry, vol 17. London: Society of Chemical Industry, 1987, pp 76–97.
18. A Warshawsky. Miner Sci Eng 5(1):35–52, 1973.
19. JMW Mackenzie, MJ Virnig, BD Boley, GA Wolfe. In: ALTA 1998 Nickel Cobalt Pressure Leaching and Hydrometallurgy Forum. Melbourne: ALTA Metallurgical Services, 1998, 34 pp.
20. MJ Nicol, CA Fleming, JS Preston. In: G Wilkinson, RD Gillard, and JA McCleverty, eds. Comprehensive Coordination Chemistry: The Synthesis, Reactions, Properties and Applications of Coordination Compounds, vol 6. Oxford: Pergamon, 1987, pp 79–842.
21. RG Bautista. In: RG Reddy, and RN Weizenbach, eds. Extractive Metallurgy of Copper, Nickel and Cobalt, vol I: Fundamental Aspects. Warrendale, PA: The Minerals, Metals and Materials Society, 1993, pp. 827–852.
22. J Preston, AC du Preez. Mintek Report No. M378, Randburg, South Africa, 1988, 30 pp.
23. WA Rickelton, D Nucciarone. In: WC Cooper and I Mihaylov, eds. Hydrometallurgy and Refining of Nickel and Cobalt. Montreal: Canadian Institute of Mining, Metallurgy and Petroleum, 1997, pp 275–291.
24. JS Preston. Hydrometallurgy 9:115–133, 1982.
25. PR Danesi, L Reichley-Yinger, G Mason, L Kaplan, EP Horwitz, H Diamond. Solv Extr Ion Exch 3:435–452, 1985.
26. JA Golding, CD Barclay. Can J Chem Eng 66:970–979, 1988.
27. WA Rickelton, RJ Boyle. Solv Extr Ion Exch 8:783–797, 1990.
28. KC Sole, JB Hiskey. Hydrometallurgy 30:345–365, 1992.
29. WA Rickelton. JOM 44(5):52–54, 1992.
30. IO Mihaylov, E Krause, S Laundry, C Luong. US Patent 5,378,262, Jan 3, 1995.
31. JS Preston. Hydrometallurgy 14:171–188, 1985.
32. K Soldenhoff, N Hayward, D Wilkins. In: B Mishra, ed. EPD Congress 1998. Warrendale, PA: The Minerals, Metals and Materials Society, 1998, pp 153–165.
33. Henkel Corporation. The Chemistry of Metals Recovery Using LIX Reagents. MID Redbook, Tucson, AZ, 1997.
34. PG Thornhill, E Wigstol, G Van Weert. J Met 3(7):13–18, 1971.
35. E Wigstol, K Froyland. Het Ingenieursblad 41:476–486, 1972.
36. G Van Weert. In: Process Metallurgy of Nickel and Cobalt Extraction, Professional Development Course. Montreal: Canadian Institute of Mining, Metallurgy and Petroleum, 1997, 86 pp.
37. EO Stensholt, H Zachariasen, JH Lund, PG Thornhill. In: GP Tyroler and CA Landolt, eds. Extractive Metallurgy of Nickel and Cobalt. Warrendale, PA: Metallurgical Society, 1988, pp 403–412.
38. GM Ritcey, AW Ashbrook. Solvent Extraction: Principles and Applications to Process Metallurgy, Part I. Amsterdam: Elsevier, 1984, pp 234–236.
39. C Bozec, JM Demarthe, L Gandon. In: International Solvent Extraction Conference ISEC '74. London: Society of Chemical Industry, 1974, pp 1201–1229.
40. RD Eliasen, E Edmunds Jr. In: Proceedings of the Third Annual Meeting of the

Hydrometallurgy Section of the Metallurgical Society of CIM. Montreal: Canadian Institute of Mining and Metallurgy, 1973, pp 50–53.

41.   GA Kordosky. In: Ammtec–Henkel Solvent Extraction Symposium, Perth, 1990.

42.   GM Ritcey. In: EJ Grimsey and I Neuss, eds. Nickel '96—Mineral to Market. Victoria: Australasian Institute of Mining and Metallurgy, 1996, pp 251–258.

43.   DDJ Clemente, BI Dewar, J Hill. In: Cobalt '80. Montreal: Canadian Institute of Mining, Metallurgy and Petroleum, 1980, paper no 7.

44.   S Nishimura. In: Extraction Metallurgy '81. London: Institution of Mining and Metallurgy, 1981, pp 404–412.

45.   A Suetsuna, N Ono, TI Iio, K Yamada. Metallurgical Society of the AIME, 1980, TMS paper selection A80-2.

46.   S Makino, N Tsuchida. In: Diversity, the Key to Prosperity. Victoria: Australasian Institute of Mining and Metallurgy, 1996, pp 139–144.

47.   N Ono, S Itasako, I Fukui. In: International Symposium on Cobalt, vol 1. Brussels: Benelux Metallurgie, 1981, pp 63–71.

48.   KR Suttill. Eng Min J 195(5):29–40, 1994.

49.   JG Reid. In: Proceedings of the Australian Bureau of Agriculture and Resource Economics Conference, Canberra, Feb 6–8, 1996, pp 1–6.

50.   JG Reid. In: Proceedings of the AusIMM Annual Conference. Victoria: Australasian Institute of Mining and Metallurgy, 1995, pp 131–135.

51.   MJ Price, JG Reid. In: Solvent Extraction in the Process Industries, vol 1. London: Society of Chemical Industry, 1993, pp 159–166.

52.   JG Reid, MJ Price. In: Solvent Extraction in the Process Industries, vol 1. London: Society of Chemical Industry, 1993, pp 225–231.

53.   IG Skepper, JE Fittock. In: DC Shallcross, R Paimin, and LM Prvcic, eds. Adding Value Through Solvent Extraction, vol 1. Melbourne: University of Melbourne Press, 1996, pp 777–782.

54.   M Mackenzie. TechNews (Henkel MID) (5):1–4, 1998.

55.   J Fittock. In: WC Cooper and I Mihaylov, eds. Hydrometallurgy and Refining of Nickel and Cobalt. Montreal: Canadian Institute of Mining, Metallurgy and Petroleum, 1997, pp 329–338.

56.   A Taylor. Min Mag 172(3):167–170, 1995.

57.   B Nyman, A Aaltonen, S-E Hultholm, K Karpale. Hydrometallurgy 29:461–478, 1992.

58.   E Pääkkönene, M Mattelmäki. In: EJ Grimsey and I Neuss, eds. Nickel '96—Mineral to Metal. Victoria: Australasian Institute of Mining and Metallurgy, 1996, pp 235–241.

59.   P Hanniala. Miner Ind Int (1):43–47, 1997.

60.   JH Kyle. In: EJ Grimsey and I Neuss, Eds. Nickel '96—Mineral to Market. Victoria: Australasian Institute of Mining and Metallurgy, 1996, pp 245–250.

61.   F Habashi. In: RG Reddy and RN Weizenbach, Eds. Extractive Metallurgy of Copper, Nickel and Cobalt, vol 1. Warrendale, PA: The Minerals, Metals and Materials Society, 1993, pp 1165–1178.

62.   PG Mason, JW Gulyas. In: B. Mishra, ed. EPD Congress 1999. Warrendale, PA: The Minerals, Metals and Materials Society, 1999, pp 585–616.

63. KN Han, X Meng. In: RG Reddy and RN Weizenbach, eds. Extractive Metallurgy of Copper, Nickel and Cobalt, vol 1. Warrendale, PA: The Minerals, Metals and Materials Society, 1993, pp 709–733.

64. E Krause, BC Blakey, VG Papangelakis. In: ALTA 1998 Nickel Cobalt Pressure Leaching and Hydrometallurgy Forum. Melbourne: ALTA Metallurgical Services, 1998, 39 pp.

65. P Dickson. In: ALTA 2000 Nickel Cobalt-6. Melbourne: ALTA Metallurgical Services, 2000, 9 pp.

66. D Burvill, D White. In: ALTA 1999 Nickel Cobalt Pressure Leaching and Hydrometallurgy Forum. Melbourne: ALTA Metallurgical Services, 1999, 18 pp.

67. JH Kyle, D Furfaro. In: WC Cooper and I Mihaylov, eds. Hydrometallurgy and Refining of Nickel and Cobalt. Montreal: Canadian Institute of Mining, Metallurgy and Petroleum, 1997, pp 379–389.

68. R Grassi, D White, T Kindred. In: ALTA 2000 Nickel-Cobalt 6. Melbourne: ALTA Metallurgical Services, 2000, 10 pp.

69. J Chadwick. Min Mag 181(4):208–216, 1999.

70. S Parker. In: Proceedings of the 14th International Ferro-Alloys Conference, Monte Carlo, 1998, 15 pp.

71. Anon. Met Bull (7):8, 2000.

72. GM Ritcey, NL Hayward, T Salinovich. Australian Patent PN0441, July 18, 1996.

73. A Taylor, DC Cairns. In: ALTA 1997 Nickel Cobalt Pressure Leaching and Hydrometallurgy Forum. Melbourne: ALTA Metallurgical Services, 1997, 54 pp.

74. GL Frampton, RD Buratto. In: ALTA 1999 Nickel Cobalt Pressure Leaching and Hydrometallurgy Forum. Melbourne: ALTA Metallurgical Services, 1999, 19 pp.

75. DS Flett. J Chem Technol Biotechnol 74:99–105, 1999.

76. PM Cole, VM Nagel. Extraction Metallurgy Africa '97. Johannesburg: South African Institute of Mining and Metallurgy, 1997, 15 pp.

77. A Pavlides, J Wyethe. In: ALTA 2000 SX/IX-1. Melbourne: ALTA Metallurgical Services, 2000.

78. A Griffin, G Becker. In: ALTA 2000 Nickel Cobalt-6. Melbourne: ALTA Metallurgical Services, 2000, 12 pp.

79. AE Norton, JJ Coetzee, SCC Barnett. In: ALTA 1998 Nickel and Cobalt Pressure Leaching and Hydrometallurgy Forum. Melbourne: ALTA Metallurgical Services, 1998, 19 pp.

80. DW Dew, DM Miller. In: Proceedings of the Biomine '97 Conference. Perth: Australian Mineral Foundation, 1997, pp M7.1.1–M7.1.8.

81. DM Miller, DW Dew, AE Norton, PM Cole, G Benetis. In: WC Cooper and I Mihaylov, eds. Hydrometallurgy and Refining of Nickel and Cobalt. Montreal: Canadian Institute of Mining, Metallurgy and Petroleum, 1997, pp 97–110.

82. VM Nagel, BM Davies, PM Cole. In: M Cox, ed. International Solvent Extraction Conference ISEC '99. London: Society of Chemical Industry, 2001.

83. V Nagel, D Jacobs, G Benetis, N de Jager, A Feather. In: ALTA 2000 Nickel Cobalt-6. Melbourne: ALTA Metallurgical Services, 2000, 21 pp.

84. T Heinzle, D Miller, V Nagel. In: Proceedings of the Biomine '99 and Water Management Conference. Perth: Australian Mineral Foundation, 1999, pp 16–25.
85. J Preston, AC du Preez. J Chem Technol Biotechnol 61:159–165, 1994.
86. J Preston, AC du Preez. J Chem Technol Biotechnol 66:295–299, 1994.
87. JS Preston, AC du Preez. Hydrometallurgy 58:239–250, 2000.
88. JS Preston. Hydrometallurgy 11:105–124, 1983.
89. GSM Becker, GD Johnson. In: EJ Grimsey and I Neuss, eds. Nickel '96—Mineral to Market. Victoria: Australasian Institute of Mining and Metallurgy, 1996, pp 265–266.
90. GD Johnson, HA Evans. In: ALTA 1998 Nickel Cobalt Pressure Leaching and Hydrometallurgy Forum. Melbourne: ALTA Metallurgical Services, 1998, 11 pp.
91. RM Berezowsky. Pressure acid leaching of nickel sulfide concentrates. Presented at Extractive and Processing Division Congress, The Minerals, Metals and Materials Society Annual Meeting, Feb 12–16, 2000.
92. D Jones, J Hestrin, R Moore. In: ALTA 1998 Nickel Cobalt Pressure Leaching and Hydrometallurgy Forum. Melbourne: ALTA Metallurgical Services, 1998, 34 pp.
93. CJ Ferron, CA Fleming, DB Dreisinger, PT O'Kane. In: ALTA 2000 Nickel Cobalt-6. Melbourne: ALTA Metallurgical Services, 2000, 17 pp.
94. A Taylor. Min Mag 174(2):100–103, 1996.
95. I Mihaylov, E Krause, DF Colton, Y Okita, J-P Duterque, J-J Perraud. CIM Bull 93(1041):124–130, 2000.
96. KC Sole, JB Hiskey. Hydrometallurgy 37:129–148, 1995.
97. WA Rickelton, I Mihaylov, B Love, PK Louie, E Krause. US Patent 5.759, 512, June 2, 1998.
98. G Motteram. In: ALTA 2000 Nickel Cobalt-6. Melbourne: ALTA Metallurgical Services, 2000, 14 pp.
99. P Mason, M Hawker. In: ALTA 1998 Nickel Cobalt Pressure Leaching and Hydrometallurgy Forum. Melbourne: ALTA Metallurgical Services, 1998, 38 pp.
100. A Mezei, R Molnar, T Todd, CA Fleming, P Mason. In: ALTA 1999 Nickel Cobalt Pressure Leaching and Hydrometallurgy Forum. Melbourne: ALTA Metallurgical Services, 1999, 27 pp.
101. G Miller. In: ALTA 1999 Nickel Cobalt Pressure Leaching and Hydrometallurgy Forum. Melbourne: ALTA Metallurgical Services, 1999, 15 pp.
102. LA Clinton, MG Bailie. In: ALTA 1999 Nickel Cobalt Pressure Leaching and Hydrometallurgy Forum. Melbourne: ALTA Metallurgical Services, 1999, 18 pp.
103. MG Bailie, GC Cock. In: ALTA 1998 Nickel Cobalt Pressure Leaching and Hydrometallurgy Forum. Melbourne: ALTA Metallurgical Services, 1998, 38 pp.
104. A Griffin. In: ALTA 1998 Nickel Cobalt Pressure Leaching and Hydrometallurgy Forum. Melbourne: ALTA Metallurgical Services, 1998, 26 pp.
105. D Parkes, B Grinbaum, R Kleinberger, B Arnall. In: ALTA 2000 Nickel Cobalt-6. Melbourne: ALTA Metallurgical Services, 2000, 11 pp.

106. MH Fox, SJ Ralph, NP Sithebe, EM Buchalter, JJ Riordan. In: ALTA 1998 Nickel Cobalt Pressure Leaching and Hydrometallurgy Forum. Melbourne: ALTA Metallurgical Services, 1998, 21 pp.
107. A Taylor, ML Jansen. In: ALTA 2000 Nickel Cobalt-6. Melbourne: ALTA Metallurgical Services, 2000, 13 pp.
108. Anon. S Afr Min Coal Gold Base Met (7):11–15, 2000.

4

Treatment of Soils and Sludges by Solvent Extraction in the United States

Richard J. Ayen

Neptune Consulting, Wakefield, Rhode Island

James D. Navratil

Clemson University, Anderson, South Carolina

I. TECHNOLOGY CONCEPTS

Solvent extraction is a separation and concentration process in which a non-aqueous liquid reagent is used to remove organic and/or inorganic contaminants from solids, waste residues, soils, sediments, sludges, raw industrial products, liquid process solutions, wastewater, and so on. In a solvent extraction process, the material to be removed from the feed stream is called the *solute* and the residual of the feed stream is called the *raffinate*. The solvent stream, after it has removed the solute from the feed, is called the *extract*. The process is based on well-documented scientific principles and chemical equilibrium separation techniques as described in chemistry and chemical engineering texts [1–3]. For many applications, the extract is retained and the raffinate is discarded. For waste treatment, the opposite occurs; the cleaned feed stream, the raffinate, is the desired product. Other industrial applications, which do not involve waste remediation, include the decaffeination of coffee, the extraction of oil from soybeans, and the separation of copper from leaching fluids. Several distinctions must be made to understand the concepts behind solvent extraction. Solvent extraction does not include soil washing but does include leaching. Soil washing is a related technology and involves the use of dilute aqueous solutions of detergents or chelating agents to remove contaminants through desorption, abrasion, or physical separation, whereas solvent extraction relies on the action of organic chemical reagents. Solvent extraction, as described in this chapter, is used to describe both leaching and extraction operations; leaching implies the removal of a contaminant from a solid, whereas extraction implies the removal of a contaminant from a liquid.

A general overview of the extraction process is depicted in Fig. 1. The contaminated soil and the solvent enter an extraction step, where they are mixed. The extract, which contains most of the solvent plus the organic contaminants, is then separated from the contaminant-free soil, the raffinate. The extract is then separated, usually by distillation, into a solvent stream and a concentrated contaminant stream. The solvent is recycled to the extraction step, usually with fresh solvent makeup added. The concentrated contaminants are then disposed of by some appropriate method, as described in the following sections of this chapter. Additional solvent is stripped from the decontaminated solids, and this solvent is also recycled to the extraction step. In the case of a soil remediation project, the solvent-free and contaminant-free solids are then generally returned to where they were located before remediation was begun.

A. Scientific Principles

In an extraction operation, the two liquid phases can be immiscible or partially miscible. Maximum separation of contaminants is effected under the following conditions:

> The solute is more soluble in the solvent phase.
> The solvent is completely immiscible with the feed.
> The solvent has a substantially different specific gravity from that of the feed.

In the case of leaching, the solute is bound to a solid substrate. The solubility of the solute in the extracting liquid competes with the solute's affinity for the substrate through low energy adsorptive binding, high energy chemisorption, or incorporation into the solid matrix. The solubility of the solute in the solvent is partially dictated by the permanent and induced dipole interactions. The stronger the interactive binding between the solute and the solid, the more difficult it will be to extract the solute from the solid.

B. Basic Concepts and Definitions for Solvent Extraction Technology

A useful concept in explaining liquid–liquid or solid–liquid solvent extraction processes is the *stage*. In a stage, feed and solvent are combined in a mixer and

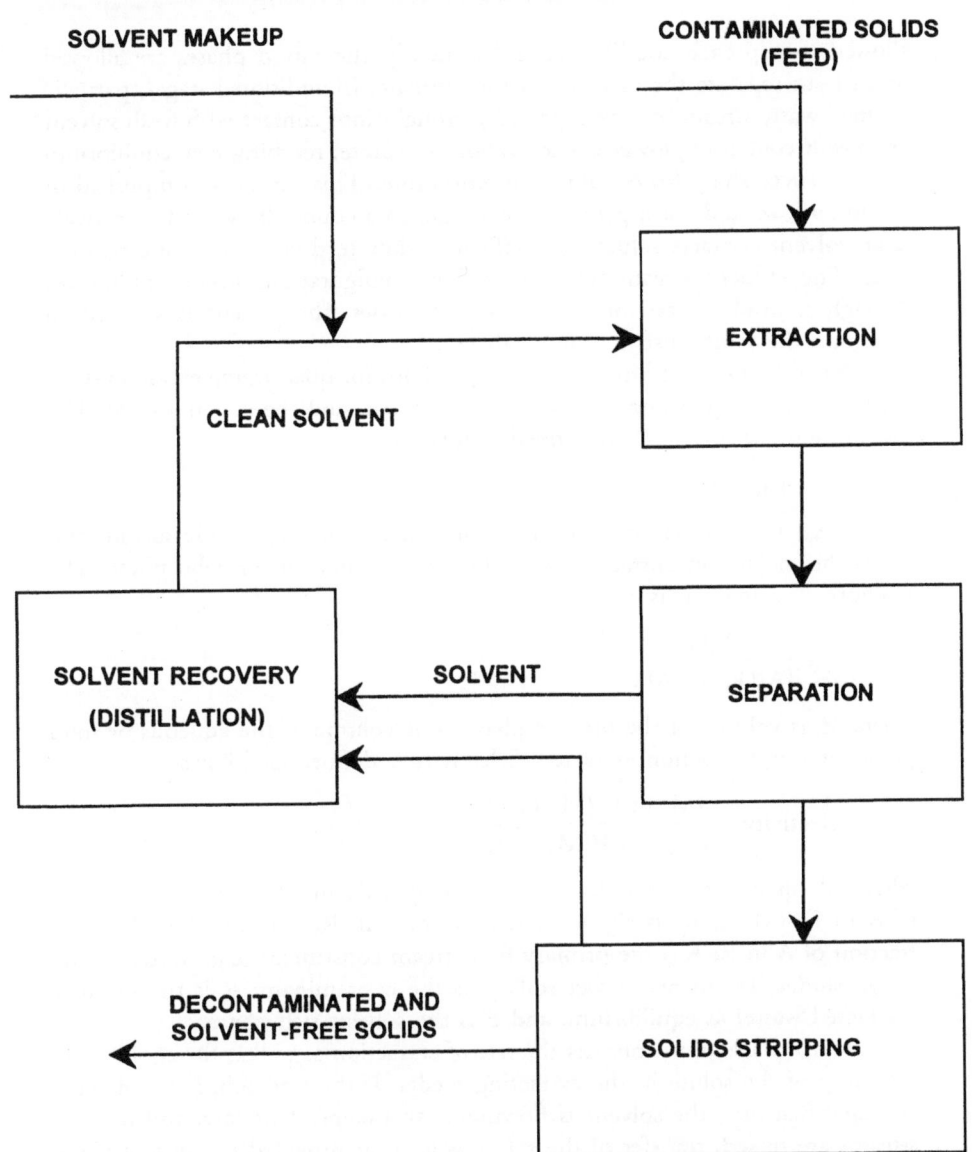

Figure 1   Generalized schematic of a solvent extraction process.

allowed to approach equilibrium. Subsequently, the mixed phases are allowed to settle to separate the extract and the raffinate. In additional stages, partially purified waste streams can be repeatedly brought into contact with fresh solvent or solvent containing lower concentrations of solute, reaching new equilibrium states at successively lower solute concentrations. This can be accomplished in a counterflow mode or a parallel flow mode. In a counterflow mode, relatively clean solvent contacts solute-lean raffinate, while feed contacts solute-rich extract. The effluent streams in parallel flow configurations reach equilibrium through multiple extraction stages. In both cases, the solvent is selected to maximize the solute distribution coefficient.

Possibly the most often utilized expressions for quantifying extractions are the distribution coefficient, the fraction extracted, and the selectivity [1]. The distribution coefficient K_D is defined as follows:

$$K_D = [A_o]/[A]$$

where $[A_o]$ is the molar concentration of solute A in the organic solvent and $[A]$ is the molar concentration of solute A in the aqueous or solid phase. The fraction extracted, F_A, is

$$F_A = \frac{[A_o]V_o}{[A_o]V_o + [A]V}$$

where V_o is volume of the organic phase, V is volume of the aqueous or solid phase, and F_A is fraction extracted. Selectivity is defined as follows:

$$\text{Selectivity} = \frac{(M_x \text{ in E})/M_A \text{ in E})}{(M_x \text{ in R})/(M_A \text{ in R})}$$

where M_x in E is the mass fraction of x in E, $(M_A$ in E) is the mass fraction of A in E, $(M_x$ in R) is the mass fraction of x in R, $(M_A$ in R) is the mass fraction of A in R; A is the primary feed stream constituent (e.g., water, wastewater, sludge, sediment, or wet soil), x is the contaminant; R is the raffinate (extracted waste) at equilibrium, and E is the solvent-rich phase.

It is also useful to consider the rate of extraction as well as the equilibrium solubility of the solute in the extracting media. If the feed solution is flowing at a specified rate, the solvent is flowing at an independent rate, and the two streams are mixed, transfer of the solute will occur provided the system is not at equilibrium. The rate of extraction at any time is proportional to the departure from equilibrium. As the liquid enters the vessel, the initial rate of extraction is given as follows:

$$\text{Rate}^0 = K_E a \Delta C_E^0 = K_R a \Delta C_R^0$$

where ΔC_E^0 is the initial concentration driving force for the solute in the extract (i.e., the concentration at the interface less the concentration in the bulk of the extract), ΔC_R^0 is the concentration driving force for the solute in the raffinate, a is interfacial area, K_E is the mass transfer coefficient for the extract, and K_R is the mass transfer coefficient for the raffinate.

As the extraction proceeds, the deviation from equilibrium is decreased ($\Delta C_E'$ and $\Delta C_R'$), and the rate of extraction is reduced: $\Delta C_E' < \Delta C_E^0$ and $\Delta C_R' < \Delta C_R^0$ so that:

$$\text{Rate}' = K_E a \Delta C_E' = K_R a \Delta C_R'$$

Theoretically, 100% extraction will never be reached, since the rate decreases: $\text{Rate}' < \text{Rate}^0$. However, a suitable extraction efficiency can usually be achieved.

A number of graphical procedures can be used to predict the number of stages needed to achieve a cleanup objective [4].

The foregoing discussion provides the fundamental concepts that describe liquid–liquid extractions. The dissolution of solutes from their mixtures with insoluble solids is called leaching or solid extraction. The solids form an open, permeable mass through which solvent percolates. There is competition between the solute's affinity for the solid phase and its solubility in the liquid phase. Since the solid phase in wastes can be rather complex, the models that describe the solid–liquid extraction are approximate, at best. In most cases, the coefficients and parameters are better estimated experimentally. For some solid–liquid systems, organics adsorbed onto activated carbon or XAD resins provide good working models. A number of XAD resins are commercially available and are commonly used for separation of organic compounds in solution. These are organic resins produced from poly(methylmethacrylate), styrene–divinylbenzene, and similar materials. The experimental variables differ for each waste type and solid sorbent, so only general concepts are described here. Graphical procedures can be used to estimate the number of theoretical, or ideal, stages required to achieve a desired extent of extraction. Procedures for doing this are given in chemical engineering unit operation textbooks such as McCabe et al. [2] and are analogous to the procedures employed for distillation, absorption, or liquid–liquid extraction. The operating line provides the same information for solid–liquid extraction as it does for liquid–liquid extraction [2]. The raffinate referred to in the operating line equation is simply the solid phase. The number of ideal stages for solid–liquid extractions is determined from construction of a McCabe–Thiele diagram [2].

One relationship characterizing adsorption is an isotherm that relates the solution phase compositions and the solid phase concentrations of adsorbable materials. For systems containing only a single adsorbable component, there is a direct correspondence between solution phase concentration and solid phase concentration. Usually, a Langmuir or a Freundlich isotherm can be used to describe the system [3,5]. Given the heterogeneity of wastes and the variable nature of the solute adsorption sites within a given waste, coefficients relating the solid and solution composition should be determined experimentally. The assumptions used in carbon adsorption of contaminant solutions are much more ideal than for the desorption of solutes from contaminant solids. Carbon is a very homogeneous substrate in comparison with waste soils, and the application of adsorption/desorption behavior for soils is better determined from experimental data and curve fitting to determine the extent of deviations from ideal behavior.

Also helpful in modeling the solid–liquid extraction process are principles of adsorption chromatography. These principles recognize the dynamic equilibrium between the solvent, solute, and adsorbent (solid phase). The equation most often used for the adsorption isotherm is the Freundlich equation [6]. From this equation, chromatographers can calculate the amount of adsorbent required to lower the concentration of solute in a solvent. In the case of extraction of contaminated solids, the opposite is desired: the removal of the solute from a contaminated solid. The equation is given as follows:

$$m = KC^{1/z}$$

where m is the amount of solute adsorbed per unit mass of adsorbent, K is the characteristic constant (experimentally determined), C is the equilibrium concentration of the solute in the solvent phase, and z is a characteristic constant with a value usually ranging from 1 to 2.

Solvent/chemical processes for remediation purposes operate in either a batch or a continuous mode, and all employ similar unit operations [7]:

Feed preparation: screening or crushing of oversized solids, chemical conditioning, pH adjustment

Extraction: removal of the contaminants from the feed using contacting devices of various kinds

Separation of solvent and extract from solids: gravity separation, filtration, centrifugation

Solvent recovery: distillation, steam stripping, pressure reduction, or phase separation to recycle the solvent

The media that can be potentially remediated by this approach include sediments, oily wastewaters, sludges, rendering wastes, biosludges, dye wastes, activated petrochemical sludge, and soils contaminated with pesticides and polychlorinated biphenyls (PCBs) [8].

Solvent extraction can occur under three processing approaches. The first approach employs two phases, in contact, at ambient pressure and temperature in which solute is exchanged between phases. Typical solvents include alkanes, alcohols, ketones, and chlorinated solvents, used either singly or in combination. Azeotropic mixtures may also be used to provide favorable boiling points (positive or negative deviations from Raoult's law). These solutions distill without change in distillate composition. It is possible to remove traces of one compound (e.g., water) by forming an azeotrope with another (e.g., ethanol). A mixture of 4% water in ethanol, by weight, boils at 78.14°C [3]. In the second approach, near-critical fluid/liquefied gas processes use butane, isobutane, propane, carbon dioxide, or other gases, liquefied under pressure, at or near ambient temperature. At the critical point, the liquid and vapor phases of the solvent form a single phase. A fluid near its critical point exhibits the viscosity and diffusivity of a gas, while also exhibiting the solvent characteristics of a liquid. Under these conditions, the solvent can very effectively penetrate the solid matrix and mobilize organic constituents. In the third approach, critical solution temperature processes utilize solvents in which the solubility can be varied by changing the operating temperature. At the lower operating temperature, the solvents are miscible, while at the upper operating temperatures, the solvents are substantially immiscible.

The types of solute extracted by solvent/chemical extractions include volatile organic compounds (VOCs), semivolatile organic compounds (SVOCs), PCBs, oils, grease, pesticides, coal tars, wood preservatives, and petroleum hydrocarbons. The solvents (or combinations) are selected based upon principles of solubility of the solute, ease and safety of material handling, and plant operational considerations (compatibility with extraction equipment).

When the process is complete, there are residue streams that can have adverse environmental impact. Air emissions can result from process leaks and/ or from venting gases. These are usually treated by carbon adsorption or other systems. Use of nitrogen blankets also minimizes emissions and reduces explosion hazards.

II. DEVELOPED SOLVENT EXTRACTION TECHNOLOGIES

Although the most detailed information is considered proprietary, seven systems were selected for discussion:

CF Systems process
Biotherm Carver–Greenfield process
IRCC BEST process
Extraksol process
Low energy extraction process (LEEP)
NuKem process
Terra-Kleen Response Group Soil Restoration Unit

These were selected because they are considered to be illustrative of recent applications of the technology and a significant amount of data was available for these processes. Five have been used commercially and four have been evaluated in the U.S. Environmental Protection Agency Superfund Innovative Technology Evaluation program [CF Systems, IRCC Basic Extractive Sludge Treatment (BEST) process, Terra-Kleen Response Group Soil Restoration Unit, and the Biotherm Carver–Greenfield process]. Key features of these seven processes are summarized in Table 1, in Section III.

A. Amine Solvent Process

Ionics Resources Conservation Company (IRCC), of Bellevue, Washington, developed the BEST process, which applies the unique miscibility properties of triethylamine [$N(CH_2CH_3)_3$; TEA] to separate sludges or oil–contaminated solids into three fractions: oil, water, and solid [9–11]. At temperatures below 18°C, TEA is fully miscible with water. Therefore cold TEA can be used to dewater solids while partially dissolving organic solutes. The organic solutes are more completely removed by extracting with TEA at temperatures above 55°C, temperatures at which TEA is not miscible with water. A predetermined amount of caustic must be added to maintain the final pH of the mixture at between 10.5 and 11. This keeps TEA in its nonionized form. Typically, the solvent/solid ratio is 3:1, and mixing times vary from 5 to 15 min per extraction. The extract is centrifuged to separate the solids from the mixture so the solids can be returned to undergo further extractions. The liquid phase (TEA, water, and oil) is heated to about 77°C. At this temperature, the TEA–water azeotrope will boil, leaving the oil phase to be removed from the extraction vessel. Since TEA and water are immiscible above 55°C, the TEA and water separate into two phases in a decanter. The TEA is steam-stripped and the residual water remains. The water would be expected to contain residual TEA and other organic contaminants found in the feed. The gaseous TEA is condensed and recycled for additional extractions. The solids are subjected to a washer/dryer, volatilizing the TEA. This process produces clean, dry solids. Steam may be injected into the washer/dryer to impart additional moisture to the solids. The residual TEA in the solids must be addressed. There are no available data to document degradation of TEA in soils [12]. When offered for disposal, concentrated TEA is ignitable [13], thus is considered to be hazardous. However, it is unlikely that the residual TEA in soils (<200 mg/kg) will meet the regulatory definition of "ignitability." Solids meeting the cleanup criteria may be returned to the land. Depending on the extent of metal or other inorganic contamination, the solvent-extracted solids may need to be treated by some other technique such as solidification/stabilization or soil washing [12,14]. Water may be acceptable for discharge to a publicly owned treatment works (POTW), although liquid phase carbon adsorption may be used to produce an aqueous discharge free of organic contaminants.

IRCC constructed a 44 kg/day BEST pilot plant that is easily transportable. This skid-mounted plant was used for treating soils contaminated with PCBs at spill sites in Ohio and New York [9]. Extensive EPA SITE testing of this system has been completed [10,12,15]. Design throughput for a full-scale, transportable BEST system is 90 metric tons/day [10]. The full-scale system can treat pumpable wastes with a particle size under 2.5 cm in diameter. The system will treat pumpable and nonpumpable solids with a particle size under 6 mm [16]. The first full-scale BEST unit was used at the General Refining Superfund site in Garden City, Georgia [17].

Since TEA is flammable, requirements of the National Fire Protection Association (NFPA) apply. Off-gases and accidental venting will release TEA to the atmosphere, resulting in a strong ammonia-like odor.

B. Supercritical Fluid/Liquefied Gas Processes

The CF Systems Corporation (CFS) of Woburn, Massachusetts (a Morrison Knudsen Company), and CF Technologies, Inc. (CFT) of Hyde Park Massachusetts, specialize in the development and application of supercritical fluid extraction processes for chemical production and hazardous waste treatment. Carbon dioxide is generally used for aqueous solutions; propane is often selected for sediment, sludges, and soil. In selecting the solvent, the solubility of CO_2 in water and the effects of pH are important considerations [18].

Contaminated sediments containing, for example, BTX (benzene, toluene, and xylene) and gasoline constituents, grease, partially oxidized hydrocarbons, and PCBs are fed top down into a high pressure contactor. Compressed propane at 12–27 bar pressure and 20–65°C temperature passes upward, counter to the solids, and dissolves organic matter. Extracted sediment (raffinate) is removed from the contactor. Propane is vaporized, recompressed, and recycled to the contactor as fresh solvent. Contaminants and natural organic matter are removed from the separation vessel and recovered for disposal or reuse.

The process has seven basic operating steps:

1. Slurried sludge is fed into a stirred-tank extractor.
2. Propane is compressed to operating pressure, condensed, and fed to the extraction vessel to dissolve the oil in the feed.
3. A mixed feed product is taken from the contactor to a decanter, in which the water/solids fraction is separated from the propane/oil fraction.
4. Water and treated solids are removed from the decanter, and the solids are de-watered so that the filter cake can be sent to landfills.
5. Propane and oil pass to a solvent recovery still.

6. Oil, collected as still bottoms, is recycled to a refinery (the BTX and gasoline fractions have reuse potential).
7. Propane is recycled as a fresh solvent.

The extraction step is a multistage process. The number of stages must be suitable to achieve concentration levels consistent with the EPA's best demonstrated available technology (BDAT) standards for refinery hazardous wastes [U.S. EPA Resource Conservation and Recovery Act (RCRA) K048 to K052]. The treated oil and raffinate from the commercial unit are reported to conform to the BDAT standards for 16 volatile and semivolatile organic compounds. Posttreatment must be considered for two of the product streams generated from the supercritical extraction process. The extract contains concentrated organic matter, and the treated sediments contain water and solids. Extracted contaminants must be contained, handled, stored, and transported off-site in a manner consistent with EPA RCRA regulations [7]. The wet product must be dewatered and the water returned to the feed stream to minimize the wastewater treatment requirements. Unless the treatment process has been determined by the EPA to result in a delisted material, a listed feed yields a listed product. After treatment, the product must be evaluated to determine the feasibility of delisting.

On a pilot scale, CFS has operated a trailer-mounted unit [19]. Feed slurry is passed through a basket filter to remove particles greater than 3 mm in diameter. The treatment capacity of the propane system at 20 bar and 32–49°C is about 23 metric tons/day. This plant reports extraction efficiencies of 80% to essentially 100% for the liquefied propane extraction system. With the supercritical CO_2 extraction systems, CFS reports extraction efficiencies of 95% to essentially 100%.

Foaming in treated sediment and extracted contaminant tanks was evident throughout the SITE project experiments [19]. The suspected cause was propane entrainment. Foaming has two adverse effects: extracted contaminants persist in the foam, and increased volume of the product stream leads to higher costs and contaminant migration. Potential sources of ignition must be eliminated in the process plants where propane and butane (and their mixtures) are used. Obviously, these hazards are mitigated when CO_2 is used.

PCBs were removed from hazardous wastes at the Star Enterprise site in 1991 and 1992 in Port Arthur, Texas [20]. Oil, grease, and polynuclear aromatic hydrocarbons (PAHs) were removed from 60,000 m^3 of soil at the United Creosoting site in Conroe, Texas [21].

C. Biotherm Carver–Greenfield Process

The Carver–Greenfield process for drying and solvent extraction treatment was developed by Dehydro-Tech Corporation (DTC) [22–24] and, since 1996, has been available from Biotherm, LLC. The original process was developed by Charles Greenfield in the 1950s at the Carver Process Laboratory [25]. The first commercial plant was installed in 1961 to treat meat-rendering wastes. The system was designed to treat nonhazardous wastes but has been adapted for use in hazardous waste remediation projects. Over 85 Carver–Greenfield plants have been licensed worldwide [25]: 53 designed to dry and de-oil slaughter-house wastes. The process separates mixtures into solids, oil, and water, while using a solvent to extract organics [26–28]. In some cases, heavy metals are removed from wastes because they form complexes with the organic solvents used to effect the extraction.

The process consists of the following steps:

1. Pretreatment by removing debris and grinding feed particles to less than 0.6 cm.
2. Feedstock slurrying by mixing the waste with a solvent, which can be an oil, usually Isopar-L, a food-grade isoparaffinic oil with a boiling point of 204°C, or isooctanol [29].
3. Evaporation and solvent extraction stages in which the waste is extracted and the water in the slurry is evaporated [22,25,30,31].
4. Condensation and oil/water separation using a decanter to separate water from immiscible components—the water is generally clean and can be treated with standard wastewater treatment technologies.
5. Separation of the solvent from the solids by centrifuging (additional extraction steps may be required when the solids are dried).
6. Removal of the solvent from the centrifuge cake by heating and stripping (with nitrogen gas) [26].
7. Distillation of solvent to separate the solvent from the indigenous oils and organics. Complete separation is not always obtained.

The process produces the following residues [32]:

Concentrated indigenous oil and organics.

Water, substantially free of solids and oils (which may be dischargeable to a POTW).

Clean, dry solids (which may contain metals requiring further treatment). Although not a residue, vent gases are produced and may require treatment by granular activated carbon.

Much of the operating data for the treatment of hazardous waste by the Carver–Greenfield process is proprietary, but some pilot and field data are available. In recent treatability tests, a pilot-scale facility had a maximum feed rate of 45 kg/h [26]. In this SITE demonstration a 10:1 ratio of solvent to feedstock was reported. In the pilot-scale demonstration, the system temperature of 65–93°C in the initial evaporation/extraction was increased to 110–

135°C to evaporate the water. At 180°C, carrier solvent was removed from the solids [22]. Material with sizes greater that 0.6 cm could not be treated by this process, but could be ground to acceptable sizes.

As is the case in many of these processes, if the feed is a listed waste under EPA RCRA regulations [33], the residue may be considered to be a listed waste (the "derived-from" rule [34]). This provision and another provision called the "mixture" rule are important considerations. The mixture rule states that under certain circumstances, mixtures of hazardous wastes and solid wastes are deemed hazardous waste. Since all the processes described here generate residues and products, the waste classification pursuant to RCRA is crucial. Merely subjecting a feed to a solvent extraction process may not ensure that the residue is nonhazardous even if the levels of contaminants that remain are very low.

D. Extraksol Process

The Extraksol process is conducted by a transportable system developed by the Sanivan Group of Anjou, Quebec, Canada. Unit operations are as follows [35]:

> Washer/dryer extraction
> Nitrogen gas blanketing
> Geomembrane filtration
> Steam generation
> Condensation
> Solvent distillation

The extraction vessel is charged with solids, purged with nitrogen gas, and filled with low boiling solvent (the composition of the solvent is proprietary). No internal agitation is used. The developer claims that the rotation of the mixing cell combined with the dewatering effect of hydrophilic solvents fractures porous solids. After the vessel's rotation has been stopped, the solids settle to the bottom. The contaminated solvent is decanted and pumped to a still for solvent recovery, thus completing one washing cycle. The solids will contain additional solvent that is removed by filtering using a geotextile filter. Although the developer does not suggest that plugging is a problem, the company does limit the application of the technology to solids with less than a 30% clay fraction. Hot nitrogen gas is used to heat the solids to remove the last of the solvent. The product is a dry, decontaminated solid. Operating in the drying cycle only, the Extraksol unit can be used as a thermal desorber to volatilize organic compounds. The contaminated solvent is distilled to remove the low-boiling components, which can be reused as clean solvent. Any water remaining may either be discharged as wastewater or be further treated. The residue contains the concentrated contaminants and requires further treatment such as incineration or dehalogenation depending on its composition.

As in all these treatment processes, a pilot-scale treatability study for Extraksol is usually recommended. The following information acquired from such a study is generic to all the processes:

Extent of agitation/rotation required
Solid/liquid ratio
Solvent type and composition
Mixing time
Evaluation of the residue
Number of washings required (developer limits to a maximum of nine)
Equipment compatibility (e.g., emulsions that plug the geotextile filter)
Processing time

The Sanivan Group has constructed two full-scale systems (1 metric ton/h) that treat PCB-contaminated soils, remove oil and grease from refinery sludges, and remove PCBs from porous gravel and activated carbon. The process was designed to accommodate large nonporous solids. However, porous solids may be accommodated if crushed to sizes smaller than 5 cm [7]. Waste streams containing more than 30% water are undesirable owing to reduced removal efficiency of organics, additional volumes of wastewater, and increased mixing times [36]. This application also uses chelating agents to mobilize the lead for additional treatment.

E. Low Energy Extraction Process (LEEP)

LEEP was developed in 1987 by a New York University research team in the course of a study funded by the EPA [37,38]. The purpose of the study was to develop a low energy, cost-effective process for removing PCBs from contaminated soils, sediments, and sludges. The process has been under commercial development by ART International since June 1988. It was accepted in the SITE program as an emerging technology in June 1989.

Contaminants are leached from the solid matrix through the use of a hydrophilic leaching solvent and then concentrated by means of either a hydrophobic stripping solvent or distillation. ART International has developed two process: PCB leaching using a mixture of acetone and a proprietary additive, and tar leaching using only acetone as the solvent. The important distinction of the LEEP process is that it utilizes two extractions: one to extract the contaminants from the soils/sludge, and the second to extract the contaminants from the leaching solvent. The first extraction generally involves acetone (or some mixture)—a hydrophilic solvent. The contaminants from the PCB extraction that were extracted into acetone are back-extracted into kerosene—a hydrophobic solvent. The contaminants from the tar plant extraction are not back-extracted, but are separated from the solvent by distillation.

The process consists of the following steps:

1. Separation and screening of contaminated materials—draining of excess (>50%) water and crushing of oversize particles to <0.6 cm
2. Extraction of the solid fraction
3. Treatment of the aqueous fraction initially separated from the waste feed (usually by carbon adsorption)
4. Treatment of the leaching solvents (acetone) by liquid–liquid extraction with kerosene or by distillation
5. Drying of the solids with reinjection of water to mitigate fugitive emissions
6. Disposal (as in the case of PCB concentrate) or reuse (as in the case of tar plant extracts)

The residues consist of clean solids, treated water, and contaminated stripping solvent. A variety of leaching solvents (as distinguished from stripping solvents) have been investigated in addition to acetone: n-butylamine, diethylamine, and methylene chloride [38].

A pilot plant, which operates at a nominal throughput of 100 kg/h, has been used for treatability studies. A full-scale facility, although not currently available, is expected to have a capacity of 12 metric tons/h. In this process, as in many of the others reviewed here, vent gases result. The methods of treatment range from carbon adsorbers to refrigeration (condensation) for minimizing hydrocarbon emissions. Particles smaller than 20 cm will be crushed to 1.3–2.5 cm by equipment located internal to the plant. Particles greater than 20 cm will have to be crushed externally.

F. NuKEM Development Process

NKD, a division of American NuKEM, developed two waste treatment processes employing solvent extraction: one for treating contaminated soils [40] and one for treating wastewaters and sludges from petroleum refineries (EPA RCRA waste codes K048 to K052) [41].

In the soil treatment process, excavated soil is screened to remove rocks and debris. The remaining soil is reduced to below 5 cm. The filtered contaminant soil is fed into a mixer and combined with a proprietary solvent and a proprietary reagent. Generally, the solvent/solid mass ratio varies from 1:1 to 2:1. The slurried soil from this pretreatment step is fed to a countercurrent extractor, where contaminants are removed. The countercurrent extractor employs mixers-settlers for the first separation. Three to five stages are usually sufficient to reach target decontamination levels. The treated soil contains residual solvent, which is removed by a solvent dryer. The recovered solvent is then condensed and recycled to the initial pretreatment step or to the last stage of the extraction unit. The contaminated solvent is withdrawn from the mixer-settler and is distilled to separate the solvent from the contaminants. While the solvent is recycled, the residual waste steam (which is now concentrated) is either destroyed on site or incinerated [40].

In the treatment of wastewaters and sludges from petroleum refineries (the K-wastes just listed), a multistage column extractor is used. The vast majority of these wastes exist as bulk water streams containing small amounts of oil and solid. Most of the aforementioned processes require dewatering prior to treatment. In general, the presence of water reduces the efficiency of most extractions because of the formation of emulsions, which decrease the solvent–waste contact. However, NuKEM claims to be able to add a proprietary reagent to the mixture to alleviate this problem. In the liquid–liquid extraction, solvent is fed at the bottom of the extraction column at a nominal flow rate of 75 L/min (proposed full-scale plant). Solids and water flow down the column, while solvent flows from the bottom to the top, extracting oil from the refinery waste as it rises. The clean solids and water from the bottom of the column are pumped to a solvent stripper. The solids, which are now solvent-free, are pumped to a filter. The water, filtered in this step, is sent to the refinery's wastewater treatment system. The filter cake is sent to a disposal site for stabilization. The extract is fed to a fractional distillation column for the recovery of solvent. The recovered oil is recycled to the refinery fractionation unit, coker, or fluidized catalytic cracker.

In August 1989, a pilot plant version of the soil decontamination unit was being operated at NKD, Houston [40]. In October 1990, a pilot-scale version of the refinery waste decontamination unit was also being operated (600 L/day) [41]. Projections for full-scale operations have been made based on pilot-scale experiments. API separator sludge would be about 9000 metric tons/year, and slop oil production would be 900 metric tons/year. Sludge would be drawn from the storage tank at a rate of about 40 L/min (for 4 days), while slop oil would be processed at a range of 4–20 L/min (on the fifth day). The extraction column needed to accommodate these streams would be 0.76 m in diameter and 10.8 m in height. Solvents would be injected into the base of this full-scale system at a rate of 80 L/min.

NKD claims an ability to handle highly aqueous wastes in a straightforward manner. However, most wastes contain numerous organic contaminants. Those with vapor pressures greater than the solvent will appear in the distillate from the solvent distillation step, while those with lower vapor pressures remain in the concentrated residue. The presence of the more volatile contaminants may necessitate venting the overhead gases to relieve excess pressure.

G. Soil Restoration Unit

Terra-Kleen Response Group, Inc. (Terra-Kleen) developed the Solvent Restoration Unit (SRU) that uses up to 14 different solvents in treating contaminated solids [42]. The process steps are the same as those in the Extraksol process: washing, drying, and solvent recovery. The SRU process is unlike a typical batch process in that solids are conveyed through an extraction unit, where they are mixed with solvent. Conveyors are used to move solids from the feed hopper to the extractor and on to the solids drying unit. Fresh solvent continuously circulates through the extractor. As the solvent mixes and extracts contaminants, fresh solvent reestablishes equilibrium and extracts additional contaminants. As in most instances, extraction is diffusion limited. The extracted solids, which contain residual solvent, are dried to produce a clean, solvent-free soil for return to the land. Solvent vapors from this drying step are condensed for reuse. Toward the end of the drying cycle, dried soil particles may become airborne. A bag filter placed ahead of the condenser collects the fugitive dust that can be discharged as a clean solid. The gas exiting the bag filter is reheated and reused as feed gas for the dryer. The water must be analyzed to determine if it can be discharged to a POTW.

From 1988 to 1990, a trailer-mounted system was used to remove organic contaminants from the Traband site in Oklahoma [42]. This system has not been dismantled. However, during the demonstration, the cleanup goal of 100 ppm PCBs was met. The initial concentrations of PCBs were 4600, 3300, and 88 ppm. Following treatment, the concentrations were 94, 47, and 4 ppm, respectively [42]. A full-scale system was constructed and used to remove PCBs and chlorobenzenes from contaminant soils at the Pinette Salvage Yard Superfund site in Washburn, Maine. The cleanup objective of 5 mg/kg PCBs was met, but the design throughput was never achieved. The system was demonstrated in the EPA SITE program during May and June 1994 at Naval Air Station North Island Site 4 in San Diego, California. Five metric tons of soil contaminated with PCBs was successfully treated [43], reducing PCB concentrations from about 150 ppm to less than 2 ppm.

Several full-scale systems were subsequently built. PCBs were removed from 9000 metric tons of soil at the Naval Air Station North Island site. PCBs were removed at a remote Air Force site in Alaska. 1,1,1-Trichloro-2,2-bis(p-chlorophenyl)ethane (DDT), 1,1-dichloro-2,2-bis(p-chlorophenyl)ethane (DDD), and 1,1-dichloro-2,2-bis(p-chlorophenyl)ethylene (DDE) were removed from soil at the Naval Communications Station in Stockton, California [43].

III. SUMMARY OF COMMERCIAL APPLICATIONS

A summary of the principal parameters of the aforementioned technologies is given in Table 1. This table lists the preferred wastes, solvent system, and contaminants. Solvent extractions have been reportedly selected in six Records of Decision to clean Superfund sites [44]. Concentration factors of 10,000:1 have been measured. Removal efficiencies exceeding 90% are generally reported for organic contaminants, with residual levels below 1 ppm in many cases.

A. Suitable Feedstocks for Solvent Extraction Technology

Solvent extraction processes have been selected by the EPA for sites contaminated with PCBs, volatile organic compounds, and pentachlorophenol. Tests have been successfully carried out on pesticides, oil and grease, dioxins, petroleum hydrocarbons, polynuclear aromatic hydrocarbons (PAHs), and a wide range of semivolatile compounds. Solid matrices tested have included soils, sludges, sediments, slurries, and drilling cuttings. Pentachlorophenol has been removed from activated carbon [43].

B. Successful Past Applications

The list of successful past applications is now extensive and includes remediation of sites contaminated with PCBs, PAHs, various pesticides, oil and grease, halogenated volatile compounds, solvents, pentachlorophenol, polychlorinated dibenzo-p-dioxins, and polychlorinated dibenzofurans [45].

C. Process Design and Commercialization

In general, commercial solvent extraction systems are based on equipment from the chemical process and hydrometallurgical industries [43]. Much of the equipment can therefore be purchased "off the shelf." The integrated systems, however, must be designed and assembled by remediation companies, and designs vary widely from company to company. An individual vendor might employ different designs and solvents for different projects depending on the nature of the contamination and site conditions.

Table 1 Summary of Various Solvent Extraction Processes

Technology	Preferred wastes	Contaminant[a]	Solvent	Capacity (commercial unit)
CF Systems	Slurried solids, waste water, aqueous systems	Hydrocarbons, aromatics, halomethanes, BTX, alcohols	Propane, butane, CO_2, (supercritical fluid)	10–100 tons/day; 5–50 gal/min
Biotherm Carver–Greenfield	High water content wastes, rendering waste	Fuel oils, PCBs, PAHs, dioxins, metals	Isopar-L, Isooctanol	5–50 tons/day
IRCC BEST	Soils, sludges, sediments	PCBs, PAHs, VOCs, SVOCs, pesticides, K044 to K052	Triethylamine $N(CH_2CH_3)_3$	100 tons/day
Extraksol	Soils, sludges, sediments	Oil and grease, PAHs, PCP, metals	Proprietary and various	24 tons/day
LEEP; META-LEEP for metals remediation	Soils, sludges, sediments, high moisture wastes	Petroleum hydrocarbons, pesticides, chlorophenols, metals	Acetone, others	168–336 tons/day
NuKEM	Soils, sludges, oily waste water	PCBs (demonstrated); VOCs, SVOCs (anticipated)	Proprietary and various	None
Terra-Kleen Soil Restoration Unit	Soils, sludges, sediments	PCBs, diesel fuel, potentially metals	Proprietary and various	48 tons/day

[a]BTX, benzene, toluene, and xylene; VOCs, volatile organic compounds as defined by RCRA; SVOCs, semivolatile organic compounds as defined by RCRA.

D. Treatment Trains

Solvent extraction systems as deployed at a remediation site generally consist of a number of unit operations. In addition to the solvent extraction operation itself, other operations may provide pretreatment of the feedstock and/or posttreatment of the treated solids, wastewater, off-gases, or concentrated extract.

E. Pretreatment and Posttreatment Process Steps

Pretreatment at a remediation site generally consists of removal of trees, stumps, and other vegetation. During excavation of the soil, any cobble, boulders, and debris must be removed. These large objects must often be cleaned to remove contaminated soil from them. Shedders and screens are then used to prepare the feedstock. Maximum particle sizes are usually in the 0.6–8 cm range. Some processes then require the addition of water, while others require water removal by mechanical dewatering or evaporation.

Posttreatment includes management of treated solids, water or air emissions, and concentrated extract. The treated solids must be analyzed to determine whether the contaminants have been sufficiently removed to meet the cleanup requirements. If not, the solids are re-treated. Any water removed must be tested to determine whether the water meets the requirements for discharge locally or to determine a suitable outlet. Treatment before discharge or other disposal may be required. Any hazardous metals present in the feedstock will not be removed by the extraction process, and it may be necessary to add stabilization reagents to the soil. Air emissions are usually minimal, but it may be necessary to employ condensers and/or activated carbon systems to clean gas streams before discharge. The extract, which contains the concentrated contaminants, is normally sent off-site for incineration or other means of disposal. Recycle of contaminants is sometimes possible.

F. Acceptance by the Regulators and the Public

Solvent extraction has been accepted to some extent by the EPA. As of November 1996, it had been accepted as the preferred technology at five sites in the United States. Reasons for acceptance include the extensive evaluation the EPA has carried out on this technology, extensive commercial-scale experience in other (nonremediation) areas, adaptability of the technology to a wide variety of site conditions, and competitiveness in cost with other remediation technologies. High removal efficiencies and low residual values of contaminants have been demonstrated for a wide range of contaminants and conditions.

Public acceptance has also been favorable, and the advantages of the technology appear to be recognized. The use of high odor (e.g., triethylamine) or flammable solvents can be a problem, and the impact of these potential problems must be minimized.

G. Cost Information

Costs vary widely because of site-specific considerations [43]. The scope corresponding to a given estimate may or may not include regulatory, permitting, and analytical activities. Effluent treatment also may or may not be included. A range of $110–$576 per metric ton has been reported. These costs were for the 1995–1997 time frame and were based on estimates from technology vendors and a model developed by the EPA.

The EPA has published detailed cost estimates for the CF Systems and Carver–Greenfield processes [43]. These estimates included a breakdown of site-specific costs. For the CF Systems process, the EPA assumed a base case of treating 800,000 metric tons of sediments contaminated with PCBs in concentrations of 580 mg/L at 450 metric tons/day over an 11.3-year period. Analytical costs were $500/day. The estimated cost was $163/metric ton ±20%, including excavation and pre- and posttreatment costs, but excluding contaminant disposal cost. Excavation and pre- and posttreatment costs were assumed to be 41% of the total cost. For a second, "hot spot," case, it was assumed that 57,000 metric tons of sediment contaminated with 10,000 mg/L of PCBs was to be treated at 90 metric tons/day over a one-year period. Analytical costs were again $500/day. The estimated cost was $492/metric ton, −30% +50%. Excavation and pre- and posttreatment costs were estimated at 32% of total cost.

For the Carver–Greenfield process, it was assumed that 21,000 metric tons of drilling mud contaminated with petroleum wastes was to be treated. The total cost was estimated at $576/wet metric ton. This includes $264/metric ton for incineration of contaminants, but excludes regulatory, permitting, analytical, effluent treatment, and disposal costs. The cost would drop to $285/metric ton if the recovered oil could be sold to a refinery for $26/metric ton.

ACKNOWLEDGMENT

The contributions of Dr. Marion R. Surgi in gathering materials for and drafting an early version of this document are gratefully acknowledged.

REFERENCES

1. J Rydberg, C Musikas, GR Chopin. Principles and Practice of Solvent Extraction. New York: Dekker, 1992.

2. WL McCabe, JC Smith, P Harriot. Unit Operations of Chemical Engineering. New York: McGraw-Hill, 1985.
3. RA Alberty, F Daniels. Physical Chemistry. New York: Wiley, 1979.
4. O'Brien & Gere Engineers, Inc. Innovative Engineering Technologies for Hazardous Waste Remediation. New York: Van Nostrand, 1995.
5. CN Haas, RJ Vamos. Hazardous and Industrial Waste Treatment. Englewood Cliffs, NJ: Prentice-Hall, 1995.
6. EW Berg. Physical and Chemical Methods of Separation. New York: McGraw-Hill, 1963.
7. US Environmental Protection Agency. Innovative Treatment Technologies Semi-Annual Status Report, 4th ed. EPA/540/R-92/001. Cincinnati, OH, EPA, October 1992.
8. US Environmental Protection Agency. A Compendium of Technologies Used in the Treatment of Hazardous Wastes. EPA/625/8-87/014. Cincinnati, OH: EPA, 1987.
9. LC Robbins. A Permanent Solution to Hazardous Wastes: The B.E.S.T. Solvent Extraction Process. Marketing information from Resources Conservation Company, Bellevue, WA, 1990.
10. MK Tose. Removal of polychlorinated biphenyl (PCBs) from sludges and sediments with B.E.S.T. extraction technology. Paper presented at 1987 Annual Meeting with Biotechnology Conference, New York, Nov 15–20, 1987.
11. LD Weimer. The B.E.S.T. solvent extraction process applications with hazardous sludges, soils and sediments. Paper presented at the Third International Conference New Frontiers for Hazardous Waste Management, Pittsburgh, 1989.
12. MC Meckes, E Renard, J Rawe, G Wahl. J. Air Waste Management Assoc 8:42, 1992.
13. US Code of Federal Regulations (CFR) 261.21.
14. US Environmental Protection Agency. Solvent Extraction Treatment. Engineering bulletin. EPA/540/2-90/013. Cincinnati, OH: EPA, 1990.
15. US Environmental Protection Agency. Resources Conservation Corporation B.E.S.T. Solvent Extraction Technology Application Analysis Report. EPA/540/AR-92/079. Cincinnati, OH: EPA, 1993.
16. US Environmental Protection Agency. Evaluation of the B.E.S.T. Solvent Extraction Sludge Treatment Technology, Twenty-Four Hour Test. EPA/600/2-88/051. Cincinnati, OH: EPA, 1988.
17. www.clu-in.org, Jan 25, 2000.
18. J Markiewicz. Solvent extraction of PCB-contaminated soils. Proceedings of Superfund XV Conference, Washington, DC, 1994, pp 650–652.
19. US Environmental Protection Agency. CF Systems Organics Extractions Process, New Bedford Harbor, Massachusetts, Applications Analysis Report. EPA/540/A5-90/002. Cincinnati, OH: EPA, August 1990.
20. www.frtr.gov/matrix/, Jan 25, 2000.
21. www.frtr.gov/matrix2/, Jan 25, 2000.
22. TD Trowbridge, TC Holcombe, EA Kollitides. Extraction and drying of superfund wastes with the Carver–Greenfield process. Paper presented at Third Forum on

Innovative Hazardous Waste Treatment Technologies: Domestic and International, Dallas, 1991.

23. US Environmental Protection Agency. Dehydro-Tech Corporation (Carver–Greenfield) process for extraction of oily waste. In: The Superfund Innovative Technology Evaluation Program: Technology Profiles, 2nd ed. EPA/540/5-89/013. Cincinnati, OH: EPA, November 1989, pp 31–32.

24. US Environmental Protection Agency. Dehydro-Tech Corporation (Carver–Greenfield) process for extraction of oily waste. In: The Superfund Innovative Technology Evaluation Program: Technology Profiles, 3rd ed. EPA/540/5-90/006. Cincinnati, OH: EPA, November 1990, pp 36–37.

25. NETAC. Carver–Greenfield Process for Oily Waste Extraction. NETAC Environmental Product Profiles. Pittsburgh: National Environmental Technology Applications Corporation, April 1991.

26. US Environmental Protection Agency. Carver–Greenfield process. Dehydro-Tech Corporation. Application Analysis Report. EPA/540/AR-92/002. Cincinnati, OH: EPA, August 1992.

27. US Environmental Protection Agency. Dehydro-Tech Corporation (Carver–Greenfield) process for extraction of oily waste. In: Innovative Treatment Technologies—Overview and Guide to Information Sources. Cincinnati, OH: EPA, October 1991, pp 5–13.

28. US Environmental Protection Agency. Dehydro-Tech Corporation (Carver–Greenfield) process for extraction of oily waste. In: The Superfund Innovative Technology Evaluation Program: Technology Profiles, 4th ed. EPA/540/5-91/008. Cincinnati, OH: EPA, November 1991, pp 58–59.

29. US Environmental Protection Agency. Literature Survey of Innovative Technologies for Hazardous Waste Site Remediation 1987–1991. Preliminary draft. Washington, DC: EPA, February 1992.

30. Process to rid oil from soil succeeds in EPA demonstration. Environ Today Jan/Feb 1992, 11.

31. Carver–Greenfield process applied to petroleum-contaminated soils and sludges. Hazardous Waste Consultant, Jan/Feb 1(7), 1991.

32. US Environmental Protection Agency. The Carver–Greenfield process. In: Superfund Innovative Technology Evaluation Program: Demonstration Bulletin. EPA/540/MR-92/002. Cincinnati, OH: EPA, February 1992.

33. US Code of Federal Regulations (CFR) 261.31,32, and 33.

34. US Code of Federal Regulations (CFR) 261.3(c)(2)(i).

35. D Mourato, J Paquin. Extraction of PCB from soil with Extraksol. Marketing information from the Sanivan Group, Montreal, Canada, 1990.

36. US Environmental Protection Agency. Synopses of federal demonstrations of innovative site remediation technologies. EPA/540/8-91/009. Cincinnati, OH: EPA, May 1991.

37. W Steiner, B Rugg. METALEEP—The total solution for on-site remediation. US EPA Abstract Proceedings: Third Forum on Innovative Hazardous Waste Treatment Technologies: Domestic and International, Dallas, June 11–13, 1991. EPA/540/2-91/016, pp 48–49.

38. W Steiner, B Rugg. L.E.E.P.—Low energy extraction process for on-site remediation of soil, sediment and sludges. Air and Waste Management Association, Kansas City, MO, June 21–26, 1992, 33.03.

39. DW Hall, JA Sandrin, RE McBride. Environm Prog 9(2): 98, 1990.

40. MJ Massey, S Darian, Extractive decontamination of soils and sludges. Paper presented at PCB Forum, EPA International Conference for the Remediation of PCB Contamination, Houston, Aug 29–30, 1989.

41. MJ Massey, S Darian, NKD process for extractive treatment of refinery oil wastes. Report to Committee on Refinery Environmental Control, American Petroleum Institute, Oct 2, 1990.

42. AB Cash. Proposal for participation in the US Environmental Protection Agency's SITE Program, Oklahoma City, 1991.

43. MJ Mann, RJ Ayen, LG Everett, D Gombert II, CR McKee, M Meckes, RP Traver, PD Walling Jr, S-C Way. Liquid Extraction Technologies. Annapolis, MD: American Academy of Environmental Engineers, 1998, pp 10.1–13.66.

44. US Environmental Protection Agency. Synopses of Federal Demonstrations of Innovative Site Remediation Technologies, 2nd ed. Unpublished draft, 1992.

45. www.frtr.gov, Jan 20, 2000.

5

The Design of Solvents for Liquid–Liquid Extraction

Braam van Dyk and Izak Nieuwoudt

Institute for Thermal Separation Technology, University of Stellenbosch, Stellenbosch, South Africa

I. INTRODUCTION

In the recovery of valuable components from organic mixtures with liquid–liquid extraction (LLE), the single most important design variable is the choice of solvent. Solvents with higher selectivities will require fewer stages and lower solvent-to-feed ratios to effect a separation and so result in lower capital and operations cost for the process equipment. Higher recoveries will reduce solvent losses and so decrease operating costs.

Other than the obvious requirement that two immiscible liquid phases form, there are a number of properties that are desirable for a solvent:

The liquid phases should be easily separable. No emulsions must form.
No unwanted azeotropes must form between the solvent and any of the feed components.
The solvent's selectivity for the key components must be as high as possible.
The solubility of the solvent in the raffinate phase should be low.

The challenge is to find solvents that meet all these requirements simultaneously. This chapter gives an overview of the solvent search methods that have been proposed and suggests a novel solvent search method. An experimental case study is used to demonstrate the power of the new method.

221

II. SOLVENT SELECTION METHODS

A. Database Searches

There are many databases of solvent properties (boiling points, freezing points, solubility, etc.) that may be searched to find a suitable solvent. Unfortunately, even the largest of these databases contain at most a few thousand solvents. Most of these are also limited to such properties of pure components as boiling and freezing points. Those that do contain phase equilibrium data (e.g., solubility data) are almost always limited to binary mixtures.

Another fact to keep in mind is that database searches will yield only solvents that have been used. There may well be much more effective solvents for the specific separation that have not yet been used.

B. Broad Screening Methods

To structure the search for solvents a two-step method is used. First, all possible solvents are screened in order to prepare a short list of candidates. These solvents are then more strictly evaluated in order to find a suitable solvent. Seader and coworkers (Seader et al., 1997) proposed a number of methods to screen possible solvents.

1. Selection by Homologous Series

Homologs of one of the key components tend to be selective for that component (Seader et al., 1997). Homologous components also tend to form near ideal solutions and rarely form azeotropes (Scheibel, 1948). This method of solvent selection is most effective if the key components do not contain similar functional groups.

A further complication is that adding a homolog of one of the key components does not always easily attain a liquid–liquid phase split, unless the mutual solubility of these components was already relatively low.

2. The Robbins Chart

The Robbins chart (Table 1) (Robbins, 1980) indicates the manner in which solutions may be expected to deviate from Raoult's law. This may be used to select a solvent that will give a positive deviation with one key component and a negative deviation with the other.

Unfortunately many molecules contain more than one functional group. It is possible that these groups will be in conflicting categories on the Robbins

Table 1 Robbins Chart: Solute–Solvent Group Interactions

Group:	Deviations from Raoult's law by group[a]											
	1	2	3	4	5	6	7	8	9	10	11	12
H-bond donor												
1 Phenol	0	0	−	0	0	−	−	−	−	−	−	−
2 Acid, thiol	0	0	−	0	−	−	0	0	0	0	−	−
3 Alcohol, water	−	−	0	+	+	0	−	−	−	−	−	−
4 Active-H on multihaloalkane	0	0	+	0	−	−	−	−	−	−	0	−
H-bond acceptor												
5 Ketone, disubstituted amide, sulfone, phosphine oxide	0	−	+	−	0	+	+	−	−	+	0	0
6 Tertiary amine	−	−	0	−	+	0	+	0	0	+	0	0
7 Secondary amine	−	0	−	−	+	+	0	0	+	0	0	0
8 Primary amine, ammonia, unsubstituted amide	−	0	−	−	−	0	0	0	0	0	−	−
9 Ether, oxide, sulfoxide	−	0	−	−	−	0	+	0	0	0	0	0
10 Ester, aldehyde, carbonate, phosphate, nitrate, nitrile, nitrite, intramolecular bonded	−	0	−	−	+	+	0	0	0	0	+	+
11 Aromatic, olefin, haloaromatic, multihaloalkane with no active-H, monohaloalkane	−	−	−	0	0	0	0	−	0	+	0	0
Non-H-bonding												
12 Paraffin, carbon disulfide	−	−	−	−	0	0	0	−	0	+	0	0

[a] +, positive; 0, none; −, negative.
Source: Robbins (1980).

chart. In such cases other methods must be used to decide which group should take preference.

To demonstrate the use of the Robbins chart, we consider the separations of a paraffin from benzene, an aromatic compound. Benzene falls into group 11 on the Robbins chart, while the paraffin selected, heptane, falls into group 12. From the Robbins chart we see that group 8 (primary amines, ammonia, and unsubstituted amides) will give a negative deviation from Raoult's law with the aromatic solute and a positive deviation with the paraffin.

Although it is very probable that an amine or amide will be effective in separating the mixture, there is no indication of whether a phase split will occur.

The Robbins chart also indicates that groups 4 (multihalogenated paraffins with active hydrogen atoms), 7 (secondary amines), and 9 (ethers, oxides, and sulfoxides) will all yield positive deviations from Raoult's law with the paraffin and no deviations with the aromatic. These solvents should also be considered as possible solvents. Again, the number of liquid phases that will form is not indicated.

3. Hydrogen Bonding

Where the formation of hydrogen bonds is possible, care must be taken that only one of the key components will form hydrogen bonds with the solvent. The formation of these bonds can greatly enhance the selectivity of the solvent. A number of charts exist that indicate the formation of hydrogen bonds between families of molecules (Ewell et al., 1944; Gilmont et al., 1961; Berg, 1969).

The use of these charts may lead to the same problem with conflicting functional groups encountered with the Robbins chart. In these cases this method, like the Robbins chart, must be used in conjunction with other selection criteria.

4. Polarity Effects

Polarity effects are among the primary causes for deviations from Raoult's law (Prausnitz and Anderson, 1961). Polarity differences are also the principal cause for the formation of immiscible liquid phases.

The use of polarity to select solvents is more effective when there is a more than a negligible difference in polarity between the key components. A solvent with higher polarity than the more polar key component or one with lower polarity than the least polar key component should be used.

Table 2 gives an indication of the relative polarities of a number of different components. This table is most accurate for molecules of similar size. For a more quantitative method, dipole and quadrupole moments may be used

Design of Solvents for LLE

225

Table 2 Relative Polarities of Functional Groups

Polarity	Species
Most polar	Water
↑	Organic acids
	Amines
	Polyols
	Alcohols
	Esters
	Ketones
	Aldehydes
	Ethers
	Aromatics
	Olefins
Least polar	Parafins

Effect of branching	
Most polar	Normal
↕	Secondary
Least polar	Tertiary

Source: Seader et al. (1997).

to give some indication as to the relative polarities. Seader (Seader et al., 1997) also recommended using the polarity contribution to the Hansen three-term solubility parameter (Hansen, 1973; Barton, 1991). Barton also gave a group contribution method for estimating these parameters.

C. Identification of Specific Solvents

Once the broad screening methods just discussed have been used to prepare a list of candidate solvents, specific solvents must be identified by means of stricter guidelines.

1. Boiling and Freezing Points

The boiling point requirement for the solvent will depend on the method of solvent recovery in the later stages of the process. If the solvent's boiling point differs significantly from that of the components that will be recovered in the solvent phase, solvent recovery should be easier.

Species with large differences in boiling points are also less likely to form azeotropes.

The freezing point of the solvent is important for practical considerations. The solvent should not solidify at the process temperatures. This should include process conditions in winter, when ambient temperatures are low.

2. Selectivities

Since selectivity is the most important property of any candidate solvent, it should be the primary criterion for selection of solvents. If experimental equilibrium data are available, the candidates may be ranked according to their selectivities. Activity coefficient models may also be used to calculate the selectivities. The availability of parameters for these models may be a problem, but in their absence group contribution methods can be used to estimate selectivities.

The accuracy of the group contribution methods may vary and is a strong function of the database the parameters were fitted on. The applicability of these methods also depends on the availability of the interaction parameters that are needed.

3. Summary

The procedure for solvent selection discussed above may be summarized as follows:

Search available databases for previously used solvents.
Find families of candidate solvents through broad screening methods: Homologs of one of the key components should be selective for that component. Tables of deviations from Raoult's law (e.g., the Robbins chart) indicate functional groups that candidate solvents should possess. Solvents that form hydrogen bonds with one of the key components will be selective for that component. Solvents that are more polar than the most polar key component or less polar that the least polar key component will be selective for those key components with corresponding polarity.
Identify specific candidate solvents from these families of candidates. Two immiscible liquid phases must form. The solvent must be liquid at the process temperature and pressure. The solvent must have a high selectivity for one of the key components. The solvent must have a low solubility in the raffinate phase.
Test final candidates in the laboratory to identify the best solvent.

III. DESIGNER SOLVENTS: THE STATE OF THE ART

In the estimation of physical properties, there are two basic problems: a forward problem and a backward problem. The forward problem involves the calculation of macroscopic properties, given the molecular structure. The backward problem is the reverse thereof. It involves the construction of an appropriate molecular structure, given the desired macroscopic properties. Computer-aided molecular design (CAMD) is concerned with solving the backward problem.

Numerous methods are available to solve the forward problem, The UNIFAC method (Fredenslund et al., 1975) may be used to estimate activity coefficients in nonideal mixtures, while Joback's methods (Joback and Reid, 1987) may be used to estimate, among other properties, the boiling and freezing points of pure solvents. There are also group contribution methods to calculate any number of physical properties, ranging from viscosity to heat capacity.

Solving the backward problem has proven to be much more complex, and a number of methods have been proposed. All have both advantages and disadvantages. Some of these methods are reviewed here.

A general problem in CAMD is the accuracy of the methods used to solve the forward problem for each candidate molecule. The use of inaccurate methods to predict physical properties will result in the design of inappropriate solvents, irrespective of the CAMD method used.

There are four main categories of CAMD algorithms:

Interactive methods
Combinatorial methods
Construct-and-test methods
Mathematical programming methods

A. Interactive Methods

Interactive methods make use of the user's knowledge and experience to find a set of functional groups that may be combined into molecules with the required properties. Once this set of functional groups has been found, an exhaustive search method must be used to combine them into a molecule or group of molecules.

Interactive methods are based on the linear nature of most group contribution methods. In linear methods, the individual contributions of all the functional groups are added to find the molecular property. This allows the molecule to be assembled group by group. Given a requirement such as a boiling point higher than 500 K, groups are added one at a time and each resulting molecular property evaluated until the constraint is met.

The groups that are added or excluded are determined solely by the user. The computer algorithm will generally only compute the properties for the molecule at each step in the design process.

When multiple properties are specified, it becomes rather more difficult to keep track of all the changing properties. For these cases, Joback proposed a graphical method (Joback and Stephanopoulos, 1989). The contribution of a functional group to each of the specified properties is written as a vector. As each functional group is added, this vector is plotted on the design space to give a graphical representation of the effect of each group towards achieving the design goal. Because of its graphical nature, it is obvious that this method will work well when only two, or at best three, properties are specified. However, many estimation techniques require more properties. Joback mentions that some estimation methods of liquid heat capacities require as many as seven fundamental physical properties (Joback and Stephanopoulos, 1989). This would render the graphical technique impractical.

Fortunately, many physical properties are highly intercorrelated. This allows one physical property to be replaced with a function of other physical properties. This is done by using techniques like principal factor analysis or principal curves, which allow us to reduce the dimensionality of the design space to a point where the contributions may again be plotted.

The group contribution methods used to calculate activity coefficients are not linear, however, and this makes it very difficult to apply the graphical design technique to the design of solvents for liquid–liquid extraction. It will not be possible to calculate contribution vectors for each functional group because the contribution of the groups will depend on the groups that are already present as well as other factors (e.g., temperature).

Although the interactive design methods are simple to implement, they are limited by the knowledge of the user. The ability of the computer to do the repetitive calculations needed for optimization and search algorithms is not utilized at all. To make effective use of the computer, the process of selection of functional groups must be automated. The methods described in the next sections were designed with this goal in mind.

B. Combinatorial Methods

In its crudest form, the combinatorial method is a brute force search method. All possible combinations of functional groups are tested until a set of groups are tested until a set of groups is found that could be used to construct molecules with the required properties. With even the simplest group contribution methods, with small sets of functional groups, this method will be so resource intensive as to be utterly impractical.

Joback and Stephanopoulos (Joback and Stephanopoulos, 1989) have proposed an automated design method that will greatly reduce the search space in which a combinatorial search must function.

The first step in their method is to select all the groups from which the molecule may be constructed. Joback gives the groups in Table 3 as an example of an initial set for designing a refrigerant. At least two groups will be needed to construct a molecule. The maximum number of groups is decided beforehand. Joback's generator now starts selecting groups from this set. First all possible combinations of two groups are constructed, then all possible combinations of three groups, and so forth, up to the maximum number of groups allowed.

Setting an upper limit of six groups and using Table 3, this will give a starting population of 5×10^8 sets (keep in mind that each group may be selected multiple times), most of which would not represent physically viable molecules. Each of these sets is represented by a vector. Ethane, for example, would be written as [2 0 0 0 0 .... 0 0]. To use the brute force approach, all the specified properties must now be calculated for each of these vectors. Each vector must also be tested for physical viability, using rules such as those in Table 4.

Clearly this process would require an enormous amount of computer resources and time. Merely storing all these vectors in the computer memory would require more than 13 gigabytes of RAM!

To reduce the size of the problem, Joback proposed the use of meta-groups and meta-molecules. Meta-groups are formed by abstracting the functional groups of Table 3 into clusters of groups. Instead of combining groups from Table 3 into molecules, these meta-groups are combined into meta-molecules. A meta-molecule that contains two members of the first meta-group will be written as [2 0 0 0 0 .... 0 0]. The representation is the same as before, but now each component of the vector represents not an individual functional group, but a set of functional groups. Each meta-molecule represents a set of molecules that may be expanded by taking all possible combinations of the individual groups from the meta-groups in each meta-molecule. The meta-

Table 3   Initial Groups for Refrigerant Design

—CH₃	>CH₂	>CH—	>C<	=CH₂	=CH—	=C<
=C≤	≡CH	≡C—	—F	—Cl	—Br	—I
—OH	—O—	>CO	—CHO	—COOH	—COO—	=O
—NH₂	>NH	>N—	—CN	=NO₂	—SH	—S—

Source: Joback and Stephanopoulos (1989).

Table 4 Structural Constraints

1. A collection of groups must contain at least two groups.
2. If a collection contains both cyclic and acyclic groups, then it must also contain mixed groups.
3. The number of groups in a collection having an odd number of free bonds must be even.
4. If the collection contains n groups with b free bonds, then $b/2 \leq \frac{1}{2}n(n-1)$.
5. If a collection contains n groups that are all acyclic, and n_i denotes the number of groups with a global valence of i, then $n_1 = 2 + n_3 + 2n_4 + \cdots + (i-2)n_i + \cdots$.

Source: Joback and Stephanopoulos (1989).

groups in Table 5 were formed by grouping the functional groups in the initial set by the number of free bonds in each group.

As an example, these meta-groups will be used to find a molecule containing two to four functional groups and possessing a boiling point of more than 300 K.

1. The first step is to generate all the possible meta-molecules that contain two to four meta-groups. This gives a total of 336 meta-molecules.
2. For each of these meta-molecules, a boiling point is computed by means of a group contribution method like that of Joback (Joback and Reid, 1987). Since the meta-molecules are actually sets of molecules, their boiling points will also be sets. As a practical measure these sets are written as closed intervals. This means that only the highest and lowest boiling points possible in each set need be stored.
3. Once the physical properties for the meta-molecules have been calculated, these are tested against the requirements in the design spec-

Table 5 Example Meta-Groups

Meta-group 1	$-CH_3$	$=CH_2$	$\equiv CH$	$-F$	$-Cl$	$-Br$
	$-I$	$-OH$	$-CHO$	$-COOH$	$=O$	$-NH_2$
	$-CN$	$-NO_2$	$-SH$			
Meta-group 2	$>CH_2$	$=CH-$	$=C=$	$\equiv C-$	$>CO$	$-COO-$
	$-O-$	$>NH$	$-S-$			
Meta-group 3	$=C<$	$>CH-$	$>N-$			
Meta-group 4	$>C<$					

Table 6 Property Values for the Meta-Molecules

Meta-molecule	Range of T_b (K)
(2 0 0 0)	[177, 536]
(2 1 0 0)	[200, 617]
(3 0 1 0)	[178, 730]
(2 2 0 0)	[222, 699]

ification. In this example, only meta-molecules with an upper bound higher than 300 K are retained, leaving 65 of the initial 336.

4.   These meta-molecules are then tested against the physical constraints listed in Table 4. After this test, only four meta-molecules remain. These are listed in Table 6.

At this stage, the search space has been reduced from an initial 637,392 possible combinations to 30,600 possible combinations, a reduction of 95%. To further reduce the number of candidates, the remaining meta-molecules are expanded in a stepwise manner. This is done by dividing the meta-groups into smaller groups. This division is based on the chemical and/or physical properties of the groups. The first meta-group may be divided into two smaller meta-groups, as seen in Table 7. This division was done on the basis of boiling point contributions.

All the meta-molecules containing meta-group 1 must be expanded as well. Again all possible combinations of meta-groups 1,1 and 1,2 are taken for each meta-molecule. For example, meta-molecule [2 0 0 0] may be expanded into three new meta-molecules: [2 0 0 0 0], [0 2 0 0 0], and [1 1 0 0 0]. Note that the dimensionality of the vectors has increased from four to five. The property intervals for each of these new meta-molecules are not calculated, and those not satisfying the design specifications are eliminated. This process is repeated for another meta-group until each meta-group contains only one functional group and each meta-molecule represents an actual molecule. Then a

Table 7 Divided Meta-Groups

	Functional groups							ΔT_b (K)
Meta-group 1,1	—CH$_3$	=CH$_2$	≡CH	—F	—Cl	—Br		[−10.5, 90.8]
	—I	—OH	—CHO	=O	—NH$_2$	—SH		
Meta-group 1,2	—COOH	—NO$_2$	—CN					[125.7, 169.1]

brute force or other combinatorial method must be used to combine the functional groups in each of these sets into actual molecules.

Joback's method greatly reduces the search space for combinatorial methods, but is very sensitive to the way in which meta-groups are divided. Neither is the best division of meta-groups obvious when multiple physical properties are specified.

Further complications arise when this method is applied to the design of solvents. The UNIFAC method that is used to calculate activity coefficients is not a purely additive method. This makes it difficult to calculate the intervals of contributions for the different meta-groups.

The method may be used to design either aliphatics or aromatics, but not simultaneously, unless the physical constraints are expanded.

C. Construct-and-Test Methods

The method developed by Gani and coworkers (Gani et al., 1983, 1991; Brignole et al., 1986; Nielsen et al., 1990) uses knowledge of the physical and chemical properties of the functional groups to limit the search space for a "construct-and-test" method.

Gani et al. classified the set of UNIFAC groups according to the number of free attachments and also according to chemical and physical considerations (see Table 8). Structures are built by combining these groups according to a set of conditions. These conditions are applied at every step, so that only feasible structures are assembled.

The primary conditions set for these molecular structures are as follows:

The final structure must have no free bonds.
Groups of categories 2–5 (Table 8) have certain restrictions on their joining with other groups.

Table 9 gives a summary of the combinations that will satisfy these conditions. There are also a number of secondary conditions:

Only groups with known parameters for the group contribution methods may be used.
Highly branched structures are to be avoided.
Structures that may exhibit proximity effects (e.g., two or more large structures bonded to adjacent carbon atoms) are to be rejected.

The method of combining functional groups requires that different types of molecular structure (e.g., straight chain aliphatics, branched aliphatics, aromatics) be considered separately.

The first step in applying this method is to select the type of molecular structure (e.g., aliphatic or aromatic) and select the functional groups that will

Table 8 Gani's Classification of UNIFAC Groups

Class	Category				
	1	2	3	4	5
1	CH_3	CH_2CN CH_2NO_2	CH_3CO $CONH_2$ $CON(CH_3)_2$	OH CHO CH_2Cl I Br F Cl CH_3COO CH_3O $C_2H_5O_2$ CH_3S	CCl_2F CH_2SH $CHCl_2$ C_4H_3S SH $C\equiv CH$ COO CCl_3 CH_2NH_2 CCl_2F CHClF
2	CH_2	$CHNO_2$	CH_2CO CH_2COO CH_2O $CONCH_3CH_2$	$CHNH_2$ CH_2NH CHCl $CONHCH_2$ $C_2H_4O_2$ CH_2S	$CH=CH$ $CH_2=C$ C_4H_3S CH_3N $C\equiv C$
3	CH		$CON(CH_2)_2$	CNHN CH_2N CCl CHO CHS	$CH=C$ CCl_2
4	C				$C=C$
5	ACH		$ACCH_2$ ACCH AC	$ACCH_3$	ACOH $ACNH_2$ ACCl ACNO2

Source: Gani et al. (1991).

be considered. The algorithm is summarized in Table 10. This algorithm will construct all possible molecular structures that are feasible under the conditions specified by Gani et al. (1991). Once these structures have been assembled, they are all tested against the property specifications. These specifications must of course include calculations of liquid–liquid equilibria.

Properties are classified into four groups:

Primary pure component properties
Secondary pure component properties

Table 9 Rules Related to the Primary Conditions

Total number of groups	Largest class of group	Groups from largest class	Maximal number of groups allowed from sum of categories				
			3	4	5	3 + 4 + 5	4 + 5
Nonaromatic compounds							
2	1	2	2	1	1	2	1
3	2	1	2	1	1	2	1
4	3	1	2	1	1	2	1
4	2	2	2	2	1	2	2
5	4	1	2	1	1	2	1
5	3	1	2	2	1	2	2
5	2	3	3	2	1	3	2
6	4	1	3	2	1	3	2
6	3	2	3	2	1	3	2
6	3	1	3	2	1	3	2
6	2	4	3	3	1	3	3
7	4	1	3	2	1	3	2
7	3	2	3	3	1	3	3
7	3	1	3	3	1	3	3
7	2	5	4	3	1	4	3
8	4	2	3	2	1	3	2
8	4	1	3	3	1	3	3
8	3	3	3	2	1	3	2
8	3	2	3	3	1	3	3
8	3	1	3	3	1	3	3
8	2	6	4	3	1	4	3
Aromatic compounds[a]							
6	5	6	0	3	1	3	3
7	5	5	1(1)	2	1	3	2
8	5	5	2(1)	2	1	3	2
8	5	5	1(2)	2	1	3	2
9	5	5	3(1)	0	0	3	0
9	5	5	1(3)	2	1	3	2
9	5	5	1(1) + 1(2)	1	1	3	1

[a]Values in parentheses for aromatic compounds indicate the number of free attachments after ring completion.
Source: Gani et al. (1991).

Table 10   The Construct-and-Test Algorithm

Variables
M_A: minimum number of groups in a molecule
M_B: maximum number of groups in a molecule
M_C: category in Table 8 to be used
M_2, M_3, M_4, M_5:   utility variables

Instructions
Step 1:
1.1 Input M_A and M_B
1.2 $M_C \rightarrow 1$

Step 2:
2.1 $M_2 \rightarrow M_B - 1$
2.2 For nonaromatic compounds:
    Choose a group from category M_C and class M_2
    If $M_2 > 4$ then
        Choose from category M_C or category 5 and class 4
        $M_2 \rightarrow 4$
    End if
2.3 For aromatic compounds:
    Choose a group from category M_C or category 5 and class 5.
2.4 For either case:
    If more than one group exists in the specified class and category then create a molecular structure for each case.

Step 3:
3.1 $M_3 \rightarrow M_B - M_2$
3.2 If $M_3 = 1$ then
    Terminate structures by choosing M_2 groups from class 1 and any category that satisfies the primary and secondary conditions (see above and Table 9). Repeat until all the structures formed in step 2 have been considered.
    End if
3.3 If $M_3 > 1$ then
    Continue with step 4.
    Else
    Go to step 6.
    End if

Step 4:
4.1 $M_4 \rightarrow M_B - M_2$
4.2 If $M_4 > 4$ then
        $M_4 \rightarrow 4$
    End if
4.3 Choose a group from class M_4 and a category that satisfies the primary and secondary conditions (see above and Table 9).

Table 10 Continued

4.4 Repeat 4.3 for all allowable categories in class M_4.
4.5 Repeat 4.3 until all the structures created in step 3 have been considered.

Step 5:
5.1 $M_5 \rightarrow M_B - M_2 + 1$
5.2 If $M_5 = 1$ then
      Go to step 3.
   Else
      Choose a group from class M5 and any category that satisfies the primary and secondary conditions (see above and Table 9).
      Repeat 4.3 until all the structures created in step 3 have been considered.
   End if
5.3 $M_2 \rightarrow M_2 + 1$
5.4 Go to step 4.

Step 6:
6.1 $M_C \rightarrow M_C + 1$
6.2 If $M_C \geq 5$ then
      Go to step 7.
   Else
      Go to step 2.
   End if

Step 7:
7.1 $M_B \rightarrow M_B - 1$
7.2 Repeat from step 2 until $M_B = M_A$.
7.3 If $M_B < M_A$ then
      Select new type of structure and repeat from step 1.
   End if

End

Primary mixture properties
Secondary mixture properties

    Primary properties may be calculated directly from group contributions without any knowledge of other physical properties of the molecule. To calculate secondary properties, other physical properties must be known. The primary properties are often needed to calculate the secondary properties.
    The properties are calculated in the order listed. This allows properties that are easier to calculate to be tested first. Only if these properties match the requirements are further properties calculated.

Like Joback's combinatorial method, this is an attempt to limit the search space of an essentially brute force method. While Joback attempts to eliminate candidates based on their physical properties, Gani et al. eliminate candidate molecules based on physical viability.

Owing to the complexity of the natural laws that govern the stability of chemical compounds, the set of rules used by Gani et al. will have some exceptions. This may cause feasible candidates not to be considered and may result in the inclusion of some nonfeasible candidates. It is up to the user to check for these eventualities. Continued research on this topic should increase the accuracy and generality of these rules.

D. Mathematical Programming Methods

A number of mathematical programming methods for CAMD have been proposed (Macchietto et al., 1990; Naser et al., 1991; Churi et al., 1996; Maranas, 1996; Vaidyanathan et al., 1996). Most of these methods use variations of mixed integer nonlinear programming (MINLP) to solve the backward problem, though Macchietto and Naser both relaxed the integer specification and employed continuous optimization algorithms in their methods.

In all these methods, the variables that are optimized are the number of structural groups from each UNIFAC group that is present in the solvent or mixture of solvents. A single goal function that includes all the required properties is defined and minimized or maximized. After sets of functional groups have been assembled, molecules must be built from these sets. These molecules are usually assembled with exhaustive search algorithms.

Constraints are set in all the above-mentioned methods to ensure that the assembled groups form viable molecules. Most of these methods were developed to design polymers and polymer blends, and therefore only certain physical viability aspects are checked. Certain UNIFAC groups like $C=C$ and $C\equiv C$ are usually excluded beforehand.

The mathematical programming methods are computationally inexpensive in comparison to other methods. Since, however, they tend to be susceptible to local minima traps, they are not guaranteed to find the global optimum. The brute force methods used to assemble molecules from the functional groups returned by the mathematical programming methods are computationally expensive but operate on a much smaller number of groups.

If the postprocessing step constructs more than one molecule from the set of functional groups, the liquid–liquid equilibria must be recalculated. This will significantly decrease the speed of the algorithm because the most time-consuming calculations must be done twice. Sets of functional groups that give good selectivities as a whole may also have to be abandoned if the solubilities of these smaller molecules in the raffinate are too high.

IV. DESIGNER SOLVENTS: AN EVOLUTIONARY APPROACH

In 1975 John Holland published a paper describing how the Darwinian principles of evolution may be applied to optimization problems (Holland, 1975). The genetic algorithms (GA) he described were found to be efficient, robust optimization techniques that work in the most difficult search spaces without problems like susceptibility to local minima traps. These algorithms have also been applied to CAMD problems, both for the design of polymers (Venkatasubramanian et al., 1994) and for the design of solvents for extractive distillation (Van Dyk, 1998; Van Dyk and Nieuwoudt, 2000). The basic principles of GA are briefly explained before we look in detail at the application of GA to CAMD.

A. A Brief Introduction to Genetic Algorithms

The basic principle of evolution through natural selection is that the individuals that are more suited to survive in their environment will have a better chance of surviving and reproducing. The genetic material that gave them this enhanced fitness will then be passed on to their offspring, who will also have a better chance to propagate. In time the entire population will have the characteristics that make them more suited to their environment.

This survival of the fittest principle is used in GA to solve optimization problems by defining a fitness function suited to the problem. This fitness function may take any form, but better solutions to the problem must have higher fitness values. Any number of parameters may be incorporated into the fitness, but a single overall fitness is assigned to each candidate solution to the problem.

1. The Encoding Scheme

Candidate solutions are encoded in linear structures called chromosomes. The elements comprising the structure are called genes.

2. The Evaluation Scheme

Once the chromosomes have been encoded, some method must exist to decode the chromosome into the parameters that are to be optimized and calculate the fitness of the individual. In CAMD this would be the group contribution methods that convert the structural information stored in the chromosome into physical properties. The fitness function must be chosen with care, since this is the function that will ultimately be optimized.

3.   The Population Scheme

Evolution does not influence the extent to which an individual is suited to its environment but instead strives to improve the average fitness of the entire population. For GA to mimic this natural phenomenon, a population scheme must be implemented. Unlike most optimization techniques, a GA does not sample the solution space at a single point at a time. Instead GAs are largely parallel. A population of chromosomes is maintained, consisting of typically hundreds to thousands of individuals. These individuals are then selected probabilistically, based on their fitness values to reproduce and create new individuals.

The two most popular types of population scheme are steady state reproduction and generational reproduction. In steady state reproduction, new individuals will replace their parents if they have higher fitness values. The total population size will thus remain constant. In generational reproduction, the current generation (the entire population) will generate offspring that are kept in a second population—the next generation. When the new generation comprises as many individuals as the parent generation, the parent generation as a whole is discarded and replaced with the new generation. Both methods have certain benefits, but some researchers claim that generational reproduction is superior to steady state reproduction (Davis, 1991).

A variation on generational reproduction is to implement an elitist strategy. In this case the best 5–10% of each generation is copied unchanged into the next generation. The rest of the generation is filled by letting offspring individuals replace parents. This usually has the effect of speeding up convergence in the GA, since good solutions are not lost to random changes.

4.   The Reproduction Scheme

The reproduction scheme determines the manner in which chromosomes reproduce to form new individuals. This is done through the application of genetic operators to chromosomes. The two most popular operators are point mutations and crossovers. During a point mutation operation, a single gene on a chromosome is randomly changed. During a crossover, two chromosomes are both divided into two segments that are then joined across to form two new individuals. The crossover point on each chromosome may be determined randomly, or the midpoint of each chromosome may be used.

Other operators include the insertion of random genes and the deletion of genes from a chromosome. The point on the chromosome where these operators will act is determined randomly. In these cases the length of the chromosome may vary. This type of GA is called a "messy" GA (Koza, 1992).

B. The Application of GA to Solvent Design

1. The Encoding Scheme

The first step in applying GA to solvent design, or any other CAMD problem, is to decide on an encoding scheme. Although binary strings would be perfectly adequate, experience shows that enumerated types are much easier to implement in CAMD problems. Using enumerated types entails assigning numbers to certain submolecular groups and then storing a list of these numbers instead of the binary string. Each of these groups represents a single gene. The structures of these groups are determined by the group contribution methods that are used.

For the design of solvents, the most important group contribution method used would be one of the versions of UNIFAC. Group contribution methods must then be found that use structures that are compatible with the UNIFAC groups. Joback's methods (Joback and Reid, 1987) may, for example, be used to estimate boiling and freezing points.

2. The Genes

Once the estimation methods have been chosen, the set of structural groups is fixed. A method must now be found to combine these groups into molecules. Since the genetic operators require that the genes combine linearly to form the chromosomes, the structural groups cannot be directly used as genes without discarding all those that do not have either one or two free bonds. This would also limit the algorithm to designing linear molecules. An effective solution to this problem is to construct genes from combinations of these structural groups.

Using these predefined genes has a number of advantages: it prevents most unrealistic and reactive combinations of functional groups. In methods like those discussed earlier, a great deal of work has gone into determining a set of rules for physical viability of candidate molecules. This is a very difficult task, and some of these rules have as many exceptions as cases in which they are applicable. Using predefined genes in linear combinations to build molecules largely circumvents this problem. It allows more complex molecules than simple linear combinations of UNIFAC groups. Genes of arbitrary complexity may be constructed. This removes most of the constrictions caused by the requirement of linear combinations and allows the mixing of aromatic and aliphatic groups. This was a problem in some of the algorithms published earlier, and as a rule aromatic and aliphatic solvents were designed in separate cycles of the algorithm. Although having predefined genes may be seen to limit the solution space, a sufficiently large set of genes will include most feasible combinations of the available structural groups.

The linear combination of genes into chromosomes requires that two sets of genes be defined. The first set will each have one free bond and will be used as the first and last genes in each chromosome. The second set will each have two free bonds and will be used to construct the middle part of each chromosome. Should cyclic molecules be desired, it is merely necessary to omit the starting and ending genes in the chromosome.

3. The Chromosomes

Once the genes have been defined, it is a simple matter to combine them into chromosomes. Because the solvents should be in the liquid phase, it is undesirable to waste processor time constructing huge molecules that will exist only in the solid phase at process temperatures. To prevent this, an arbitrary limit is placed on the length of a chromosome. It has been found that a maximum length of eight genes is more than adequate. This will however depend on the specific gene set that is used.

Each chromosome will consist of a start gene and an end gene, each with one free bond, and up to six middle genes with two free bonds each. This is illustrated in Fig. 1. To accommodate cyclic molecules in this structure, an empty gene is defined as a part of the start/end set. This also allows the molecular UNIFAC groups (e.g., H_2O) to be handled with the same data structures as the other chromosomes. In these cases an empty group will be used as the end gene, while the molecular UNIFAC group will serve as the start gene. No middle genes are used in these cases.

4. The Evaluation Scheme

As discussed earlier, the evaluation scheme entails the choice of fitness functions. The fitness function gives the algorithm an indication of the degree in which the candidate solvent fulfills the specifications set. These specifications were given in Section I. The more important of these may be summarized as follows:

> Two liquid phases must form.
> The selectivity of the solvent for one of the key components must be high.

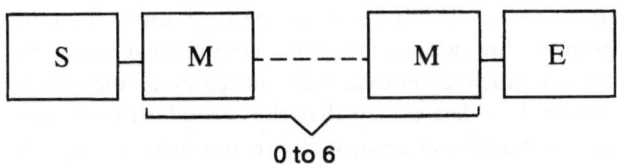

Figure 1 A typical chromosome.

The recovery of the solvent in the solvent phase must also be maximized. No unwanted azeotropes must form between the solvent and any of the feed components.

All these properties require the use of an activity coefficient model like UNIFAC. To determine whether two phases form, a flash calculation must be done. The selectivities and recoveries may be calculated from the results of the flash calculation. It is possible to check for the formation of azeotropes by calculating binary vapor–liquid equilibrium curves for each of the feed components with the solvent. Alternatively one may specify the minimum boiling point of the solvent to be far from that of the feed components. Species with widely differing boiling points tend not to form azeotropes.

Other specifications may require the use of additional estimation methods to calculate, for example, the density and surface tension of each of the liquid phases to determine the ease of separation.

Unfortunately, group contribution methods are not always very accurate. When the only method available to estimate a property is of dubious accuracy, it may be better not to specify requirements for that property. The final set of candidate solvents can then be tested experimentally against these requirements.

A very simple and efficient way to combine the multiple specifications into a single fitness value is to calculate a fitness for each property and use a weighted mean as the overall fitness of the chromosome. The weights assigned to each property will be determined by the specific problem under consideration.

All the foregoing requirements are one-sided; that is, for a candidate to be acceptable, a property value must be above or below a certain level. It would also be convenient to have fitness values within the interval $[0,1]$, since this would make calculation of the weighted mean simpler and thus faster. There are two types of function that would be well suited to all these requirements: a step function (or Boolean function) and a sigmoidal function.

For requirements like the formation of two liquid phases, the Boolean function would be perfectly suited, since the mixture either splits into to two phases (at the process temperatures) or does not. There is no middle ground. For other properties, such as selectivity, choosing the function type is not so simple.

Suppose a minimum value for the selectivity is specified and a Boolean function is used as the property fitness function. Now consider two chromosomes. The first has a selectivity much lower than the required value; the other's is just below the requirement. It is fair to say that the second chromosome must contain genetic material (functional groups) that could help accomplish the desired separation, whereas the first probably does not. Unfortunately, the Boolean fitness function will assign the same low fitness to each chromosome.

The desired genetic material of the second chromosome will have no better chance of being incorporated into the next generation than the undesired genetic material of the first chromosome. This failure to identify desired traits in candidates could prove detrimental to the algorithm.

Consider now the same situation, but with a sigmoidal fitness function instead. The second chromosome will now have a high fitness than the first, because its selectivity is higher. Its desired genetic material will thus have a better chance of surviving into the next generation. This allows the algorithm to identify favorable traits in candidates and concentrate these traits in future generations.

In Eq. (1), an example of a property fitness function that may be used,

$$F = \frac{1}{1 + \exp[-\beta(P_i - P_r)/P_r]} \tag{1}$$

F is the property fitness, P_i the property value for the ith chromosome, P_r the required value for the property, and β a gradient parameter. This function has the following properties (see Fig. 2):

At the required value, the fitness is 0.5.
For values higher than the requirement, the fitness approaches unity.
For values lower than the requirement, the fitness approaches zero.
The slope of the function may be controlled with the β parameter.
The function is smooth and continuous.

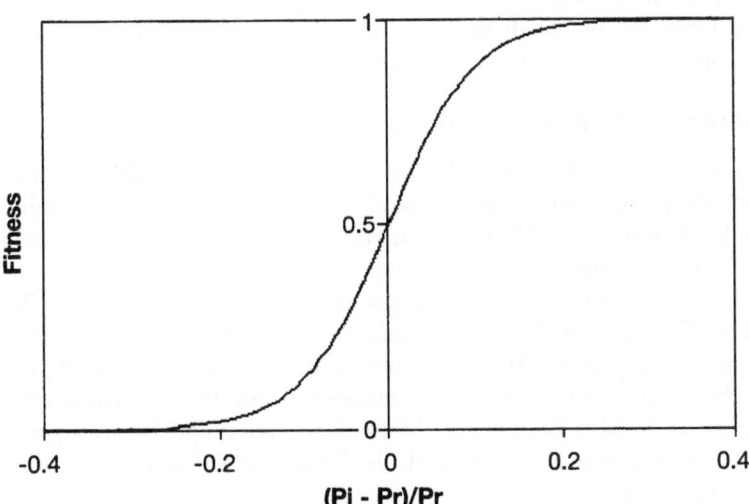

Figure 2 The sigmoidal fitness function.

5. The Population Scheme

In most applications of GA the initial population is random. As stated earlier, generational reproduction is considered to be superior to steady state reproduction. An elitist policy is also implemented to help speed conversion by preventing the loss of good candidates to random changes. A policy in which the best 10% of candidates are retained has been found to give good results.

6. The Reproduction Scheme

All the operators discussed earlier (point mutation, crossover, insertion, and deletion) are applicable to CAMD problems. The use of the insertion and deletion operators is limited by the restrictions in the structure of a chromosome. Start and end genes may not be inserted or deleted. The maximum number of middle genes is also fixed.

When the crossover operator is applied, two new chromosomes are created. This is allowed only if there is enough space for both in the generation that is being constructed. The chromosomes are cut as close as possible to their centers to prevent complications with the length restrictions placed on the chromosomes. The exact crossover point is determined randomly within an interval, around the center.

7. The Basic Algorithm

Once the evaluation, encoding, population, and reproduction schemes have been decided upon, the algorithm may be implemented. The basic GA is summarized in Table 11. Although this algorithm will give good results in most cases, there are a number of ways in which it can be improved.

8. Missing Interaction Parameters

One of the primary problems experienced in CAMD is the availability of parameters for the group contribution methods that are used. A good example of this is the large number of UNIFAC interaction parameters that are not yet available in the literature.

In some publications these missing parameters are taken to be zero (Reid et al., 1987). This often leads to gross errors in the calculations for selectivity, recoveries, and phase splits. Another option would be to eliminate all chromosomes that require these missing parameters. This method could cause "good" genetic material to be eliminated if interactions required by other groups in the same chromosome were not available. This would slow convergence in the evolutionary process.

A less harsh method of solving the problem of missing interactions is to penalize chromosomes for each missing interaction by subtracting a small

Table 11 The Basic Algorithm

---

1. Initialize a population of chromosomes.
2. Evaluate each chromosome in the population.
   2.1. Estimate the properties.
   2.2. Calculate the property fitness values.
   2.3. Calculate the global fitness.
3. Choose best 10% and copy them to the new generation.
4. Create new chromosomes.
   4.1. Choose an operator.
   4.2. Choose parent chromosome(s).
   4.3. Apply the operator to the parent chromosome(s).
5. Copy the new chromosomes to the new generation until this generation has been filled.
6. Replace the current generation with the new generation.
7. If time is up, go to step 8, else repeat from step 2.
8. Evaluate each chromosome in the population.
9. Return the best solution.

---

amount from its fitness. This will cause all groups with missing interaction parameters to be eliminated through the process of natural selection, usually within a few generations. This allows the "good" genes also contained in these "bad" chromosomes some time to be replicated into other individuals that do not require missing interaction parameters.

9. Biased Gene Selection

In the state-of-the-art section of this chapter (Section III), we discussed a number of heuristic methods for broad screening. These methods are based on years of experience that should not be discarded lightly. It is possible to incorporate these heuristics into the GA by biasing gene selection toward functional groups that are known through experience to perform well. This will in many cases speed up convergence by allowing the algorithm to spend more time in the areas of the solution space that are suspected to contain the optimal solutions.

The Robbins (Robbins, 1980) chart in particular can very easily be incorporated into a GA. Robbins lists 12 functional groups and the deviation from Raoult's law expected for binary mixtures containing one of these functional groups in each species. If the genes used in the GA are also classified according to the Robbins categories into which their constituent functional groups fall, it is a trivial matter to find the expected deviations from Table 1. For each deviation in the desired direction, the selection probability of the gene is increased and for each deviation in the other direction, it is decreased.

The use of selection probabilities for the functional groups also enables the user to "switch off" any of the genes by setting its selection probability to zero. If, for example, a solvent is required to separate a mixture containing amines, genes containing carbonyl groups should not be used because the solvent might react with the mixture components.

10. Physical Realism

Although the use of predefined genes eliminates most unrealistic combinations of functional groups, it is still possible that certain combinations of genes may lead to molecules that are not physically viable. To check for such combinations, a set of rules similar to those defined by Joback and Stephanopoulos (Table 4) or Gani et al. (Table 9) may be used. The number of rules and their nature will be determined by the gene set that is used.

As is the case with missing UNIFAC interactions, chromosomes that violate any of these rules should penalized by a small subtraction from their fitness values for each violation. These combinations will then be eliminated through natural selection.

11. Fine-Tuning the Parameters

Like all numerical methods, genetic algorithms have certain parameters that influence their performance. While the performance of an ODE integrator could be adjusted by setting the step size, a GA is tuned by setting the selection probabilities of its operators. There are two ways in which this could be done: through a sensitivity analysis or through auto-tuning implemented in the algorithm.

The sensitivity analysis is done by running the algorithm with a test case, using different values for the various parameters. Through this method an optimal set of parameters may be found for both speed of convergence and the fitness of the final generation. The effect of these parameters on the performance of the algorithm can be seen in Fig. 3. The different sets of parameters used in this test are listed in Table 12, which shows that parameter set D not only caused the fastest initial climb in the fitness but also produced the best individual after 40 generations.

When a sensitivity analysis is used to determine the parameter set, it is important that the test case be representative of the type of system the algorithm will be applied to. The optimal parameters for finding solvents for the separation of acetone and methanol through extractive distillation are not necessarily the optimal parameters when a solvent is required to separate ethanol from water with liquid–liquid extraction.

A more flexible method for finding the optimal parameter set is by implementing auto-tuning of the parameters. Whenever a chromosome is created

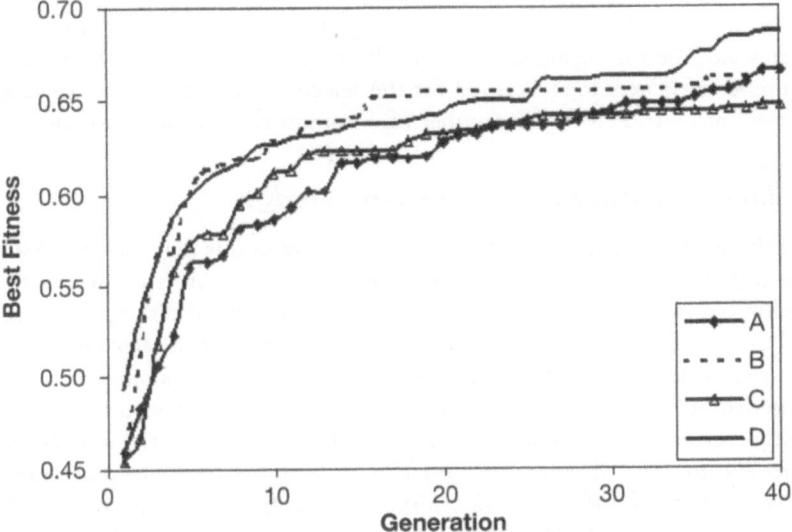

Figure 3 The performance of the algorithm with different parameters.

during the evolution process, the tuning algorithm must note which operator was used and also the fitness of the parent chromosome. After the new generation has been evaluated, the fitness of each chromosome is compared with that of its parent. If the new chromosome has a higher fitness, the selection probability of the operator that created it is increased by a small amount. If the new chromosome has a lower fitness than its parent, the operator's selection probability is decreased. The overall effect of auto-tuning is to find the optimal parameter set for the specific problem "on the fly." Best results are achieved if the initial parameter set does not differ greatly from the optimal set. The sen-

Table 12 The Different Parameters for the Sensitivity Analysis

Parameter	A	B	C	D
β	10	20	10	20
Point mutation, %	35	35	40	40
Crossover, %	30	30	20	20
Insert, %	17.5	17.5	20	20
Delete, %	17.5	17.5	20	20

sitivity analysis method described above may be used to find a good initial parameter set. The tuning algorithm must also be given enough time to execute. The number of generations required for the selection probabilities to converge is inversely proportional to the number of chromosomes in the population.

12. Incorporating the User's Knowledge

One of the primary advantages of the interactive method is that the knowledge and experience of the user may be used to design the solvent. It is also possible to employ the user's knowledge to aid the GA. This is done through seeding.

If a solvent is known or suspected to perform well for the separation that is to be performed, it may be seeded into the initial random population. Since these seeded solvents will usually have much higher fitness values than the randomly created chromosomes that make up the rest of the initial population, their genetic material is sure to be used in at least the first few generations. The seeding process allows the algorithm to identify interesting areas of the solution space at an early stage. More time will then be spent exploring these areas.

13. Evolving Fitness Function

Although the sigmoidal fitness function is well suited to most CAMD problems, there is at least one disadvantage to using it. If the required value for a property is high, it often happens that none of the chromosomes in the initial, random population come close to this requirement. Since the fitness function is flat in both extremes, there will be very little difference in the property fitness values of chromosomes that lie at respectively -10 and -100 on the abscissa (see Fig. 2). In these cases the sigmoidal fitness function suffers from the same problems as the Boolean fitness function. The result of this problem is that the algorithm will find it very difficult to identify better solutions and to use their genetic material to improve the next generation. It will in fact become dependent on random mutations to find better solutions to work with. This situation is very unsatisfactory. Fortunately there is a simple solution to the problem: evolving fitness functions.

Instead of using the specified value of the property, P_t in Eq. (1), to calculate the property fitness, a lower value is initially used. This moves the middle of the sigmoidal curve toward the area occupied by the initial population, where the curve is much steeper, making it easier to identify the better individuals. This adjusted requirement is then increased with subsequent generations until it is equal to the requirement specified by the user. The biological equivalent of this process is the change in the environment with time and the continuous adaptation of species to suit this changing environment.

V. A CASE STUDY IN EVOLUTIONARY SOLVENT DESIGN

To illustrate the power of CAMD in process design, we will discuss an application of solvent design using SolvGen, an implementation of the genetic algorithm (Van Dyk, 1998). The gene set used by SolvGen consists of 531 start/end genes and 368 middle genes. Each gene may consist of up to six types of UNIFAC group, and each chromosome may contain up to eight genes. This gives a total solution space of the order of 10^{20} molecules. Each generation contains 10,000 chromosomes.

A problem of some industrial importance is the separation of phenols from so-called neutral oils. In practice such mixtures would comprise dozens of species. To simplify matters, only a mixture of phenol and aniline will be considered. The object of the case study is to find a solvent to recover phenol with a high selectivity and recovery.

The use of co-solvents and anti-solvents is a common practice for liquid–liquid extraction processes. In this case water and hexane will be used as the co-solvent and anti-solvent, respectively. A water-to-feed ratio of 1:1 and a hexane-to-feed ratio of 4:1 were decided upon. These ratios are mass-based. The composition of the feed stream is listed in Table 13.

The GA solvent design algorithm was allowed to run for 15 generations (~10 min). The mole fraction of solvent was set at 0.15 with a temperature of 313 K. Some of the chromosomes of the final generation are listed in Table 14, along with the predicted selectivity for each of these. These values were calculated by using the modified UNIFAC (Dortmund) model. The selectivity was calculated according to Eq. (2).

$$\alpha_{ij} = \frac{[x_i^1/x_i^2]}{[x_j^1/x_j^2]} \tag{2}$$

with α_{ij} being the selectivity for component i over component j, x_i^1 the mass

Table 13 The Feed Composition (Combined with Water and Hexane)

Component	Mass fraction	Mole fraction
Phenol	0.148	0.084
Aniline	0.019	0.011
Water	0.167	0.492
Hexane	0.666	0.413

Table 14 Some Candidate Solvents

Solvent structure	Selectivity $\alpha_{phenol/aniline}$
7-Hydroxo-2,5,9,12,15-pentaoxahexadecane	8.32
1,5-Dihydroxo-3,6,9,12-tetraoxatetradecane	8.45
1-Hydroxo-3,6,9,12-tetraoxatridecane	7.65
1,5-Dihydroxo-3,7,10-trioxaundecane	8.18

or mole fraction of component i in phase 1, x_i^2 the mass or mole fraction of component i in phase 2, and similarly for component j.

The resulting optimal solvents are all very similar to polyethylene glycols that are readily procured. Thus they will be much cheaper than solvents that must be specially synthesized.

Using the interactive design mode of the SolvGen, we find that the predicted selectivity for triethylene glycol, using the given solvent mole fraction, is 7.475, for tetraethylene glycol it is 7.840, and for pentaethylene glycol it is 8.141.

In a detailed study of the separation of phenolic compounds from neutral oils conducted by Venter and Nieuwoudt (1998), more complex feed streams were used, to accurately represent the compositions of industrial feed streams. The authors reported that the polyethylene glycols, combined with water and hexane, are indeed very effective solvents for this separation. Triethylene glycol is recommended as the preferred solvent, based on their experimental results, some of which are given in Fig. 4. The selectivities graphed in this figure were also calculated according to Eq. (2). The feed composition for these equilibrium measurements is given in Table 15.

Table 15 Feed for Liquid–Liquid Equilibrium

Component	Mass fractions
Phenol	0.710
Aniline	0.067
Benzonitrile	0.067
Mesitylene	0.090
Ethyl methyl pyridine	0.067
	Mass ratios
Hexane/feed	5 : 1
Water/solvent	0.3 : 1

Source: Venter and Nieuwoudt (1999).

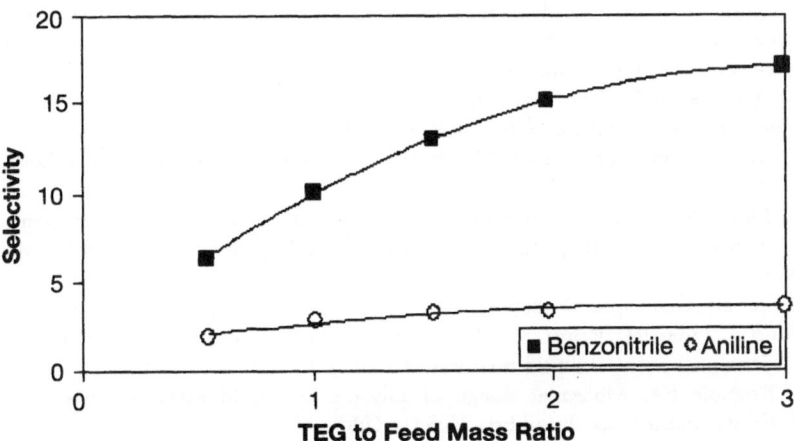

Figure 4 Selectivity of triethylene glycol (TEG) for phenol over neutral oils at 40°C. (From Venter and Nieuwoudt, 1999.)

VI. CONCLUSIONS

Computer-aided molecular design (CAMD) is a very powerful tool for process designers. As our case study showed, effective solvents for liquid–liquid extraction processes may be found within minutes. Very little experience or prior knowledge is required of the designer.

Nor are the algorithms limited to designing solvents. The number of possible applications is nearly limitless, including, for example, the design of polymers and refrigerants. As computers rapidly become more powerful and affordable, computer-based methods such as these will become part of the standard toolbox of all design engineers.

The common weakness in the CAMD algorithms discussed here is the accuracy of the group contribution methods they depend upon. Many researchers are working to improve the existing methods and develop group contribution models for other properties. As these efforts gain momentum, the accuracy and applicability of CAMD will increase simultaneously.

REFERENCES

Barton AFM. CRC Handbook of Solubility Parameters and Other Cohesion Parameters. Boca Raton, FL: CRC Press, 1991.

Berg L. Selecting the agent for distillation processes. Chem Eng Prog 65(9):52, 1969.

Brignole EA, et al. A strategy for the design and selection of solvents for separation processes. Fluid Phase Equilibria 29:125, 1986.

Churi N, Achenie LEK. Novel mathematical programming model for computer aided molecular design. Ind Eng Chem Res 35:3788, 1996.

Davis L, ed. Handbook of Genetic Algorithms. New York: Van Nostrand Reinhold, 1991.

Derr EL, Dean CH. Analytical solution of groups: Correlation of activity coefficients through structural group parameters. Inst Chem Eng Symp Ser (Lond) 3:40, 1969.

Ewell RH, et al. Azeotropic distillation. Ind Eng Chem 36:871, 1944.

Fredenslund A, Jones RL, Prausnitz JM. Group-contribution estimation of activity coefficients in non-ideal liquid mixtures. AIChE J 21(6):1086, 1975.

Gani R, Brignole EA. Molecular design of solvents for liquid extraction based on UNIFAC. Fluid Phase Equilibria 13:331, 1983.

Gani R, et al. A group contribution approach to computer-aided molecular design. AIChE J 37(9):1318, 1991.

Gilmont R, et al. Correlation and prediction of binary vapour–liquid equilibria. Ind Eng Chem 53:223, 1961.

Hanson CM. In: RW Tess, ed. Solvent Selection by Computer in Solvents: Theory and Practice. Washington, DC: American Chemical Society, 1973.

Holland JH. Adaptation in Natural and Artificial Systems. Ann Arbor: University of Michigan Press, 1975.

Joback KG, Reid RC. Estimation of pure-component properties from group-contributions. Chem Eng Commun 57:223, 1987.

Joback, KG, Stephanopoulos G. Designing molecules possessing desired physical property values. Presented at FOCAPD'89, Snowmass CO, 1989, 383.

Koza JR. Genetic Programming: On the Programming of Computers by Means of Natural Selection. Cambridge, MA: MIT Press, 1992.

Macchietto S, et al. Design of optimal solvents for liquid–liquid extraction and gas absorption processes. Trans. Inst Chem Eng 68(A):429, 1990.

Maranas CD. Optimal computer-aided molecular design: A polymer design case study. Ind Eng Chem Res 35:3403, 1996.

Naser SF, Fournier RL. A system for the design of an optimum liquid–liquid extractant molecule. Comput Chem Eng 15(6):397, 1991.

Nielsen B, et al. Computer-aided molecular design by group contribution method. Comput Appl Chem Eng 227, 1990.

Prausnitz JM, Anderson R. Thermodynamics of solvent selectivity in extractive distillation of hydrocarbons. AIChE J 7:96, 1961.

Reid RC, Prausnitz JM, Poling BE. The Properties of Gasses and Liquids, 4th ed. New York: McGraw-Hill, 1987.

Robbins LA. Liquid–liquid extraction: A pre-treatment process for wastewater. Chem Eng Prog 76(10):58, 1980.

Scheibel EG. Principles of extractive distillation. Chem Eng Prog 44(12):927, 1948.

Seader JD, et al. Distillation. In: RH Perry and DW Green, eds. Perry's Chemical Engineer's Handbook, 7th ed. New York: McGraw-Hill, 1997.

Vaidyanathan R, El-Halwagi M. Computer-aided synthesis of polymers and blends with target properties. Ind Eng Chem Res 35:627, 1996.

van Dyk B. The design of solvents for extractive distillation. MSc Ing thesis, University of Stellenbosch, Stellenbosch, South Africa, 1998.

van Dyk B, Nieuwoudt I. Design of solvents for extractive distillation. Ind Eng Chem Res 39:1423, 2000.

Venkatasubramanian V, et al. Computer-aided molecular design using genetic algorithms. Comput Chem Eng 18(9):833, 1994.

Venter DL, Nieuwoudt I. The separation of *m*-cresol from neutral oils with liquid–liquid extraction. Ind Eng Chem Res 37:4099, 1998.

Venter DL, Nieuwoudt I. The separation of phenolic compounds and neutral oils, International Solvents Extraction Conference, Barcelona, Spain, July 1999.

6

Extraction Technology for the Separation of Optical Isomers

André B. de Haan

Twente University, Enschede, The Netherlands

Béla Simándi

Budapest University of Technology and Economics, Budapest, Hungary

I. INTRODUCTION

Chirality is a key characteristic of all molecules with biological relevance. Very often the two mirror images (enantiomers) of a molecule will show different interactions with a biological receptor or enzyme, hence giving rise to different biological effects. As a result, in many cases only a single enantiomer is responsible for the desired biological activity of a compound. In such cases the other enantiomer represents a 50% impurity that often leads to unwanted (harmful) side effects or, in the case of agrochemicals, to environmental pollution. Well-known examples are anti-inflammatory drugs such as ibuprofen (**1**), ketoprofen, and flurbiprofen, whose anti-inflammatory activity predominantly resides in the (*S*)-iosomers. In food applications (*S*)-limonene (**2**) has a lemon odor, the (*R*)-isomer an orange one; (*R*)-asparagine (**3**) has a sweet taste, while the (*S*)-form has an insipid taste [1−5].

(**1**)	(**2**)	(**3**)
Ibuprofen	Limonene	Asparagine

255

Historically, synthetic products for application in medicine, agriculture, and the food industry were developed and marketed as racemates (50:50 mixtures of enantiomers) because suitable single-isomer manufacturing technology was not available. This is the main reason that most of the drugs, agrochemicals, flavors, and fragrances used nowadays are marketed as racemic mixtures [5–8]. In the past decade, however, the pressure on companies to develop and produce enantiomerically pure chiral products increased dramatically. Evidence for this increased pressure is the recent introduction of increasingly rigorous regulations by several regulatory bodies [9–11] and increased public concern over health and environmental problems. This trend is supported by the fact that of the synthetic chiral drugs now in the development pipeline, 40% are developed in the single enantiomer form, in comparison to only 10% of the commercially available drugs [5–8]. As a result, the efficient and economical production of enantiomerically pure chiral products has become one of the major challenges facing the modern chemical, pharmaceutical, agrochemical, and flavor and fragrance industries.

II. CURRENT CHIRAL MANUFACTURING AND SEPARATION TECHNOLOGY

There are two main routes to the production of optical pure enantiomers: resolutions and stereoselective syntheses. Asymmetric synthesis involves the conversion of an achiral starting material into a chiral product by stereoselective reaction with a chiral reagent or catalyst. In principle, asymmetric synthesis is the most cost-effective method for producing single-enantiomer products, because the entire precursor is converted to the desired enantiomer. Despite this obvious advantage, stereoselective synthesis can be particularly difficult because the stringent requirements for enantiomeric purity require exceptional enantioselectivity (>100) from the catalyst system, which is often difficult to achieve and maintain throughout the manufacturing process [12]. Therefore, many of the developed asymmetric synthesis routes require a final purification step to remove the undesired enantiomer from the final product.

Because of this inherent limitation of asymmetric synthesis, resolution methods are currently used for the production of the majority of the synthetic single enantiomer products. In a resolution process, a racemic mixture is synthesized and separated into its two enantiomers. The popularity of the resolution route is highlighted by the fact that 65% of all synthetic enantiomeric drugs are made by resolution of racemic drugs or their intermediates [6]. Resolution techniques use the competitive advantage of the relative straightforwardness of synthesizing a racemic mixture that can always be resolved into a product with the desired enantiomeric purity. The main disadvantage of resolution methods is the limitation to 50% yield unless they are combined with a racemization operation or asymmetric synthesis.

The industrially applied resolution methods can be divided into three categories: selective crystallization, selective conversion, and chromatography [1–5]. Diastereomeric salt crystallization is widely applied as a cost-effective way to separate enantiomers. However, it is relatively inflexible and requires detailed investigation of system behavior before use. In addition, the cost of the chiral crystallization agents can be high, depending on their availability. Conversion of one of the enantiomers from a racemic mixture by an enantio-selective chemical or biological (enzyme) catalyst is another way to create a difference in molecular properties that allows resolution of the racemate. The required selective catalyst makes selective conversion suffer from the same dis-advantages as those of asymmetric synthesis, as well as requiring additional unit operations for the separation, back-conversion of the modified enantiomer, and racemization of the undesired enantiomer.

In the past 10 years the application of preparative chromatography for the resolution of pharmaceutical end products has become the chief competi-tion to these classical methods [13–15]. This development was particularly stimulated by the introduction of the simulated moving bed concept and the increased availability of less expensive chiral stationary phases. For pharmaceu-tical companies, the main advantages driving the application of preparative chromatography to pharmaceutical end products are the ability to scale up directly from analytical conditions and the capability of preparing larger amounts of testing material for clinical tests on short notice. For larger-scale production however, although chromatography provides highly pure products, it still suffers from low productivity, high capital costs, and a significant dilution of the product, requiring additional operations for solvent removal and recycling.

Several attempts have been made to develop new competing technologies for industrial-scale chiral separations based on alternative technological concepts such as extraction (liquid, supercritical, aqueous two-phase), liquid membranes, functionalized membranes [16–22], and selective complexation to large sub-strates combined with ultrafiltration. Of these alternatives, extraction-based technologies may become an industrially attractive option. This is mainly be-cause, although the selectivity of a single extraction is usually small, staging of multiple extractions is easily achieved and well established. In addition, the capacity of extraction is often much greater than that of chromatography, which is also difficult to operate in a continuous way. Countercurrent and fraction-ating extraction processes are easily operated continuously. In this chapter we explore recent developments in various extraction technologies for the separa-tion of chiral components.

III. PRINCIPLES OF ENANTIOSELECTIVE EXTRACTION

A. Introduction

Dashkevic [23] was the first to demonstrate the principle of enantioselective extraction. He showed that selective partitioning of ($\pm$)-mandelic acid enantiomers (4) between two liquids requires the incorporation of enantioselective recognition in one or both of the liquid phases. Enantioselectivity originates from bonding interactions between the racemate and the resolving medium that are more favorable for one enantiomer than the other through geometrical configuration differences. Although chiral solvents are the simplest way to achieve enantioselective recognition, most methods make use of a chiral additive (chiral selector, carrier, reagent) that is added to an achiral solvent. An obvious advantage of these systems is the possibility of tailoring the selector molecules to the applications. Chiral separation is achieved by the formation of diastereomeric complexes (host–guest, HG) between the chiral selector (host, H) and racemic solute (guest, G):

$$H(+) + G(+) \overset{K^+}{\Longleftrightarrow} HG(+, +) \tag{1}$$

$$H(+) + G(-) \overset{K^-}{\Longleftrightarrow} HG(+, -) \tag{2}$$

with

$$K^+ = \frac{[HG(+, +)]}{[H(+)][G(+)]} \quad \text{and} \quad K^- = \frac{[HG(+, -)]}{[H(+)][G(-)]} \tag{3}$$

(4)
Mandelic acid

Enantioselectivity, β, arises from the difference in the equilibrium constants K^+ and K^- of both diastereomeric complexation reactions and is expressed as follows:

$$\ln \beta = \ln \frac{K^+}{K^-} = -\frac{\Delta\Delta G}{RT} \tag{4}$$

where $\Delta\Delta G$ is the difference in the Gibbs energy of complexation for both diastereomeric complexation reactions. The obtained enantiomeric purity is often expressed as enantiomeric excess, ee, which is defined as the excess of the predominant enantiomer expressed as a percentage:

$$ee = \frac{(\text{predominant enantiomer}) - (\text{minor enantiomer})}{(\text{predominant enantiomer}) + (\text{minor enantiomer})} \times 100 \qquad (5)$$

For a racemic mixture, the ee can be directly related to the selectivity:

$$ee = \frac{\beta - 1}{\beta + 1} \times 100 \qquad (6)$$

Chiral discrimination requires at least three points of interaction between the selector and one of the guest enantiomers, of which at least one must be stereo-selective [24,25]. This so-called three-point model is schematically illustrated in Fig. 1. In this model, the three interaction points of one enantiomer G-(+) cannot be situated in the same locations as the corresponding three points of the other enantiomer G-(−) and therefore cannot be recognized in the same way. The forces between the contact points may be based on multiple inter-molecular host–guest interactions, including the following:

Hydrogen bonding and/or dipole–dipole interaction
$\pi-\pi$ Interactions
Hydrophobic van der Waals–type interactions
Steric attraction/repulsion

These interactions may either be attractive or repulsive. Two attractive and one repulsive interaction (e.g., by steric hindrance) can also be used to achieve chiral recognition [26]. The simplest way to divide host compounds is to distinguish them by their size and the shape of their binding sites. The first class consists of relatively small molecules with convex binding sites that interact with a limited portion of the guest surface (complexation). The second class consists of relatively large molecules that typically bind a guest by encapsulating it within a cleft or pocket (inclusion). Their interaction with each other is usually at many different points over the surface of the guest.

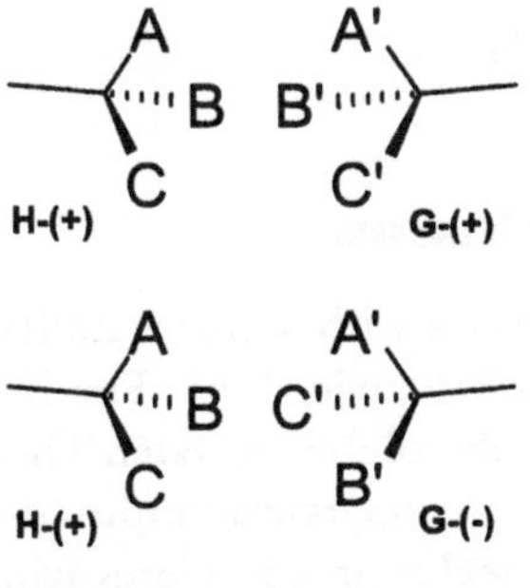

Figure 1 Schematic of three-point interaction model.

B. Chiral Separation Based on Complexation (Chiral Solvating Agents)

Chiral solvating agents were originally developed to determine the purity and absolute configurations of enantiomers by NMR spectrometry [25]. Addition of a chiral solvating agent or lanthanide chiral shift agent separates the NMR signals of the target enantiomers by the formation of temporary diastereomeric complexes. Among the most widely employed chiral solvating agents are aryl-trifluoromethylcarbinols and 1-arylethylamines. Both agents bind to the solutes via two-point hydrogen bonding and rely on the proximity of the aryl group to induce magnetic anisotropy. Aryltrifluoromethylcarbinols have been found to be effective for solutes such as carboxylic esters, lactones, ethers, and aryl-amines. 1-Arylethylamines have been found to be effective for solutes such as alcohols, carboxylic acids, and amides.

In extraction studies, most of the selectors used are based on derivatives from naturally occurring optically active compounds. Examples are alkyl esters of tartaric acid (5) [27–29] and various amino acids such as L-proline and L-hydroxyproline (6) [30]. A relatively long hydrocarbon tail is attached to make them soluble in the solvent phase. To improve enantioselectivity, the selectors are often employed as transition metal complexes with Cu(II), Co(II), Ni(II), or various lanthanides [30–33]. The main function of the metal ion is to create an optically active complex that can be used for enantioselective extraction through ligand exchange. The enantioselectivity of these selectors originates mainly from the chirality of the chosen starting compound but is enhanced through the three-dimensional structure of the transition metal complex.

(5)
Dialkyl tartrate

(6)
N-n-Alkylhydroxyproline

C. Chiral Separations Based on Inclusion

Cyclodextrins (Fig. 2) are cyclic, natural oligosaccharides with a shape similar to a truncated cone, containing a relatively hydrophobic cavity [34]. The hydrophilic nature of the outside surface makes them easily soluble in water. They are composed of six, seven, or eight optically active glucopyranose units (α-, β-, or γ-cyclodextrins) and have the ability to form inclusion complexes with guest molecules. When an optically active guest is included in a cyclodextrin

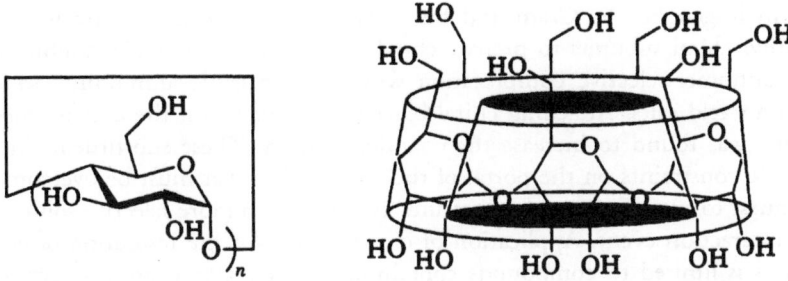

Figure 2 Schematic of cyclodextrin structure. (Reproduced with permission from Ref. 36.)

cavity, the strength of complex formation and chiral recognition is determined by a complex combination of hydrophobic interactions in the cavity, point interactions with the hydroxyl groups, and steric effects [26]. Cyclodextrins have found widespread application as the active component in HPLC and GC stationary phases and in capillary electrophoresis for the analytical separation of chiral components [6,35]. However, so far their use in extraction is limited to only a very few examples. Besides cyclodextrins, linear polysaccharides such as maltodextrins and maltooligosaccharides were also found to be useful as chiral selectors in electrophoresis [35]. Both are composed of α-(1,4)-linked D-glucose units. Although hydrogen bonds and dipole–dipole interactions are assumed to be the main interactions, the helical structure of dextrins may often provide additional enantioselectivity. Application of linear polysaccharides in extraction might also be effective in incorporating the required enantioselectivity into one of the two phases.

A new approach may become the use of calixarenes, which are the artificial analogs of natural cyclodextrins. They also form bucketlike structures with hydrophobic pockets that can form inclusion complexes. Although at present only little is known about their application for chiral separations, it can be anticipated that calixarenes may become a new class of efficient chiral selectors for various applications [36].

Another group of synthetic chiral resolution agents that exhibit stereoselective inclusion complexation are chiral crown ethers that consist of macrocyclic polyethers [37,38]. The main interactions are assumed to be hydrogen bonds between the three hydrogens attached to the nitrogen of the analyte and the dipoles of the oxygens of the macrocyclic ether. Chirality is introduced by attaching two halves of the macrocyclic ether ring to chiral side groups, such as tartaric acid derivatives and dinaphthyl compounds. Enantioselectivity is obtained by a combination of the strength and the high degree of directionality of the hydrogen bonds. Cram and his coworkers [37] were the first to use binaphthyl chiral subunits to prepare chiral active crown ethers (7) exhibiting high enantiomer-selective complexation with chiral organic ammonium salts and amino acid salts. Attaching chiral barrier substituents on the crown ether backbone was found to increase the enantioselectivity. These substituents impose steric constraints on the portal of the crown ether that must be overcome by the guest to enter or leave the host and also promote a more selective solute–carrier interaction event. Application of crown ethers for the resolution of enantiomers is limited to compounds containing primary amino groups such as amino acids, amino alcohols, and amines.

(7)
Binaphthyl crown ether

D. Other Chiral Selector Systems

Analogous to cyclodextrins, aqueous micellar aggregates are also able to provide a pseudophase with a hydrophobic interior in which chiral recognition of enantiomers may occur through properly oriented interactions. Chiral recognition by micelles has been shown possible by using either chiral surfactants (8) or comicellar systems in which chiral selectors are solubilized in the hydrophobic interior formed by achiral surfactants [39]. An illustrative example of micelles based on chiral surfactants is the use of anionic n-dodecanoyl-L-valine sodium salts (9) that appeared to be capable of resolving amino acid derivatives. The hydrogen bonding affinity of chiral amides incorporated into these surfactants appeared to be effective for the chiral recognition of enantiomers inside the hydrophobic micellar interior. On the other hand, the comicellar N-N-didecyl-L-alanine-copper(II) complex/achiral sodium dodecyl sulfate (SDS) surfactant system was found to be effective for the separation of N-dansyl-amino acids. Another example of the use of comicellar systems is the separation of amino acids by L-5-cholesterol glutamate–copper(II) complex solubilized by the achiral surfactant Serdox NNP [40]. The basis for separation in this system is the greater preference for binding D-phenylalanine than the L-counterpart.

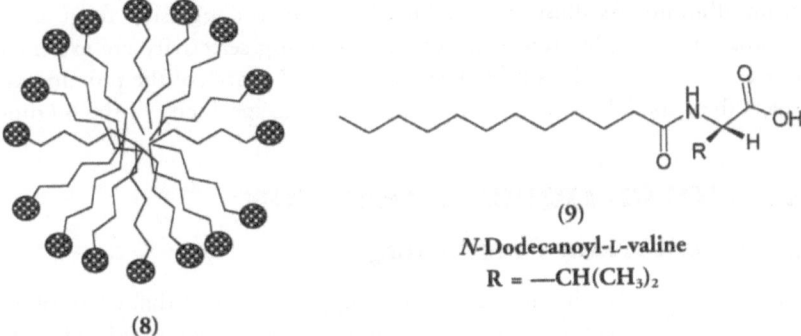

(8)
Micelle with chiral ligands

(9)
N-Dodecanoyl-L-valine
R = —CH(CH₃)₂

Dendrimers are hyperbranched macromolecules with a tailor-made structure that can be modified to enable chiral recognition by attaching chiral groups to the surface of the dendritic structure (**10**) and/or by introducing chiral branches in the dendritic matrix [41,42]. Both modifications are essentially based on the use of available optically active components. Various amino acids and chiral sugar units have been studied for surface modification. Chiral branches were introduced by incorporating tartrate units into the dendritic matrix. The obtained chiral surface and internal chiral cavities make chiral dendrimers in principle applicable to enantioselective complexation and/or encapsulation of guest molecules. Although not yet demonstrated, application of chiral dendrimers in extraction can be expected in the near future, especially because their large size enables easy separation and subsequent enantiomer recovery by, for instance, membrane filtration.

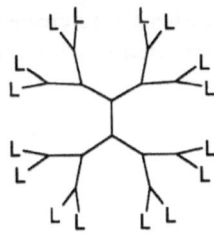

(10)
Dendrimer with chiral ligands

Proteins are widely employed in chromatography and capillary electrophoresis for analytical chiral separations [35]. Typical proteins used include bovine serum albumin, human serum albumin, cellulase, ovomucoid, and casein. Although their diverse nature should also make these substances useful as chiral selectors in extraction, this has been demonstrated only for the complexation of amino acid [43,44] and methamphetamine [45] enantiomers by bovine serum albumin. As illustrated by Fig. 3, the main conclusion from all studies was that the complexation constant and resulting selectivity are extremely dependent on the pH. This is because the electrical charge of the proteins strongly affects their spatial conformation and thereby the specificity of the binding site.

IV. LIQUID–LIQUID EXTRACTION

A. Equilibrium Partitioning

Dashkevic [23] was the first to demonstrate the principle that enantiomers can be separated by enantioselective partitioning between two liquids. He used several optically active organic solvents to selectively extract one of the enantiomers from racemic aqueous ($\pm$)-mandelic acid solutions. In most cases a partial resolution resulted. The most favorable separation factor, 1.055, was obtained with ($-$)-menthone. To our knowledge, Dashkevic [23] and Bowman [27] were the only ones to use an optically active solvent to obtain the desired enantioselectivity. In most other studies reported until now, the enantioselectivity is obtained by dissolving a suitable chiral selector in an organic solvent. In most cases an organic solvent with minimal physical solubility of the enantiomers is chosen. These selector solutions have until now mainly been used to study the resolution of racemates that possess two potential hydrogen bonding sites. This is not completely surprising, since two simultaneous hydrogen bond interactions can provide relatively high enantioselectivity. The main classes that have been studied are amino acids, amino alcohols, and organic acids. Of these three groups most of the reported work is related to the extractive resolution of amino acids. Takeuchi [30] used N-n-alkyl-L-proline–copper(II) complexes for the resolution of DL-leucine (**11**). The use of copper(II) is essential to create a selector that is capable of an enantioselective ligand exchange reaction with amino acids (AA):

$$AA_{aq} \leftrightarrow AA_{org} \tag{7}$$

$$AA_{org} + Cu-L_{2,org} \leftrightarrow AA-Cu-L_{org} + L_{org} \tag{8}$$

(11)
Leucine

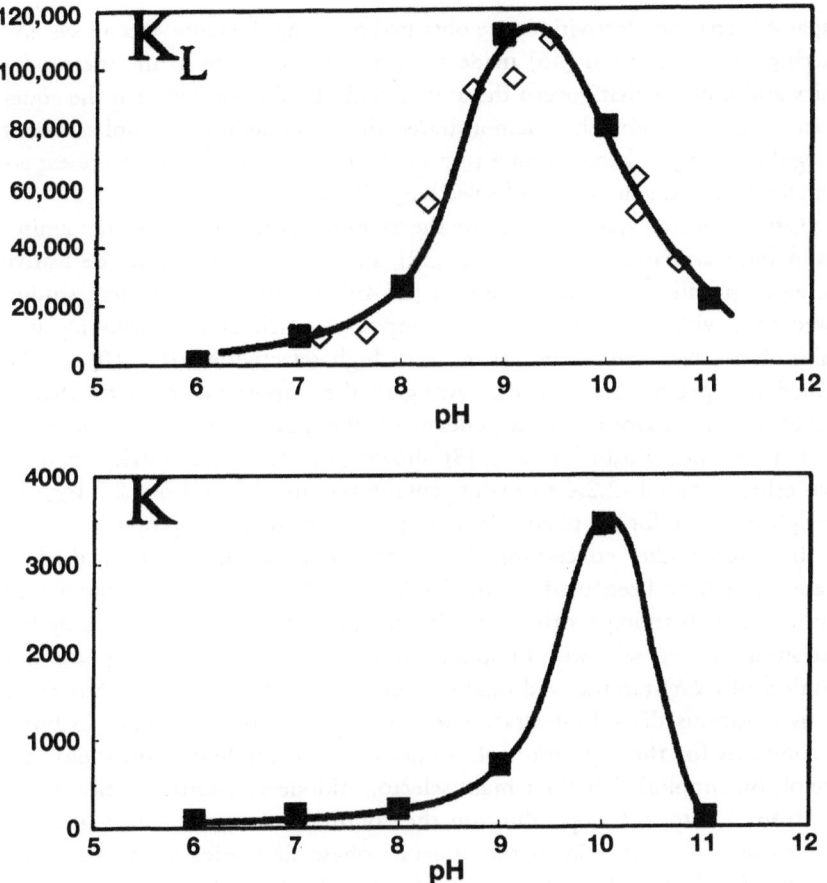

Figure 3 Influence of pH on the complexation constant of tryptophan with bovine serum albumin protein. (Reproduced with permission from Ref. 44.)

According to this mechanism, the overall distribution ratio is then determined by the combined solubility of the physically dissolved and complexed amino acid:

$$D = \frac{[AA]_{org} + [AA\text{---}Cu\text{---}L]_{org}}{[AA]_{aq}} \tag{9}$$

For DL-leucine the highest enantioselectivity (1.5–1.9) was obtained with *n*-alcohols as the solvent. The selectivity increased with increasing alkyl chain length until a constant value was obtained from eight carbon atoms and higher. In the case of other amino acids (valine, norvaline, norleucine, and isoleucine) comparable enantioselectivities were obtained by using the same selector system. Pickering and Chaudhuri [46] made a more detailed study of the thermodynamics and kinetics that govern the amino acid distribution between the aqueous and organic phase. They demonstrated that a semiempirical application of the regular solution theory is able to provide a good description of the experimentally observed solubility and selectivity effects.

Other selector systems used for the enantioseparation of racemic amino acids include lanthanide complexes [31,32], a carbomoylated quinine derivative [47], and optically active crown ethers [48,49]. Comparison of these studies showed that with inclusion complex compounds such as the carbomoylated quinine derivatives and crown ethers, very high selectivities ($\beta > 10$) can be obtained for specific amino acids. Owing to the importance of steric effects, however, the selectivity is very dependent on the amino acid side chain structure. For instance Kozbial et al. [48] showed the enantioselectivity in their crown ether–ethanol–2,2,4-trimethylpentane system to vary from 2.8 for DL-phenylglycine, 4.1 for DL-phenylalanine up to 11.6 for DL-tryptophan.

In most studies concerning the separation of amino alcohols, dialkyl-tartrate esters have been used as chiral selectors. Prelog et al. [28] found that α-amino alcohols (norephedrine, ephedrine, and others) can be resolved by the partitioning of their salts with lipophilic anions between an aqueous phase and a solution of (R,R)-tartaric acid dialkyl esters in 1,2-dichloroethane. Abe et al. [29] used various dialkyl-L-tartrate esters in combination with aqueous boric acid solutions for the separation of amino alcohols (pindolol, propanolol, alprenolol, bucumolol). For their main selector, didodecyl-L-tartrate, selectivities range from 1.6 to 2.9, depending on the conditions and the amino alcohol. The presence of boric acid in the aqueous phase dramatically enhanced the selectivity. As illustrated in Fig. 4, the selectivity for the extraction of pindolol (12) increased at lower temperatures [29]. This temperature effect originates from the stronger hydrogen bonding interactions between the amino alcohol and the selector at lower temperatures. Prelog [28] reached the same conclusion, obtaining a maximum selectivity around 1.7 for the resolution of ($\pm$)-norephedrine (13) at 4°C. The absence of selectivity for α-methylbenzylamine salts confirms that at least two hydrogen bond interactions are necessary for enantioselective differentiation of amino alcohols with dialkyl-L-tartrate esters. In these systems the highest distribution coefficients are obtained with polar solvents such as alcohols. The length of the ester alkyl moiety appeared to have only limited effect on the selectivity. As shown in Fig. 5 for pindolol, the selectivity reached an approximately constant value for chain lengths exceeding two carbon atoms. The same conclusion was reached by Prelog [28] for the separation of ($\pm$)-norephedrine. Prelog showed that although the chain length and structure of the alcohol in the tartaric acid ester has a significant influence

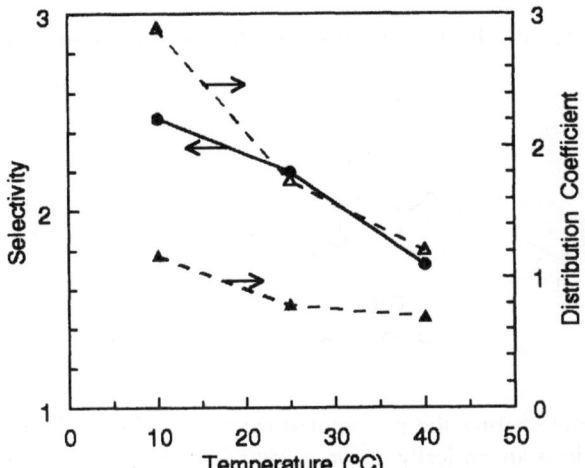

Figure 4   Effect of temperature on the extraction of (±)-pindolol. (Data from Ref. 29.)

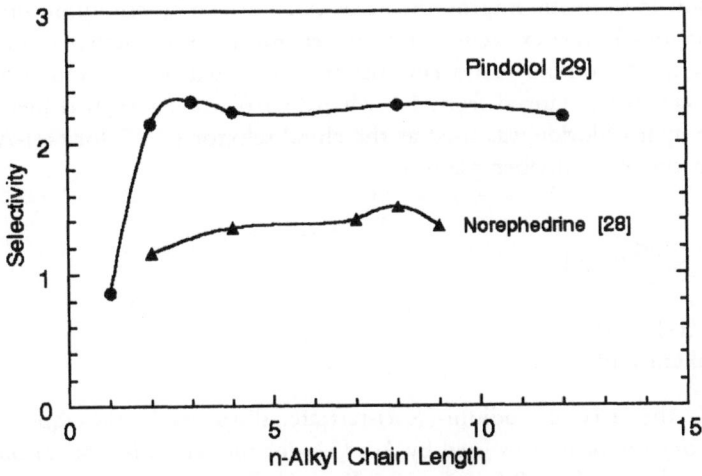

Figure 5   Effect of tartrate ester alkyl chain length on the selectivity of (±)-pindolol and (±)-norephedrine extraction. (Data from Refs. 28 and 29.)

on the distribution coefficient, the selectivity is affected to only a minimal extent.

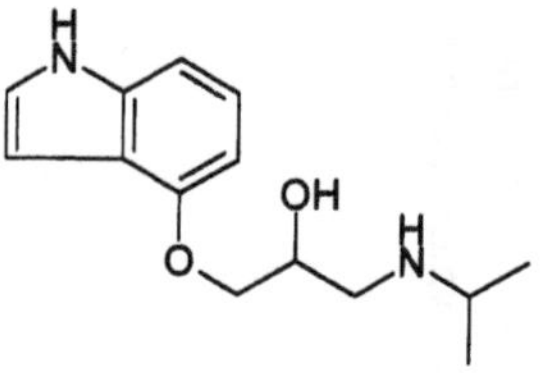

(12)
Pindolol

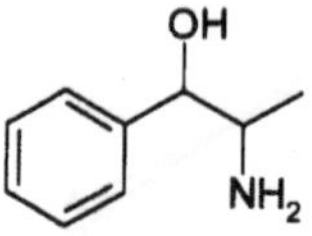

(13)
Norephedrine

The first study demonstrating the preferential extraction of organic acid enantiomers from water into an optically active solvent is that of Bowman et al. [27]. They proposed that the selective extraction of L-camphoric acid (14) by D-diisoamyl tartrate and D-diisopropyl tartrate is related to the formation of hydrogen bonds between the selector and chiral solute molecules. Again, the highest selectivities were obtained with racemates that are capable of forming two hydrogen bonds. The most efficient resolution was obtained with D-diisoamyl tartrate because the very low water solubility in the selector phase strongly reduces the nonselective physical dissolution of DL-camphoric acid. Very important is their finding that adding L-tartaric acid to the aqueous phase results in a significant improvement of the resolution of camphoric acid. This result indicates that the combined use of counteracting selectors in the two phases can result in dramatic improvements of extractive chiral resolution processes. The enantioselective extraction from water into trichloromethane of an organic acid salt, (±)-sodium mandelate, has been reported by Romano et al. [50]. Typical selectivities ranged from 1.1 when (−)-*N*-(*n*-octadecyl)-α-methylbenzylamine hydrochloride was used as the chiral selector to 1.3 for (−)-*N*-(1-naphthyl)methyl-α-methylbenzylamine.

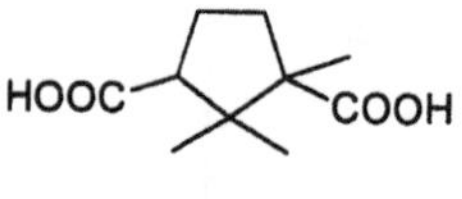

(14)
Camphoric acid

A system with the selector, sodium-(*R,R*)-tartrate, dissolved in the aqueous instead of the organic phase was used by Acs [51] for the separation of various secondary amine bases. Using 0.5 equiv of sodium-(*R,R*)-tartrate in the aqueous phase, they observed that the *R*-base extracted into the aqueous phase while the *S*-base was retained in the benzene phase:

R,S-base + Na-hydrogen-(R,R)-tartrate

$$\uparrow\downarrow \qquad\qquad\qquad\qquad (10)$$

R-base · Na · (R,R)-tartrate (aq) + S-base (benzene)

The recovered amount and purity of the two enantiomers were optimized by the ratio of the aqueous to the benzene phase. Rabai [52] demonstrated that other enantiopure bases such as quinine, quinidine, cinchonine, and cinchonidine also are able to separate the optical isomers of organic acid group containing components.

B. Multistage Extraction

The characteristic feature of enantiomer resolution is that the mixture must be separated into two fractions consisting of the individual nearly pure enantiomers. Because of the low selectivities (mostly 1.2–4) obtained in the equilibrium studies, a series of extraction stages is required to obtain high optical purities. Extraction with many stages is a relatively well-established industrial and laboratory technique that was used for the first time for the full separation of optical isomers by Bauer et al. [53], they resolved 1,2-(2-oxytetramethylene) ferrocene enantiomers with a (+)-diethyl tartrate solution in cyclohexane by the Craig countercurrent distribution technique. Around the same time, the Craig countercurrent extraction apparatus was also applied to obtain full separation of (±)-sodium mandelate isomers by Romano et al. [50]. The apparatus for Craig extraction involves several hundred, typically 200, separatory funnel-like tubes that can be gently shaken [54]. The system contains (n + 1) tubes, all but one of which are partially filled with solute-free heavy phase. The "zeroth" tube contains initially solute dissolved in the light phase and pure heavy phase. After equilibration, the light phase from the "zeroth" tube is transferred to the first and new light phase is added to the "zeroth." Again, the concentrations are allowed to equilibrate, light phases are moved to the next tube, and fresh light solvent is added to the first tube. This sequence is repeated hundreds of times. The result of these transfers is that the solute develops a concentration profile distributed across the tubes. In spite of automation however, the Craig technique is too complex to prevent application for preparative-scale extractions.

For larger-scale preparative extractions, a more efficient extraction can be obtained when the two liquid phases are contacted countercurrently. However, as shown in Fig. 6, classical countercurrent extraction suffers from the disadvantage that only one of the two enantiomers can be obtained with sufficient purity by removing the other one more or less selectively from the raffinate. For most chiral separations the recovery will be low because the extract composition is limited by the equilibrium between the racemate feed and the chiral selector containing solvent. To combine complete separation and high recovery

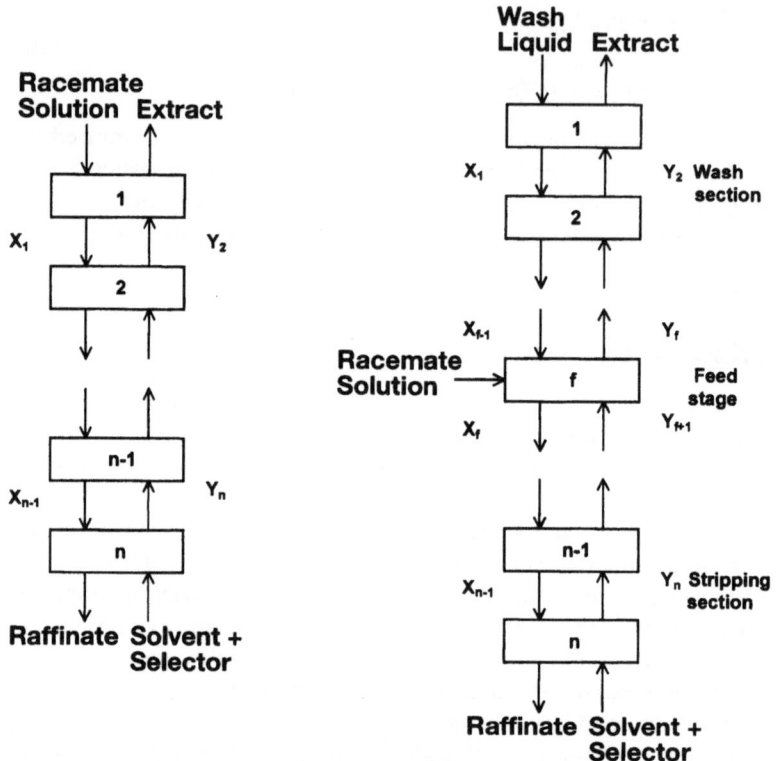

Figure 6   Schematic of countercurrent and fractional extraction.

of the enantiomers in two optically pure fractions, fractional extraction is the most promising liquid–liquid separation technique [55]. Figure 6 illustrates that two solutes can be separated nearly completely by isolating one solute in the extraction solvent and the other solute in the raffinate by applying a wash solvent. The lower section of the process is much the same as a stripping extractor, with the solvent entering the bottom and stripping one of the components from the raffinate. As the extract travels above the feed stage, it is contacted countercurrently with a wash solvent, which scrubs the unwanted solute out of the extract. Abe and coworkers [56] simulated this through an ingenious but labor-intensive arrangement of separatory funnels for the enantioseparation of ($\pm$)-propanolol (**15**) with didodecyl-L-tartrate. As in chromatography, the extent of purification depends on the number of stages or transfer units. The ratio of wash solvent to extraction solvent can be optimized to give the lowest number of stages required. Takeuchi et al. [57] used an *N-n*-dodecyl-

L-hydroxyproline–copper(II) complex dissolved in *n*-butanol to resolve racemic valine (**16**) by fractional extraction in rotating segmented glass columns (Fig. 7). They demonstrated that with single-stage selectivity of 2.86, full separation of both valine enantiomers was achieved with their setup, corresponding to the equivalent of at least 24 theoretical stages. However, scale-up of their equipment can be expected to be a difficult challenge.

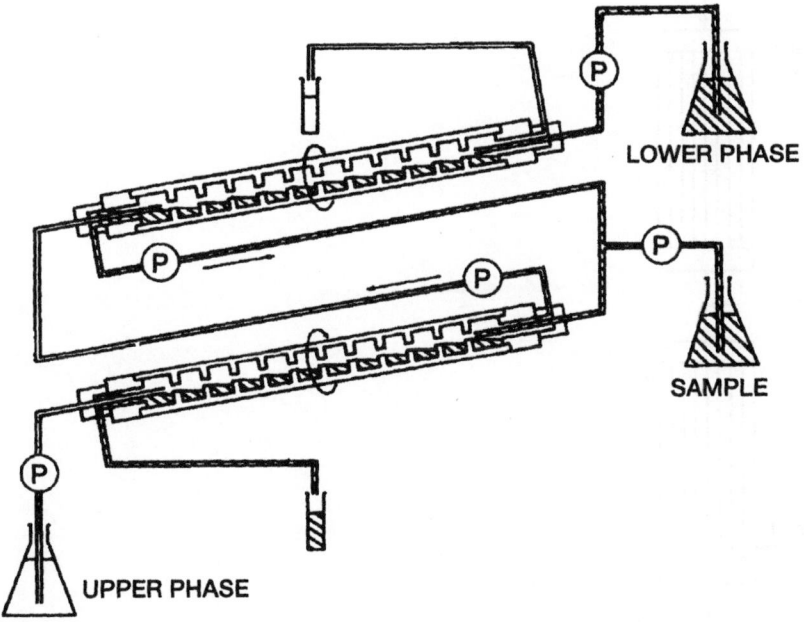

(15)
Propanolol

(16)
Valine

(17)
Isoleucine

The same holds for the analytical vertical rotating column setup developed by Nishizawa et al. [58]. These researchers demonstrated that fractionating extraction with two columns in series was sufficient to achieve full (>99.9%)

LOWER PHASE

SAMPLE

UPPER PHASE

Figure 7   Schematic representation of the countercurrent extraction setup used by Takeuchi et al. [57]. (Reproduced with permission from Ref. 57.)

separation of (±)-mandelic acid with *N-n*-dodecyl-L-proline–copper(II) complex in butanol. Yokouchi et al. [59] demonstrated that the same selector could also be used for the full separation of racemic isoleucine (**17**) in a countercurrent fractional extraction device based on a cascade of 21 small mixer-settler-like units. In both examples, water was used as the wash fluid.

   N-n-Dodecyl-L-hydroxyproline was used in octanol by Ding et al. [60,61] to separate DL-leucine into its pure enantiomers. These investigators used polypropylene microporous hollow-fiber modules in which the pores were filled with a crosslinked polyvinyl alcohol gel as the contacting device. Because of their large surface area per volume, these contactors can achieve the same performance as conventional equipment in a significantly smaller volume. As illustrated in Fig. 8, two separate membrane modules were used for extraction and stripping, and the feed was injected in between these two modules. Figure 9 demonstrates that complete separation could be obtained, provided the water and solvent flow are low enough in velocity to provide sufficient contacting stages. All three examples [58–61] demonstrate that full separation and recovery

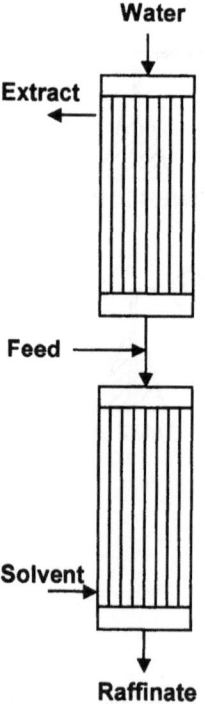

Figure 8 Fractional extraction with hollow-fiber membrane modules.

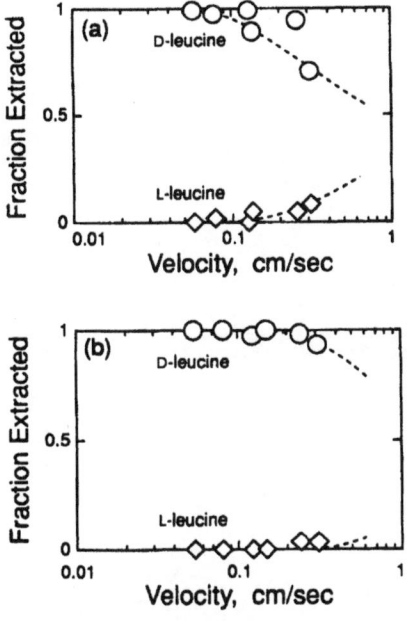

Figure 9 Separation of DL-leucine in hollow-fiber membrane extraction using an N,n-dodecyl-L-hydroxyproline solution in octanol for two modules: (a) 12 cm or (b) 32 cm. (Reproduced with permission from Ref. 60.)

of pure enantiomers from racemic mixtures can be achieved when fractional extraction is used.

V. LIQUID MEMBRANES

An attractive alternative for the separation of optical isomers could be liquid–membrane separations [62,63]. Compared to liquid–liquid extraction, the main features of this approach are high mass transfer rates and efficient selector utilization, resulting in low solvent and selector inventories. Compared to polymeric membranes, liquid membranes can offer considerably higher transport rates and selectivities, thereby reducing staging requirements markedly. Enantioselective transport through the membrane is possible when chiral carriers are used that preferentially form a diastereomeric solute–carrier complex with one of the two enantiomers. The most common configurations are bulk, emulsion, and supported or contained liquid membranes, as shown in Fig. 10.

The first example of enantioselective transport through a bulk liquid membrane was reported by Cram and coworkers [37,64]. They used chiral

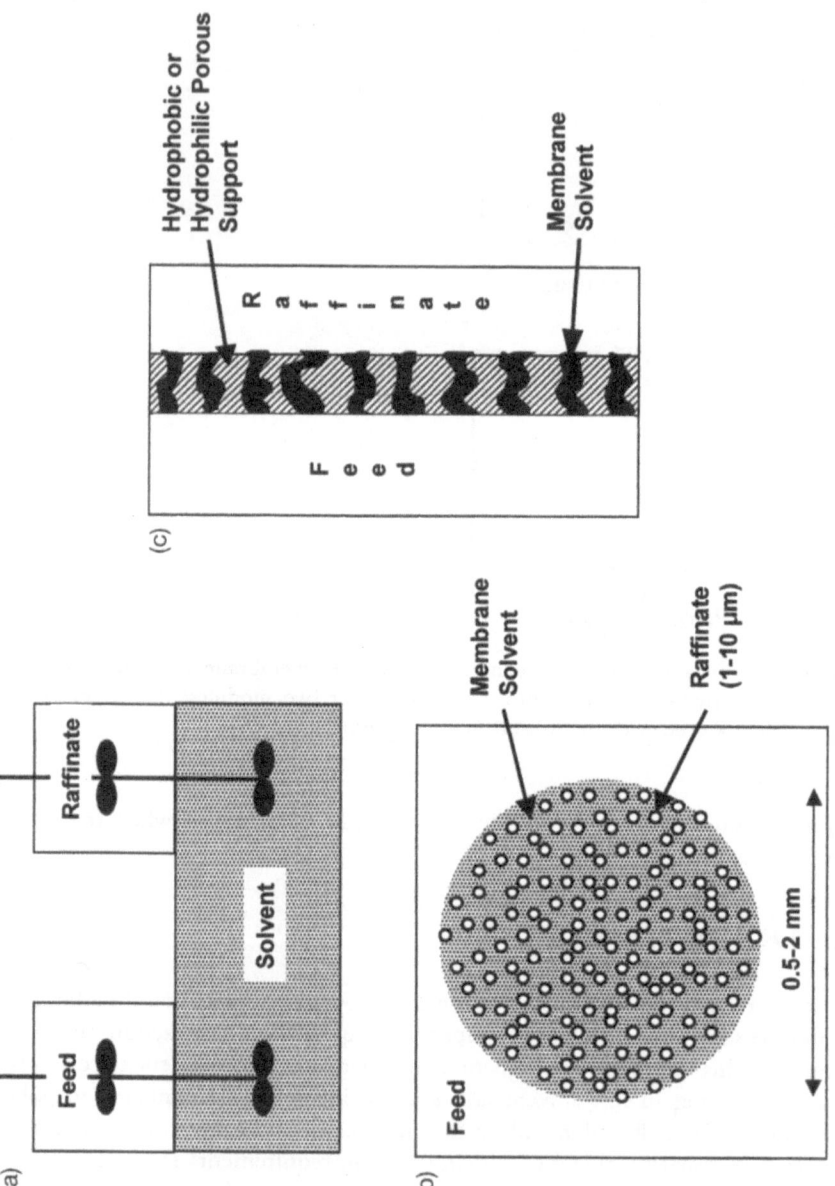

Figure 10 (a) Bulk, (b) emulsion, and (c) supported liquid membrane configuration.

crown ethers dissolved in chloroform for the preferential transport of protonated ($\pm$)-α-substituted benzylamines enantiomers. The liquid–membrane apparatus was assembled by placing chloroform in the bottom of a U-tube and layering the aqueous feed and receiving solutions above the organic phase in each arm of the tube. A logical extension studied by Cram is a machine that separates enantiomers by simultaneously transporting each from the source phase into two different receiving phases [37]. Obviously, the machine requires an enantiopure sample of each enantiomer of the carrier. Conceptually, the resolving machine simply combines two U-tube experiments run simultaneously, each using a different enantiomer of the carrier but feeding from the same source.

Kozbial et al. [48] compared the enantioselectivity of bulk liquid membranes with liquid–liquid extraction for the separation of amino acids with a chiral crown ether. Their results indicated a significantly lower enantioselectivity of the bulk liquid membrane system compared with the liquid–liquid extraction system for all amino acids studied. This result could be explained by assuming that the release step from the membrane to the receiving phase is the rate-determining step. Water-based bulk liquid membranes with cyclodextrin carriers were studied by Armstrong and Jin [65] for the separation of various organic, water-insoluble enantiomers. These investigators demonstrated that cyclodextrins are capable of enantioselective transport of solutes from an ether feed solution through the aqueous membrane to an ether receiving solution. However, in bulk liquid membranes the membrane phase is a bulk, flowing, or well-mixed phase that provides only a low interfacial surface area, hence, slow mass transfer rates. This disadvantage is overcome to a large extent in emulsion and supported liquid membrane systems.

A. Emulsion Liquid Membranes

In emulsion liquid membranes, the receiving phase is stirred with a surfactant to generate an emulsion, which is then mixed with the feed phase that contains the material to be extracted [62,63]. After phase separation, a de-emulsification step is required to separate the membrane and the internal receiving phase, to allow recovery of the extracted material. Pickering et al. [66,67] studied emulsion liquid membranes for the separation of racemic phenylalanine (**18**). They used copper(II)–N-decyl-L-hydroxyproline as the carrier, dissolved in a decane–hexanol mixture as the membrane phase. Phenylalanine was introduced in the internal emulsion droplets and allowed to travel to the external receiving phase that contained perchloric acid or MES(2-[N-morpholino]ethanesulfonic acid)–NaOH buffer as stripping agent. The selectivity for D-phenylalanine (1.68–1.56 at pH 4–6) appeared to increase with decreasing pH of the internal source phase at the expense of a strong decrease in distribution coefficient of both enantiomers. This implies that the higher enantioselectivity at low pH can be exploited only by using significantly higher amounts of organic solvent and chiral extractant. The presence of a pH gradient over the membrane phase enhanced the extraction rate, which was explained by the change in ionic charge on the phenylalanine enantiomers on entering the receiving phase. Unfortunately, loss of the D-isomer from the source phase combined with an increase in its concentration in the receiving phase led to a reduction in the enantioselectivity during the course of extraction. Compared with solvent extraction (ee 25%), the initial enantiomeric excess appeared to be somewhat higher, with a value around 40%. Skrzypinski et al. [68] applied liquid emulsion membranes for the stereoselective extraction of dipeptides and phosphonodipeptides with a carbon tetrachloride membrane containing 18-crown-6 as the chiral carrier. The obtained stereoselectivity ranged from 1.1 to 1.8 and appeared to be almost independent of membrane thickness, indicating that the transport through the membrane is the rate-determining step.

(18)

Phenylalanine

Both studies support the more general conclusion that the emulsion liquid membrane process is not sufficiently selective to achieve full separation in one stage. Even if more selective carriers are available, the strong depletion of the enantiomer in the feed phase will have a dramatic effect on the enantiomer ratio in the receiving phase. Staging is further complicated by the required emulsification and de-emulsification steps, which make emulsion liquid membrane processes more complex and very difficult to stage.

B. Supported and Contained Liquid Membranes

In addition to bulk and emulsion–liquid membranes, related selectors have been employed in supported and contained liquid membranes. In supported liquid membranes, the membrane phase is retained within the pores of a porous polymer film by capillary forces. Chiral crown ethers [69–73] have been used as carriers for the separation of amino acids. Good separations with an optical resolution ratio of 18.9 for L-phenylglycine, 15.0 for L-glutamic acid, and 14.7 for L-arginine were reported. As expected, the stability of the supported liquid membrane appeared to be the main issue. After a certain period of time, loss of membrane phase causes a sudden decrease in enantioselectivity that indicates the breakdown of the supported liquid membrane. The main cause appeared to be the solubility of the membrane solvent in the feed and receiving phase. Presaturation of the feed and receiving phase with the membrane solvent resulted in a drastic increase of the supported liquid membrane lifetime. Instead of chiral crown ethers, Bryjak et al. [74] used water-insoluble chiral alcohols, nopol and (2S)-(−)-methyl-1-butanol, for the enantioselective transport of various amino acid hydrochlorides. The obtained enantioselectivities were low and ranged from 0.4 to 1.5.

Although supported liquid membranes have numerous potential advantages, they have not found applications at industrial level. This is mostly because their main problems, stabilization and solubilization of the membrane phase, have not yet been sufficiently mitigated. Long-term stability is easily achieved in a contained liquid membrane system. This is essentially a bulk liquid membrane system in which the mass transfer rates are greatly increased by circulating the membrane liquid between an absorption and a desorption hollow-fiber unit. Formation and dissociation of the chiral complex takes place at different locations, allowing the choice of appropriate conditions for both stages. Pirkle and Bowen [75] used selectors based on N-(1-naphthyl)leucine to separate the enantiomers of amino acid derivatives. As with the emulsion liquid membranes, the initially high enantioselectivity dropped quickly because the feed phase was depleted in the desired enantiomer. Cyclodextrins were applied in a contained aqueous liquid membrane setup for the separation of various water-insoluble isomeric mixtures by Mandal et al. [76]. Although their setup was not tested for optical isomers, the chiral recognition capabilities of cyclodextrins should make this arrangement capable of separating optical isomers, as well.

VI. SUPERCRITICAL FLUID EXTRACTION

A. Introduction

Supercritical fluid chromatography (SFC) using a chiral stationary phase was developed in the 1990s for the analytical and preparative [77] separation of the enantiomers of important chiral compounds [78–80] and other biologically active molecules [81,82]. The efficiency of SFC suggested that extraction under supercritical conditions might also be applicable for optical resolution on a preparative and industrial scale. The well-known reasons for choosing supercritical fluids, particularly carbon dioxide, over other solvent systems are as follows [83,84]:

> Unique solvation and favorable mass transfer properties, which can be varied simply by adjustment of the system pressure and/or temperature.
> Carbon dioxide is an ideal solvent because it is nontoxic, nonexplosive, readily available, and easily removed from the products.

Possibility of eliminating or to reducing environmental hazards associated with toxic organic solvents.

Low critical temperature of carbon dioxide (31°C), which allows operation close to ambient temperature, eliminating decomposition or racemization of thermally labile substances.

Enantioselective extraction by supercritical fluids is possible by resolving a racemic acids mixture with less than one equivalent of a resolving agent [85]. In this way a complex equilibrium is established involving both enantiomers of the free acid and two diastereomeric salts. The enantioselectivity of extraction arises from two distinct phenomena, the second one being specific to supercritical fluid extraction (SFE):

1. Higher stability of one of the diastereomeric salts resulting in its preferred formation. Consequently the free acid becomes optically active.

2. Poor solubility of the diastereomeric salts in supercritical CO_2, permitting the selective extraction of the free acid.

These two phenomena are interdependent because the dissociation constant of the diastereomeric salts is also dependent on the state of the solvent (pressure and temperature). A series of SFE experiments with different racemic organic acids and basic resolving agents [86,87] (see Table 1) has shown that most of the diastereomeric complexes are insoluble in CO_2. Only with (R)-$(+)$-1-phenylethylamine and (S)-$(+)$-2-benzylamino-1-butanol did some of the diastereomeric complexes appear to be slightly soluble in CO_2 and to contaminate the extracts. Table 2 records the configuration and ee values of the major extracted enantiomers. These results illustrate clearly that the obtained enantioselectivity depends strongly on the exact enantiomer–selector combination. Ex-

Table 1   Racemic Compounds and Resolution Agents

Racemic compounds
1. $(\pm)$-2-(4-Isobutyl-phenyl)-propionic acid
2a. $(\pm)$-*cis*-Chrysanthemic acid
2b. $(\pm)$-*trans*-Chrysanthemic acid
3a. $(\pm)$-*cis*-Permetric acid
3b. $(\pm)$-*trans*-Permetric acid
Resolution agents
4. (R)-$(+)$-1-Phenylethylamine
5. (R,R)-$(-)$-1-*p*-Nitrophenyl-2-aminopropane-1,3-diol
6. (S)-$(+)$-2-Amino-1-butanol
7. (S)-$(+)$-1-Phenyl-*N*-methylisopropylamine
8. (S)-$(+)$-2-Benzylamino-1-butanol

Table 2 Configuration and Enantiomeric Excess (%) of Extracts and Raffinates (for the free acid)

Acid		Base								
	4		5		6		7		8	
	Conf	ee	Conf	ee	Conf	ee	Conf	ee	Conf	ee
1 Extract	S	33	R	8						
Raffinate	R	42	S	11						
2a Extract			R	2	R	4			R	43
Raffinate			S	2	S	4			S	76
2b Extract									S	3
Raffinate									R	3
3a Extract	R	40	R	15			S	2		
Raffinate	S	48	S	38			R	2		
3b Extract	S	9	S	13					S	27
Raffinate	R	3	R	10					R	31

tension of the investigations to the separation of racemic mandelic acid, phenylpropionic acid, and phenylbutyric acid with (R)-(+)- or (S)-(−)-methylbenzylamine as a resolution agent provided remarkable partial resolutions (61–95% ee) on an analytical scale [88]. During the SFE resolution of the racemic base (±)-6-phenyl-2,3,5,6-tetrahydroimidazo-[2,1-b]thiazole with (2R,3R)-O,O'-dibenzoyltartaric acid monohydrate, the (+)- and (−)-enantiomers were enriched in the extract and raffinate, respectively [89].

B. Effect of Process Conditions

Extractions at various molar ratios of the resolving agent to the racemic mixture (0.25–0.75) have shown that an increase of the molar ratio results in enhanced enantiomeric excess but reduced yields [86–89]. Characterization of this combined effect through the modified Fogassy parameter indicated optimum molar ratios between 0.25 and 0.5. Consequently, in standard experiments the racemic acid was allowed to react with 0.5 equiv of the chiral bases.

(19)
Chrysanthemic acid

(20)
S-(+)-2-Benzylamino-1-butanol

Dissociation of the diastereomeric salts and the dissolving power of CO_2 are influenced by extraction temperature and pressure. During the resolution of $(\pm)$-*cis*-chrysanthemic acid (**19**) with (S)-$(+)$-2-benzylamino-1-butanol (**20**) as the resolving agent, the pressure of CO_2 (in the range of 90–150 bar) was found to have a direct bearing on yield and enantiomeric excess [87]. In addition to the higher yield, the enantiomeric excess of the extracts increased from 41% to 62% by raising the pressure from 90 bar to 150 bar. The effect of temperature was investigated only in the supercritical state at a constant pressure of 100 bar. The ee values of the extracts obtained after a single extraction decreased from 65% (32°C) to 25% (50°C). This reduction of selectivity with increasing temperature was also found in the liquid–liquid extraction studies but seems more pronounced in case of SFE with CO_2. For a better understanding, ee values of the total extracts obtained from a single extraction were plotted against the density of the fluid (Fig. 11). Below 0.6 g/cm³ enantiomeric excess is nearly linear with CO_2 density and approaches a constant value in the higher density region. In this density region (0.6–0.85 g/cm³) the ee values are significantly higher at lower temperatures (32–35°C). A more detailed quantification of this combined pressure and temperature effect through experimental design indicated that a combination of a high temperature (47°C) and high pressure (200 bar) might be optimal for the production of pure $(+)$- and $(-)$-*cis*-permetric acid enantiomers [90].

C. Purification of Partially Resolved Mixtures

Further purification of the partially resolved $(\pm)$-*cis*-chrysanthemic acid was possible by repeated extractions after the addition of base (S)-$(+)$-2-benzylamino-1-butanol equimolar to the content of $(-)$-*cis*-chrysanthemic acid until a satisfactory resolution was achieved (90% ee). Since the rate constants for diastereomeric association of the enantiomers are different, an increased selectivity can also be achieved by partial salt formation with an achiral substance followed by the extraction of the free enantiomer [91]. This procedure was used to increase the separation of $(\pm)$-6-phenyl-2,3,5,6-tetrahydroimidazo-[2,1-*b*]thiazole enantiomers by SFE [89]. Addition of hydrochloric acid in quantities close to the equivalent amount of the racemic fraction of the mixture increased the enantioselectivity considerably, and nearly complete resolution of the $(+)$- and $(-)$-enantiomers was achieved in a single SFE step.

From the foregoing it can be concluded that supercritical fluid extraction may offer a promising alternative for the resolution of racemic acids and bases. Published results indicate that SFE might be more selective for the separation of enantiomeric mixtures than conventional extraction because differences in solubility, dissociation constants, and stability are larger in supercritical carbon dioxide than traditional solvents. One of the main challenges in making the

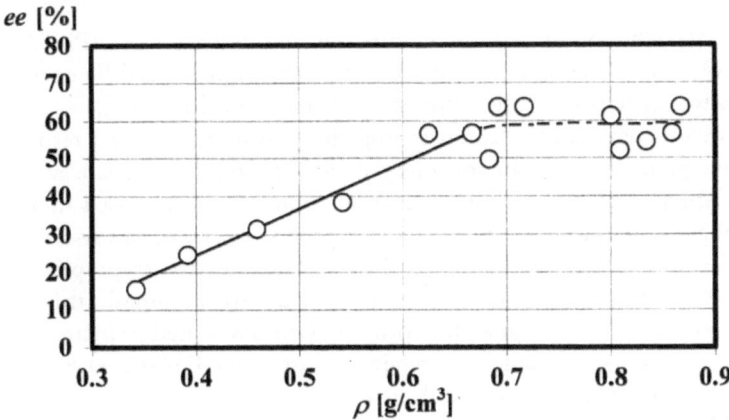

Figure 11 Effect of the CO_2 density on the enantiomeric excess of the total extracts obtained by single extraction.

process applicable to industrial use is to provide an easy way of staging to achieve full resolution in one single operation. From currently running commercial-scale SFE units for the decaffeination of green coffee beans or tea and the extraction of hops, spices, and tobacco [83,84], sufficient experience is available to support the design of commercial installations and upscaling to establish such facilities.

VII. AQUEOUS TWO-PHASE EXTRACTION

Aqueous two-phase systems result from the incompatibility between aqueous solutions of two polymers [e.g., polyethylene glycol (PEG) and dextran] or between one polymer and an appropriate salt (e.g., potassium phosphate) when certain threshold concentrations are exceeded. Since the bulk of both phases consists of water (60–90%), aqueous two-phase systems often suffer from the minor differences in partition behavior for many molecules [92–95]. Introduction of chiral differentiation in such systems requires the incorporation of affinity-based interactions. As with enantioselective micelles, this can be accomplished by the introduction of a selector that partitions into one phase or by the attachment of a chiral ligand to one of the polymers.

One of the first reports of chiral resolution in combination with an aqueous two-phase system is that of Kuhlman et al. [96]. They used a biochemical reaction, catalyzed by an acylase enzyme, followed by extraction for the resolution of amino acids. The distribution coefficients of the substrate *N*-acetyl-

D,L-methionine and the product L-methionine in the system used (PEG/ K_2HPO_4/KH_2PO_4) were 11.2 and 1.54, respectively. This significant difference permits an effective separation of the substrate and the formed product.

The first quantitative study demonstrating the possibility of chiral resolution by selective partitioning in an aqueous polymer–polymer two-phase system is that of Ekberg et al. [97]. Bovine serum albumin (BSA) was used in a dextran–PEG phase system to separate D- and L-tryptophan (21). In this system 95% of the BSA was in the dextran-rich phase, while the free tryptophan partitioned more equally between the phases (D = 1.2). An automatic thin-layer countercurrent distribution apparatus with 60 cavities was used to increase the separation [93]. The L-enantiomer was retained more in the BSA-containing dextran phase than the D-enantiomer, and an overall separation factor of 3.1 was obtained. No resolution of racemate was obtained when extraction was carried out without BSA. The method was extended by the same group for separation of enantiomeric and diastereomeric mixtures [98]. The chiral components capable of resolving racemates were proteins (e.g., BSA), carbohydrates (e.g., cyclodextrin, cellulose), amino acids (e.g., D- or L-proline), and derivatives thereof. The experimental results indicated that besides D,L-tryptophan, R,S-methylsulfinyl benzoic acid and R,S-terbutaline could be successfully resolved.

| (21) | (22) |
| **Tryptophan** | **Kynurenine** |

Sellergren et al. [99] have demonstrated that only a few countercurrent extractions were required for a preparative-scale separation of D- and L-kynurenine (22) by a dextran–PEG two-phase system with BSA. The yield and optical purity of the pooled fractions are given in Table 3. Preparative separation of D,L-kynurenine was performed by countercurrent chromatography and mixer-settlers using a PEG–disodium hydrogen phosphate phase system with BSA [100,101]. Shinomiya et al. [101], for example, used 6% BSA as the chiral selector in an aqueous two-phase system comprising 10 wt% PEG 8000 and 5 wt% disodium hydrogen phosphate. The selective partitioning of BSA provided sufficient enantioselectivity to enable full separation of both enantiomers. An eight-extraction procedure, similar to that reported by Sellergren [99], was used for partial resolution of some drug enantiomers [102]. The phase systems were prepared with dextran, PEG, and protein [BSA and ovomucoid (OVM)]. Ofloxacin (an antimicrobial agent) enantiomers were well recognized by the BSA system, whereas the enantiomers of carvedilol (a β-adrenergic blocker) and DG-

Table 3 Yield and Optical Purity of D,L-Kynurenine in the Pooled Fractions

Fraction	Yield (%)	Configuration	Optical purity (%)
2–4	22	D	90
5	13	D	83
6	35	D	55
7, 8	30	L	90

Source: Ref. 99.

5128 (an oral hypoglycemic drug) were hardly recognized by the OVM system. Although the distribution coefficients of both ofloxacin enantiomers increased at lower pH, the enantioselectivity decreased. After eight transfers in separatory funnels, the *R*-enantiomer was recovered in 62% optical purity at pH 9.

A similar method to resolve optical isomers was disclosed by Hsu [103]. Separation of enantiomers was achieved by partitioning between a water-soluble polymer phase and a second aqueous phase containing the chiral selector, such as monosaccharides, disaccharides, peptides, proteins, antibodies, and optically active amino acids. Additives such as salts, buffers, sugars, and polymers may be added to enhance physical phase separation and separation factor. The illustrative examples are summarized in Table 4. The separation of enantiomers can be improved by attaching chiral receptors chemically to the phase-forming polymers. In this case the resulting procedure is called affinity partitioning, which has the intrinsic advantage of ensuring its primary distribution to one phase in the system. Although extensively described for nonchiral applications (biomolecules, metals) with affinity ligands attached to polymers such as PEG

Table 4 Separation Factors of Enantiomers in Aqueous Two-Phase Systems

Racemate	System	β
D,L-Phenylalanine	L-lysine–PEG8000–water	1.11
β-Lactoglobulin A and B	L-lysine–PEG8000–water	1.94
D,L-Tryptophan	L-lysine–PEG8000–water	1.24
D,L-Tryptophan	L-serine–PEG8000–water	1.19
D,L-Tryptophan	L-proline–PEG8000–water	1.06
D,L-Tryptophan	L-proline–PEG8000–water–CuSO₄	1.18
D,L-Tryptophan	Sucrose–PPG425–water	0.90
D,L-Tryptophan	PEG–potassium phosphate–L-lysine–water	1.03

Source: Ref. 103.

or Dextran [92], the use of such receptors in the separation of enantiomers has not yet been reported. An indication that attaching chiral ligands to polymers might make them useful for chiral separations is the enantioselective permeation that can be obtained by immobilizing chiral ligands such as L-menthol or L-phenylglycine in the pores of ultrafiltration membranes [104–107].

VIII. MEMBRANE FILTRATION ASSISTED EXTRACTION

An alternative way to keep the selector molecules in the extract phase separated from the raffinate phase may be the use of membrane filtration. To be effective, the membrane pores should retain the selector molecules and selector–enantiomer complex, while the free enantiomer molecules can freely exchange phases. Complete retention in the extract phase requires sufficiently large selector molecules or aggregates such as enantioselective micelles, dendrimers, proteins, or cyclodextrins. Micellar enantioselective complexation has been used for the separation of amino acids with nonionic micelles containing an amphiphilic chiral Cu(II)–amino acid derivative selector molecule. Several studies describe the separation of racemic phenylalanine with L-5-cholesteryl-glutamate (**23**) as the selector in micelles formed with a nonionic surfactant Serdox NNP 10 [40,108,109]. The selector can form ternary complexes with a Cu(II) ion and the D- or L-amino acid.

(23)
L-5-Cholesteryl glutaminate

Analyses of the complexation behavior demonstrated that competitive multicomponent Langmuir isotherms are sufficient for an accurate description of the DL-phenylalanine complexation [109]. Characterization of the enantioselectivity was done through the operational selectivity obtained from micellar-enhanced ultrafiltration (MEUF) experiments. As schematically illustrated in Fig. 12, an ultrafiltration membrane retains the micelles with the preferentially bound enantiomer, which is permeable to the uncomplexed enantiomer. It is clear that the determined operational selectivities, defined as the ratio of the distribution

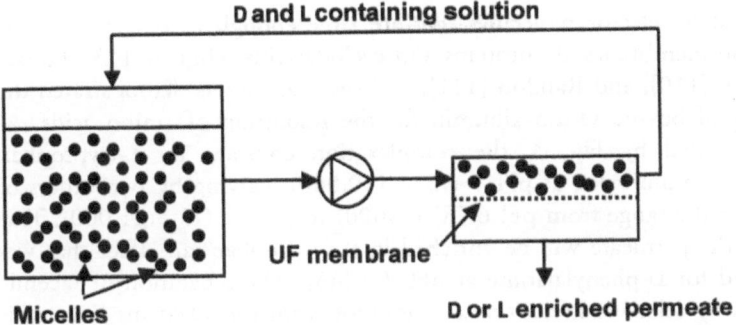

Figure 12 Principle of micellar-enhanced ultrafiltration (MEUF).

coefficients D of D- and L-phenylalanine between the permeate and retentate, may include some mass transfer and kinetic effects. The selectivity was maximal with short surfactant chain length and a pH around 7 [108]. At pH below 6, no complexation occurred and at higher pH, enantioselective interactions are dominated by nonchiral ionic interactions. The highest enantioselectivity (4.2) was obtained when the chiral cosurfactant (L-5-cholesteryl-glutamate) was present in a 2:1 excess over racemic phenylalanine, and the cosurfactant to Cu(II) ratio was 1:1. The purity of L-phenylalanine in the permeate reached a maximum of 73% under these optimal conditions. It is obvious that the obtained selectivities (1.5–4) are not sufficient to obtain both enantiomers at purity exceeding 99% with a single-stage separation process. Thus, a multistage MEUF process as shown in Fig. 13 is required to achieve full separation of the amino acid enantiomers. Experimental data indicate that the productivity of multistage process can be improved by reducing the affinity of the micelles for the enantiomers [108].

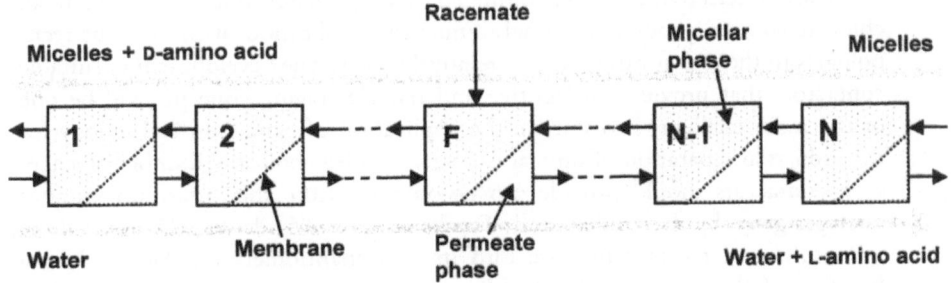

Figure 13 Cascaded countercurrent system for the separation of racemic amino acid mixtures with micellar-enhanced ultrafiltration.

Alternative selector molecules that are large enough to be separated by ultrafiltration membranes are proteins and cyclodextrins. Higuchi [43], Poncet [44], Garnier [110], and Randon [111], and their coworkers demonstrated the applicability of bovine serum albumin for the resolution of amino acids. As already illustrated by Fig. 3, the complexation constant for L-tryptophan reached a maximum value at pH 9 (K_L = 110,000), varying by two orders of magnitude in the range from pH 6 (K_L = 1000) to pH 11 (K_L = 21,000). This means that the permeate will be enriched in D-tryptophan, an effect that was also observed for D-phenylalanine at pH 7.0 [43]. The separation of racemic amino acids by selective inclusion complexation with cyclodextrins in combination with ultrafiltration is described by Kokugan et al. [112,113]. For threonine and phenylalanine, the rejection of the L-isomer was much stronger, indicating that D-isomer will again be enriched in the permeate. This is induced by the difference in the inclusion binding strength of the amino acid enantiomers. Figure 14 illustrates that the studied α-cyclodextrin tended to bind threonine more strongly, while phenylalanine was more strongly bonded to the β-cyclodextrin.

IX. FUTURE OUTLOOK

The overview presented illustrates that extraction-based technologies offer interesting opportunities for the resolution of enantiomer mixtures. Their main limitation is the limited selectivity of extraction, which is especially important for applications in the multipurpose environment in which the majority of enantiopure products are made. The variety of products requiring to be resolved calls for a technology concept that will be able to separate whole classes of products (amino acids, organic acids, amines, alcohols, etc.) with a single unit. Although a high selectivity might be reached for a specific product, the multipurpose environment requires this technology concept to remain applicable when lower selectivities are encountered. This problem, found in virtually all chiral separation processes, strongly limits the application of the various technologies to those that can be staged relatively easily. The development of suitable contactors that provide competitive and reliable staging concepts will be one of the main challenges to initiate the application of extraction technology for the industrial separation of optical isomers. An idea about the number of stages such contactors should provide can be obtained from the minimum number of stages given by the Underwood–Fenske relation [55]. In Fig. 15 this is done for resolving a racemic mixture into its two enantiomers (>99% pure) as a function of the selectivity. It can be seen that for most of the reported selectivities (1.2–4), minimally 10–40 stages are required. This means that under practical operating conditions one can expect to need 20 up to 80 stages.

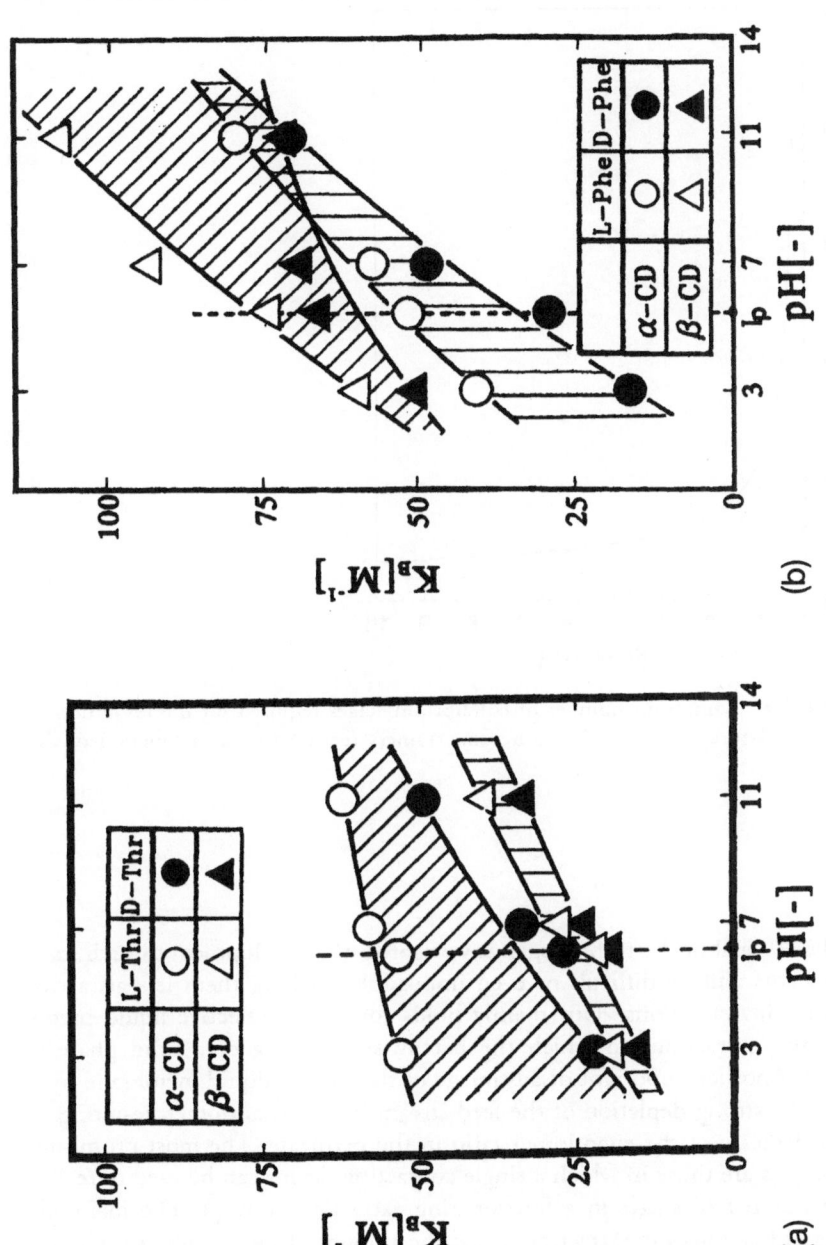

Figure 14 Effect of pH on binding constant for threonine (a) and phenylalanine (b) with α and β-cyclodextrin. (Reproduced with permission from Ref. 111.)

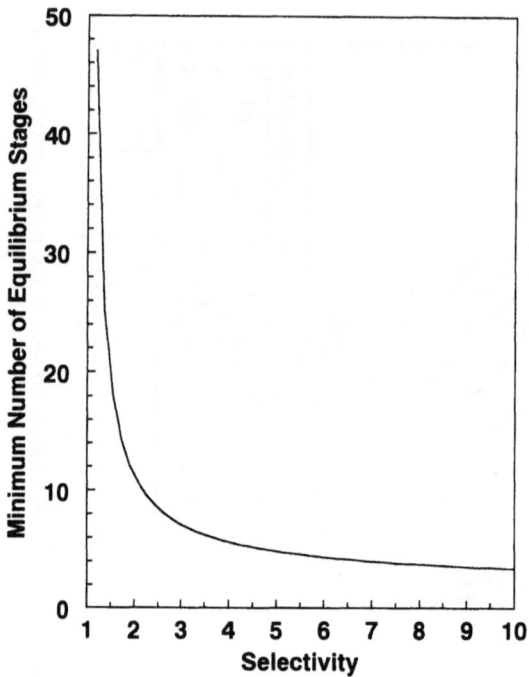

Figure 15   Minimum number of equilibrium stages required for the resolution of a racemic mixture into its >99% pure enantiomers (ee > 98%) as a function of the selectivity.

For several of the technologies investigated, the development of such staging concepts will be difficult or even impossible, making them less attractive for future investigations. This mainly holds for enantioselective liquid membranes and extractions in which the batchwise handling of a solid phase is involved. Another important disadvantage of the various liquid membrane concepts is the strong depletion of the feed stream in one enantiomer, resulting in a rapid decrease of the enantiomer ratio in the permeate. The most promising technologies are those in which a single contacting device can be used to realize a large number of stages in a fractionating extraction concept. The main advantage of fractionating extraction is that the amount of solvent and wash liquid used is adjustable from product to product. This ensures that in spite of low and varying selectivities, the desired purity of extract and raffinate can be achieved if sufficient contacting stages are provided. Various setups based on rotating columns [57,58], mixer-settlers [59], and membrane contactors [60,61,114–116] have been proposed in the literature and were discussed in this chapter. It is well known that mixer-settler systems with many stages, high efficiencies, and a wide range of capacity are relatively easy to design and construct. However, an important drawback to their application in chiral extraction will be the requirement for a relatively large inventory of expensive selectors. Although not yet applied to chiral extraction, centrifugal contactors and extraction columns are promising alternatives that largely circumvent the disadvantage of high selector inventory and are readily designed with significant amounts of stages.

Until now, most of the studies have focused on extractions under analytical conditions, to demonstrate the enantioselectivity of the used selectors. Because of the low racemate concentrations in the feed and low selector concentrations, usually very low overall productivities are obtained. To make extraction technology applicable for industrial systems, studies must be conducted at conditions under which high productivities can be achieved. This means high feed and selector concentrations applied in a fractionating extraction concept. Such studies should also include a more detailed characterization and modeling of the used systems to facilitate future scale-up to industrial contactors. The preferred contactors for such studies should be easily scalable from laboratory tests to production. This is a primary requirement of most enantiopure product producers, since most of them employ a multipurpose custom-manufacturing environment in which only very limited time is allowed for process development. Therefore a need exists to compare contacting devices under industrial operating conditions for their efficiency in realizing a large number of stages and ease of scale-up and design.

For many food and pharmaceutical applications, special attention needs to be given to the choice of solvent. The solvents used in those applications should be product compatible, nontoxic, harmless to consumers, and environmentally friendly. When other solvents fail to meet these conditions, aqueous two-phase and supercritical fluid extraction might be viable alternatives. Both techniques may offer cleaner products and equivalent or better recoveries than conventional technologies. Disadvantages are the relatively high need for chemicals to form aqueous two-phase systems and the relatively high pressures needed for supercritical fluid extraction. When both techniques have been transformed in accordance with a multistage fractionating extraction concept, their application might become economically feasible for specific products.

In general, it can be concluded that although extraction technology has proven its technical feasibility for the resolution of racemates, much work still needs to be done to transform the analytical concepts studied into an industrially applicable multipurpose technology. In such process development studies, the feasibility borders of chiral extraction technology can be substantially moved by changing the chiral selector system. Therefore a strong need remains for a continued search to improved chiral selectors. In addition to attaining improved selectivities, however, it is important to pay particular attention to selector solubility, ease of regeneration, and stability.

ACKNOWLEDGMENTS

We express our gratitude to M. Steensma, G. W. Meindersma, N. J. M. Kuipers (University of Twente), G. Kwant (DSM-Research), and J. Sawinsky and E. Fogassy (Budapest University of Technology and Economics) for their suggestions during the preparation of this manuscript.

REFERENCES

1. EL Eliel, SH Wilen. Stereochemistry of Organic Compounds. New York: Wiley, 1994.
2. J Jacques, A Collet, SH Wilen. Enantiomers, Racemates and Resolutions. New York: Wiley, 1981.
3. RA Sheldon. Chirotechnology, Industrial Synthesis of Optically Active Compounds. New York: Dekker, 1993.
4. AN Collins, GN Sheldrake, J Crosby. Chirality in Industry, The Commercial Manufacture and Applications of Optically Active Compounds. Chichester: Wiley, 1992.
5. AN Collins, GN Sheldrake, J Crosby. Chirality in Industry II, The Commercial Manufacture and Applications of Optically Active Compounds. Chichester: Wiley, 1997.
6. S Ahuja. Chiral Separations. Washington, DC: American Chemical Society, 1997.
7. SC Stinson. Chem Eng News 18:44–74, 1995.
8. SC Stinson. Chem Eng News 21:83–104, 1998.
9. Pharmaceutical Manufacturers Association. Pharm Technol 46: May 1990.
10. FDA Policy Statement. Chirality 4:338–342, 1992.
11. M Gross. Drug Inf J 27:453, 1993.
12. H-U Blaser, M Studer. Chirality 11:459–464, 1999.
13. MJ Gattuso, B McCulloch, JW Priegnitz. Chem Tech Eur 27–30, 1996.
14. M Schulte, JN Kinkel, RM Nicoud, F Charton. Chem Ing Techn 69:670–683, 1996.
15. ER Francotte. Chimia 51:717–725, 1997.
16. T Masawaki, S Matsumot, S Tone. J Chem Eng Jpn 27:4 517–522, 1994.
17. T Aoki, S Tomizawa, E Oikawa. J Membr Sci 99:117–125, 1995.
18. S Tone, T Masawaki, T Hamada. J Membr Sci 103:57–63, 1995.

19. M Yoshikwa, J-I Izumi, T Kitao, S Koya, S Sakamoto. J Membr Sci 108:171–175, 1995.
20. T Aoki, A Maruyama, K-I Shinohara, E Oikawa. Polym J 27:5, 547–550, 1995.
21. S Tone, T Masawaki, K Eguchi. J Membr Sci 118:31–40, 1996.
22. T Aoki, K-I Shinohara, T Kaneko, E Oikawa. Macromolecules 29:4192–4198, 1996.
23. LB Dashkevic. Tr Leningrad, Khim-Farm Inst 6:29, 1959.
24. S Topiol, M Sabio. Enantiomer 1:251–265, 1996.
25. TH Webb, CS Wilcox. Chem Soc Rev 22:383–395, 1993.
26. K Kano. J Phys Org Chem 10:286–291, 1997.
27. NS Bowman, GT McCloud, GK Schweitzer. J Am Chem Soc 90:3848–3852, 1968.
28. V Prelog, Z Stojanac, K Kovacevic. Helv Chim Acta 65:377–384, 1982.
29. Y Abe, T Shoji, M Kobayashi, W Qing, N Asai, H Nishizawa. Chem Pharm Bull 43:262–265, 1995.
30. T Takeuchi, R Horikawa, T Tanimura. Anal Chem 56:1152–1155, 1982.
31. H Tsukube, J Uenishi, T Kanatani, H Itoh, O Yonemitsu. J Chem Soc Chem Commun 477–478, 1996.
32. H Tsukube, S Shinoda. Bol Soc Chil Quim 42:237–245, 1997.
33. H Tsukube, S Shinoda, J Uenishi, T Kanatani, H Itoh, M Shiode, T Iwachido, O Yonemitsu. Inorg Chem 37:1585–1591, 1998.
34. J Szejtli. Cyclodextrin Technology. Amsterdam: Kluwer Academic, 1988.
35. K Verleysen, P Sandra. Electrophoresis 19:2798–2833, 1998.
36. E Weber. In: DM Ruthven, ed. Encyclopedia of Separation Technology. New York: Wiley, 1997.
37. DJ Cram. Angew Chem Int Ed Engl 27:1009–1020, 1988.
38. K Naemura, Y Tobe, T Kaneda. Coord Chem Rev 148:199–219, 1994.
39. A Dobashi, M Hamada. J Chromatogr A 780:179–189, 1997.
40. AL Creagh, BBE Hasenack, A van der Padt, EJR Sudholter, K van't Riet. Biotechnol Bioeng 44:690–698, 1994.
41. YM Chen, CF Chen, F Xi. Chirality 10:661–666, 1998.
42. HF Chow, TKK Mong, CW Wan, ZY Wang. Adv Dendritic Macromol 4:107–133, 1999.
43. A Higuchi, T Hashimoto, M Yonehara, N Kobota, K Watanabe, S Uemiya, T Kojima, M Hara. J Membr Sci 130:31–39, 1997.
44. S Poncet, J Randon, JL Rocca. Sep Sci Technol 32:2029–2038, 1997.
45. Y Yanagisawa, K Nakazato, T Nagai. Chirality 10:742–746, 1998.
46. PJ Pickering, JB Chaudhuri. Chirality 11:241–248, 1999.
47. KH Kellner, A Blasch, H Chmiel, M Lämmerhofer, W Lindner. Chirality 9:268–273, 1997.
48. M Kozbial, M Pietraszkiewicz, O Pietraszkiewicz. J Inclusion Phenom Mol Recognit Chem 30:69–77, 1998.
49. SC Peacock, DJ Cram. J Chem Soc Chem Commun 282–284, 1976.
50. SJ Romano, KH Wells, HL Rothbart, W Rieman. Talanta 16:581–590, 1969.
51. M Acs, D Kozma, E Fogassy. ACH Mod Chem 132:475–478, 1995.
52. J Rabai. Angew Chem Int Ed Engl 31:1631–1633, 1992.

53. K Bauer, H Falk, K Schlögl. Monatsh Chem 99:2186–2194, 1968.
54. LC Craig, D Craig. In: A Weissberger, ed. Techniques of Organic Chemistry, vol 3, Part 1: Separation and Purification, 2nd ed. New York: Wiley, 1956, p 149.
55. PA Schweitzer. Handbook of Separation Techniques for Chemical Engineers, 3rd ed. New York: McGraw-Hill, 1997.
56. Y Abe, T Shoji, S Fukui, M Sasamoto, H Nishizawa. Chem Pharm Bull 44: 1521–1524, 1996.
57. T Takeuchi, R Horikawa, T Tanimura. Sep Sci Technol 25:941–951, 1990.
58. H Nishizawa, K Tahara, A Hayashida, Y Abe. Anal Sci 9:611–615, 1993.
59. Y Yokouchi, Y Ohno, K Nakagomi, T Tanimura, Y Kabasawa. Chromatography 19:374–375, 1998.
60. HB Ding, PW Carr, EL Cussler. AIChE J 38:1493–1498, 1992.
61. EL Cussler. Proceedings of Chiral '95 Europe, pp 133–138.
62. JW Frankenfeld, NN Li. In: RW Rousseau, ed. Handbook of Separation Process Technology. New York: Wiley, 1987, pp 840–861.
63. L Boyadzhiev, Z Lazarova. In: RD Noble and SA Stern, eds. Membrane Separation Technology, Principles and Applications. Amsterdam: Elsevier, 1995, pp 283–339.
64. M Newcomb, RC Helgeson, DJ Cram. J Am Chem Soc 96:7367–7369, 1974.
65. DW Armstrong, HL Jin. Anal Chem 59:2237–2241, 1987.
66. PJ Pickering, JB Chaudhuri. Chirality 9:261–267, 1997.
67. PJ Pickering, CR Southern. J Chem Technol Biotechnol 68:417–424, 1997.
68. W Skrzypinski, E Sierleczko, P Plucinski, B Lejczak, P Kafarski. J Chem Soc Perkin Trans 689–693, 1990.
69. T Yamaguchi, K Nishimura, T Shinbo, M Sugiura. Maku (Membrane) 10:178, 1985.
70. T Yamaguchi, K Nishimura, T Shinbo, M Sugiura. Chem Lett 1549, 1985.
71. T Yamaguchi, K Nishimura, T Shinbo, M Sugiura. Bioelectrochem Bioenerg 20: 109–123, 1988.
72. T Shinbo, T Yamaguchi, K Sakaki, H Yanagishita, K Kitamot, M Sugiura. Chem Express 7:781–784, 1992.
73. T Shinbo, T Yamaguchi, H Yanagishita, K Sakaki, D Kitamoto, M Sugiura. J Membr Sci 84:241–248, 1993.
74. M Bryjak, J Kozlowski, P Wieczorek, P Kafarski. J Membr Sci 85:221–228, 1993.
75. WH Pirkle, WE Bowen. Tetrahedron Asymmetry 5:773–776, 1994.
76. DK Mandal, AK Guha, KK Sirkar. J Membr Sci 114:13–24, 1998.
77. G Fuchs, L Doguet, D Barth, M Perrut. J Chromatogr 623:329, 1992.
78. P Macaudiére, M Caude, R Rosset, A Tambute. J Chromatogr 405:135, 1987.
79. CR Lee, JP Porziemsky, MC Aubert, AM Krstulovic. J Chromatogr 539:55, 1991.
80. P Biermanns, C Miller, V Lyon, W Wilson. LC-GC Int 11:744, 1993.
81. Z Juvancz, K Markides. LC-GC Int 5:44, 1993.
82. K Anton, J Eppinger, L Frederiksen, E Francotte, TA Berger, WH Wilson. J Chromatogr 666:395, 1994.

83. M McHugh, V Krukonis. Supercritical Fluid Extraction, Principles and Practice. Stoneham: Butterworths, 1986.
84. G Brunner. Gas Extraction, An Introduction to Fundamentals of Supercritical Fluids and the Application to Separation Processes. New York: Springer, 1994.
85. WJ Pope, SJJ Peachey. J Chem Soc 75:1066, 1899.
86. E Fogassy, M Ács, T Szili, B Simándi, J Sawinsky. Tetrahedron Lett 35:257, 1994.
87. B Simándi, S Keszei, E Fogassy, J Sawinsky. J Org Chem 62:4390–4394, 1997.
88. R Bauza, A Rios, M Valkárcel. Anal Chim Acta 391:253, 1999.
89. S Keszei, B Simándi, E Székely, E Fogassy, J Sawinsky, S Kemény. Tetrahedron Asymmetry 10:1275–1281, 1999.
90. B Simándi, S Keszei, E Fogassy, S Kemény, J Sawinsky. J Supercrit Fluids 13:331–336, 1998.
91. E Fogassy, F Faigl, M Ács. Tetrahedron 41:2841, 1985.
92. RD Rogers, MA Eiteman. Aqueous Biphasic Separations, Biomolecules to Metal Ions. New York: Plenum Press, 1995.
93. P-Å Albertsson. Partition of Cell Particles and Macromolecules, 2nd ed. New York: Wiley, 1971.
94. H Walter, D Brooks, D Fisher. Partitioning in Aqueous Two-Phase Systems: Theory, Methods, Uses and Applications to Biotechnology, Orlando, FL: Academic Press, 1985.
95. M-R Kula. Bioseparation 1:181–189, 1990.
96. W Kuhlmann, W Halwachs, K Schügerl. Chem-Ing-Tech 52:607, 1980.
97. B Ekberg, B Sellergren, P-Å Albertsson. J Chromatogr 333:211–214, 1985.
98. B Sellergren, B Ekberg, P-Å Albertsson, K Mosbach. US Patent 4,960,762 (1990).
99. B Sellergren, B Ekberg, P-Å Albertsson, K Mosbach. J Chromatogr 450:277–280, 1988.
100. S Yukihisa, S Kazufusa, K Yozo. Nippon Kagaku Kaishi 12:1067–1071, 1994.
101. K Shinomiya, Y Kabasawa, Y Ito. J Liq Chromatogr Relat Technol 21:135–141, 1998.
102. T Arai, H Kuroda. Chromatographia 32:56–60, 1991.
103. JT Hsu. US Patent 4,980,065 (Dec. 25, 1990).
104. T Masawaki, M Sasai, S Tone. J Chem Eng Jpn 25:33–39, 1992.
105. T Masawaki, S Matsumoto, S Tone. J Chem Eng Jpn 27:517–522, 1995.
106. S Tone, T Masawaki, T Hamada. J Membr Sci 103:57–63, 1995.
107. S Tone, T Masawaki, K Eguchi. J Membr Sci 118:31–40, 1996.
108. PEM Overdevest, JTF Keurentjes, A Van der Padt, K Van't Riet. In: JF Scamehorn and JH Harwell, eds. Surfactant-Based Separations: Science and Technology, ACS Symp Ser 740. Washington, DC: American Chemical Society, 1999, pp 123–138.
109. PEM Overdevest, A van der Padt. CHEMTECH 29:12, 17–22, 1999.
110. F Garnier, J Randon, JL Rocca. Sep Purif Technol 16:243–250, 1999.
111. J Randon, F Garnier, JL Rocca, B Maisterrena. J Membr Sci 175:111–117, 2000.

112. T Kokugan, Kaseno, T Takada. Sen-I Gakkaishi 50:581–586, 1994.
113. T Kokugan, Kaseno, T Takada, M Shimizu. J Chem Eng Jpn 28:267–273, 1995.
114. JTF Keurentjes, LJWM Nabuurs, EA Vegter. J Membr Sci 113:351–360, 1996.
115. JTF Keurentjes. PCT Int Pat Appl EP93 02715 (1993).
116. CPMG A'Campo, MS Leloux. PCT Int Pat Appl WO96/11894 (1996).

7

Regularities of Extraction in Systems on the Basis of Polar Organic Solvents and Use of Such Systems for Separation of Important Hydrophobic Substances

Sergey M. Leschev

Belarusian State University, Minsk, Republic of Belarus

I. INTRODUCTION

Solvent extraction is a promising method for the separation, purification, and isolation of various organic and inorganic substances. It is characterized by simplicity and efficiency, and the possibilities for control, by changing the nature and composition of the liquid phases of the extraction systems, are wide. In this connection, the choice of optimal extraction system is of great importance, but choosing well is hard or even impossible without knowledge of the main regularities of the extraction process. So far, "classical" extraction systems consisting of organic solvents or their mixtures and water and aqueous solutions of mineral salts have been the ones most widely investigated. The main regularities of extraction of organic substances in these systems are sufficiently well generalized and systematized with the use of methods featuring group increments [1,2] and correlation equations [3,4].

It is well known that extraction systems consisting of an aliphatic hydrocarbon and polar organic solvents and their aqueous mixtures were first successfully used in petrochemistry for the separation of near-boiling aliphatic and aromatic hydrocarbons. The further use of such systems in various fields of

295

petrochemistry, connected mainly with hydrocarbon extraction [5], remains wide, and numerous investigations in this direction are still forth coming [6]. On the other hand, these systems are sporadically used for the isolation, separation, and purification of other hydrophobic organic substances that are important for chemical industry, such as higher homologs of alcohols, amines, and carboxylic acids [7,8]. Some theoretical aspects of the extraction of various organic substances by such systems have been discussed in the literature [9], but the main regularities of extraction in these "nonaqueous" systems are still not generalized in reviews.

Therefore, the main goals of present chapter are as follows:

1. To show the main regularities of extraction processes in systems consisting of an aliphatic hydrocarbon and a polar phase, which may be made up of polar organic solvents, their aqueous mixtures, or solutions of zinc halides in polar organic solvents.
2. To establish the main differences of such systems from "classical" aqueous extraction systems.
3. To illustrate the methods of choosing and subsequent use of systems on the basis of polar organic solvents for the separation, purification, and isolation of some synthetic and natural hydrophobic organic substances that are important for chemical industry.

A systematic experimental study of the extraction of various classes of organic nonelectrolytes (oxygen-, nitrogen-, sulfur-, and halogen-containing substances, unsaturated and aromatic hydrocarbons) as well as of some organic salts in systems on the basis of polar organic solvents has been carried out at the Belarus State University [10–20] and forms the basis for this chapter.

II. METHODOLOGY

The experimental methods for "nonaqueous" extraction systems have been reported [10–20]. In the majority of cases, n-octane was used as the standard nonpolar hydrocarbon phase, whereas lower alcohols (from methanol to propanol and 2-propanol), lower glycols, N,N-dimethylformamide (DMF), dimethyl sulfoxide (DMSO), acetonitrile (AN), 2-methoxyethanol, and aqueous mixtures of these solvents were used as the polar phase. Aqueous organic phases were characterized by the volume fraction of the organic solvent (φ), varying from 0.2 to 1.0. Solutions of anhydrous zinc halides in various polar solvents with the salt concentration from 1 to 7 M (mol/dm^3) were also used as the polar phase.

The extraction of various classes of model and natural organic nonelectrolytes was investigated. The model substances were oxygen containing—primary alcohols, ethers, esters, ketones, and carboxylic acids [10–12]; nitrogen containing—primary, secondary, and tertiary amines [13] and nitroalkanes [14]; sulfur containing—sulfides, disulfides, sulfoxides, sulfones, and mercaptanes [15,16]; and halogen containing—alkyl halides [10]. All these model substances involved linear hydrocarbon radicals with 1 to 12 carbon atoms. Unsaturated and aromatic hydrocarbons were used as model substances, too [17–19]: the former included lower homologs of ethylene, acetylene, and butadiene with aliphatic and aromatic substituents. Aromatic hydrocarbons included benzene and its derivatives and some polycyclic aromatic hydrocarbons (PAHs). Natural organic substances included the terpenoid alcohol linalool [20], the bicyclic diterpene glycol sclareol [21], the water-insoluble vitamin α-tocopherol [22], petroleum sulfides and sulfoxides [23], various oil fractions [23,24], and turpentine [25].

Partition constants, P, of model nonelectrolytes between octane and the polar phases were determined at $20 \pm 1°C$. For the suppression of the dissociation of amines and carboxylic acids in the polar phase, small additions of mineral alkalies and strong acids were used. The partition constants were obtained at high dilutions that corresponded to the ideal region of concentrations of the distributed substances, where self-association was practically absent [10–18]. The partition constants were, therefore, close to the thermodynamic values.

Extraction constants, K_{ex}, of some organic salts [in the majority of cases, halides of higher quaternary ammonium salts (QAS) and higher amines salts with molar masses up to 1000 g/mol] were also obtained at $20 \pm 1°C$. Thermodynamic extraction contants of these substances in various extraction systems were calculated [26,27], but the procedure for the determination of these constants was much more complicated than for organic nonelectrolytes. The upper concentration limit for ideal behavior of nonelectrolytes in octane varied from 0.01 M for primary alcohols to 10^{-4} M for higher quaternary ammonium salts, while for low polarity organic substances this could reach 1–2 M [17,18]. The concentration activity coefficients γ of some investigated substances were calculated from the dependence of their partition coefficients D on their concentrations in the hydrocarbon or polar phases [26,27].

The investigators studied the dissociative extraction of higher amines, containing up to 36 carbon atoms in their aliphatic hydrocarbon radicals, and of carboxylic acids, containing 10–18 carbon atoms. The process of dissociative extraction was described by the log D values that were linear functions of the pH of the polar phase [28,29]. These functions were used for the calculation of P values of higher amines from the experimentally determined values of D and their protonation constants in the polar phase [28,29]. This permitted the determination of P values that were extremely high even for systems with polar organic or aqueous organic phases and could not be determined experimentally.

The extraction of amine salts from the polar phase to octane was taken into account, and methods of its control were proposed [28,29].

The extraction of some natural complicated mixtures of compounds of various structure and molar mass was described by means of the partition coefficients. Typical examples were oil sulfoxides [23], oil mercaptans [24], and turpentine carboxylic acids [25].

The determination of concentrations of these substances in the phases of the extraction systems was carried on by various methods as required. Amines were determined by nonaqueous titration, polycyclic aromatic hydrocarbons by UV photometry, and higher QAS by extraction photometry [26]. Higher sulfoxides and sulfides were determined by specially developed highly sensitive and selective methods of extraction photometry using acidic dyes [30,31]. Concentrations of volatile substances were determined by means of gas chromatography.

The logarithms of thermodynamic partition and extraction constants were used for the calculation of group increments as described elsewhere [10–20]. It is necessary to emphasize that these increments were calculated from thermodynamic values of P and K_{ex} of substances that ensure the absence of intramolecular interactions of groups in their molecules. Otherwise, special indexes were used for noting such interactions. For instance, methylene group increments were calculated from the difference of logarithms of partition constants of two homologs with linear hydrocarbon chains:

$$I_{CH_2} = \frac{\log P_{n+m} - \log P_n}{m} \tag{1}$$

where $n + m$ and n are the numbers of carbon atoms in molecules of two homologs: for instance, n-butylamine ($n = 4$) and n-octylamine ($n + m = 8$) [13]. Generally, homologs with sufficiently large differences in carbon atom numbers were taken for decreasing the error in I_{CH_2} values. Methylene group increments may be calculated from the data on aliphatic hydrocarbon solubilities in the polar phases of the extraction systems [32], and these values were found to be close to values calculated from Eq. (1). The error in I_{CH_2} values did not exceed ± 0.02.

Increments of functional groups connected to aliphatic linear hydrocarbon radicals were calculated as follows:

$$I_f = \log P_i - nI_{CH_2} - iI_H \tag{2}$$

where i is the number of replacements of hydrogen atoms for functional groups and I_H is the increment for the hydrogen atom, which was experimentally determined to be equal to 0.5 I_{CH_2} [33].

Increments of polar functional groups, connected to an aromatic substituent, were marked by subscript "arom" and were calculated as follows:

$$I_{f\,arom} = \log P - I_{C_6H_5} \qquad (3)$$

where $I_{C_6H_5}$ is the increment of a phenyl group in $\log P$ that can be calculated from the data on benzene or toluene distribution [18]. Increments in $\log K_{ex}$ of functional groups of quaternary ammonium salts were calculated according to this procedure, too [27].

The error in I_f of organic nonelectrolytes did not exceed ± 0.1. Because of the large numbers of carbon atoms in molecules of the higher QAS, the error in increments of QAS functional groups was $\pm 0.5-1.0$.

III. PHENOMENOLOGICAL DESCRIPTION OF THE EXTRACTION PROCESS

Experimental distribution and extraction data for concrete representatives of various classes of organic substances were reported in the systematic study mentioned in Section I [10–20]. Typical results of the distribution of organic nonelectrolytes in systems with aqueous organic polar phases are described in this section.

The dependencies of the partition constants of nonelectrolytes on the nature and composition of the polar phase are rather complicated. For instance, partition constants of some polar low molecular mass substances grow with increasing organic solvent content φ in the polar phase but may have an extremum. On the other hand, partition constants of nonpolar organic substances and higher homologs of polar substances diminish with increasing φ in the polar phase [10–20] as shown in Fig. 1.

The appearance of an extremum in the dependence of the thermodynamic extraction constant on φ is typical for organic electrolytes [20], and this indicates a common nature for the driving forces of extraction processes with the participation of organic electrolytes and polar nonelectrolytes.

Systems with polar phases involving solutions of zinc halides in polar solvents also show a relatively complicated dependence of the partition constants on salt concentration in the polar phase [11] (Fig. 2). The partition constants of polar and electron donor organic nonelectrolytes fall with increasing salt concentration in the polar phase, whereas the reverse is typical for nonpolar and weakly polar substances. The slopes of the falling or increasing partition constants depend on the number of carbon atoms in the hydrocarbon radicals of these substances.

It is necessary to emphasize that these results cannot be explained on the basis of stoichiometric approaches as used for "classical" extraction systems, in particular, the involvement of various solvates of constant composition formed in the polar phase [34]. The foregoing results show the influence of the nature

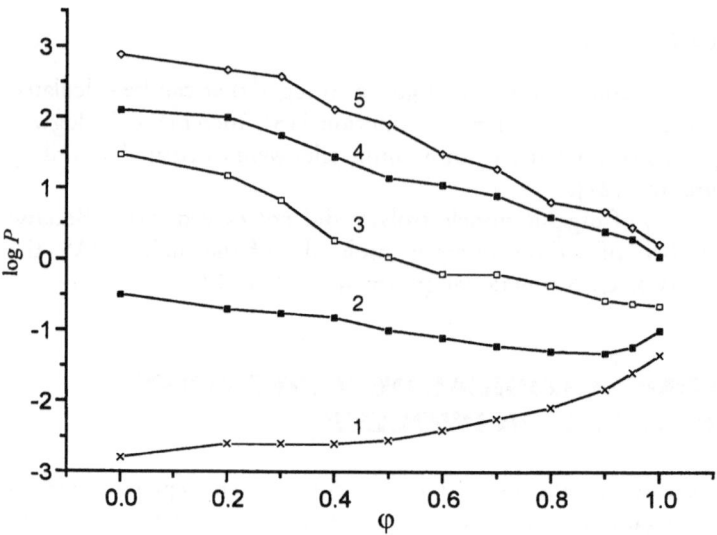

Figure 1   Dependencies of log P of some organic nonelectrolytes on φ in octane–methanol–water mixtures: 1, methanol; 2, n-butylamine; 3, n-octanol; 4, diethyl sulfide; 5, toluene. (From Refs. 11, 13, and 19.)

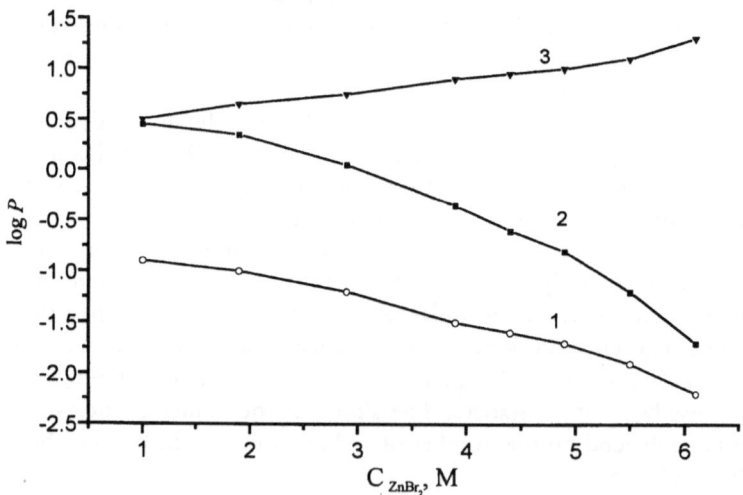

Figure 2   Dependencies of log P of some organic nonelectrolytes on salt concentration in octane–zinc bromide solutions in methanol: 1, n-octanol; 2, diethyl sulfide; 3, toluene. (From Ref. 36.)

and composition of the polar phase on extraction to be much more complicated than that of the nature and composition of the organic phase of "classical" systems.

IV. GROUP INCREMENTS AND THEIR USE FOR THE DESCRIPTION AND PREDICTION OF THE EXTRACTION OF VARIOUS SUBSTANCES

The dependence of group increments on the nature and composition of the polar aqueous organic phase is simple (Tables 1 and 2). The increments of the

Table 1 Group Increments in Systems Consisting of Octane and Aqueous Methanol Mixtures

φ	I_{CH_2}	$I_{C_6H_5}$	$I_{C\equiv CH}$	$I_{C\equiv CH\ arom}$	I_I	I_{Br}	I_{Cl}
1.00	0.10	0.0	-0.3	-0.3	-0.1	-0.3	-0.4
0.95	0.14	0.2	-0.4	-0.4	-0.1	-0.2	-0.14
0.90	0.18	0.3	-0.4	-0.4	0.0	-0.2	-0.4
0.80	0.24	0.6	-0.3	-0.1	0.0	-0.2	-0.4
0.70	0.30	1.0	-0.3	0.0	0.1	-0.2	-0.3
0.60	0.33	1.2	-0.1	0.1	0.3	-0.1	-0.2
0.50	0.37	1.4	0.0	0.3	0.4	-0.1	-0.2
0.40	0.41	1.6	0.05	0.4	0.4	0.0	-0.2
0.30	0.49	1.8	0.05	0.5	0.4	0.0	-0.1
0.20	0.54	1.9	0.05	0.6	0.5	0.0	-0.1
0.00	0.61	2.0	0.05	0.7	0.6	0.1	-0.1

φ	I_{NH_2}	I_{NH}	I_N	I_{OH}	I_O	I_S	I_{SO}
1.00	-1.5	-1.4	-1.2	-1.5	-0.6	-0.4	-2.1
0.95	-1.9	-1.8	-1.5	-1.8	-0.8	-0.4	-2.7
0.90	-2.1	-1.9	-1.7	-2.1	-0.8	-0.5	-3.1
0.80	-2.4	-2.2	-2.0	-2.4	-1.0	-0.6	-3.7
0.70	-2.6	-2.5	-2.2	-2.7	-1.2	-0.6	-4.2
0.60	-2.6	-2.6	-2.4	-3.0	-1.3	-0.6	-4.3
0.50	-2.7	-2.7	-2.5	-3.1	-1.3	-0.7	-4.4
0.40	-2.7	-2.7	-2.5	-3.1	-1.5	-0.6	-4.7
0.30	-3.0	-3.1	-3.1	-3.3	-1.7	-0.7	-5.5
0.20	-3.1	-3.3	-3.3	-3.4	-1.9	-0.7	-5.7
0.00	-3.2	-3.6	-3.5	-3.7	-2.2	-0.9	-6.4

Source: Refs. 11, 13, 16, and 17.

Table 2 Group Increments in Systems Consisting of Octane and Zinc Bromide Solutions in Methanol

C_{ZnBr_2} (M)	I_{CH_2}	$I_{C_6H_5}$	I_i	I_{OH}	I_O
6.1	0.26	0.8	0.1	−4.4	−2.5
5.5	0.24	—	—	−3.9	−1.8
4.9	0.22	0.7	0.0	−3.6	−1.7
4.4	0.22	—	—	−3.5	−1.5
3.9	0.21	0.6	0.0	−3.3	−1.4
2.9	0.19	—	—	−2.8	−1.0
1.9	0.17	—	—	−2.4	−0.8
1.0	0.15	0.3	−0.05	−2.2	−0.7

C_{ZnBr_2} (M)	I_{CO}	I_S	I_{SS}	I_{SH}	I_{SO}
6.1	−3.2	−3.0	−0.9	−2.0	−7.4
5.5	−2.8	−2.4	−0.7	−1.8	−7.0
4.9	−2.6	−1.9	−0.5	−1.6	−6.7
4.4	−2.5	−1.7	−0.5	−1.5	−6.4
3.9	−2.3	−1.4	−0.4	−1.3	−6.4
2.9	−2.0	−0.9	−0.3	−1.0	−5.2
1.9	−1.7	−0.5	−0.2	−0.7	−4.3
1.0	−1.5	−0.3	−0.3	−0.6	−3.6

Source: Refs. 35 and 36.

nonpolar methylene group and those of the majority of low-polarity groups fall with increasing φ, whereas increments of polar groups become larger. This can be readily explained by the destruction of the spatial structure of the polar phase and the reduction of its solvating ability [13] when water is replaced by an organic solvent. Another reason for the growth of polar group increments is the increasing solubility of the organic solvent in the octane phase. The influence of this factor on the extraction of polar substances, especially of organic electrolytes, is discussed in more detail later.

Zinc halides play a role that is analogous to that of water in aqueous solvent mixtures. It "structures" the polar phase and increases its solvating ability. The increase of the methylene group increments indicates "structuring" of the polar phase with increasing salt concentrations, whereas the fall of the increments of electron donor and polar groups indicates growth of its solvating ability. The latter phenomenon is in some ways unusual, since in aqueous extraction systems mineral salts are usually used for increasing the extraction from water (i.e., for salting-out) [3]. This phenomenon is attributable to the ability of zinc ions to form complexes with electron donor substances and to the coordinative unsaturation of zinc halides, especially in concentrated solutions [35]. This fact is confirmed by the practical disappearance of salting-in (into the polar phase) when there is conjugation of electron pairs or when electron pairs become delocalized in molecules of the distribuent. For instance, the disulfide group and the thiophene ring hardly salt in at all [36]. Zinc halides are among the few salts able to salt in various polar and electron donor organic substances convenient for practical purposes [36]. Even changing the halide anions in zinc halides for oxygen-containing anions leads to practical loss of the salting-in ability because most of the coordination sites of the zinc cation are saturated by the oxygen atoms of the anions [36].

For practical purposes, salt-containing extraction systems with participation of concentrated solutions of zinc halides in some octane-miscible polar solvents, such as propanol and butanol, have the following very important property: very low values of methylene group increments and, at the same time, highly negative increments of polar and electron donor groups [36]. This fact allows us to isolate effectively high molecular mass polar and electron donor organic substances from hydrocarbon solutions (see later), and, at the same time, distinguishes salt-containing systems from the usual ones.

Group increments can be effectively used for the description of observed regularities of extraction of various substances. All observed phenomena can be explained by the effects of oppositely directed and different slopes of polar and nonpolar group increment dependencies on φ or on zinc halide concentration. In particular, for polar low molecular mass substances, either the increase of the slope for polar group increments with growing φ exceeds the falling slope of hydrocarbon radical increment or these slopes are comparable. On the other hand, for higher homologs of polar substances and for nonpolar and low-polarity substances, the slope of hydrocarbon radical changes provides a dominant contribution to the dependence on increasing φ or on salt concentration in the polar phase.

The dependence of polar group increments on the nature of the polar organic solvent can be explained by the chemical nature of the group and the structure and solvation ability of the polar solvent, but special corrections for the solubility of the polar solvent in octane must be made [37].

Another important use of group increments is for the prediction of partition and extraction constants of substances of simple molecular structure with any molar mass in given extraction systems (Table 3).

The term "simple molecular structure" refers to substances with linear hydrocarbon radicals or with branching far from the functional group. In other words, these substances are characterized by the practical absence of significant intramolecular interactions of groups. This limits the variety of substances, the extraction of which can be effectively described and predicted by means of group increments. Still, a large number of organic substances important for the

Table 3 Comparison of the Logarithms of the Partition and Thermodynamic Extraction Constants of Some Substances in Octane–Aqueous Polar Organic Solvent Systems and Their Aqueous Mixtures, Determined Experimentally and Calculated with the Use of Group Increments

| | Solutes | | | | | |
| | Dinonylamine | | Dibutyldecylamine | | Triheptylamine | |
Polar phase	Exp	Calc	Exp	Calc	Exp	Calc
Propanol, $\varphi = 0.70$	1.3	1.4	1.4	1.5	1.8	1.9
Propanol, $\varphi = 0.60$	1.8	1.8	1.9	1.8	2.3	2.4
Propanol, $\varphi = 0.50$	2.4	2.3	2.3	2.3	2.9	2.9

| | Solutes | | | | | |
| | Tridecylsolutyl-ammonium sulfide | | Trinonyloctadecyl-ammonium sulfide | | Tetrakis(decyl)am-monium chloride | |
Polar phase	Exp	Calc	Exp	Calc	Exp	Calc
Methanol, $\varphi = 1.00$	0.3	0.5	1.2	1.5	0.8	0.9
Methanol, $\varphi = 0.95$	−0.1	0.1	—	—	0.7	0.7
DMF, $\varphi = 1.00$	−1.4	−1.2	−0.4	−0.2	−0.8	−0.5
DMF, $\varphi = 0.95$	−1.0	−0.9	−0.2	−0.2	−0.8	−0.6
DMF, $\varphi = 0.90$	−0.8	−0.6	—	—	—	—
Acetonitrile, $\varphi = 1.00$	−1.4	−1.4	0.0	0.1	−0.5	−0.5

Source: Refs. 27 and 28.

chemical industry belong to the category of substances of "simple molecular structure," including primary aliphatic alcohols, ketones, esters, carboxylic acids, amines, and unsaturated hydrocarbons. However, the efficiency of the use of the group increments method depends on the limitation of the fragments into which the extracted molecule can be conceptually divided, or on special corrections used for the procedure of increment summing [1,2]. Intramolecular interactions in molecules of the extracted substances strongly influence the increments of various groups when water is the polar phase of the extraction system, and in this case group increments have the widest range (Table 1). For systems with polar organic and aqueous organic phases, the role of these interactions becomes much less significant. For instance, increments of aliphatic and aromatic ethynyl groups are practically equal in extraction systems with large values of φ, whereas for water the increment of an aromatic ethynyl group strongly exceeds that of aliphatic one [17]. Furthermore, extraction of benzene derivatives with phenyl substituents in their molecules in systems with a polar organic phase can be satisfactorily described and predicted by using phenyl group increments, in contrast to the octane–water system [38]. On the other hand, extraction of condensed aromatic hydrocarbon derivatives containing phenyl substituents in their molecules cannot be described by using group increments in these systems [39]. Table 3 shows that generally, the prediction error is comparable to the experimental errors. This allows the consideration of group increments as objective and full characteristics of extraction systems.

Group increment analysis shows the main difference between the extraction systems discussed here and the classical aqueous ones. For the former, all the increments of polar and nonpolar groups, and in particular, methylene group increments, are functions of the nature and composition of the polar phase (see Tables 1 and 2). In comparison, for the overwhelming majority of classical extraction systems, the methylene group increment is almost independent of the nature of the organic solvent [40]. For the latter systems with nonpolar or low-polarity organic phases, the methylene group increment varies from 0.58 to 0.64. This fact depends on the practical absence of spatial structure of these organic solvents and their relatively low solubility in water. A similar picture is sometimes observed for some other low-polarity groups, such as phenyl and sulfide groups, in classical systems [1,2,15,18].

Another important feature of systems with polar organic, aqueous organic, and salt-containing phases is a reduced value of the methylene group increments in comparison with classical systems. Even small additions ($\varphi = 0.05$) of polar organic solvents to water reduce the methylene group increment and the hydrophobic effect [41].

Group increments can characterize some specific and important propeties of polar organic solvents, their aqueous mixtures, and solutions of mineral salts in polar solvents. Such properties include the propensity for intermolecular interactions and solvating ability with respect to various substances. For instance, the liquid structure formed by the polar solvent as a result of its self-association is a very important property that determines extraction and the ability of the solvent to dissolve hydrophobic substances. The methylene group increment was proposed to represent the energetic strength of the spatial structure of the solvent and a measure of its solvophobic effect [42]. The factors determining the methylene group increments for various solvents have been discussed [42,43]. They include the polyfunctionality of the solvent molecule, its polarity, and the donor–acceptor ability of its functional groups. Some data on methylene group increments for polar solvents, nonmiscible and miscible with aliphatic hydrocarbons, are given in Tables 4 and 5, respectively.

Water is outstanding among the polar solvents, including those that are more polar than water. Even sulfuric acid, which has several structure-forming centers in its molecule and is more polar than water, has a methylene group increment value inferior to that of water. Methylene group increments have so

Table 4 Methylene Group Increments for Octane–Polar Organic Solvent Systems and Gibbs Energies of Transfer of the Methylene Group from the Solvent to Octane

Solvent	I_{CH_2}	$-\Delta G°_{CH_2}$ (J/mol)
Water	0.61	3430
Sulfuric acid	0.42	2340
Nitric acid	0.14	840
Glycerol	0.35	1970
Ethylene glycol	0.27	1510
1,3-Propanediol	0.21	1170
1,4-Butanediol	0.19	1070
2,5-Hexanediol	0.14	800
Diethylene glycol	0.20	1130
Tetraethylene glycol	0.16	880
Furfuryl aldehyde	0.13	710
Formic acid	0.26	1460
Propylene carbonate	0.20	1130
Monoethanolamine	0.26	1460
Ethylene diamine	0.18	1000
Formamide	0.31	1740
N,N-Dimethylformamide	0.11	630
Dimethyl sulfoxide	0.22	1230
Nitromethane	0.17	960
Acetonitrile	0.13	710
Acetone cyanhydrine	0.20	1130

Source: Ref. 43.

far been determined for more than 160 solvents, including those that are miscible with aliphatic hydrocarbons. A special procedure was proposed for the determination of methylene group increments for solvents of the latter kind [43]. These values can be used for the estimation of the solvophobic effect for the dissolution of various hydrophobic substances in such solvents.

The quantitative description of the dependencies of group increments on the nature and composition of the polar phases allows a drastic reduction in the amount of experimental data required for the determination of partition and extraction constants. Numerous attempts have been directed at such a description of partition constants and group increments, using physicochemical properties of the distributed substances and the solvents [3,4,44–47]. For the systems under discussion, however, attempts to show dependencies of group increments—in particular, methylene group increments [42]—on the nature and composition of the polar phases in terms of such physicochemical param-

Table 5 Methylene Group Increments for Hypothetical Octane–Solvent Extraction Systems That Are Miscible with Aliphatic Hydrocarbons and Gibbs Energies of Transfer of the Methylene Group from Octane to the Solvent

Solvent	I_{CH_2}	$-\Delta G^\circ_{CH_2}$ (J/mol)
Ethanol	0.078	435
n-Propanol	0.068	380
i-Propanol	0.074	415
n-Butanol	0.064	360
n-Pentanol	0.058	325
n-Hexanol	0.055	310
n-Octanol	0.048	265
2-Ethylhexanol	0.037	210
Methyl acetate	0.068	380
Ethyl acetate	0.058	325
Propyl acetate	0.043	245
Pentyl acetate	0.030	165
Diethyl ether	0.008	45
Dihexyl ether	0.015	85
Acetone	0.083	465
2-Butanone	0.056	325
Cyclopentanone	0.056	190
2-Octanone	0.034	445
Butanenitrile	0.079	335
Octanenitrile	0.060	330
Benzonitrile	0.059	—

Source: Ref. 43.

eters as the solubility parameter, surface tension, dipole moment, molar volume, and other parameters have not been successful. Such dependencies are valid only for limited groups of similar polar solvents and extracted substances. On the other hand, dependencies of group increments on φ for systems with aqueous organic polar phases and on salt concentration for systems with salt-containing polar phases may be described by adjusting empirical correlation equations [36,48]. Since these equations generally have no physical basis, they cannot be used for extraction systems of other types.

In conclusion, the group increment method is the most convenient method for the generalization of data on extraction of substances with simple molecular structure in the extraction systems reviewed. Because of the high importance of substances of complicated structure, the extension of this method for use for such substances is very desirable.

V. MUTUAL SOLUBILITY OF PHASE COMPONENTS AND ITS INFLUENCE ON GROUP INCREMENTS

Since immiscible liquids in contact dissolve each other to some extent, a great deal of mutual solubility data for water and organic solvents is reported in the chemical literature [1,49]. Minimal mutual solubilities of the components of the phases are typical for extraction systems comprising an aliphatic hydrocarbon and water. On the other hand, systems involving an aliphatic hydrocarbon with polar organic solvents and their aqueous and mineral salt mixtures show elevated mutual solubility compared with the standard octane–water system [50] (Table 6).

The solubility of polar organic solvents, in particular alcohols, in the octane phase can reach 2 M. Since alcohols are strong solvating agents for various polar substances [51], they should have a great influence on the solubility of polar solvents in octane in the extraction process. It was found that in the great majority of cases, the solubility of octane in the polar phase does not exceed 10% by volume, and its influence on the group increments is negligible. Furthermore, the solubility of polar solvents in octane, up to 1 M, influences neither the methylene group increment nor, generally, that of such low-polarity groups as phenyl and ethynyl radicals, and halogen atoms. A marked influence of such solubility, however, is typical for high-polarity groups of nonelectrolytes and especially for functional groups of organic salts.

The modeling of the solvating action of polar low molecular mass solvents dissolved in the octane phase is based on their higher homologs, dissolved in the octane phase of octane–water [37] and octane–ethylene glycol [52] systems. The large hydrocarbon radicals of these modeling substances make them practically insoluble in water or ethylene glycol, and thus these systems are chosen as standard "clean" ones.

The solvation of the polar group of the extracted substance by the polar solvent in the octane phase leads to increasing polar functional group increments (Figs 3 and 4). The action by the modeling additives on the polar group increments becomes appreciable from some definite value of their concentration in octane. This value depends on the nature of the polar group and the additive and characterizes the strength of the solvation interactions between the polar group and the modeling additive, as does also the slope of the dependence of the polar group increment on the concentration of the additive. Figures 3 and 4 also show that the increases of some polar group increments can reach 3 units per 1 mol/dm^{-3} of polar solvent solubility in octane, and that alcohols are the strongest solvating agents for such groups, especially non-self-associated alcohols.

Table 6 Solubilities (mol/dm³) of Polar Organic Solvents in Octane (Oct) and Hexadecane (Hex) Phases of Extraction Systems with Polar Organic or Aqueous Organic Phases

	Solvents					
	Methanol		Ethanol		n-Propanol	
φ	Oct	Hex	Oct	Hex	Oct	Hex
1.00	1.00	0.34	—	1.1	—	—
0.95	0.44	0.19	3.0	0.39	—	1.8
0.90	0.25	0.13	1.1	0.25	—	0.87
0.80	0.15	0.085	0.52	0.14	1.8	0.59
0.70	0.10	0.066	0.31	0.093	1.4	0.45
0.60	0.080	0.062	0.22	0.075	1.3	0.41
0.50	0.060	0.050	0.15	0.070	1.2	0.34
0.40	0.045	0.038	0.11	0.060	1.0	0.34
0.30	0.030	0.024	0.081	0.050	0.73	0.26
0.20	0.020	0.016	0.050	0.040	0.21	0.15
0.00	0.0037[a]	0.0030[a]	—	—	—	—

	Solvents					
	DMF		Acetonitrile		DMSO	
φ	Oct	Hex	Oct	Hex	Oct	Hex
1.00	0.34	0.24	0.41	0.30	0.050	0.045
0.95	0.23	0.16	0.31	0.21	0.038	0.030
0.90	0.16	0.095	0.31	0.21	0.027	0.023
0.80	0.11	0.080	0.30	0.19	0.018	0.016
0.70	0.072	0.060	0.29	0.18	0.009	0.008
0.60	0.044	0.035	0.27	0.17	0.005	0.004
0.50	0.028	0.022	0.27	0.17	0.002	0.002
0.40	0.019	0.015	0.27	0.17	—	—
0.30	0.012	0.010	0.22	0.15	—	—
0.20	0.007	0.006	0.15	0.12	—	—

[a]Literature data on water solubility in hydrocarbons.
Source: Ref. 50.

The effect of increasing water or ethylene glycol concentration in the octane phase as the concentration of the modeling additive grows is taken into account. The hydration and solvation numbers of most strongly solvating modeling additives at concentrations up to 1 M did not exceed 0.1 [52], and such

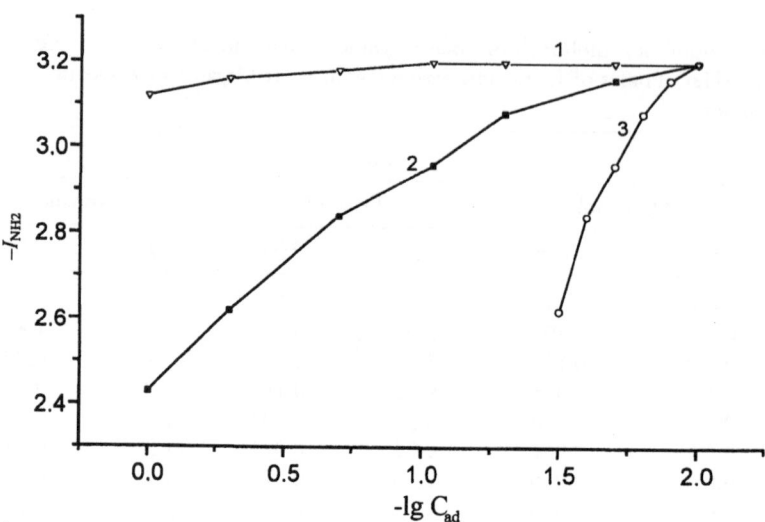

Figure 3   The influence of modeling additions on primary amino group increment in an octane–water system: 1, dioctylformamide; 2, n-octanol; 3, nonassociated octanol. Concentrations of nonassociated octanol are calculated on the basis of the concentration dependence of the octanol partition constant between octane and water. (From Ref. 37.)

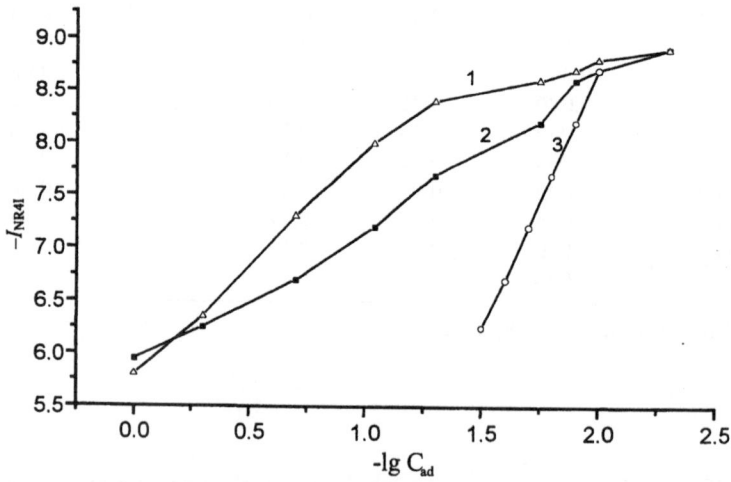

Figure 4   The influence of modeling additions on the functional group increment of iodide quaternary ammonium salts in an octane–ethylene glycol system: 1, dioctyl-formamide; 2, n-dodecanol; 3, nonassociated dodecanol. (From Ref. 52.)

amounts of water or ethylene glycol cannot have significant effect on the solvating ability of the additive.

The dependencies shown in Figs. 3 and 4 are used for the determination of corrections for the increments of polar groups, ΔI_f, for the calculation of group increments in hypothetical "clean" extraction systems, $I_{f\,cl}$:

$$I_{f\,cl} = I_f - \Delta I_f \tag{4}$$

where ΔI_f is the increase of the polar group increment at a value of modeling additive concentration that is equal to the solubility of the polar solvent in the octane phase, and I_f is the polar group increment for the actual extraction system.

Group increments in clean systems for primary amines are presented in Table 7. These systems are characterized by more negative values of the polar group increments than real extraction systems involving aqueous alcohols as the polar phases. Systems with participation of DMF and DMSO are characterized by higher apparent affinity to amino groups than systems with an aqueous propanol phase, which again differ from systems with aqueous ethanol and especially methanol. However, in clean systems all aqueous alcohol mixtures are stronger solvating agents for amines than aqueous DMF and DMSO mixtures, and the apparent differences among the aqueous alcohol mixtures are due to alcohol solubility in octane rather than to the nature of the alcohol, especially for small values of φ.

Increments of functional groups in clean extraction systems may be considered to be a measure of the affinity of various groups to the polar phase of the extraction system at comparable affinities to the octane phase. The group affinity to the octane phase is mainly a function of the size or the molar refraction of the group [53], which can vary within wide limits, hence can influence strongly the apparent affinity of the group to the polar phase. The solvation Gibbs energy of any substance or group, which can be calculated from liquid–vapor equilibrium data, is the proper measure of its affinity to the individual or mixed solvent [54].

VI. DETERMINATION OF GIBBS ENERGIES OF SOLVATION OF ORGANIC NONELECTROLYTES BASED ON THEIR DISTRIBUTION

The standard Gibbs energy of the distribution of a nonelectrolyte between octane and a polar phase, $\Delta_{distr} G^\circ$, can be expressed by the equation:

Table 7 Comparison of Group Increments of Primary Amine Groups in Real and "Clean" Extraction Systems

	Solvents					
	Methanol		Ethanol		n-Propanol	
φ	Real	"Clean"	Real	"Clean"	Real	"Clean"
1.00	−1.5	−2.3	—	—	—	—
0.95	−1.9	−2.5	—	—	—	—
0.90	−2.1	−2.6	−1.5	−2.3	—	—
0.80	−2.4	−2.7	−1.7	−2.3	−1.0	−1.9
0.70	−2.6	−2.8	−2.0	−2.5	−1.2	−2.1
0.60	−2.6	−2.8	−2.1	−2.5	−1.3	−2.1
0.50	−2.7	−2.8	−2.1	−2.5	−1.3	−2.0
0.40	−2.7	−2.8	−2.2	−2.5	−1.3	−2.1
0.30	−3.0	−3.1	−2.4	−2.6	−1.5	−2.3
0.20	−3.1	−3.1	−2.8	−2.9	−2.1	−2.4

	Solvents			
	DMF		DMSO	
φ	Real	"Clean"	Real	"Clean"
1.00	−0.9	−0.9	−1.1	−1.1
0.95	−1.2	−1.2	−1.2	−1.2
0.90	−1.5	−1.5	−1.3	−1.3
0.80	−1.9	−1.9	−1.5	−1.5
0.70	−2.1	−2.1	−1.8	−1.8
0.60	−2.2	−2.2	−2.0	−2.0
0.50	−2.2	−2.2	−2.5	−2.5
0.40	−2.4	−2.4	−2.7	−2.7
0.30	−2.5	−2.5	−2.9	−2.9

Source: Ref. 37.

$$\Delta_{distr} G° = \Delta_{solv} G°_{oct} - \Delta_{solv} G°_{pol} \tag{5}$$

where $\Delta_{solv} G°_{oct}$ is the solvation Gibbs energy by octane and $\Delta_{solv} G°_{pol}$ is that by the polar solvent. Although $\Delta_{distr} G°$ can be readily determined experimentally, the determination of $\Delta_{solv} G°_{oct}$ and $\Delta_{solv} G°_{pol}$ often requires complicated and difficult vapor pressure measurements above the solutions of the nonelectrolyte in the respective solvents. These values can be more easily estimated on the basis of the distribution and of group increments. The following procedure was recommended for this purpose [55]. The Gibbs energies of distribution are determined for the "clean" octane–water system and are described by the equation:

$$\Delta_{distr}G° = \Delta_{solv}G°_{oct} - \Delta_{hydr}G° \tag{6}$$

where $\Delta_{hydr}G°$ is the Gibbs energy of hydration of the nonelectrolytes. In some cases this value can be taken from the chemical literature [54]. Gibbs energies of hydration were calculated from saturated vapor pressure data of various classes of nonelectrolytes and their solubilities in water [55]. Data for the calculation were selected in such a way that vapor pressures and solubilities were small, and the gaseous and aqueous phases were close to ideality. Equation (6) was used for the calculation of $\Delta_{solv}G°_{oct}$. Comparison of $\Delta_{solv}G°_{oct}$ of some nonelectrolytes derived from the vapor pressure above their octane solutions [56] and those calculated from distribution data [Eq. (6)] confirmed the validity of that equation. The terms of Eq. (6) for some organic nonelectrolytes are shown in Table 8.

Subsequently, the use of Eq. (5) to calculate $\Delta_{solv}G°_{pol}$, with $\Delta_{distr}G°$ representing the distribution of Gibbs energy in the "clean" extraction system,

Table 8 Solubilities of Some Organic Nonelectrolytes in Water, Their Saturated Vapor Pressures, and Their Standard Molar Gibbs Energies of Hydration, Distribution and Solvation by Octane at 20°C

Nonelectrolyte	Solubility, s (mass %)	Saturated vapor pressure, p (kPa)	Gibbs energies (kJ/mol)		
			$-\Delta_{hydr}G°$	$-\Delta_{distr}G°$	$-\Delta_{solv}G°_{oct}$
Benzene	0.14	10.11	3.68	12.80	16.48
Styrene	0.013	0.56	4.02	17.28	21.30
Thiophene	0.32	8.38	5.90	10.67	16.57
n-Hexanol	0.59	0.093	18.41	1.42	19.83
2-Butanone	1.75	1.21	14.27	6.15	20.42
Propyl acetate	2.3	3.33	12.43	5.40	17.82
n-Butylamine	—	—	18.66[a]	-2.76	15.90
Di-n-butylamine	0.50	0.20	15.02	10.67	25.69
Nitromethane	9.5	3.72	16.90	-4.77	12.13
Nitroethane	4.5	2.13	16.02	-1.34	14.69
Diethyl sulfide	0.22	7.18	5.15	12.05	17.20
Dimethyl disulfide	0.21	3.06	7.11	11.30	18.41
Ethyl mercaptan	1.5	55.86	5.86	4.73	10.59

[a]Data from Ref. 54.
Source: Ref. 55 except as noted.

presented no further problems. For substances of simple molecular structure $\Delta_{solv}G^{\circ}_{oct}$, $\Delta_{solv}G^{\circ}_{pol}$, and $\Delta_{hydr}G^{\circ}$, as well as $\Delta_{distr}G^{\circ}$ are additive, permitting the use of the group increment method for the generalization of Gibbs energies of solvation of nonelectrolytes [47].

VII. THE SOLVOPHOBIC EFFECT OF THE POLAR PHASE AND ITS INFLUENCE ON EXTRACTION

The solvophobic effect is a very important property of any solvent. Many methods of its estimation are described in the chemical literature. A very convenient measure of the solvophobic effect is the methylene group increment for the Gibbs energy of distribution of nonelectrolytes between octane and various solvents [42] or for the Gibbs energy of transfer of solid paraffins from solvents miscible with aliphatic hydrocarbon to octane [57]. This measure considers octane to be a standard nonpolar solvent with a zero value of the solvophobic effect. However, since this measure corresponds to one methylene group or to a linear aliphatic hydrocarbon radical, it cannot be directly used for the estimation of the solvophobic effect for branched, aromatic, and unsaturated low-polarity organic molecules, which also show typical solvophobic behavior [17,18]. Moreover, it is useful to deal separately with the influence of the octane phase and the solvophobic effect of the polar phase. It is also interesting to separate the solvophobic and solvation terms of the Gibbs energy of the solute in the polar phase. The Gibbs energy of solvation of a nonelectrolyte can be presented as the sum of two terms: the Gibbs energy of cavity formation in the solvent structure for accommodation of the solute molecule, $\Delta_{cav}G^{\circ}$, and the Gibbs energy of its interaction with its surrounding, $\Delta_{interact}G^{\circ}$ [58]. The following estimation of the solvophobic effect for various solvents and solutes was attempted [59]. The Gibbs energy of hydration of the methylene group, determined from numerous experimental data, was taken as the starting point of the estimation procedure of the solvophobic effect:

$$\Delta_{hydr}G^{\circ} = \Delta_{cav}G^{\circ} + \Delta_{interact}G^{\circ} = 0.75 \text{ kJ/mol} \tag{7}$$

The division of the solvation Gibbs energy into solvophobic and interaction terms is a complicated problem, and some attempts to solve it have been proposed [60]. This work allows us to clarify the nature and driving forces of solvation interactions in polar solvents and to estimate the driving forces of extraction processes.

The most important term in Eq. (7) is $\Delta_{cav}G^{\circ}$, which was estimated by various authors to vary from 2.26 [60] to 5.31 kJ/mol [61]. It was also estimated as follows [59]. Octane was chosen as the standard solvent with a zero value for the solvophobic effect, because aliphatic hydrocarbons have the least ability to participate in intermolecular interactions. The $\Delta_{interact}G^{\circ}$ term for methylene group was then estimated from data on the distribution of water between its saturated solution in octane and its saturated vapor where it is present mainly as separate molecules. The Gibbs energy of solvation of individual water molecules by octane was -2.64 kJ/mol [59]. On the other hand, the enthalpies and Gibbs energies of solute solvation in hydrocarbon solution are known to be proportional to their molar refractions [53]. Hence the Gibbs energy of methylene group solvation by hypothetical destructed water is estimated to be $-2.64:0.78 = -3.38$ kJ/mol, 0.78 being the ratio of the molar refraction of water and that of a methylene group. The Gibbs energy of methylene group solvation by hydrocarbons at 20°C, derived from numerous data [54], is -2.76 kJ/mol however [44], and this indicates the existence of additional inductive interactions, -0.88 kJ/mol, of the methylene group with water in comparison with hydrocarbons. Therefore, $\Delta_{cav}G^{\circ}$ for methylene group accommodation in the water structure is $0.75 - (-2.76) - (-0.88) = 4.39$ kJ/mol.

For any substance, $\Delta_{cav}G^{\circ}$ is determined by the surface area of its molecule. The surface areas calculated by various methods of various molecules and groups, including that of the methylene group, have been presented in the chemical literature [62,63]. The following approximate value was deduced for the solvophobic effect for any particle with a surface area, S, in units of that of the methylene group [59]:

$$\Delta_{cav}G^{\circ} \sim 7.1 S I_{CH_2} \tag{8}$$

where I_{CH_2} is the methylene group increment for extraction in the octane–polar phase system. The Gibbs energy of solvation of any particle by an individual or mixed polar solvent "destructured" to the octane level, ΔG°_{destr}, can then be expressed by [59]:

$$\Delta G^{\circ}_{destr} \sim \Delta_{solv}G^{\circ}_{pol} - 7.1 S I_{CH_2} \tag{9}$$

The solvophobic (cavity) and interaction terms of solvation Gibbs energies of various molecules and groups by polar solvents, calculated in this work, are presented in Table 9.

All the values of ΔG°_{destr} are seen to be much less than $\Delta_{solv}G^{\circ}_{oct}$, owing to the specific solvation interactions of solutes with polar solvents in comparison with octane. The larger the solute molecule, and the more solvation centers are present in the solute molecule, the nature of which, is also very important, the greater the differences between $\Delta_{solv}G^{\circ}_{oct}$ and ΔG°_{destr}.

From all these results it can be concluded that the solvophobic effect drives the transfer of nonpolar and low-polarity molecules from the polar phase to gaseous or nonpolar liquid phases. It is easy to see that in the absence of a

Table 9 Cavity and Solvation Terms of the Solvation Gibbs Enegy (kJ/mol) of Groups and Molecules by Various Polar Organic Solvents

| | Groups | | | |
| | CH$_2$ | | CH=CH$_2$ | |
Polar solvent	Cav	Solv	Cav	Solv
Methanol	0.71	−2.93	1.51	−6.36
2-Methoxyethanol	0.63	−2.89	1.38	−7.36
DMF	0.84	−3.05	1.67	−8.20
1,3-Propanediol	1.42	−3.05	—	—
Diethylene glycol	1.42	−3.05	3.05	−7.32
DMSO	1.42	−3.05	3.05	−7.82
Ethylene glycol	1.84	−3.10	4.06	−7.20
Glycerol	2.38	−3.18	—	—
Water	4.39	−3.72	9.12	−8.37
Octane[a]	—	−2.76	—	−5.40

| | Groups | | | |
| | Benzene | | Naphthalene | |
Polar solvent	Cav	Solv	Cav	Solv
Methanol	2.9	−18.8	4.6	−33.5
2-Methoxyethanol	2.5	−18.4	4.2	−33.9
DMF	3.4	−20.9	5.0	−36.8
1,3-Propanediol	5.9	−19.3	8.8	−35.2
Diethylene glycol	5.9	−20.5	8.8	−36.4
DMSO	5.9	−22.2	8.8	−39.8
Ethylene glycol	8.0	−19.7	12.1	−36.0
Glycerol	10.5	−20.5	15.5	−36.4
Water	18.4	−21.8	27.2	−37.2
Octane[a]	—	−16.3	—	−29.3

[a]Data given for comparison of the solvating ability of octane and polar solvents.
Source: Ref. 59.

solvophobic effect, the Gibbs energies of distribution of all substances would be positive, even for a methylene group, and the mere existence of extraction systems is determined by the solvophobic effect. It is not surprising, therefore, that water, characterized by the maximal solvophobic effect, yields a very large number of extraction systems with various organic and inorganic liquids.

The solvophobic effect plays an important role in processes of PAH solubility in various polar solvents, including even those that are miscible with aliphatic hydrocarbons [43]. Table 10 illustrates some actual and hypothetical PAH solubilities, the latter in the absence of solvent solvophobic effects, calculated by means of Eq. (9).

Accounting for the solvophobic effect allows the ready interpretation of unexpected regularities of PAH solubilities in some ethers and esters. In particular, the decreasing of PAH solubilities from diethyl to dibutyl ether and from pentyl to ethyl acetate is determined by the practical absence of a solvophobic effect in ethers, whereas esters show an appreciable solvophobic effect, the reduction of which dominates the diminished solvation ability of the solvent when an ethyl radical is replaced by an amyl one in the ester molecule. The maximal solvating ability with respect to PAHs is typical for solvents with polarizable double bonds in their molecules. Water is a very strong solvating agent for PAHs because the concentration of polar groups in it is very high. DMSO, however, which has a highly positively charged sulfur atom in its molecule with free d-electron orbitals, is able to form π complexes with PAH molecules and shows an even higher solvating ability than water.

The solvophobic effect plays an especially important role in the extraction of substances with very large hydrocarbon radicals—for instance, in the case of higher homologs of polar and low-polarity organic nonelectrolytes, as well as in the case of high molecular mass organic salts. Table 11 illustrates the influence of the solvophobic effect of the polar phase on extraction of some

Table 10 Real and Hypothetical Solubilities (mol/dm^{-3}) of Some Polycyclic Aromatic Hydrocarbons in Various Solvents

		Solubilities (mol/dm^3)			
		2,4,6-Triphenylbenzene		Anthracene	
Solvent	I_{CH_2}	Real	Hypothetical	Real	Hypothetical
n-Octane	0.000	0.0092	0.0092	0.0092	0.0092
Methanol	0.110	0.0016	0.14	0.0051	0.071
n-Propanol	0.068	0.0038	0.060	0.0081	0.041
Ethyl acetate	0.058	0.15	1.6	0.040	0.16
Pentyl acetate	0.030	0.23	0.78	0.045	0.090
Diethyl ether	0.008	0.064	0.090	0.029	0.035
Di-n-butyl ether	0.012	0.027	0.044	0.018	0.024
Acetone	0.083	0.079	2.3	0.046	0.33
Water	0.610	3.7×10^{-10}	19	2.0×10^{-7}	0.38

Source: Ref. 43.

Table 11 Extraction Constants of Some Hydrophobic Organic Substances[a]

Polar phase	I_{CH_2}	$(C_9H_{19})_2C_{18}H_{37}N^+I^-$	$(C_9H_{19})_2C_{18}H_{37}N$	$C_9H_{19}I$
DMF	0.11	0.4 (0.01)	5×10^3	4.0
Acetonitrile	0.13	1.0 (0.04)	3×10^4	7.0
Ethylene glycol[b]	0.27	3×10^3 (3×10^2)	4×10^8	3×10^2
Water[b]	0.61	5×10^{13} (5×10^{12})	3×10^{19}	2×10^6

[a]The distribution coefficients of trinonyl octadecylammonium iodide (0.1 M in the polar phase) are given in parentheses.
[b]Data calculated from the group increments.
Source: Ref. 64.

hydrophobic substances [64]. It shows a steep fall of the distribution constants and coefficients of up to 15 powers of 10, when water is replaced completely or even partially by polar solvents. At times the sign of the Gibbs energy of extraction changes from negative to positive (see the QAS extraction). In such cases, the solvophobic effect practically determines the extraction of such highly hydrophobic substances. An effective control of the polar phase solvophobic effect can always take place by changing the nature and content of the polar solvent in the polar phase. Unfortunately, only a limited group of polar organic solvents can be used for the preparation of stable extraction systems with aqueous organic polar phases. These solvents were listed earlier. On the other hand, solvents miscible with water, such as acetone, dioxane, and tetrahydrofuran, pass in a very high degree into the octane phase, which makes any such extraction system useless for practical purposes.

An original possibility of the control of the solvophobic effect takes place in salt-containing extraction systems. A large number of octane-miscible polar solvents, such as higher alcohols, ketones, ethers, and esters, give stable extraction systems when salt having a concentration of 1 M or higher is added to the polar solvent [36]. Control of the solvophobic effect of the polar phase can then be carried out by varying the nature of the solvent and the concentration of the salt.

In conclusion, the use of the group increments method is a very convenient way of investigating the systems reviewed. Since the extraction processes in these systems are more complicated than classical systems and have not yet been widely investigated, an extension of the bank of group increments, which now contains some 3000 experimental points for more than 250 extraction systems [65], is the optimal way of studying such systems.

VIII. EXTRACTION, SEPARATION, ISOLATION, AND PRECONCENTRATION OF NATURAL AND SYNTHETIC HYDROPHOBIC ORGANIC SUBSTANCES

Now we turn to a practical illustration of the efficiency of the group increment method for the choice of extraction systems. Mainly we shall consider results of the extraction-based separation, isolation, and purification of various classes of hydrophobic organic substances, obtained at the Analytical Chemistry Chair of Belarus State University.

Natural and synthetic hydrophobic organic substances are of great importance for the chemical industry and for the national economy. They are effectively used as extraction agents, surfactants, monomers, composition materials, and raw materials for industrial and preparative organic synthesis. These substances are represented by higher homologs of various classes of organic substances and by lower nonpolar organic compounds. Such substances very often exist as complicated mixtures in natural products or form complicated mixtures in the process of their synthesis. Typical examples are oil products, as well as reaction mixtures of higher amines and quaternary ammonium salts synthesized from ammonia and higher alkyl halides. The purity of such substances may in some cases determine the efficiency of their use [66].

The separation of such mixtures and isolation of individual hydrophobic substances from them with the use of the traditional methods of organic chemistry are generally problematic or even impossible. In particular, vacuum distillation is not effective, either because of the high boiling temperatures of high molecular mass substances and the possibility of their decomposition or because azeotropic mixtures form. For instance, many oil product components, including sulfur–organic impurities, are distilled together. Recrystallization is limited by formation of gels and stable solutions. This is typical for substances with bulky hydrocarbon chains or for mixtures of homologs of high molecular mass substances. Classical extraction systems of the organic solvent–water type cannot be used because of the extremely low solubility of the hydrophobic substances in water and their quantitative transfer from water to the organic solvent (Table 11).

On the other hand, the present extraction systems, characterized by efficiency, simplicity and mild process conditions, afford a promising method of separation and purification of various substances in industry and in laboratory practice. Extraction systems with aqueous polar organic and hydrocarbon phases have been used episodically for the separation of some hydrophobic substances, but the main regularities of the extraction process in such systems have not been completely studied and generalized. Therefore, the use of such "nonaqueous" systems in the chemical industry as well as in various fields of chemistry still has not achieved its potential.

On the basis of group increments, it is possible to calculate separation factors f and optimal distribution constants and coefficients P_{opt}, D_{opt} of various substances [65]:

$$f = \frac{P_1}{P_2}; \quad \log P_1 = \Sigma I_{i_1}; \quad \log P_2 = \Sigma I_{i_2}; \quad f = 0.4343 \, \exp[\Sigma I_{i_1} - \Sigma I_{i_2}]$$

(10)

where indexes 1 and 2 desginate substances and I_i are group increments. The following conditions pertain to optimal extraction separation:

$$P_1(\text{or } D_1) P_2(\text{or } D_2) \approx 1; \quad \text{then} \quad \text{if } P_1(\text{or } D_1) > 1, \, P_2(\text{or } D_2) < 1 \quad (11)$$

$$P_{1 \, opt}(\text{or } D_{1 \, opt}) = f^{1/2}; \quad P_{2 \, opt}(\text{or } D_{2 \, opt}) = f^{-1/2} \quad (12)$$

Calculations of separation factors for the majority of hydrophobic substances show that maximal separation factors are typical for the octane–water system because of the maximal absolute values of the group increments. However, conditions (11) and (12) are then not met because of the tremendous values of P and D of the substances.

Replacement of water by polar organic solvents and their aqueous and mineral salt mixtures is the only way for conditions (11) and (12) to be met.

The most general way of estimating possibilities of specific extraction systems to be used for separation of various organic substances as well as for the choice of optimal systems is the following [65]. A specific and high value of f (e.g., 100), should be taken, and optimal P or D should be calculated according to condition (12). This value of f can be conditionally assumed to be realizable. The minimal and maximal carbon atom numbers in molecules of optimally separated substances are then estimated.

$$n_{1 \, min} \geq \frac{\log P_{1 \, opt} - I_{f_1}}{I_{CH_2}}$$

(13)

$$n_{2 \, max} \leq \frac{\log P_{2 \, opt} - I_{f_{21}}}{I_{CH_2}}$$

(14)

where I_{CH_2} is the methylene group increment and I_f the functional group increment. Systems characterized by moderate and especially low values of I_{CH_2} are least sensitive to the size of the hydrocarbon radical and, at the same time, they are most effective for the isolation of hydrophobic substances from hydrocarbon solutions. But it is necessary to remember that separation factors of the overwhelming majority of hydrophobic substances decrease with growing organic solvent content in the polar phase, and that is why in some cases aqueous mixtures with $\varphi = 0.80-0.95$ are more effective for separations than individual polar organic solvents [27–29].

As the result of these investigations, many new, simple, and effective extraction methods of separation, preconcentration, and purification of natural and synthetic hydrophobic substances have been developed.

These methods, which can be divided into two large groups—methods based on molecular extraction and methods based on dissociative extraction—are discussed in Sections VIII.A and VIII.B, respectively. In these sections, values of $P \ll 1$ mean that the distributed substance strongly prefers the polar phase, whereas $P > 1$ means that it prefers the nonpolar hydrocarbon phase.

A. Methods Based on Molecular Extraction

1. Separation of Higher Hydrocarbons and Alcohols, of Higher Alcohols and Diols, and of Higher Alkyl Halides and Alcohols, Containing up to 16 Carbon Atoms in the Radical [65–68]

The separation factors of a hydrocarbon and alcohol and of an alcohol and diol can be described by the equation:

$$\log f = I_H - I_{OH} \tag{15}$$

Similarly, the separation factor of an alkyl halide and an alcohol can be described by the equation:

$$\log f = I_{Hal} - I_{OH} \tag{16}$$

Systems with salt concentrations about 6 M in the polar phase, characterized by high negative values of I_{OH} and near zero values of I_{Hal}, as well as by low and moderate values of I_{CH_2}, were found to be most effective for this purpose. The realized f values are between 5000 and 10,000, hydrocarbon and alkyl halide P values are between 100 and 150, and alcohols have P from 0.01 to 0.003 and diols from 10^{-4} to 10^{-7}. In these systems the preconcentration factors of higher alcohols, related to 95% extraction of the alcohols by the polar phase, can reach 10–20 [64].

An alternative method of separating the above-mentioned substances is based on a two- or three-step extraction of higher alcohols by lower alcohol–aqueous mixtures. This method is much less effective (realized f values 30–100) but much more convenient, since regeneration of the polar phase by distillation of the lower alcohol is easy.

2. Separation of Nitroalkanes and Alkyl Nitrites, Radicals of Which Contain from 4 to 12 Carbon Atoms [14]

Mixtures of nitroalkanes and alkyl nitrites are formed as the result of the reaction of higher alkyl halides and sodium nitrite in N,N-dimethylformamide (DMF) media. The separation factor in this case is described by:

$$\log f = I_{ONO} - I_{NO_2} \tag{17}$$

where I_{ONO} and I_{NO_2} are the functional group increments of alkyl nitrite and nitroalkane. The realized f value of these substances for the aliphatic hydrocarbon–DMF extraction system is about 5. Complete separation of these substances may be reached by multistep extraction. Nitroalkanes are concentrated in the polar phase, while alkyl nitrites are in the hydrocarbon one.

3. Separation of Near-Boiling Unsaturated Hydrocarbons [17]

Typical examples are the separation of ethylstyrenes and divinylbenzenes and the ethylbenzenes and ethyl-ethynylbenzenes derived from them. The separation factor of ethylstyrenes and divinylbenzenes, independently on the place of the substituents in the benzene ring, can be described by the equation:

$$\log f = 2.5 I_{CH_2} - I_{CH=CH_2 \, arom} \tag{18}$$

The separation factor of ethyl-ethynylbenzene and diethynylbenzene can be described by a similar equation:

$$\log f = 2.5 I_{CH_2} - I_{C\equiv CH \, arom} \tag{19}$$

These methods are based on extraction of divinylbenzenes and diethynylbenzenes by acetonitrile, dimethyl sulfoxide, N,N-dimethylformamide, and their aqueous mixtures. The realized f value of ethylstyrenes and divinylbenzenes is about 1.6, for ethyl-ethynylbenzenes and diethynylbenzenes it is 10–20. It is possible to obtain concentrates of diethynylbenzenes containing 80 to 95% of the main substance with the yield of 70–90% while initial mixtures contain only about 40% of diethynylbenzenes [18].

4. Isolation of the Terpenoid Alcohol Linalool from Coriander Oil [20]

Another isolation is based on extraction of linalool from oil by acetontrile and its aqueous mixtures. Partition constants of linalool between the oil and the polar phases in this case are about 0.15–0.20, but the oil solubility in the polar phase is insignificant. It is possible to isolate about 95% of linalool by means of a three-step extraction with ready regeneration of the polar phase by distillation.

5. Isolation of Sclareol from the Wastes of Musk Sage Processing [21]

Sclareol is a water-insoluble bicyclic diterpene glycol $C_{20}H_{34}(OH)_2$ that is an important raw material in perfumery and medicine. It was found that sclareol can be effectively extracted from aliphatic hydrocarbons by DMF and DMSO, which are characterized by low and moderate values of methylene group increments but also by highly negative values of oxy group increments [11]. The partition constant of sclareol in the octane–DMF system is 0.075 and in the octane–DMSO system it is 0.053. Still, DMSO is proposed to be the more effective and selective extraction agent of sclareol because its methylene group increment has a sufficiently higher value than that of DMF. Sclareol can be isolated from DMSO solution by two- to threefold dilution of the DMSO by water and subsequent extraction by an aliphatic hydrocarbon.

6. Separation and Purification of Dihexylphosphoric Acid and Trihexyl Phosphate [69]

Trihexyl phosphate (THP), used as an extraction agent and plasticizer, usually contains impurities of dihexylphosphoric acid (DHP). DHP is used as an ionophor and extraction agent of cationic species. Owing to the acidic properties of DHP and its relatively large hydrocarbon radical, basic solvents, characterized by moderate values of the methylene group increment, were proposed for the extraction isolation of DHP. Aqueous N,N-dimethylformamide mixtures, containing 10–20% of water by volume, were proposed as most effective for DHP isolation from hydrocarbon solution and for the extraction separation of THP and DHP. Partition constants of DHP under these conditions vary from 0.013 to 0.028, while the partition constant of THP exceeds 50. DHP can be isolated from the polar phase by two- and threefold dilution by water, with recovery of more than 98% of the main substance. The THP content does not exceed 1%.

7. α-Tocopherol Purification from Isofitol Impurities [22]

Isofitol is a higher tertiary alcohol used for the synthesis of α-tocopherol. Isofitol impurities are effectively extracted by concentrated zinc halide solutions in lower alcohols, while α-tocopherol extraction is poor. Partition constants of α-tocopherol in systems consisting of an aliphatic hydrocarbon–salt solution are about 20–30; those of isofitol are 0.1–0.15.

8. Separation, Fractionation, and Purification of Higher Aliphatic Sulfides and Sulfoxides in Oil Sulfoxides [70]

The separation factor of sulfides and sulfoxides is described by the equation:

$$\log f = I_S - I_{SO} \tag{20}$$

The separation of higher sulfides and sulfoxides is based on the extraction of the sulfoxide molecules containing up to 16 carbon atoms by polar organic solvents and their aqueous mixtures, while the sulfides are extracted by the hydrocarbon phase. The practically realized f values of these substances are about 100–1000, the partition constants of sulfides vary from 10 to 1000, and the partition constants of sulfoxides vary from 0.02 to 1.0. Analysis of functional group increments of sulfur–organic substances [16] also shows that extraction separation of sulfones and sulfides and of sulfones and sulfoxide can be easily realized.

Purification and fractionation of oil sulfoxides is based on the extraction of nonsulfoxide impurities and sulfoxides of higher molecular mass by hydrocarbons from polar organic solvents and their aqueous mixtures [70].

Zinc halides solutions are the most effective for extraction of sulfoxides from hydrocarbon solutions, but they also extract other classes of polar organic substances and require a difficult operation of salt regeneration after extraction.

9. Isolation of Oil and Synthetic Sulfides from Hydrocarbon Solutions and in Particular the Desulfuration of Hydrocarbon Raw Material [23]

Concentrated sulfuric acid with small additions of water is usually used for the isolation of oil sulfides from various oil fractions [71], to take advantage of the protonation of organic sulfides by sulfuric acid. This method is not selective, on the other hand, sulfuric acid is highly aggressive with respect to many components of hydrocarbon raw materials. Methods were developed based on the high affinity of the aliphatic sulfide group to concentrated solutions of zinc halides (see Table 2). It is interesting to find that such solutions are the only possible extraction agents of the higher sulfides, because there are no polar organic solvents able to give strong complexes with organic sulfur. Sulfoxides have an even higher affinity to salt solutions, but they are effectively extracted by polar organic solvents and their aqueous mixtures, in contrast to the higher sulfides. The degree of extraction of the latter by zinc halide solutions from high-boiling oil fractions, such as heavy vacuum gas oils, can reach 90–95%. In some cases a severalfold preconcentration of sulfides from hydrocarbon solutions is possible [64]. Thiophene and its derivatives are poorly extracted by salt solutions [36], and this fact can be used for the separation of higher sulfides, sulfoxides, and thiophene derivatives. Aromatic hydrocarbons are poorly extracted by salt solutions, too [67].

B. Methods Based on Dissociative Extraction

1. Separation of Higher Mono-, Di-, and Trialkylamines Containing from 6 to 12 Methylene Groups in Their Hydrocarbon Chains [28,29]

In the case of higher mono-, di-, and trialkylamines, the differences in dissociation ability of higher amines in polar solvents, such as lower alcohols or DMF and their aqueous mixtures, are taken into account. Values of the protonation constants of higher amines in polar solvents and their aqueous mixtures are roughly as follows: monoalkylamines > dialkylamines > trialkylamines. The difference between the logarithms of the protonation constants of mono- and trialkylamines is about 1.0–2.1, while those between mono- and dialkylamines are 0.2–0.4, but in aqueous solutions these differences are very small. The separation factor of higher amines can be described by the equation:

$$\log f = (i_1 - i_2)I_H + (n - m)I_{CH_2} + \Delta I_{N_i} + \Delta \log K_{pr} \tag{21}$$

where i is degree of amine substitution (mono to tri), n and m are carbon atom numbers in the amine radicals, ΔI_{N_i} is the difference of increments of amine groups and $\Delta \log K_{pr}$ is the difference of the logarithms of protonation constants of the higher amines in the polar phase.

The methods of separation are based on extraction of amines of lower molecular mass and substitution by the acidified polar solvents such as lower alcohols and N,N-dimethylformamide and their aqueous mixtures. For instance, monoalkylamines, having a minimal number of carbon atoms in the molecule and maximal values of the protonation constant, are extracted first, and trialkylamines are extracted last [28,29]. The optimal pH of extraction for the fixed extraction system is a function of amine molar mass, and substitution and can be calculated with the use of group increments and protonation constants of higher amines [28,29]. The optimal pH varies from 5 to 7 for the separation of tri- and dialkylamines and from 7 to 9 for the separation of di- and mono-alkylamines. Distribution coefficients of higher amines in this case vary from 0.01 to 100. The maximal realizable f value for trialkyl- and dialkylamines is about 10,000 for a single-step extraction [29]. Owing to the high values of $\log K_{pr}$ differences of amines of various substitution in some polar solvents and their aqueous mixtures, separation of isomeric higher primary and tertiary amines is possible with realized f values up to 100 [28]. The possibility of such high realizable values of f is determined by the dramatic fall of extraction of higher amine salts by hydrocarbons, beside the high differences of the logarithms of protonation constants [28,29], when water is replaced by polar solvents and their aqueous mixtures, which in turn increase as the hydrocarbon radical of the higher amine becomes larger [28,29]. These methods are used to remove amine and nonamine impurities from higher amines as well as for control of reaction mixtures in the synthesis of higher amines [72].

2. Separation of Higher Quaternary Ammonium Salts (QAS) Containing up to 60 Carbon Atoms in the Molecule, and Substances Used for QAS Synthesis [64]

The separation factor of amine and QAS can be described by the equation:

$$\log f = -2 \log \gamma_{\pm} - \log C_{QAS} - (i - 4)I_H + (n - m)I_{CH_2} + I_{N_i} - I_{N_A}$$

$$(22)$$

where $\gamma_{\pm}$ are the mean ionic activity coefficients of QAS ions in the polar phase [26], I_{N_A} are increments in logarithms of thermodynamic extraction constants of QAS functional groups, and C_{QAS} is the QAS concentration in the polar phase. The other quantities are as defined for Eq. (21). Being strong electrolytes, quaternary ammonium salts are quantitatively extracted by polar organic solvents and their aqueous mixtures, while higher nonelectrolytes, including higher amines, are quantitatively extracted by the aliphatic hydrocarbons when the pH of the polar phase exceeds 10 (Table 3). When the QAS solutions in the polar solvents are sufficiently concentrated, the realized f value of about 0.1–0.2 M, is about 10^5–10^7 according to Eq. (22). Other nonelectrolytes, such as alkyl halides and hydrocarbons, can be effectively removed by extraction with aliphatic hydrocarbons (Table 11).

These methods allow us to produce higher QAS of high purity (≤99.9%), when the initial products usually contain less than 60–70% of the main substance. The methods are also very effective for purification of QAS of unsymmetrical structure, which cannot be effectively recrystallized from organic solvents, even from aliphatic hydrocarbons. Typical examples of such QAS are various anionic forms of trialkyloctadecylammonium [27,73] and QAS that are synthesized from ethylenediamine [52,73]. A highly sensitive method for the determination of long chain amine microimpurities in higher QAS was developed, too [72]. It is based on a preliminary quantitative extraction, removing the long chain amine impurity from the QAS matrix dissolved in DMF, by the aliphatic hydrocarbon and subsequent sensitive extraction–photometric determination of the amine with acidic dyes. Detection limits of higher amines impurities is about 3×10^{-4}%, the standard deviation being 1–2%.

3. Highly Effective Extraction and Preconcentration of Higher Sulfoxides, Including Oil Sulfoxides, from Hydrocarbon Solutions and from Oxidized Hydrocarbon Raw Materials [74]

Our next separation is based on the use of sulfuric acid solutions in acetonitrile. The protonation constants of higher sulfoxides in acetonitrile are approximately 10^5 times larger than in water and in lower alcohols, owing to the aprotic nature of acetonitrile. On the other hand, acetonitrile forms very stable extraction systems with aliphatic hydrocarbons. The distribution coefficients of the higher sulfoxides, containing up to 16 carbon atoms in the molecule, at molar ratios of sulfoxide to acid of $1:1.2-1.4$, are about $0.001-0.02$. This allows us to concentrate 5–10-fold the higher sulfoxides containing up to 16 carbon atoms in their radicals and also to purify them from nonsulfoxide impurities.

4. Highly Effective Extraction and Preconcentration of Mercaptans from Hydrocarbon Solutions and Low-Boiling Hydrocarbon Raw Materials [24]

Another extraction is based on the use of alkaline solutions in mixtures of lower alcohols and polar aprotic solvents, such as DMSO, DMF, and acetonitrile. Because of the high activity of hydroxide ions in such mixtures, these solutions are in 100–200 times more effective for mercaptan extraction than alkaline solutions in lower alcohols and aqueous alcohol mixtures, which are usually used for the removal of mercaptans from hydrocarbon raw materials [75]. The distribution coefficients of mercaptans are about $0.001-0.0001$ at molar ratios of mercaptan to alkali of $1:1.2-1.4$, allowing complete removal of mercaptans from hydrocarbon raw materials. In particular, the required volume ratio of hydrocarbon raw material to alkaline solution for complete removal of the mercaptan (>99%) can be about $10-50:1$. For 95% extraction by the polar phase, the preconcentration of the mercaptans can reach 500. Such extraction systems can, therefore, be used for decreasing the detection limit of mercaptans in hydrocarbon raw materials.

5. Extraction of Higher Carboxylic Acids from Hydrocarbon Solutions and Purification of Various Materials from Carboxylic Acids Impurities [25,65]

Our final method is based on the use of alkaline solutions in lower alcohols, glycols, and their aqueous mixtures. Distribution coefficients of carboxylic acids, containing up to 16 carbon atoms in their radicals, are about 0.001–0.03 at pH values of the polar phase of 9.5–10. This allows us to remove carboxylic acid impurities from such nonelectrolytes as oils, fats, higher nitriles, and turpentine.

IX. CONCLUDING REMARKS

All the methods described in Section VIII include a procedure for the isolation of the extracted substances and solvent regeneration. This procedure is simplest when low-boiling solvents such as the lower alcohols and acetonitrile are used, and it can be realized by solvent vaporization. When high-boiling solvents, like DMSO, glycols, and DMF, are used, their regeneration cannot be effectively realized by distillation. In this case the polar solvent is diluted two- to threefold by water and extracted by a low-boiling nonpolar solvent for the isolation of the extracted substance. Subsequently, the polar solvent can be completely or partly dried by evaporation of the water. The salt-containing polar phase can similarly be regenerated by its two- to threefold dilution with water and extraction by an aliphatic or aromatic hydrocarbon. After that, water and the polar solvent are distilled and the salt is dried at about 300°C.

These methods are effectively used in scientific investigations at the Department of Analytical Chemistry of the Belarus State University. They are used for producing high-purity extraction agents, surfactants, and ionophores that are delivered to various scientific organizations. The latest investigations have shown the possibility of extractive isolation and extractive separation of polycyclic aromatic hydrocarbons [39] and some sulfur-containing aromatic substances [76] present in oil, in coal tar, and in various environmental objects. In particular, partition constants of isomeric PAH of linear and angular structure in some extraction systems can differ by 1.5–2.0 times [39], providing an opportunity to separate them by means of multistep extraction.

These results show that extraction systems based on polar solvents can be effectively used for solving of numerous problems of separation, purification, isolation, and preconcentration of various classes of synthetic and natural substances. It is possible that they will be very useful in the technology of numerous other organic water-insoluble substances, such as lipids and their derivatives, oil-soluble vitamins, and biologically active substances.

REFERENCES

1. A Leo, C Hansch, D Elkins. Chem Rev 71:525, 1971.
2. RF Rekker, HM Kort. Eur J Med Chem—Chim Therm 14:479, 1979.
3. Ya I Korenman. Ekstraktsia Fenolov Moscow-Gor'ky, 1973.

4. Y Marcus. J Phys Chem 95:8886, 1991.
5. GI Bittrich, AA Gaile, D Lempe, VA Proskuriakov. Razdelenie Uglevodorodov s Ispolzovaniem Selectivnich Rastvoritelei. Leningrad: Khimia, 1987, pp 1–192.
6. GK Ziganshin, AA Osintzev, FI Serdjik, KG Ziganshin. Proceedings of Sovremennie Problemi Khimii i Teknologii Ekstraktsii, Moscow, 1999. Part 2, p 99.
7. PL Bocca, R Caggiano, G Variali. Nuova Chim 50:33, 1974.
8. US Patent 4,208,334. CO7 D 311172 (1980).
9. H Watanari, N Suzuki. Bull Chem Soc Jpn 53:1848, 1980.
10. GL Starobinets, EM Rakhman'ko, SM Leschev. Zh Fiz Khim 52:2284–2287, 1978.
11. SM Leschev. Izv Vuzov SSSR Ser Khim Khim Tekhnol 34:42, 1991.
12. EM Rakhman'ko, SF Furs, GL Starobinets. Akad Nauk BSSR 4:39, 1983.
13. GL Starobinets, EM Rakhman'ko, SM Leschev. Akad Nauk BSSR 2:30, 1979.
14. SM Leschev, PM Malasko. Zh Prikl Khim 65:449, 1992.
15. SM Leschev, Yu I Denisenko. Neftekhimia 29:850, 1989.
16. SM Leschev, IV Melsitova. Izv Vuzov SSR Ser Khim Khim Tekhnol 35:65, 1991.
17. SM Leschev, VI Onischuk. Neftekhimia 32:367, 1992.
18. SM Leschev, VI Onischuk. Neftekhimia 33:82, 1993.
19. SM Leschev, VI Onischuk. Neftekhimia 29:356, 1989.
20. SSSR Patent 1,440,907. Bjullet Izobret 44 (1988).
21. VV Egorov, LV Koleshko, GL Starobinets. Izv Akad Nauk BSSR 3:37, 1984.
22. SSSR Patent 1,154,279. Bjullet Izobret 17 (1985).
23. SM Leschev, SF Furs, Yu I Denisenko. Neftekhimia 30:423, 1990.
24. SM Leschev, IV Melsitova. Neftekhimia 33:185, 1993.
25. AI Lamotkin, GA But'ko, SM Leschev, AN Pronevich. Gidroliz Lesokhim Promyst 5:8, 1991.
26. SM Leschev, NF Sharamet. Zh Fiz Khim 66:1220, 1992.
27. SM Leschev. Zh Prikl Khim 66:176, 1993.
28. SM Leschev, EM Rakhman'ko, GL Starobinets, I Yu Rumiantsey. Izv Akad Nauk BSSR 2:8, 1988.
29. SM Leschev, EM Rakhman'ko. Zh Prikl Khim 63:129, 1990.
30. SSSR Patent 1,427,259. Bjullet Izobret 39 (1988).
31. SSSR Patent 1,425,530. Bjullet Izobret 35 (1988).
32. VV Egorov, GL Starobinets. Izv Akad Nauk BSSR 61:56, 1981.
33. SM Leschev, VI Onischuk. Zh Prikl Khim 61:1796, 1988.
34. Ya I Korenman. Zh Fiz Khim 48:653, 1974.
35. GL Starobinets, EM Rakhman'ko, TK Haletskaya, SF Furs. Dokl Akad Nauk BSSR 28:243, 1984.
36. SM Leschev, SF Furs, EM Rakhmanko, I Yu Rumiantsev. Zh Fiz Khim 72:1218, 1998.
37. EM Rakhman'ko, GL Starobinets, SM Leschev, I Yu Rumiantsev. Dokl Akad Nauk BSSR 29:724, 1985.
38. SM Leschev, AV Sinitsina. Neftekhimia 37:56, 1997.
39. SM Leschev, AV Sinitsina. Neftekhimia 37:552, 1997.
40. SS Davis. Sep Sci 10:1, 1975.
41. SM Leschev, VI Onischuk. Zh Strukt Khim 32:146, 1992.
42. SM Leschev, EM Rakhman'ko. Zh Strukt Khim 30:136, 1990.

43. SM Leschev, NP Novik, AV Sinkevich, TP Novik. Proceedings of Sovremennye problemy khimii i tekhnologii ekstraktsii, Moscow, 1999. Part 1, p 173.

44. S Srebrenic, C Cohen. J Phys Chem 80:996, 1976.

45. WJ Murray, LH Hall, LB Kier. J Pharm Sci 64:1978, 1975.

46. GL Starobinets, EM Rakhman'ko, SM Leschev. Zh Fiz Khim 53:2720, 1979.

47. GL Starobinets, SM Leschev, EM Rakhman'ko. Izv Akad Nauk BSSR 1:124, 1980.

48. EM Rakhman'ko, SM Leschev, GL Starobinets, I Yu Rumiantsev. Izv Akad Nauk BSSR 6:27, 1983.

49. Y Marcus. The Properties of Solvents. Chichester: Wiley, 1998.

50. SM Leschev, IV Melsitova. Zh Prikl Khim 66:562, 1993.

51. VV Egorov, GL Starobinets. Izv Akad Nauk BSSR 1:52, 1976.

52. GL Starobinets, EM Rakhman'ko, I Yu Rumiantsev, SM Leschev. Izv Akad Nauk BSSR 6:28, 1981.

53. BN Solomonov, AI Konovalov. Usp Khim 60:45, 1991.

54. A Ben-Naim, Y Marcus. J Chem Phys 81:2016, 1984.

55. SM Leschev, IV Melsitova, VI Onischuk. Zh Fiz Khim 67:2383, 1993.

56. AM Bogomolny. Spravochnik Ravnovesije Zhydkost-par. Leningard: Khimia, 1987.

57. SM Leschev, NP Novik. Zh Strukt Khim 40:514, 1999.

58. O Sinanoglu. Molecular Associations in Biology. New York: Academic Press, 1968, 427 pp.

59. SM Leschev. Zh Fiz Khim 73:74, 1999.

60. MH Abraham. J Am Chem Soc 101:5477, 1979.

61. VP Belousov, M Yu Panov. Termodinimika Vodnykh Rastvorov Nnejelectrolitov. Leningard: Khimia, 1983.

62. A Bondi. J Phys Chem 68:441, 1964.

63. RS Pearlman, SH Yalkowsky, S Banerjee. J Phys Chem Ref Data 13:555, 1984.

64. SM Leschev, EM Rakhman'ko, SF Furs, I Yu Rumiantsev. Zh Anal Khim 46: 1286, 1991.

65. SM Leschev. Proceedings of the International Solvent Extraction Symposium, Moscow, 1998, p 367.

66. VV Egorov, LV Koleshko, GL Starobinets. Zh Anal Khim 47:2011, 1992.

67. SM Lechev, SF Furs, I Yu Rumiantsev. Zh Prikl Khim 65:1864, 1992.

68. SSSR Patent 1,432,046. Bjullet Izobret 39 (1988).

69. VV Egorov, GL Starobinets, LV Koleshko. Zh Prikl Khim 60:886, 1987.

70. SM Leschev, Yu I Denisenko. Neftekhimia 33:76, 1993.

71. NK Lyapina. Khimia i Fizikokhimia Seraorganicheskikh Soedinenij Neftyanikh Distillyatov. Moscow: Nauka, 1984.

72. SM Leschev, I Yu Rumiantsev. Proceedings of the International Organic Substances Solvent Extraction Conference, Voronezh, 1992. Vol 2, p 294.

73. EM Rakhman'ko, GL Starobinets, VV Egorov, AL Gulevich, SM Leschev, ES Borovsky, AR Tsyganov. Fresenius Z Anal Chem 335:164, 1989.

74. SM Leschev, IV Melsitova. Neftekhimia 35:164, 1995.

75. GF Bol'shakov. Seraorganicheskie Soedinenija Nefti. Novosibirsk: Nauka, 1986.

76. IV Melsitova, SM Leschev. Neftekhimia 36:274, 1996.

8

Developments in Dispersion-Free Membrane-Based Extraction–Separation Processes

Anil Kumar Pabby

PREFRE Plant, Nuclear Recycle Group, Bhabha Atomic Research Centre, Tarapur, Maharashtra, India

Ana-Maria Sastre

Universitat Politècnica de Catalunya, Barcelona, Spain

I. INTRODUCTION

Research and developments in membrane separation technologies are rapidly expanding all over the world, and new membrane separation technologies are overcoming the commercialization barrier [1]. Moreover, the development of dispersion-free membrane extraction (DFME) has reached significance for use in separation, purification, and analytical applications in areas like hydrometallurgy, biomedical technology, and wastewater treatment [2–9]. It is interesting to note that future trends of DFME and nonsupported liquid membrane (LM) technology are quite encouraging; they continue to command significant attention, as indicated by the large number of papers published between 1990 and mid-2000 (based on Current Content Search®) (Fig. 1). Among LM techniques, nondispersive solvent extraction (DFME) with microporous hollow-fiber modules is a relatively new technology in comparison to conventional liquid–liquid extraction, but it has appeared to be a very competitive alternative in many applications such as the extraction and separation of products in biotechnology [10,11], the extraction of organic compounds [12,13], and the extraction of metal ions [14–19]. These applications, among others, show the suitability of hollow-fiber modules as phase contactors and the great potential

331

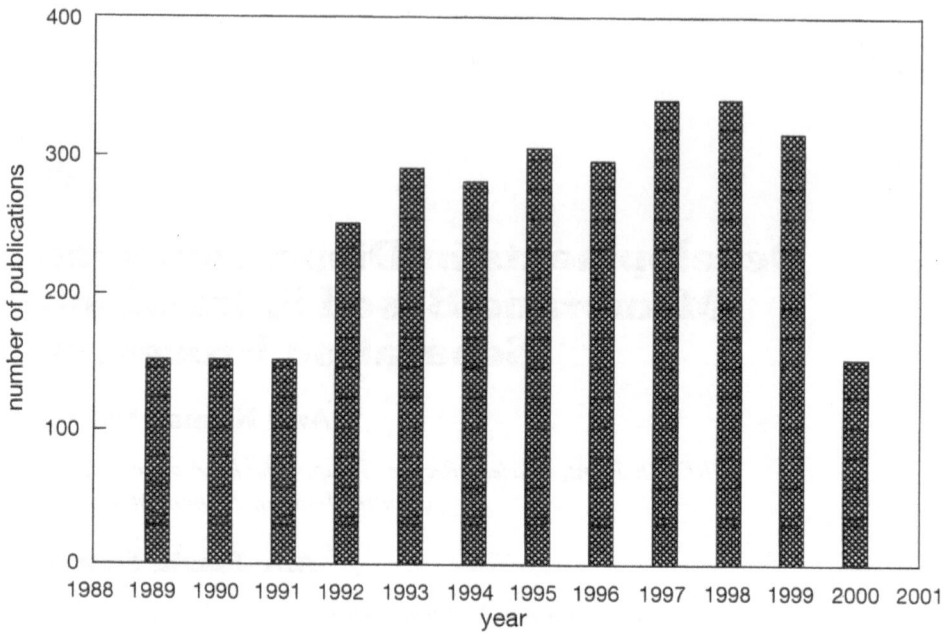

Figure 1 Number of publications pertaining to supported (DFME) and nonsupported (BLM, ELM, etc.) liquid membrane techniques in various journals between 1989 and June 2000.

of this new technology. Reed et al. [20] successfully carried out pilot-scale evaluations of microporous membrane–based solvent extraction of a wide range of organic contaminants from industrial wastewater at two industrial plants sites in the Netherlands. Process economics appeared to be quite competitive with other conventional technologies. Very recently, a nondispersive solvent extraction pilot plant for removal of Cr(VI) from galvanic process wastewaters was successfully tested by Alonso et al. [21a] in Spain, which certainly brought this technique to a level of commercialization. Kathios et al. [22] have demonstrated the utility of membrane-based solvent extraction modules for the extraction of actinides. The performance of this technique was stated to be promising. Valenzuela et al. [23–25] tested the DFME technique for real hydrometallurgical solutions using LIX 860 (5-dodecylsalicyclaldoxime) for copper recovery from acid leach residual solutions, as well as the mine water of El Teniente mine (Chile). Similarly, removal of gold cyanide from synthetic hydrometallurgical solution was accomplished successfully with nondispersive membrane extraction, and the entire process was filed as a patent [26,27].

The applications of the DFME technique for determining trace level concentrations of metal ions or organics are gaining paramount importance owing to their high selectivity and ability to concentrate aqueous streams [28–36]. In another application, Ortiz et al. [37] used DFME to separate Ni and Cd from concentrated solutions. Recent studies performed by Kralj and coworkers [38] indicate the potential of DFME applications for the selective dissolution of copper oxalate from a suspension of copper, calcium, and cadmium oxalate using LIX 84–kerosene and Durapore as a polymeric porous support. Alhusseini [39] explored the technique of membrane pulsing as a novel method of enhancing the performance of contained liquid membranes. Sirkar et al. [40–42] have successfully demonstrated hollow-fiber contained liquid membranes (HFCLMs) for efficient metal transport with better stability and reproducibility for the various chemical separation schemes. In 1994 Schlosser and Rothova published the results of a new type of hollow-fiber pertractor, in effect a flowing liquid membrane, which was reported to be similar to HFCLMs [43].

Among the potential uses of membrane contactors, a novel extraction technique was developed by performing emulsion–liquid membrane (ELM) extraction in hollow-fiber contactors, which allowed emulsion to flow in hydrophobic hollow-fiber tubes and feed into the shell side. This technique retains the advantages of ELM, namely, simultaneous extraction and stripping, while eliminating problems encountered in dispersive contacting methods, such as swelling and leakage of the liquid membrane [44].

The most important theoretical aspect of any chemical process is successful modeling, which helps researchers to understand a system better for design and performance purposes. In the last several years, detailed comprehensive reviews have been published covering the modeling of DFME [4]. Similarly, modeling of HFCLM [45,46] and DFME [47,48] was presented to focus attention on the importance of equilibrium parameters. Interestingly, the literature concerning DFME, nonsupported LMs [bulk liquid membrane (BLM) and ELM] is extensive, and several books and comprehensive reviews are available which of course, should be useful for everyone from beginners to field experts (Table 1).

In this chapter, we present a brief overview of membrane-based liquid–liquid extraction (MLLE) techniques in terms of their potential and their different applications for the separation/removal of metal cations from a range of diverse matrices. We also discuss the design of ion-selective carriers for DFME studies, their analytical importance, and stability aspects, as well as the modeling of these processes. In general, DFME studies dealing with separation and removal of organic compounds and gases are not included owing to space limitations, but a few cases of organic removal are discussed, and important pub-

Table 1 Summary of Important Literature on Supported (DFME) and Nonsupported Membrane Extraction (ELM, BLM)

Books	Ref.	Important reviews	Ref.
Cussler (1984)	49	Lonsdale (1982); Marr and Kopp, 1982; Danesi (1984–85)	55, 56, 57
Araki and Tsukube (1990)	50	Nobel and Way (1987); Noble et al. (1988); Noble et al. (1989)	6, 58, 59
Baker et al. (1991)	51	Galan et al. (1998); Baker and Blume (1990)	37, 60
Mulder (1991)	52	Kopunec and Mahn (1994)	61
Osada and Nakagawa (1992)	53	Visser et al. (1994); Shukla et al. (1996)	62, 63
Ho and Sirkar (1992)	7	Kemperman et al. (1996); Elhassadi and Do (1999)	64, 65
Bartsch and Way (1996)	3	Sastre et al. (1998); Gyves and San Miguel (1999)	4, 8
Cussler (1995)	54	Gableman and Hwang (1999)	2

lished papers on modeling and evaluating the efficiency of hollow-fiber membrane contactors to test their performance are described. In addition, potential uses of hollow-fiber contactors are briefly described in different fields such as semiconductors, human gills, and the beer industry.

II. DEVELOPMENTS IN DISPERSION-FREE MEMBRANE EXTRACTION

Dispersion-free contact between immiscible liquids can be obtained by the aid of a porous solid support such as a membrane contactor, a supported liquid membrane (SLM), or a contained or flowing liquid membrane, or simply membrane-based liquid–liquid extraction [15,40]. In SLM mode, the carrier is absorbed in the microporous wall of the polymeric support shaped as a flat sheet or tiny hollow tubes (i.d. 0.5–1 mm). Usually the feed solution is circulated through the lumen and the strip solution on the shell side of the hollow fiber. In such a case, extraction and back-extraction of the metal species take place simultaneously in one hollow-fiber (HF) module. The details of the SLM mode, its various configurations, different applications, potentials, advantages and disadvantages, latest improvements, and current status are described elsewhere [4]. Therefore, this chapter does not furnish details of SLM techniques. In a DFME mode, aqueous and organic solutions flow continuously as just described, with both phases coming into contact through the pores of the fiber wall. Here, only one separation is realized, either extraction or back-extraction. To obtain extraction and stripping together, an integrated process is used with two HF modules, one for extraction and one for stripping.

III. MEMBRANE-BASED LIQUID–LIQUID EXTRACTION (MLLE)

The capacities of performing selective metal extraction and treating dilute solutions make the DFME technique an attractive alternative to solvent extraction. In comparison to solvent extraction, DFMEs are characterized by rapid separation, high efficiency, low power consumption, and adaptability to diverse uses. Membrane-based extraction is carried out by using microporous membranes to immobilize the aqueous–organic interface with the porous structure. The solute is transported from the feed to the extractant phase, through the membrane, without phase dispersion, thus avoiding the formation of stable emulsions that may hinder the process. In this technique, it is well established [7,18] that to keep the interface within the pores of a hydrophobic membrane, it is necessary to maintain a higher local pressure in the aqueous phase (Fig.

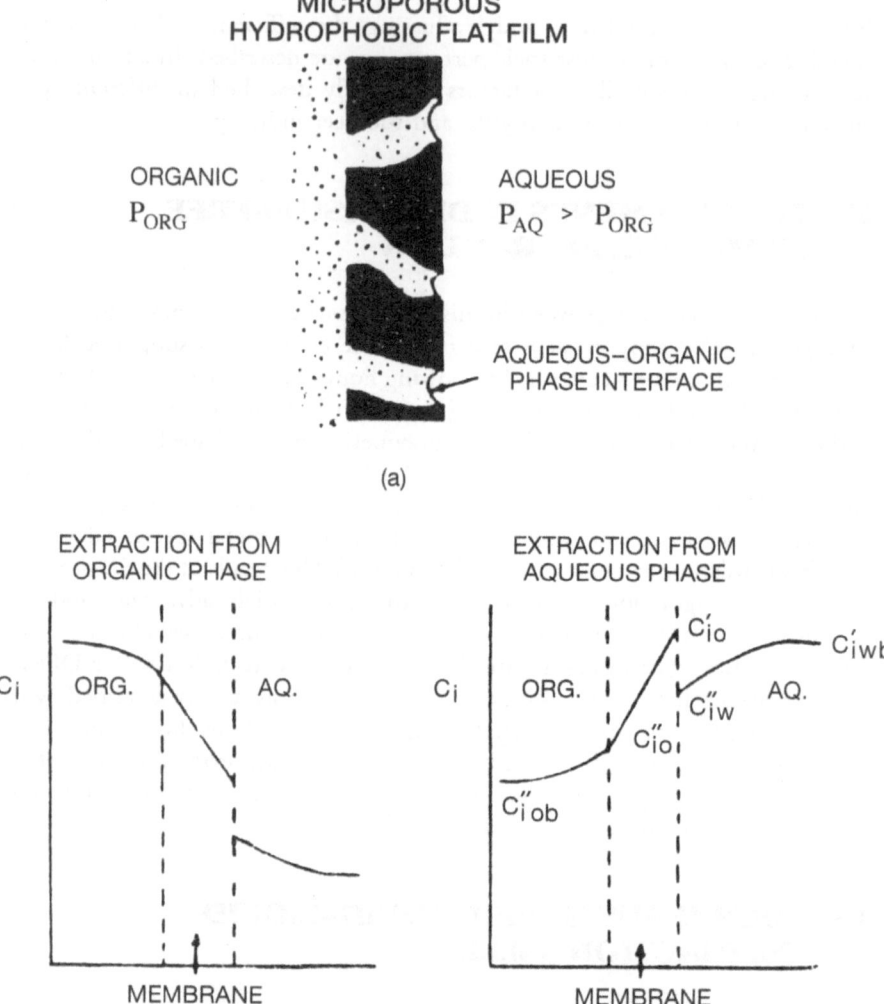

Figure 2 Solute concentration profiles for an aqueous–organic interface immobilized in the pores of (a) a hydrophobic membrane and (b) a hydrophilic membrane. (From Ref. 7.)

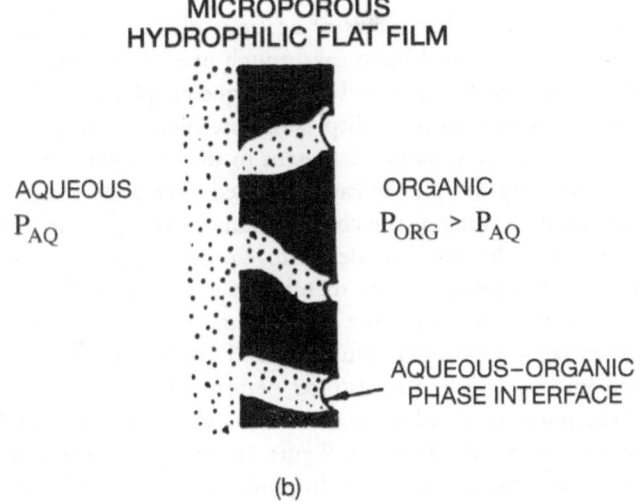

MICROPOROUS
HYDROPHILIC FLAT FILM

AQUEOUS
P_{AQ}

ORGANIC
$P_{ORG} > P_{AQ}$

AQUEOUS–ORGANIC
PHASE INTERFACE

(b)

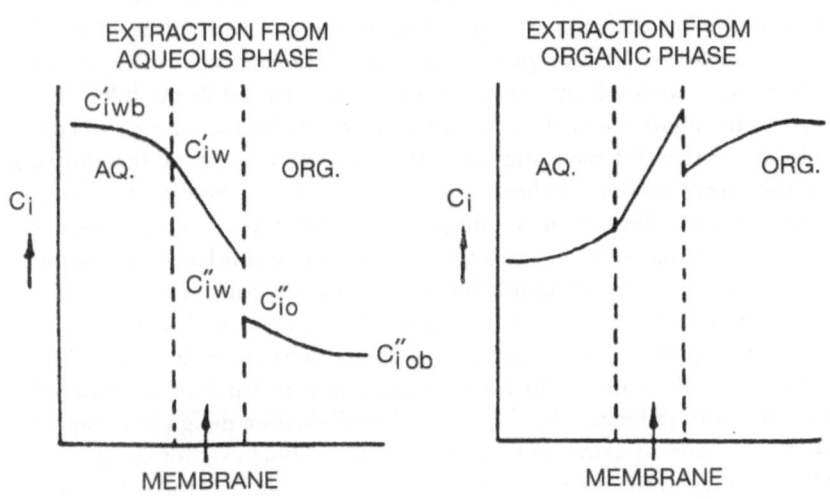

EXTRACTION FROM
AQUEOUS PHASE

C'_{iwb}

C'_{iw}

AQ.   ORG.

C_i

C''_{iw} C''_{io}

C''_{iob}

MEMBRANE

EXTRACTION FROM
ORGANIC PHASE

AQ.   ORG.

C_i

MEMBRANE

2a). It is important to note that excess of aqueous phase pressure should not exceed a critical value Δp_{cr}, called the breakthrough pressure, so that the organic phase in the pores will not be displaced by the aqueous phase [2,7].

Extraction is achieved without dispersing one phase as drops in another phase. No mechanical moving parts are present in the extractor. The flow rates of both aqueous and organic phases can be varied over wide ranges without flooding or overloading as long as the correct relative phase pressure conditions are maintained. Systems having a tendency to emulsify spontaneously can be handled without emulsification. If the membrane is in hollow-fiber form, the extractor can have very high membrane area per unit equipment volume, leading to very high extraction rates per volume [2,4,7]. Such dispersion-free, membrane-based solvent extraction can also be performed by using hydrophilic membranes whose pores are filled with an aqueous phase preferentially wetting the hydrophilic membrane. As shown in Figure 2b, the appearance of the aqueous solution on the organic side of the hydrophilic membrane and its dispersion as drops in the organic solvent can be prevented by maintaining the organic phase pressure higher than the aqueous phase pressure. The aqueous–organic interface is immobilized at each pore mouth on the organic side of the membrane. Membrane-based solvent extraction can be carried out in modules with microporous membranes in flat or hollow-fiber form. In the case of hollow fibers, Ho and Sirkar [7] have suggested that module may be of the simple shell-and-tube form without baffles, or it may have a cross-flow pattern, tangential flow filtration devices, or rotating cylindrical devices. It is most important to remember that solvent extraction is an equilibrium-based separation process, and membrane-based liquid–liquid extraction as considered here continues to be an equilibrium-based process regardless of the complications introduced by a microporous membrane. Although the separation that ultimately can be achieved is therefore limited by the conditions of equilibrium between the extract and the raffinate, the membrane-based exterior design is going to be based on the rates of extraction, as in any conventional exterior design.

The advantages membrane contactors offer over columns and other conventional mass transfer equipment include the following [2]:

1. Membrane contactors can also be used to increase conversion with equilibrium-limited chemical reactions in general. That is, by circulating the reactor contents through the contactor against a suitable extraction solvent, the products are removed and reaction equilibrium shifts to the right.

2. The available surface area remains the same at high and low flow rates because two fluid flows are independent. This property is useful in applications having a required solvent/feed ratio that is very high or very low. In contrast, columns are subject to flooding at high flow rates and unloading at low ones.

3. Modular design also allows a membrane plant to operate over a wide range of capacities. That is, small or large capacities can be obtained simply by using few or many modules.

4. Interfacial area is known and is constant, which allows performance to be predicted more easily than with conventional dispersed-phase contactors. On the other hand, interfacial area is quite difficult to determine in dispersive contactors because the bubble or droplet size distribution depends on operating conditions and fluid properties. That is why mass transfer coefficient and interfacial area are usually lumped together in mass transfer calculations. With packed columns, the interfacial area may be known, but it is often difficult to determine the loading (i.e., the fraction of the available surface that is actually used). Substantially higher efficiency, as measured by the height of a transfer unit (HTU), is achieved with membrane contactors. Also, solvent holdup is low, and this is an attractive feature when expensive solvents are employed. Membrane contactors have no moving parts like mechanically agitated dispersed-phase columns.

5. Scale-up is more straightforward with membrane contactors. That is, membrane operations usually scale linearly, so that a predictable increase in capacity is achieved simply by adding membrane modules (subject to the limitations of support equipment such as transfer pumps and piping). On the contrary, scale-up with conventional equipment is not nearly as straightforward.

On the other hand, membranes also have some disadvantages, including the following.

1. The membrane introduces another resistance to mass transfer not found in conventional operations. However, the resistance is not always important, and steps can be taken to minimize it.

2. Membrane contactors are subject to shell-side bypassing, which results in a loss of efficiency. Often bypassing is not a problem in the laboratory but becomes an issue upon scale-up to larger contactors. Fortunately, several design improvements have been proposed to address this problem.

3. Membranes are subject to fouling, although this tends to be a more serious problem with pressure-driven devices than with concentration-driven ones such as membrane contactors.

4. Membranes have a finite life, which means that the cost of periodic membrane replacement needs to be considered. Further, the potting adhesive (e.g., epoxy) used to secure the fiber bundle to the tube sheet may be vulnerable to attack by organic solvents.

These relatively few disadvantages are often outweighed by the numerous advantages cited earlier. For this reason, membrane contactors have attracted the attention of many interested parties from both academia and industry for a diverse range of applications. In our further discussions, we cite the extensive literature on DFME to summarize the advances taking place in this field.

A. Case Studies of Laboratory- or Pilot-Scale Membrane Extraction with Hollow-Fiber Modules

In the last decade, HF contactors have drawn great attention in the field of hydrometallurgy. Also, the commercial availability of hollow-fiber modules and membranes alike is important in proving the potential of this technology. With the aim of improving the performance of hollow-fiber membrane-based liquid–liquid extraction (HFMLLE), most of the research work published in the last decade deals with the study of several parameters covering all aspects of hydrodynamic and chemical parameters (Table 2). In a broad sense, three parameters —namely, permeation/mass transfer, stability, and selectivity—can be used to characterize membrane efficiency. HF modules used for this purpose are shown in Figure 3. Table 3 presents the characteristics of hollow-fiber modules commercially available for HFMLLE. Table 4 summarizes the characteristics of commercially available hollow-fiber membranes. Typical values for surface area per volume supplied by various contacting devices are listed in Table 5 [66]. Also, in relation to the support, such fundamental aspects as membrane structures, membrane fabrication (material selection), and characteristics of commercially available membranes are thoroughly discussed by Kulkarni and coworkers [67], Tsujita [68], and Feng and Huang [69]. Important applications of the HFMLLE technique in different fields are summarized in Table 6. To design a successful separation scheme with HFMLLE, one must have prior knowledge of the extraction coefficient, partition coefficient, and other solvent extraction data of the system selected, type of extractant, and source and availability of the same. Therefore, we have summarized important details of different types of extractant (acid extractants, basic extractants, chelating extractants, and solvating extractants), various applications, sources, and commercial availability in Table 7a–d.

In view of the great importance of hydrometallurgical processes and considering the compatibility of HF contractors for this purpose, several studies were performed to check the feasibility of using DFME to deal with almost all the metallic elements in the periodic table [48]. This section of the chapter covers all such important studies performed by various researchers, either using commercially available hollow-fiber contactors or constructed in laboratory by the author. In the same line, San Miguel and Gyves [70] kinetically separated indium(III) from iron(III) from nitrate media. They used a nondispersive solvent extraction technique with a hollow-fiber contactor, with Cyanex 272/ n-heptane as an extractant. An ultrafiltration hollow-fiber cartridge Xampler UFP-5-E3A [polysulfone, 5000 NMWC, 1 mm i.d. (d), 33.7 cm length (l)] from A/G Technology Corporation served as support to perform the dispersion-free solvent extraction experiments. Masterflex 7520-10 pumps with Viton

Table 2 Application-Oriented Studies for Metal Separation by HFMLLE

Ions	Operation mode	Feed	Strip	Carrier/Diluent	Ref.
Ag(I)	Extraction/back-extraction separately by one HF module	NaCN, pH 9	0.4–1 M NaOH	LIX 79/n-heptane	85
Au(I), Cu(I), Fe(II), Ni(II), Zn(II), Ag(I)	Simultaneous extraction/back-extraction with two HF modules	NaCN, pH 9–12	0.4–1 M NaOH	LIX 79/n-heptane	78
Au(I)	Extraction/back-extraction separately by one HF module	NaCN, pH 9–12; Cu(I), Fe(II), Ni(II), Zn(II), Ag(I)	0.4–1 M NaOH	LIX 79/n-heptane	79
Au(I)	Extraction/back-extraction separately by one HF module	NaCN, pH 9–12; Cu(I), Fe(II), Ni(II), Zn(II), Ag(I)	0.4–1 M NaOH	LIX 79 + TOPO/n-heptane	82
Cd(II)	Extraction module, polymeric HF; back-extraction module, ceramic multichannel	30% P_2O_5, from 85% H_3PO_4	4 M HCl	Cyanex 302/kerosene	88
Cd(II), Cu(II)	Extraction	H_2SO_4, pH 5.5		D2EHPA/WBC-15 (isododecane)	73
Ni(II), Zn(II), Am(III), Eu(III)	Extraction	Acetate buffer, pH 5		Cyanex 301/WBC-15 (isododecane)	
Cd(II), Ni(Ii), Zn(II)	Extraction using HF modules of different sizes	Sulfate, pH 4.8–5.5		DEPA/isododecane	18

Table 2 Continued

Ions	Operation mode	Feed	Strip	Carrier/Diluent	Ref.
Cr(VI)	Simultaneous extraction/back-extraction; two HF modules	CrO_4^-, Cl^-	1 M NaCl	Aliquat-336/kerosene	21b
Cr(VI)	Simultaneous extraction/back-extraction; two HF modules	CrO_4^-, Cl^-	1 M NaCl	Aliquat-336/kerosene	165
Cr(VI)	Extraction/back-extraction	CrO_4^-, Cl^-	1 M NaCl	Aliquat-336/kerosene	47
Cr(VI)	Extraction	CrO_4^-, Cl^-	1 M NaCl	Aliquat-336/kerosene	17
Cr(VI)	Simultaneous extraction/back-extraction; two HF modules	Cl^-, NO_3^-, SO_4^-, pH 5.7–7.1	1 M NaCl	Aliquat-336 kerosene	21a
Cd(II)	Simultaneous extraction/back-extraction; two HF modules	Zn, Pb, Ni	Water	Aliquat-336/kerosene	77
Cr(VI), Hg(II)	Simultaneous extraction/back-extraction; contained LM, two sets of HF modules	Cr: 0.1 M H_2SO_4, pH 2.5 Hg: 0.1 M HCl	Cr: 0.5 M NaOH; Hg: 1 M NaOH + 4 M NaCl	TOA/xylene	142
Cu(II)	Extraction	pH 1.5–6.0		E-HNBPO	16b
Cu(II)	Extraction	Cu(II), Mo(VI), Fe(II), Re(VII), pH 0.5–1.0		LIX 860	24
Cu(II)	Simultaneous extraction/back-extraction; two HF modules	Cu(II), Mo(VI), pH 0.6		LIX 860	23
Cu(II)	Extraction	Cu: pH 2.95–4.51		LIX 84/n-heptane	19
Cu(II)	Extraction/back-extraction	pH > 2.0	H_2SO_4	LIX 84/n-decane	48c
Cu(II)	Extraction/back-extraction	pH 2.95–4.51	6 N H_2SO_4 (inside emulsion droplets)	LIX 84/tetradecane surfactant ECA 5025 or Paranox 106	44

Metal ions	Process	Feed (aqueous)	Strip	Extractant/diluent	Ref.
Cu(II), Ni(II)	Extraction	pH 4.5, NaAc, HAc		D2EHPA/kerosene	48a
Cu(II), Cd(II)	Extraction	Nitrate, pH 3.8–4.0		Cyanex 272/n-heptane	80
Zn(II) Au(I)		pH 2.0–2.5 pH 10.0–10.5		LIX 79/n-heptane	15
Cu(II), Zn(II)	Extraction/back-extraction; two HF modules	0.1 M acetate buffer solution and/or HCl	HCl	PC-88A/n-heptane	70
In(III), Fe(III)	Extraction; one HF module	Nitrate, pH 1.8		Cyanex 272/n-heptane	15b
Mo(VI) Cd(II), Ni(II)	Extraction	HNO_3, pH 0–4.5		PC-88A/n-heptane	37
Nd(III), Ho(III)	Extraction/back-extraction; one HF module	$(Na, H)NO_3$, pH 2.1	1 M HNO_3	D2EHPA/dodecane	122
Nd(III)	Extraction/back-extraction; one HF module	2 M HNO_3	0.01 M HNO_3	DHDECMP and CMPO/diisopropylbenzene	22
Pb(II), Cd(II)	Extraction/back-extraction; two HF modules	$NaNO_3$/NaCl with MES and Tris buffer, pH 8		Kelex 100/n-heptane	76a
Rare earth	Extraction/back-extraction	Sodium acetate–HNO_3, $NaNO_3$–HNO_3	HNO_3	PC-88A/n-heptane	16a
Rare earth	Extraction; one HF module	Er, Ho, Y/Cl^-, pH 2.5–4.0		Oct[4]CH_2COOH	127
Zn(II)	Extraction/back-extraction	Sulfate, pH 5.3–5.7	1 M H_2SO_4	DEPA/isododecane	74

Parallel Contactor

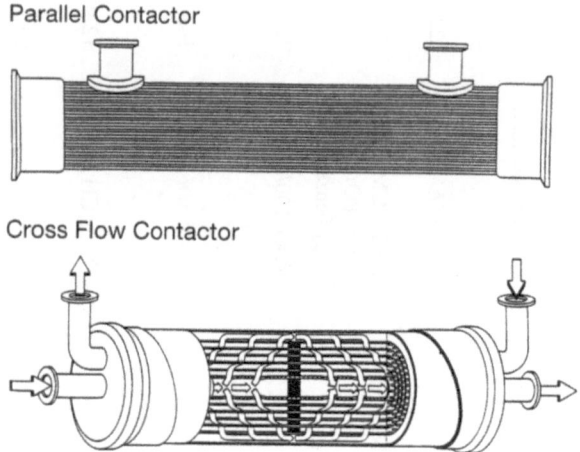

Cross Flow Contactor

Figure 3 Commercial hollow-fiber membrane extractor. (From Ref. 66.)

tubes ($\frac{1}{4}$ in. i.d., $\frac{5}{16}$ in. o.d., Cole Palmer) fed the solutions through the system. Air pressure was supplied with a diaphragm-type Gast pump. The effect of pressure differences between phases was studied at constant flow rates (Q_{org} = 4 and Q_{aq} = 60 cm^3/min) and total extractant concentration (C_{cyanex} = 0.2 M). In the case of Fe(III), the kinetics of extraction was much slower. Consequently it was possible to predict a kinetic separation of these two metals. In fact, for a flow rate of 60 cm^3/min and a period of time of 30 min, 80% of In(III) and 5% of Fe(III) could be extracted, making their separation feasible. The content of Fe(III) after acid leaching of zinc-based minerals is 10–100 times greater than that of indium. The study of the influence of the concentration of this element in the separation was very important because it has been observed that as the Fe(III) concentration increases, Y(III) [71] and In(III) [72] transport diminishes across a flat-sheet SLM (FSSLM). Figure 4 shows the separation factor defined by:

$$F_{Fe}^{In} = \frac{[In(III)_{org}][Fe(III)_0]}{[In(III)_0][Fe(III)_{org}]} \tag{1}$$

as a function of time and three different aqueous/organic flow rate ratios. The separation is accomplished better for small contact times, and it depends on the ratio of the phase flow rates. Moreover, this optimal ratio depends on the iron concentration. For an aqueous/organic flow rate ratio of 15, F_{Fe}^{In} = 52 for iron and indium concentrations of 0.1 mM, and F_{Fe}^{In} = 4.2 when a hundredfold excess of iron existed in the aqueous phase, both factors taken for a 20 min contact time. To enhance the In/Fe separation, the authors suggested that the

Table 3 Characteristics of Commercially Available Hollow-Fiber Modules

Sample no.	Fiber diameter (μm)		Number of lumina	Shell dimensions (cm)		Area per unit	Hollow-fiber membrane material	Source/supplier
	i.d.	o.d.		Diameter	Length			
1	405	464	900	1.9	15.8	46.3	Polypropylene	Liqui-Cel (Hoechst)
2	244	298	7500	4.7	24.1	40.4	Polypropylene	Liqui-Cel
3	405	464	3200	4.7	24.1	26.8	Polypropylene	Liqui-Cel
4	244	298	7500	4.7	24.1	40.4	Polypropylene	Liqui-Cel
5	240	300	2100	2.5	20.0	— $A = 0.23$ m^3	Polypropylene	Liqui-Cel
6	405	464	3200	4.7	54.6	26.8	Polypropylene	Liqui-Cel
7	240	300	10000	8	28	29.3	Polypropylene	Liqui-Cel
8	240	300	32500	10	71	$A = 1.4$ m^2 36.4 $A = 19.3$ m^2	Polypropylene	Liqui-Cel
9	240	300	—	26.4	71	$A = 193$ m^2	Polypropylene	Liqui-Cel
10	250–3000	—	—	—	17.8–182.9	$A = 0.019$ – 69.7 m^2	Polysulfone polyacrylonitrile	Koch Membrane Systems (Wilmington, DE)
11	200–5500	—	—	—	25–304.9	0.02–25 m^2	Polypropylene sulfonated polyether sulfone, polyethylene, regenerated cellulose	Microdyne Technologies (Wuppertal, Germany)
12	500–1100	—	—	—	—	0.03–5	Polysulfone with polypropylene fiber wrap	Millepore (New Bedford, MA)

Table 4 Characteristics of Commercially Available Hollow-Fiber Membranes

Membrane	Pore size (μm)	Fiber porosity (%)	Wall thickness (μm)	Material	Source
Celgard X-10	0.03 i.d. = 100	20	25	Polypropylene	Hoechst-Celanese
Celgard X-20	0.03 i.d. = 240 or 400	40	25	Polypropylene	Hoechst-Celanese
Celgard X-30	0.03 i.d. = 240	30	30	Polypropylene	Hoechst-Celanese
Accurel PP 50/280	0.2 i.d. = 600 ± 90	—	50 ± 10	Polypropylene	Enka, Germany
Accurel PP Q3/2	0.2 i.d. = 280	—	200 ± 45	Polypropylene	Enka, Germany
Accurel PP S6/2	0.2 i.d. = 600	60–65	450 ± 70	Polypropylene	Enka, Germany
Accurel PP V8/2HF	— i.d. = 5500 ± 300	—	1550 ± 150	Polypropylene	Enka, Germany

Table 5 The Specific Surface Area of Different Contactors

Contactor	Surface area per unit volume (m²/m³)
Free dispersion columns	3–30
Packed/trayed columns	30–300
Mechanically agitated columns	150–500
Membranes	1500–6000

Source: Ref. 66.

operation of the system be explored in the once-through mode with a greater membrane area and short contact times.

Further, Gyves et al. [73] used HFMLLE to recover Pb(II) and Cd(II) from aqueous media with two modules using Kelex 100 dissolved in heptane as the carrier. Experiments were performed with the same experimental setup

Table 6 Summary of Applications

Application	Comments	Selected references
Liquid–liquid extraction	Membrane contactors have been employed in a variety of chemical systems covering a broad range of m_i values. A number of membrane geometries, materials of construction, and flow configurations have been investigated.	8, 10, 22, 34, 59, 91, 93, 96, 101, 109
Wastewater treatment	Recovery from aqueous streams of a number of troublesome pollutants, including 2-chlorophenol, benzene, nitrobenzene, trichloromethane, tetrachloromethane, and acrylonitrile has been demonstrated. Bubble-free aeration offers several advantages over conventional approaches.	1, 2, 25, 28, 75, 83, 91, 93, 96, 111, 113–115, 152, 153
Metal ion extraction	The efficacy of the technology has been demonstrated for recovery of a variety of ions from aqueous wastewaters. Some authors have reported the use of dual contactors operating in chelating and stripping mode, respectively. Numerous applications using chemical reactions to facilitate ion removal have been described.	20, 57, 96, 148, 150, 151
Semiconductors	Membrane contactors are used in semiconductor manufacturing operating for the production of ultrapure water and for ozonation of cleaning water.	23, 31, 45, 143

Table 7   Hollow-Fiber-Based Liquid–Liquid Extractants

Type	Examples	Manufacturers	Commercial uses
		a. Acid extractants	
Carboxylic acids	Naphthenic acid Versatic acid	Shell Chemical Co.	Cu/Ni separation, Ni extraction, yttrium extraction
Alkylphosphoric acid	Monoalkylphosphoric acids	Mobil Chemical Co. (MEHPA, DEHPA mixture) Zeneca (Acorga SBX50)	Fe removal Sb, Bi removal from copper electrolytes
	Dialkylphosphoric acids and sulfur analogs	Daihachi Chem. Ind. Co. Ltd. (DP-8R, DP-10R, TR-33, MSP-8) Bayer AG (BaySolvex D2EHPA, D2EHPTA VP A1 4058) Hoechst (PA216, Hoe F 3787)	Uranium extraction Rare earth extraction Cobalt/nickel separation Zinc extraction, etc.
Alkylphosphinic acids	2-Ethylhexylphosphinic acid, 2-ethylhexyl ester, and sulfur annalogs	Daihachi Chem. Ind. Co. Ltd. (PC88A) Albright and Wilson Americas (Ionquest 801) Bayer AG (BaySolvex VP-AC 4050 MOOP)	Cobalt/nickel separation Rare earth separation
Alkylphosphinic acids	Dialkylphosphinic acid and sulfur analogs.	Daihachi Chem. Ind. Co. Ltd. (PIA-8) Cytec Inc. (Cyanex 272, 302, 301)	Cobalt/nickel separation Zinc and iron extraction
Arylsulfonic acids	Dinonylnaphthalenesulfonic acid	King Industries Inc. (Synex 1051)	Magnesium extraction

b. Chelating acid extractants

Hydroxyoxime derivatives	α-Alkarylhydroxines (LIX 63)	Henkel Corp. (various: e.g., LIX 860)	Copper extraction Nickel extraction
	β-Alkylarylhydroxyoximes (LIX 860) (M5640)	Zeneca Specialties (various: e.g., M5640) Inspec (MOC reagents)	
8-Hydroxyoxime derivatives	Kelex 100, 120	Witco Corp.	Gallium extraction Proposed for copper extraction
	LIX 26	Henkel Corp. (PA216, Hoe F 3787)	
β-Diketone derivatives	LIX 54 Hostraex DK16	Henkel Corp.	Copper extraction from ammonical media
Alkarylsulfonamides	LIX 34	Henkel Corporation	Development reagent
Bisdithiophosphornamide derivatives	DS 5968, DS 6001 (withdrawn)	Zeneca Specialties	Zinc extraction Cobalt/nicke/mangese separation
Hydroxamic acids	LIX 1104	Henkel Corporation	Nuclear fuel reprocessing Iron extraction Sb, Bi extraction from copper refinery liquors

Table 7 Continued

Type	Examples	Manufacturers	Commercial uses
		c. Basic extractants	
Primary amines	Primene JMT, Primene 81R	Rohm and Haas	No known commercial use
Secondary amines	LA-1, LA-2	Rohm and Haas	Uranium extraction
	Adogen 283	Witco Corp.	Proposed for vanadium and tungsten extraction
Tertiary amines	Alamines (e.g., Alamine 336)	Henkel Corporation	Uranium extraction
		Witco Corp.	Cobalt from chloride media
	Adogens		Vanadium and tungsten extraction, etc.
Quaternary amines	Aliquat 336	Henkel Corporation	Vanadium extraction
	Adogen 464	Witco Corp.	Possible chromium, tungsten, uranium extraction
Quaternary amine + nonyl phenol	LIX 7820	Henkel Corporation	Anionic metal cyanide extraction
Mono-N-substituted amide			Iridium separation from rhodium
Trialkylguanidine	KIX 79	Henkel Corporation	Gold extraction from cyanide media

d. Mainly solvating extractants and chelating nonionic extractants

Carbon–oxygen donor reagents	Alcohols, (decanol) Ketones (MIBK) Esters, ethers, etc.	Various chemical companies	Niobium/tantalum separation Zirconium/hafnium separation
Phosphorus–oxygen donor reagents and phosphorus–sulfur donors	Phosphoric esters Phosphonic esters Phosphinic esters Phosphine oxides and sulfur analogs	Albright and Wilson Americas Daihachi Chem. Ind. Co. Ltd. Cytec Industries Bayer AG Hoechst (TBP, DBBP, TOPO) (Cyanec 921, 923, 471X) (Hoecsht PX324, 320) BaySolvex VP-AC 4046 (DBBP), VP-AC 4014 (DPPP), VP-A1 4059 (DEDP)	U_3O_8 processing Iron extraction Zirconium/hafnium separation Niobium/tantalum separation Rare earth separation Gold extraction
	Dialkylphosphoric acids and sulfur analogs		
Sulfur–oxygen donors	Sulfoxides Sulfides	Daihachi Chem. Ind. Co. Ltd. (SFI-6) Hoechst (Hoe F 3440) Others	Palladium extraction in PGM refining
Nitrogen donors	Biimidazoles and bibenzimidazoles Pyridine dicarboxylic ester	Zeneca Specialties (ZNX 50) Zeneca Specialties (CLX 50)	Zinc extraction and separation from ion in chloride media Copper extraction from chloride media

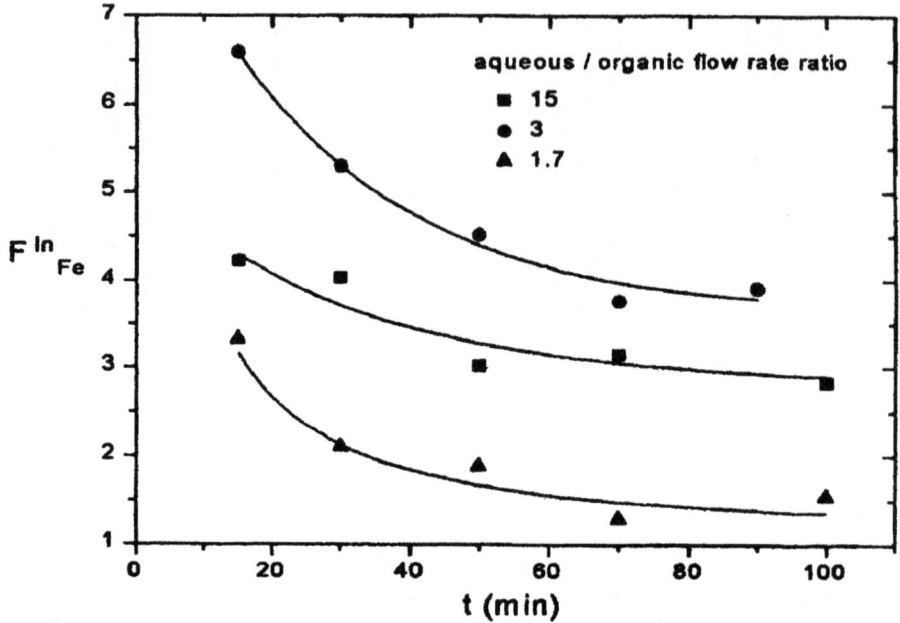

Figure 4 Separation factor as a function of time and three different aqueous/organic flow rate ratios in the presence of a 100-fold excess of Fe(III) with respect to In(III). (From Ref. 70.)

described in Reference 70. Experiments were carried out with the FSSLM technique to predict behavior in order to obtain some guidelines for further HFMLLE experiments. With FSSLM, permeabilities of Cd(II) and Pb(II) in a solution containing 0.01 mM of each metal ion in an aqueous chloride solution (I = 0.01 M) and using 1% v/v of Kelex 100 were determined at pH values of 8.0, 7.0, and 6.0. Separation factors were 0, 0.47, and 3.14, respectively, after 160 min. Consequently, a good separation between Cd(II) and Pb(II) is obtained at pH 6.0. According to the results obtained in the FSSLM experiments, two different total concentrations of Cd(II) were tested in the two-module HFMLLE setup. The phases were circulated at 20 cm³/min, ΔP in the extraction module was 2 kPa, ΔP in the back-extraction module was 21 kPa, and 1% v/v of Kelex 100 was used. At pH 8.0 in a chloride aqueous medium, 0.01 mM Cd(II) was quantitatively extracted but not back-extracted. However, when 1 mM Cd(II) was introduced in an aqueous nitrate feed solution (I = 0.01 M) at pH 6.8, 50% of Cd(II) was extracted from the feed solution and 24% was back-extracted in the strip solution in 80 min.

Geist et al. [74] reported separation of multication mixtures with a hollow-fiber module. Experimental results on the coextraction of divalent cations (Cd^{2+}, Cu^{2+}, Ni^{2+}, Zn^{2+}) into bis(2-ethylhexyl)phosphoric acid (D2EHPA) showed fine agreement with calculated results. Extraction efficiency was increased with decreasing aqueous flow rate, Q_{aq}, owing to the longer residence time. At a relatively low complexant concentration (0.05 M), zinc was readily extracted (from 100 mg/L to less than 0.1 mg/L, depending on flow rate), whereas cadmium is removed only to a small extent (e.g., from 10 mg/L to ca. 7 mg/L). This effect results from proton release due to zinc extraction, which impedes cadmium extraction. Simply by increasing complexant concentration to 0.5 M, both zinc and cadmium can be extracted to approximately the same extent. Similar calculations were performed for the separation of a mixture of four cations (Zn^{2+}, Cd^{2+}, Cu^{2+}, Ni^{2+}) (Fig. 5a): metals are removed according to their extractability with D2EHPA ($Zn^{2+} > Cd^{2+} > Cu^{2+} > Ni^{2+}$), with nickel remaining practically unaffected.

In another part of the study, the authors dealt with the separation n of Eu(III) and Am(III) using Cyanex 301, which in a purified form is able to separate Am^{3+} from Eu^{3+} with a high separation factor [75]. Stirred-cell experiments indicated that the extraction rate was limited by diffusion, a linear increase in flux, $j_{t=0}$, with stirrer speed n_{aq} was observed. Figure 5b shows the result of the calculations of Am^{3+}/Eu^{3+} separation in a hollow-fiber module when purified Cyanex 301 was used. At a flow rate of 5 L/h, 99.9% of Am^{3+} is transferred to the organic phase (i.e., a decontamination factor of 1000), whereas only about 0.3% of Eu^{3+} (not noticeable in Fig. 5b) is coextracted.

Experimental studies carried out recently by Schöner et al. [76a] proved the usefulness of hollow-fiber modules in environmental engineering. Figure 6a shows the process scheme of their zinc recovery plant. The organic (extraction) phase was a solution of bis(2-ethylhexyl)phosphoric acid (HDEPA, HX) in isododecane, WBC-15. In module I zinc was removed (extracted) from the wastewater into the organic phase as $[ZnX_2(HX)]_{org}$. Subsequently, in module II, zinc was stripped by sulfuric acid with regeneration of the complexing agent $(HX)_{2\,org}$. Liqui-Cel cross-flow modules were used with an active length of 15 cm and a diameter of 5 cm. For details of the equipment, see Schöner et al. [76b]. The aqueous phase contains 100 mg/L zinc (as zinc sulfate). The initial pH value varied between 5.3 and 5.7. The regeneration (stripping) of the organic phase was achieved with 1 M H_2SO_4. For most applications, optimization means achieving the limiting value for the zinc concentration in the purified wastewater $[Zn_{end,ex,aq}]$ and the maximum value for the concentration of zinc $[Zn_{end,reex,aq}]$ in the acid. A typical example of the calculation according to the proposed model together with experimental results is shown in Figure 6b.

The authors designed an optimization strategy directed to the maximum enrichment factor for the fluxes F_{acid} and F_{org} as variables. The most important

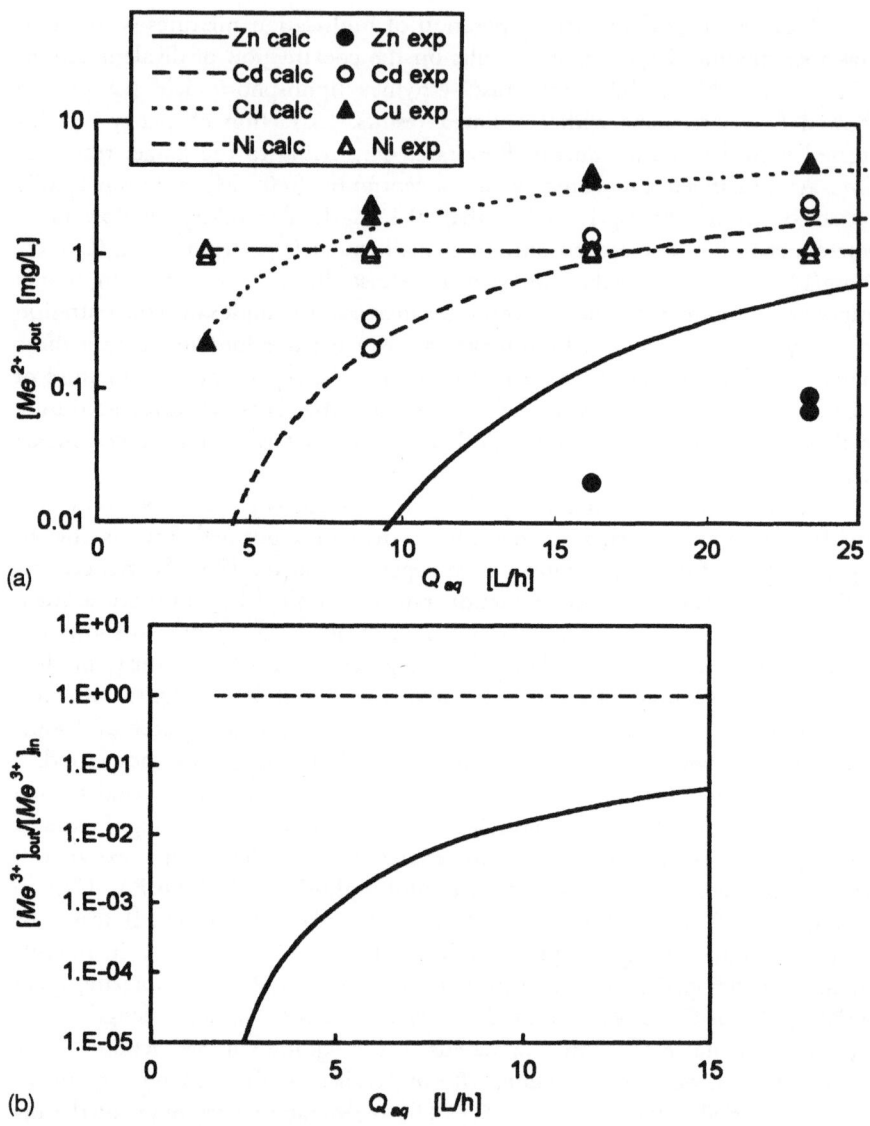

Figure 5 (a) Zinc/cadmium/copper/nickel separation in an HFM: $[Zn^{2+}]_{in}$ = $[Cd^{2+}]_{in}$ = $[Cu^{2+}]_{in}$ = 11 mg/L; $[Ni^{2+}]_{in}$ = 1.1 mg/L; pH_{in} = 5.5; $[(D2EHPA)_2]$ = 0.05 M; Q_{org} = 1.6 L/h. (From Ref. 74.) (b) Calculated americium/europium separation with purified Cyanex 301 in an HFM: $[Am^{3-}]_{in}$ = 10 mg/L; $[Eu^{3+}]_{in}$ = 1000 mg/L; pH_{in} = 4.2; [Cyanex 301] = 1.0 M; Q_{org} = Q_{aq}. Solid curves, Am^{3+}; dashed curves, Eu^{3+}. (From Ref. 74.)

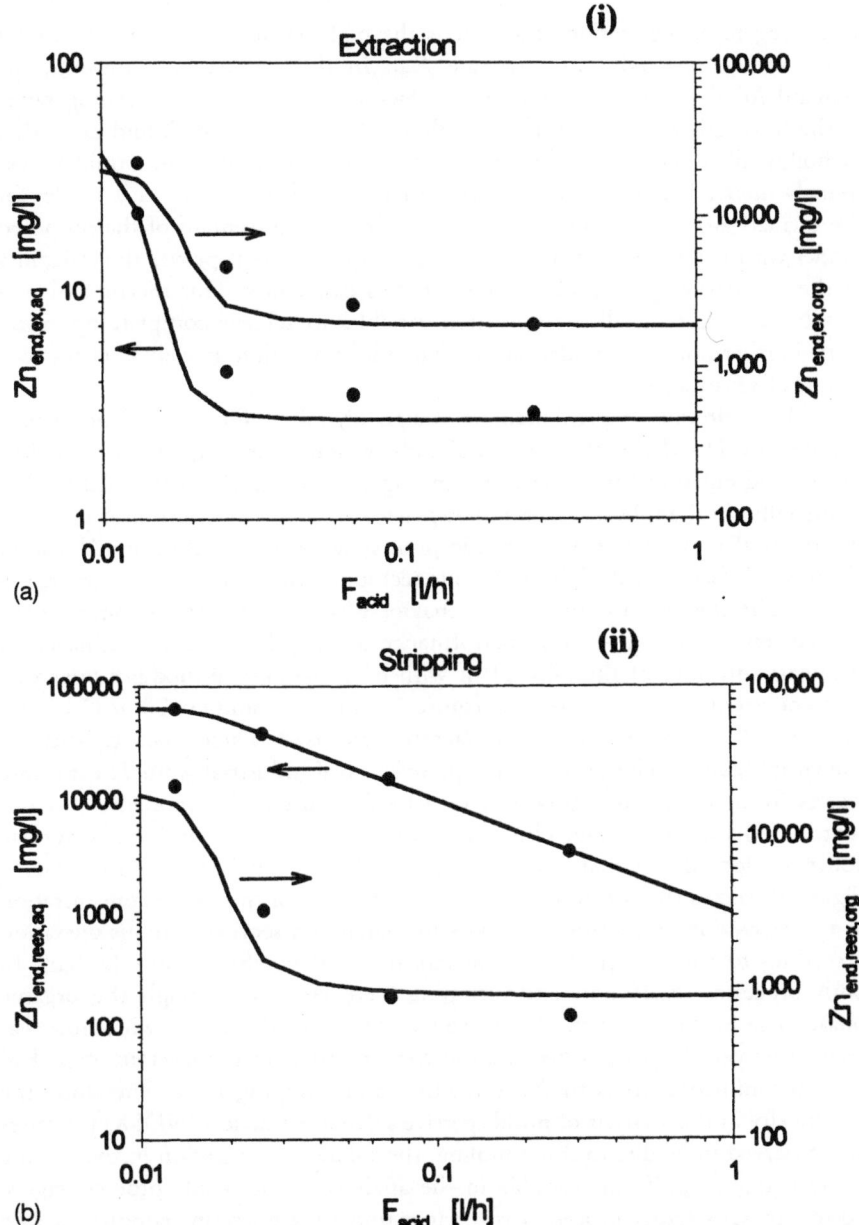

Figure 6 (a) Schematic view of the continuous extraction and stripping process for Zn recovery. (From Ref. 76a.) (b) Variation of the acid flow rate in a 15 cm module, $Zn_{0,ex,aq} = 100$ mg/L, $[H_2SO_4] = 1$ M, $[(DEPA)_2] = 0.5$ M, $F_{aq} = 10$ L/h, $F_{org} = 1$ L/h. Points, experimental data; lines, calculated results. (From Ref. 76a.)

aspect regarding separation efficiency is the high sensitivity of the enrichment factor to the selected acid flow rate F_{acid}. In this connection, however, the demand for the maximum enrichment factor means a narrow opening range in the flow rate F_{org}. The authors emphasized that for liquid–liquid extraction in hollow-fiber modules, an empirical (black box) optimization would be extremely ineffective because of the multiplicity of variables (viz., 11 variables for the extraction/stripping mode). Therefore the good agreement of the proposed model with the experimental data is important because it proves the reliability of the calculation procedure, as done in this case study. This means that it is possible to deal with all the operation variables to achieve complete optimization. Furthermore the model allows dynamic simulation as well as characterization of sensitivities.

A continuous improvement in design of hollow-fiber contactors manufactured by Hoechst Celanese (Liqui-Cel) is noted, leading to better performance and enhanced mass transfer. Among the few studies performed for hydrometallurgical applications, a nondispersive solvent extraction pilot plant for the removal of Cr(VI) from galvanic process wastewaters tested by Alonso et al. in Spain (see Ref. 21a) brought this technique to the level of commercialization. The theoretically predicted behavior was checked against some experimental results, and satisfactory performance of the pilot plant was achieved. A schematic diagram of the pilot plant with both membrane modules operating in a cocurrent mode is shown in Figure 7a. In these studies, Liqui-Cel Extra Flow 10 × 70 cm membrane contactors were used. Their characteristics are shown in Table 8. The process was applied to real industrial waste stream rinse waters from the galvanic processes of a local industry (Componentes y Conjuntos, SA) with an average chromium concentration of 64 mg/L. The average concentration of the main components in this stream is shown in Table 9. Figure 7b shows the theoretical evolution with time of the solute concentration in the extraction and stripping phases for batch and semicontinuous operation according to the model and parameters reported by the authors [21b,c]. In both processes, extraction and stripping were coupled through the organic phase. The evolution of the solute concentration in this phase with time was seen to be very important because the concentration in the organic phase had a strong influence on both the extraction and stripping rates. Therefore, the complexity in the analysis of nondispersive solvent extraction (NDSX) processes was observed to be due to this coupling, the solute concentration in the organic phase being a significant variable in the analysis of the whole process. Figure 7d reports satisfactory process performance under the operating conditions. The concentration in the extraction solution at the outlet is always less than 9.6×10^{-6} M. The organic concentration recovers its initial value after 120 h, and the stripping concentration reaches a value of around 0.140 M, higher than the minimum required value.

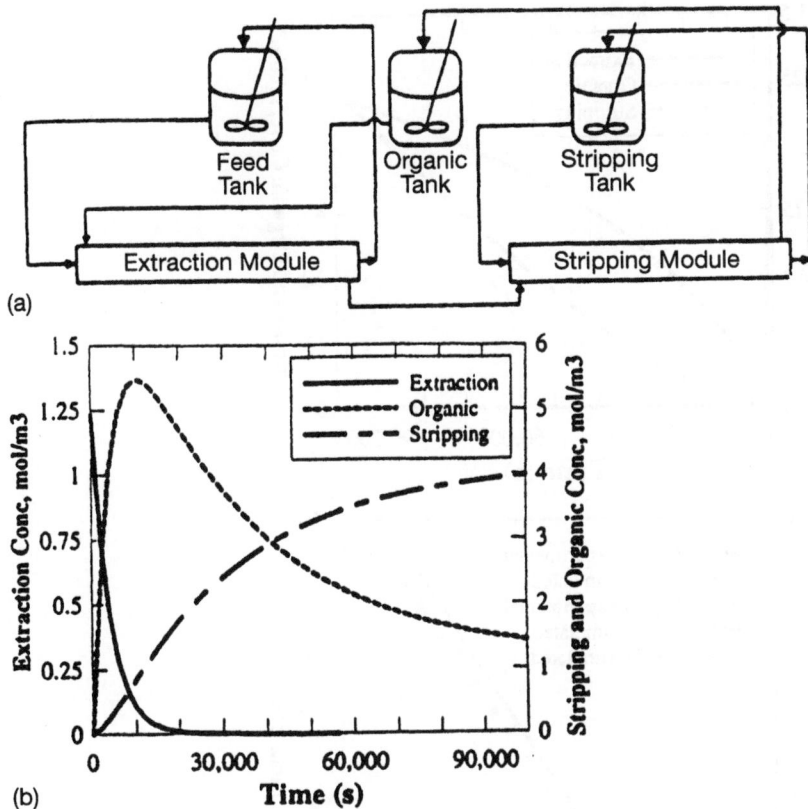

(a)

(b)

Figure 7   (a) Schematic diagram of a nondispersive solvent extraction pilot plant for chromium recovery from galvanic waste. (From Ref. 21a.) Theoretical performance of the DFME pilot plant in (b) batch mode and (c) semicontinuous mode under the following conditions: $C_T = 0.600$ M, $C_{Cl} = 1.234$ mM, $c_{Fe} = 2.22 \times 10^{-5}$ m^3/s, $F_o = 1.11 \times 10^{-5}$ m^3/s, $F_s = 2.5 \times 10^{-5}$ m^3/s, $V_e = 9 \times 10^{-2}$ m^3, $V_o = 1.5 \times 10^{-2}$ m^3, $V_s = 2.25 \times 10^{-2}$ m^3. (From Ref. 21a.) (d) Performance of a pilot plant for chromium recovery running in a semicontinuous mode. (From Ref. 21a.)

Another study [77] reported an integrated process based on the application of two separation technologies for the selective recovery of the uranium and the cadmium contained in wet-process phosphoric acid. The process is based on the removal of uranium by ion exchange using the Purolite S940 resin as ion-exchange resin followed by selective removal of Cd(II) by means of membrane-assisted solvent extraction with Aliquat 336 as the selective extractant. Operational conditions for the removal of U and Cd, respectively, are summarized in Tables 10a and 10b. Since the viability of the process for re-

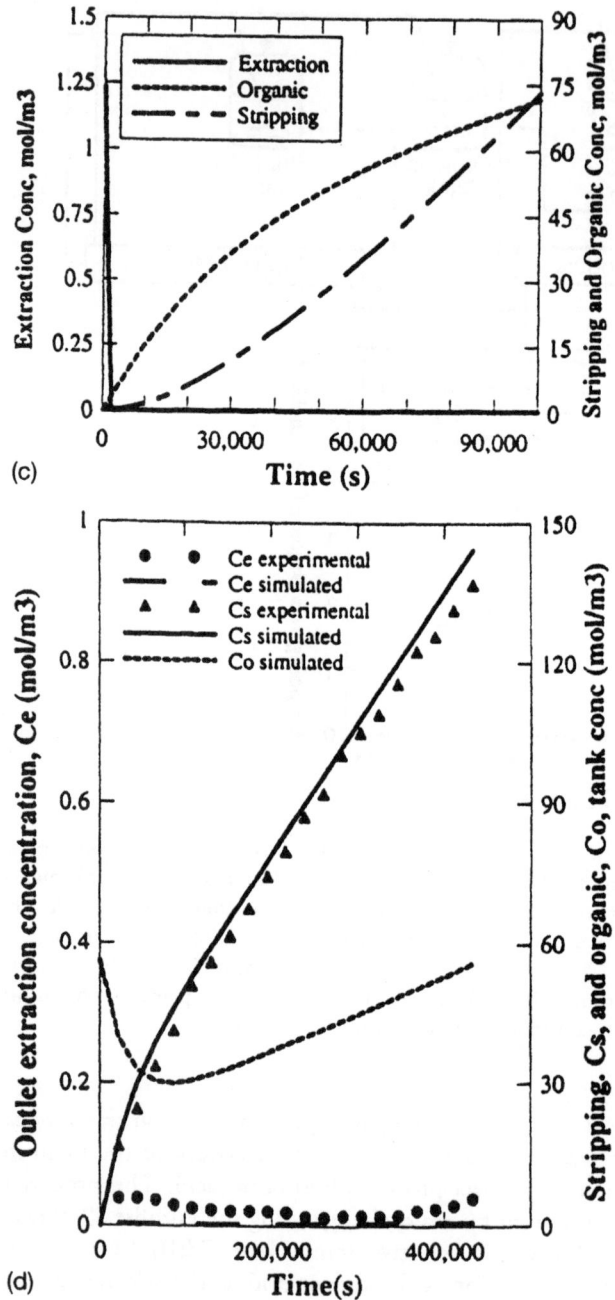

(c)

(d)

Figure 7 Continued

Table 8 Membrane Module Characteristics

Cartridge dimensions ($D \times L$)	10 cm $\times$ 71 cm
Number of fibers	32,500
Effective surface area	19.3 m^2
Effective area/volume	3641 m^2/m^3
Effective length	63 cm
Fiber potting material	Epoxy
Cartridge material	PP (stripping)–stainless steel (extraction)
Fiber type	Celgard X10 polypropylene
Inner diameter	24 $\times$ 10^{-3} cm
Thickness	30 $\times$ 10^{-3} cm
Pore size	0.05 μm
Porosity	30%

Source: Ref. 21a.

ducing the concentration of U and Cd below 2 mg/L was reported, the overall performance of this process was claimed to be satisfactory. The flow diagram of the integrated process for the removal of Cd and U from wet-process phosphoric acid is shown in Figure 8a. Figure 8b compares the simulated and experimental results for runs performed at different conditions (see Ref. 77 for more details) for the two models presented in this paper. The authors observed that a model that considers a homogeneous concentration of cadmium in the shell side of the module agrees more satisfactorily with the experimental data than the distributed concentration model.

In the present authors' laboratory, we have achieved dispersion-free membrane extraction of gold cyanide from aqueous alkaline cyanide media with LIX 79 (*N,N'*-bis(2-ethylhexyl)guanidine, RH) and stripping of gold from organic complex simultaneously using two microporous hydrophobic polypropylene hollow-fiber contactors each for extraction and stripping [26,78–80]. A sche-

Table 9 Chemical Composition of a Typical Industrial Waste Stream

Solute	Concentration (mg/L)	Solute	Concentration (mg/L)
Cl$^-$	140	Lead	1.1
NO$_3^-$	288	Nickel	1.92
SO$_4^-$	122	Chromium	64
Zinc	160	—	—

Source: Ref. 21a.

Table 10   Operational Characteristics for Removal of Heavy Metals

Operational	Unit/Symbol	Value
a. Uranium		
Concentration inlet/ outlet	mg/L	200/2
Mass action law constant	K	1
Mass transfer coefficient	k_a, cm^3 of solution/cm^3 of bed	0.02
Particle size range	mm	0.355–0.500
Average particle diameter	mm	0.4275
Number of columns	unit	2
Void fraction of the bed	e	0.28
Bulk density of the resin	g/cm^3	0.358
Reynolds number		0.0854
b. Cadmium: experimental runs in NDSX experiments		
Feed flow rate		9.10×10^{-4}–7.2×10^{-3} m^3/s
Organic flow rate		3.1×10^{-3}–4.2×10^{-3} m^3/s
Stripping flow rate		6×10^{-4}
Equilibrium flow rate		3×10^{-2}
Organic volume		718×10^{-6} m^3
Equilibration volume		1×10^{-3} m^3
Inlet feed concentration		0.437–0.495 mol/m^3
Inlet stripping cadmium concentration		0 mol/m^3
Initial organic cadmium concentration		0 mol/m^3
Run time		26–65 h

Source: Ref. 77.

Figure 8   (a) Flow diagram of the integrated process for the removal of Cd and U from wet-process phosphoric acid. (From Ref. 77.) (b) Comparison of the simulated and experimental results for experimental run I: feed flow rate, 9.0×10^{-4} m^3/h; organic flow rate, 3.1×10^{-4} m^3/h; inlet feed concentration, 0.495 mol/m^3; initial organic Cd conc, 0.0 mol/m^3; run time, 65 h. (From Ref. 77.)

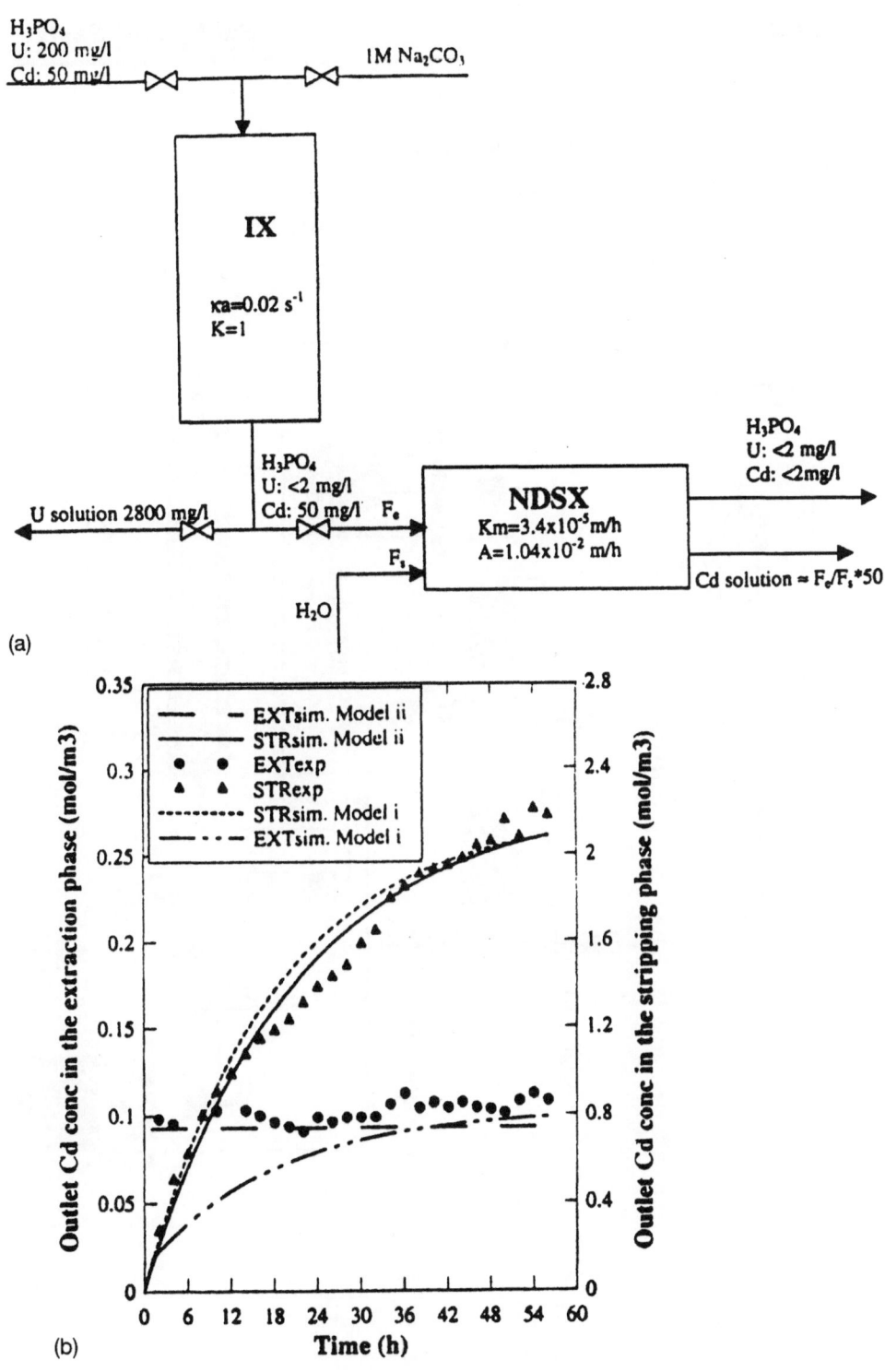

(a)

(b)

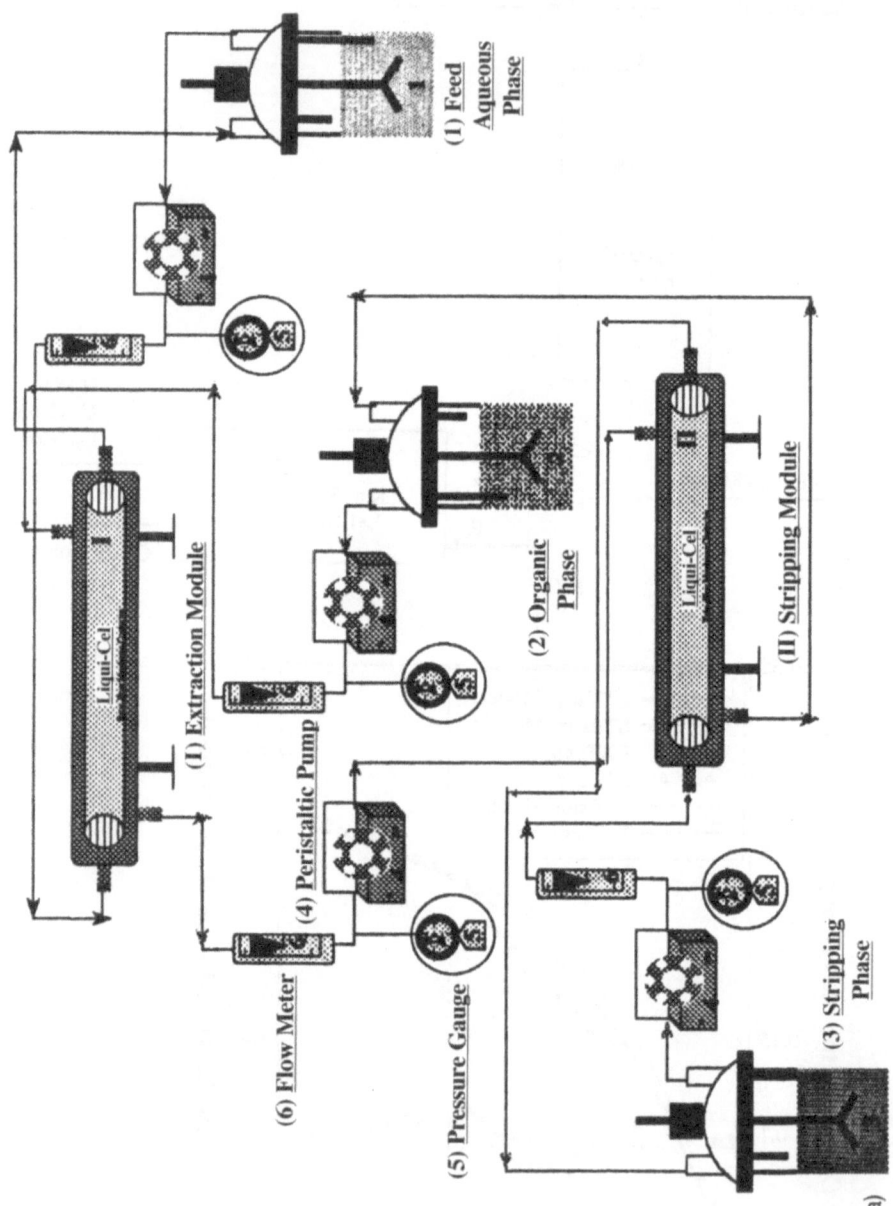

(I) Extraction Module

Liqui-Cel

(1) Feed Aqueous Phase

(2) Organic Phase

(II) Stripping Module

Liqui-Cel

(3) Stripping Phase

(4) Peristaltic Pump

(5) Pressure Gauge

(6) Flow Meter

(a)

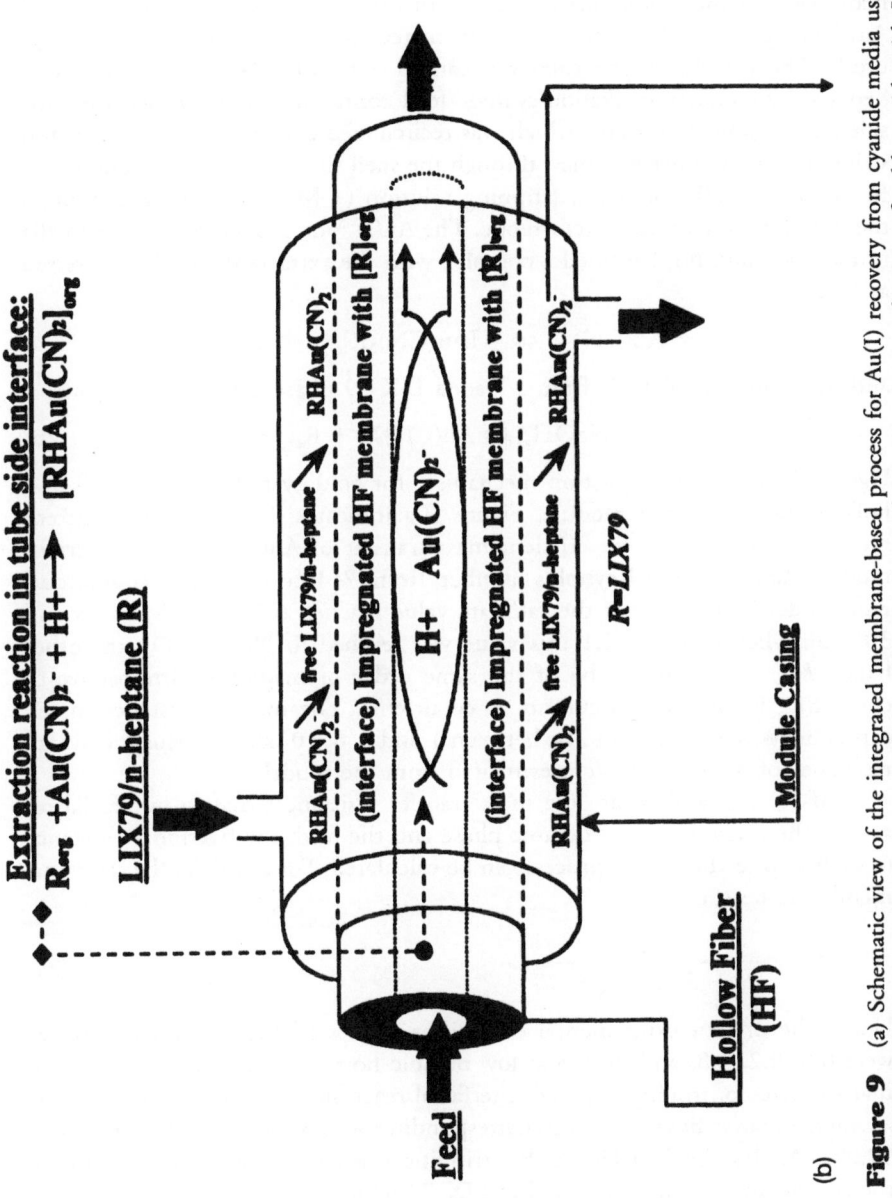

Figure 9 (a) Schematic view of the integrated membrane-based process for Au(I) recovery from cyanide media using two hollow-fiber contactors. (From Ref. 79.) (b) Schematic view of the extraction mechanism of gold cyanide with LIX 79 in a hollow-fiber contactor. (From Ref. 78.)

matic view of the membrane-based solvent extraction and stripping process of Au(I) using two hollow-fiber contactors in recirculation mode is shown in Figure 9. The HFMLLE operation was carried out with 12–18% LIX 79 in *n*-heptane by contacting alkaline cyanide feed containing gold through the tube side and organic extractant, which was recirculated between the extraction and stripping hollow-fiber modules, through the shell side, in countercurrent mode. In the second HF contactor, stripping solution (1 M NaOH) flowed through the tube side in countercurrent mode. The Au(I) ions in alkaline cyanide media (present as $Au(CN)_2^-$) formed a complex with the extractant LIX 79, expressed as follows:

$$R_{org} + H_{aq}^+ + Au(CN)_{2aq}^- \Leftrightarrow [HAu(CN)_2R]_{org} \qquad K_{ex} \qquad (2a)$$

and the stripping of Au(I) from a loaded LIX 79 phase is shown as follows:

$$[HAu(CN)_2R]_{org} + NaOH_{aq} \Leftrightarrow Au(CN)_{2aq}^- + R_{org} + H_2O + Na_{aq}^+ \qquad (2b)$$

Figure 9b shows the extraction mechanism for gold cyanide with LIX 79 in a hollow-fiber extraction module. Figure 10a indicates that the countercurrent mode is more suitable for efficient mass transfer of Au(I) than the cocurrent mode. The increase in Reynolds number, from *Re* 1.46 to 1.84 in countercurrent mode, indicated that the raffinate value reached 6×10^{-6} M in around 30 min, whereas for *Re* 1.1 this value was reached in 90 min. On the other hand, K_{Au}^E was found to be of the same order of magnitude irrespective of contacting mode (countercurrent or cocurrent). The overall resistance in the experiments was observed to be between 1 and 4×10^5 s/cm, versus the overall resistance of $3–5 \times 10^5$ s/cm estimated from the model.

If R_i, R_m, and R_f are the mass transfer resistances due to interfacial reaction, the membrane, the organic phase and the feed, the fractional resistance of each step to the overall process can be calculated. For example, the fractional membrane resistance is

$$R_m^o = \frac{R_m}{R_i + R_m + R_s + R_f} \qquad (2c)$$

Under the present experimental conditions, values of R_m^o, R_f^o, and R_s, and R_i were 0.5, 0.2, 4.8, and 94.5% at low organic flow rates. This clearly indicated that the rate-controlling step was interfacial reaction on the membrane surface. At high organic flow rates, the corresponding values were 0.5, 0.2, 0.3, and 99.0%. As described in Figure 9a, stripping was performed simultaneously in the second module by recirculating LIX 79 in the shell side of the extraction and stripping modules while NaOH solution flowed through the tube side in the second module. The overall mass transfer resistance was between 2.1 and 2.6×10^4 s/cm, versus the overall resistance calculated from Eq. (2): $2.5–3.0 \times 10^4$ s/cm. The results showed that resistance due to the organic phase is

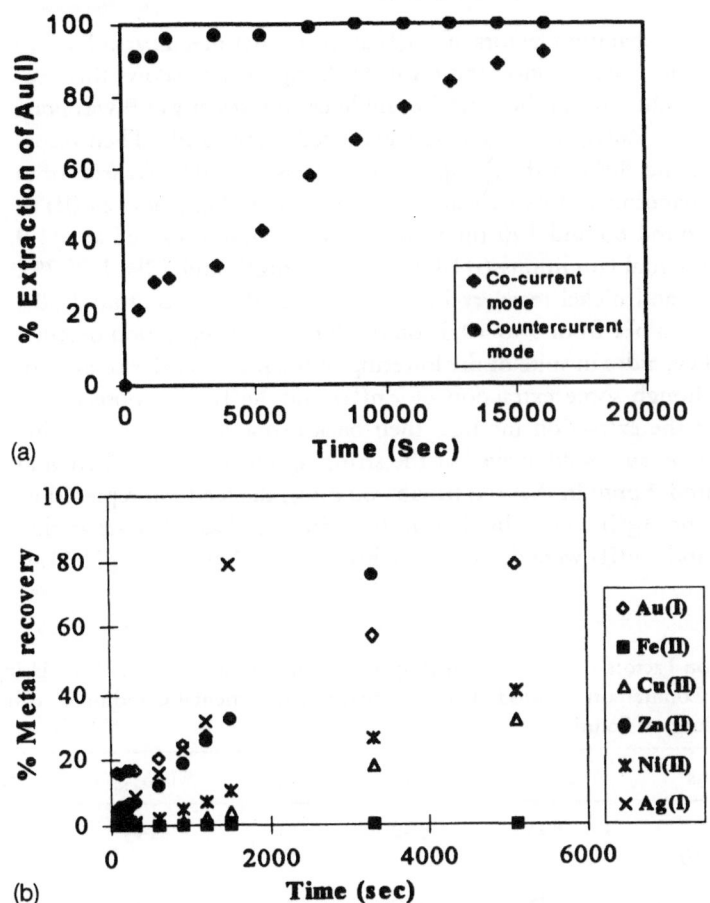

Figure 10 (a) Effect of mode of contact on extraction of Au(I) with LIX 79. (From Ref. 78.) (b) Separation of Au(I) from Fe(II), Cu(II), Ag(I), Ni(II), and Zn(II) with 18% v/v LIX 79 at pH 9.0. Composition of feed: Au(I), 10 mg/L; Fe(II), 30 mg/L; Cu(II), 30 mg/L; Ni(II), 10 mg/L; Ag(I), 3 mg/L; and Zn(II), 5 mg/L; stripping solution, 1.0 M NaOH; NaCN = 5000 ppm. (From Ref. 79.)

dominant at low organic flow rates. The fractional resistance of each step to the overall process was calculated, and R_m^o, R_s^o, and R_{st} were 48, 51, and 1%, respectively, at high flow rates and 40, 59, 1% at low flow rates. Mass transfer resistance in the organic solution and membrane was dominant at high flow rates, whereas resistance due to the organic solution was enhanced slightly at low flow rates.

Table 11 lists separation factors of Au(I) against other metal cyanides with variation of pH and NaCN concentration in feed. Figure 10b shows that separation of gold cyanide from other metal cyanide complexes at pH 9 was practically impossible, as Zn(II) and Ag(I) were extracted significantly. Their recovery rates were around 80% in the stripping phase (1 M NaOH). As seen from Table 11, the separation factors of Au(I) with respect to Ag(I) and Zn(II) at pH 10.6 were around 85 and 190 times, respectively, those obtained at pH 9 in similar experimental conditions ([NaCN] = 5000 mg/L and 18% LIX 79). Moreover, copper and nickel recovery in the stripping phase was around 30–40%. An increase in pH from 9 to 10.6 could afford better separation of Au(I) over other metal cyanides in spite of the lowering of the mass transfer coefficient at pH 10.6. Although some extraction of Zn(II) and Ag(I) were observed in this pH range in the extraction module, their back-extraction in the stripping phase was slower because gold arrived at the stripping phase faster. When gold remained at around 2 ppm in the feed (in 45–60 min), the feed was replenished to avoid Zn(II) and Ag(I) contamination in the stripping phase. The separation factors of Ag(I) and Zn(II) were decreased slightly at pH 11 (raffinate pH 11.4)

Table 11 Separation Factors S of Gold with Respect to Other Cyanide Metal Salts Using LIX 79/n-Heptane in Countercurrent Mode Under Different Experimental Conditions with a Synthetic Hydrometallurgical Solution

Experimental conditions[a,b]	$S_{Au/Cu}$	$S_{Au/Fe}$	$S_{Au/Zn}$	$S_{Au/Ni}$	$S_{Au/Ag}$
LIX 79 12% NaCN = 1000, pH = 10.6	895	359	23	1144	76
LIX 79 18% NaCN = 1000, pH = 10.6	775	370	9	203	16
LIX 79 18% NaCN = 2000, pH = 10.6	573	295	11	97	12
LIX 79 18% NaCN = 5000					
pH = 9	6	2136	0.1	0.2	0.5
pH = 10.5[c]	2534	713	19	87	44
pH = 11[c]	352	283	16	37	21

[a]Feed composition: Fe(II) (30 mg/L), Cu(I) (30 mg/L), Ni(II) (10 mg/L), Ag(I) (3 mg/L), and Zn(II) (5 mg/L) were tested in the form of a mixture with Au(I) (10 mg/L).
[b]Stripping phase, 1 M NaOH; feed flow rate, 6.94 cm^3/s.
Organic flow rate, 4.72 cm^3/s; stripping solution flow rate, 6.94 cm^3/s.
[c]LIX 79 and NaCN concentrations are the same as in the pH 9 experiment.
Source: Ref. 79.

because the effective gold extraction rate in the feed phase was reduced and took more time for 70–80% recovery in the stripping phase, which ultimately increased the extraction of Ag(I) and Zn(II) and reduced the separation factors (Table 11). In the presence of 1000–2000 mg/L NaCN, extraction of Au(I) was between 90 and 95% (45 min), whereas recovery of Au(I) in the stripping phase varied between 75 and 85% (60 min). At a high NaCN concentration (5000 mg/L), separation factors of gold with respect to Ag(I) were improved slightly. This was possibly due to the formation of $Ag(CN)_3^{2-}$ and $Ag(CN)_4^{3-}$, which were not extractable in significant amounts because the lower complexes, $Ag(CN)_2^-$, were preferred. Based on these results, the following order of selectivity with 18% LIX 79 was obtained at pH 10.6 and 5000 mg/L NaCN in the feed (Table 11):

$$Au(CN)_2^- > Zn(CN)_4^{2-} > Ag(CN)_2^- > Ni(CN)_4^{2-} > Fe(CN)_6^{4-} > Cu(CN)_4^{3-}$$

Also, the extraction of these complexes strongly depends on the metal coordination number; from results shown in Table 11, it is seen that the extraction order follows the series $Me(CN)_2^- > Me(CN)_4^{n-} > Me(CN)_6^{n-}$.

The separation factor of each base metal was calculated by the following equation. For example, $SF_{Au,/Cu}$ is defined as:

$$SF_{Au,/Cu} = \frac{[Au(I)]_{0,strip}}{[Au(I)]_{0,feed}} \frac{[Cu(I)]_{0,feed}}{[Cu(I)]_{0,strip}} \tag{3}$$

Also, to concentrate Au(I), the feed was continuously replaced and the stripping solution was maintained the same throughout these experiments. The latter flowed through the tube side at 6.94 cm^3/s, and the organic solvent was passed through the shell side at 4.72 cm^3/s. The gold cyanide was concentrated up to 400 mg/L. In this separation process, the value of K_{Au}^E was 5.0–7.1 × 10^{-6} cm/s and K_{Au}^S was 3.8–4.8 × 10^{-5} cm/s.

The removal from wastewater of toxic heavy metals, such as Cu(II), Cd(II), and Zn(II), was reported by Kumar and Sastre [81] from the environmental point of view, using a dispersion-free hollow-fiber membrane extraction process. The organic extractant used for copper extraction was 0.1 M Cyanex 272 in n-heptane. Both Cd(II) and Cu(II) were extracted together from aqueous solutions between pH 3.8 and 4.0 and Zn(II) in the pH range of 2.0–2.5. The extraction of copper and cadmium ranged between 90 and 95% in 4 to 5 h, while that of Zn(II) was in the range of 85–90%. The extractant concentration affected the exit concentration of the metal ions, and 0.1 M was found to be suitable for the removal of these metal ions. A flow rate of the organic solvent greater than 25 L/h affected the contact time between the aqueous and organic phases, resulting in poor extraction of the metal ions. The mass transfer coefficients are presented in Table 12. The pressure difference was between 0.2 and 0.6 kg/cm^2.

Table 12 Removal of Cu(II), Cd(II), and Zn(II) Using Nondispersive Extraction Technique as a Function of Extractant Concentration[a]

Cyanex 272 concentration (mol/dm³)	Removal of metal (%)			Mass transfer coefficients, k_A (cm/s $\times$ 10^{-7})		
	Cu(II)	Cd(II)	Zn(II)	Cu(II)	Cd(II)	Zn(II)
0.05	65	55	65	0.7	0.5	0.7
0.10	90	85	90	7.2	6.5	7.0
0.15	95	90	92	8.2	7.9	8.0
0.20	95	91	95	8.3	8.0	8.5

[a] Feed: Cu(II), Cd(II), and Zn(II) = 100 ppm in 0.1 mol/dm³ NO_3^- media, pH 3.8–4.0; Cu(II), Cd(II) for Zn(II) 2.0–2.5; extractant: Cyanex 272 in *n*-heptane; flow rate (organic): 23 dm³/h; flow rate (aqueous): 18 dm³/h.
Source: Ref. 81.

In view of the experience with the recovery of heavy metals and gold from cyanide media, silver recovery from them was considered to be equally important. Therefore, HFMLLE studies were carried out to design a flow sheet to recover silver from alkaline cyanide media with LIX 79/*n*-heptane [82]. The extraction of silver as a function of such variables as flow rate, the LIX 79 concentration, the concentration of the stripping solution, and feed pH was optimized. The recovery of silver was carried out at pH 9–11, and around 60–80% Ag(I) was recovered in the stripping phase. Further studies are in progress.

In the earlier studies, dealing with the extraction of gold from cyanide media [78–80] with LIX 79 alone, poor selectivity against cyanides of zinc and silver [80] resulted. The addition of organophosphorus compounds such as TBP, TOPO, or DBBP to the LIX 79 improved the amine-based extraction system [83,84]. In view of the Lewis basicity of the organophosphorus compounds (phosphate < phosphonate < phosphinate < phosphine oxide), TOPO or Cyanex 921 should be selected as the modifier. Therefore, HFMLLE studies were carried out by Kumar et al. [85] to evaluate the performance of the modified extractant to recover gold from hydrometallurgical alkaline cyanide solutions. At a concentration of 4% LIX 79 + 4% TOPO, the mass transfer coefficient was six times higher than that obtained with 12% LIX 79 alone. Addition of the organic phosphine oxide modifier caused improved solvation of the amine salt, and its protonation occurred at high pH values. These changes caused limited hydration of the gold complex, which promoted its partition into the organic phase. As seen from Figure 11, the gold recovery was 95–100% in 5–10 min with the modified extractant, whereas recovery took 30–40 min under similar experimental conditions without modifier, indicating that the kinetics

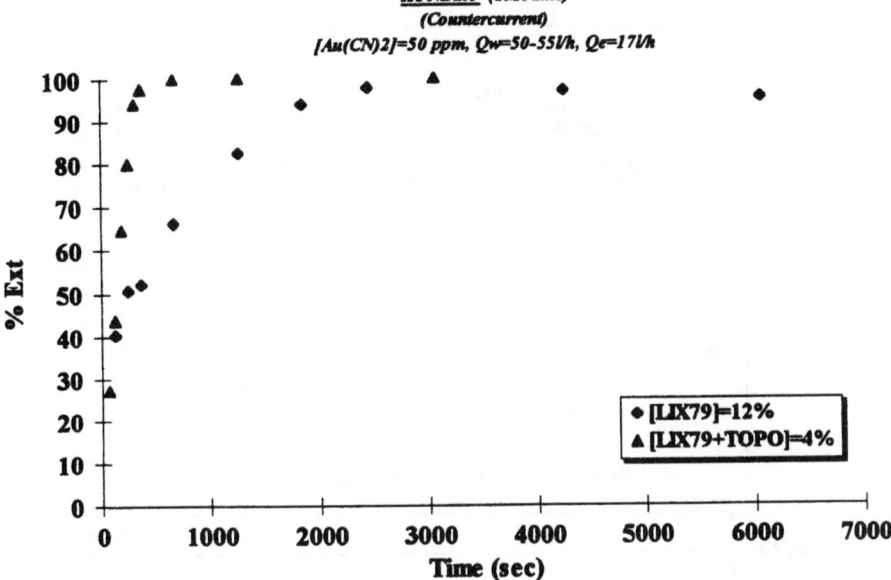

Figure 11 Recovery of gold(I) from alkaline cyanide media with modified extractant (LIX 79 + TOPO). (From Ref. 85.)

of the gold extraction system was considerably improved as well. The selectivity of Au(I) with respect to Zn(II) and Ag(I) cyanide salts was improved too. It was possible to recover Au(I) in the presence of other metal cyanide salts of Fe(II), Cu(I), Ni(II), Ag(I), and Zn(II), proving the potential of such modified extractants.

Daiminger et al. [18] performed hollow-fiber membrane extraction of Cd(II), Ni(II), and Zn(II) with bis(2-ethylhexyl)phosphoric acid (HDEPA) in isododecane, achieving a decrease in metal concentration in the aqueous phase of two to four orders of magnitude by extraction in a single-phase flow mode. The hollow-fiber contactor had a central shell-side baffle to minimize shell-side bypassing and provide a velocity component normal to the membrane surface. The concentration range of HDEPA was between 0.05 and 0.5 M. For stable operation of the HF module, the authors, like other researchers, were careful to maintain higher pressure (0.2 bar) in the aqueous phase than in the organic phase to ensure that the organic carrier stayed in the pores of the membrane. The concentrations of Cd(II), Ni(II), and Zn(II) were in the range of 10^{-5}–10^{-4} M and the aqueous phase initial pH was 4.8–5.5. The results of the experiments with an initial concentration of zinc of 100 mg/L (Fig. 12a) re-

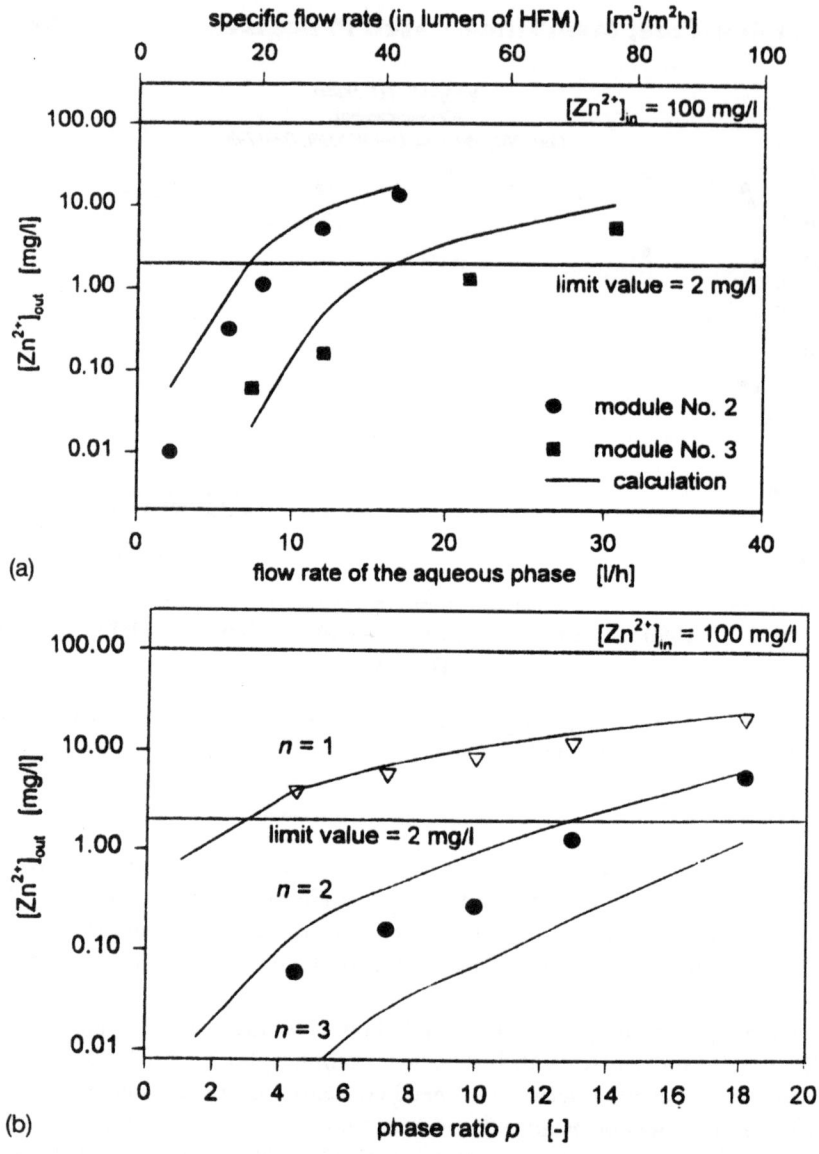

Figure 12 (a) Influence of the flow rate of the aqueous phase on the exit concentration of zinc ("100 ppm concept"): [(DEPA)$_2$] = 0.05 M, module 2: V_o' = 1.76 L/h, pH$_{in}$ = 4.9; module 3: V_o' = 1.73 L/h, pH$_{in}$ = 4.8. (From Ref. 18.) (b) Comparison of the efficiency of HFM ([DEPA)$_2$] = 0.025 M, [Zn^{2+}]$_0$ = 63 mg/L, pH$_0$ = 2.8, V_{aq} = 30 L/h, V_o = 10 L/h) and pulsed sieve plate column (with same concentrations and flows; column characteristics: height, 6 m; diameter, 38 mm; number of plates, 61; hole diameter, 2 mm; number of holes, 66, triangular array; pulse frequency, 0.45 s^{-1}, amplitude, 7.0 mm. (From Ref. 18.)

vealed that high depletion of zinc concentration was possible. With increasing aqueous flow rate, a decrease in efficiency of extraction was noted, which could be due to the shorter contact time between the phases. Hence, longer modules could allow greater throughputs for equal exit concentration of metal. Experiments conducted with wastewater demonstrated the possibility of a thousand-fold depletion of zinc starting from 100 mg/L. Similarly, metal ion concentration in the aqueous phase was depleted by two orders of magnitude for cadmium and nickel extraction in HF contactors. The initial metal concentration was set to 10 mg/L, because of the lower environmental limit values for cadmium and nickel (0.1 and 0.5 mg/L, respectively) [86]. Experiments conducted with an initial zinc concentration equal to 1000 mg/L indicated that it was possible to recover more than 98% of the metal in one passage for a small flow rate and a long residence time (13 min).

Therefore, in hydrometallurgical applications the uptake of metal ions into the organic phase is strongly dependent on the organic flow rate for a given concentration of complexing agent owing to the limited loading capacity of the organic phase. In back-extraction, in one passage of phases through a module having 9000 fibers and 1.1 m^2 membrane area, it was possible to back-extract 50–80% of the zinc, depending on its initial concentration in the organic phase, ranging from 21 to 0.21 g/L. Under conditions close to those of technical applications (i.e., high maximum throughput V'_{aq} = 180 L/h and high phase ratio) the organic flow rate is small (V'_{org} = 3.2 L/h). At least 80% of the zinc could then be easily reextracted from the organic phase in the cross-flow module (31,000 fibers, 2.2 m^2 membrane area) by maintaining $[Zn^{2+}]_{o,in}$ = 4.08 g/L and V'_{org} = 3.6 L/h. The concentrations of all species after several stages were calculated on the basis of batch simulation of an ideal continuous countercurrent cascade, to estimate the number of theoretical stages in the HFM for countercurrent extraction. The results of the calculations are shown in Figure 12b. For given conditions, the number of theoretical stages varied between two and four for the two modules used (9000 fibers, membrane area 0.49 m^2; 31,000 fibers, membrane area 2.2 m^2), which corresponded to HETS (height equivalent to a theoretical stage) values between 13 and 25 cm. It was possible to obtain a similar extraction efficiency in HFMs of 25 cm (9000 fibers, 0.49 m^2 membrane area) or of 54 cm active length (9000 fibers, 1.1 m^2 membrane area) as in an extraction column 6 m long. This could be explained by the very high specific area of the HFMs (ca. 100 m^2/m^3) [87]. Although the studies show the potential of using hollow-fiber module for environmental applications, long-term use of these modules and leaching of the extractant to the aqueous streams need to be tested to check the pollutant concentration in the wastewater.

Cadmium removal from phosphoric acid using HFM modules was reported by Alonso et al. [88]. The authors' objective was to reduce cadmium levels in fertilizers, which are a serious environmental problem. The batch nondispersive solvent extraction of Cd(II) using Cyanex 302 in kerosene was examined with two modules—a hollow-fiber module for extraction and a ceramic module for the stripping step (due to high acidity of the stripping solution)—together with three stirred tanks for homogenization of the fluid phases. The aqueous feed that passed through the extraction cell (shell side) contained Cd(II) in the range of $4.4–5.6 \times 10^{-3}$ M. The concentration of the carrier, Cyanex 302, was 5 vol% in kerosene. The flow rates of aqueous feed and organic carrier solution in the extraction module were 1.8×10^{-2} and 2.34×10^{-2} m^3/h, respectively. In the back-extraction module, the loaded organic solvent and the aqueous stripping agent (6 M HCl) circulated though the channels in the ceramic support at the rate of 2.34×10^{-2} and 2.16×10^{-2} m^3/h, respectively. A mass transfer model of cadmium was presented, and the experimental values were in accord with the simulated values. The mass transfer coefficients reported were $K_{me} = 8.33 \times 10^{-8}$ m/s and $K_{ms} = 3.33 \times 10^{-8}$ m/s, $p = 2$ (aggregation number of Cyanex 302), and $K'_e = 6103$ mol^{-2}L^2. The simulation of a batch process to reach 95% separation efficiency of the cadmium-containing phosphoric acid was carried out, and the results were reported to be promising.

The separation of nickel and cadmium from highly concentrated solutions by means of nondispersive solvent extraction was studied by Galan et al. [37]. Extraction and back-extraction processes were carried out simultaneously in a batch mode, using two parallel modules with the organic phase flowing in a closed circuit. Starting with a concentration of 0.37 M Cd and 0.37–0.68 M Ni in the aqueous phase feed, 1 M H$_2$SO$_4$ in the back-extraction module, and D2EHPA as extractant, the separation–concentration of Cd from the mixture was confirmed. Two commercial modules manufactured by Hoechst-Celanese, listed in Table 3, were used in this work; for the characteristics of the HF polypropylene membrane (Celgard X-10), see Table 4. The kinetics results presented in Figure 13 for extraction and back-extraction indicated that 90% of the cadmium had been extracted after 30 h, whereas only 20% of the nickel was coextracted. In the back-extraction stage, the separation was even higher, since the concentration of cadmium after 60 h was about 1.1 M while the concentration of Ni was only about 0.02 M, for the working conditions (0.22 M of Cd in the organic phase) heavily favored the back-extraction of Cd. The presence of Ni in the aqueous phase affected the extraction of Cd in comparison to experiments performed with CdSO$_4$ solutions alone, since the extraction flux decreased to 1.5×10^{-6} mol/m^2/s, which represents more than a tenfold decrease. The purity of the Cd and Ni products was 98.5 and 93.5%, respectively. In a batch process, separation selectivity is defined as the ratio of the differential concentration of Cd, ΔC_{Cd}, divided by the differential concentration of nickel, ΔC_{Ni}, and, thus equal to the ratio of the separation rates of both metals. Taking into account that the evolution of Ni and Cd concentrations in the stirred tank took place with constant rates for the BEX process (Fig. 13), the selectivity factor ρ was

$$\rho = \Delta C_{Cd}/\Delta C_{Ni} \tag{4}$$

During the whole experiment 67 mol of Cd appeared in the BEX tank per mole of Ni. This technology could be applied to the separation of highly concentrated mixtures of Cd and Ni, simulating the conditions that can be found in the leaching step of shredded nickel–cadmium batteries, and defining a process with a constant separation selectivity.

Juang and Huang [89] studied the extraction of copper(II) from an equimolar solution of ethylenediamine tetraacetic acid (EDTA) across microporous hollow fibers to an organic phase containing Aliquat 336/kerosene (a quaternary amine). One HF module manufactured by Hoechst-Celanese (sample 7 Table 3) was used in this work, and the characteristics of the HF polypropylene membrane (Celgard X-30) are presented in Table 4. This study was important

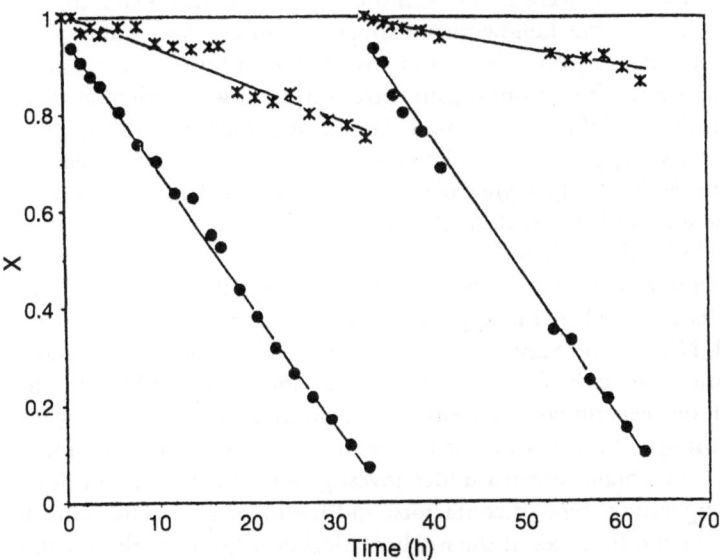

Figure 13 Experimental dimensionless Cd(II) (circles) and Ni(II) (asterisks) concentrations in the BEX aqueous phase vs time for experiment III: number of cycles, 2; initial/final Cd concentration in the feed aqueous phase, 0.42/0.029 M or 0.36/0.035 M; initial/final Ni concentration in the feed aqueous phase, 0.38/0.28 or 0.355/0.31 M; initial/final Ni concentration in the BEX aqueous phase, 0/0.002 or 0.002/0.02 M; pH 3.5. (From Ref. 37.)

because temperature was one of the variables that influenced the extraction rate of Cu in the HF contactor. Figure 14a shows a typical time profile of the organic phase concentration of the chelated anions at different temperatures. The measured extraction rate or metal flux, J_{CuL}, is

$$J_{CuL} = \frac{V_o}{A} \frac{d[CuL]_o}{dt} \tag{5}$$

The apparent activation energy of the overall process, obtained from the temperature dependence of the extraction rate (Fig. 14b), is calculated to be only about 4.7 kJ/mol. Hydrophobic complex formation reactions in LLE processes generally have an activation energy greater than 40–80 kJ/mol [89]. The comparatively low activation energy obtained indirectly indicated that the increase in J_{CuL} with increasing temperature mainly results from diffusion processes or that the contribution of the interfacial chemical reaction to the overall mass transfer process is small. A mass transfer model was presented, taking into account all diffusion processes and assuming that chemical reaction equilibrium holds at the interface. Close agreement between the measured and modeled results was obtained (12% average SD). On the basis of the fractional resistance of each diffusion step, the hollow-fiber extraction process was found to be governed by combined aqueous layer and membrane diffusion ($\Delta_a + \Delta_m >$ 0.91). the resistance resulting from organic layer diffusion was negligibly small. The role of membrane diffusion was more significant at higher $[CuL^{2-}]_a$, lower pH_o, and lower $[NR_4Cl]_o$. On the other hand, Δ_a becomes larger under the conditions of lower $[CuL^{2-}]_a$, higher pH_o, and higher $[NR_4Cl]_o$. In this work Δ_a tends to increase and Δ_m to decrease when the extraction experiment was started as then $[NR_4Cl]_o \gg [CuL^{2-}]_a$.

In interesting recent work, Urtiaga et al. [90] and Galan et al. [91] compared the results obtained in the application of three liquid membrane technologies (ELM, NDSX, and SLM) in a batch mode for the removal of cadmium from phosphoric acid with the thiophenic extractant Cyanex 302. Detailed descriptions of the experimental systems and conditions are given elsewhere [88,92,93]. Although the chemistry of the extraction system was the same in the three liquid membrane systems under investigation, the diversity in liquid membrane configuration, type of contactors, and hydrodynamic conditions introduced substantial differences in the mathematical description of the systems. The simulated curves of the concentration of Cd with time in the feed tanks

Figure 14 (a) Time profiles of concentrations of the chelated anions in the organic phase at different temperatures. (From Ref. 89.) (b) Temperature dependence of the extraction rate of the chelated anions. (From Ref. 89.)

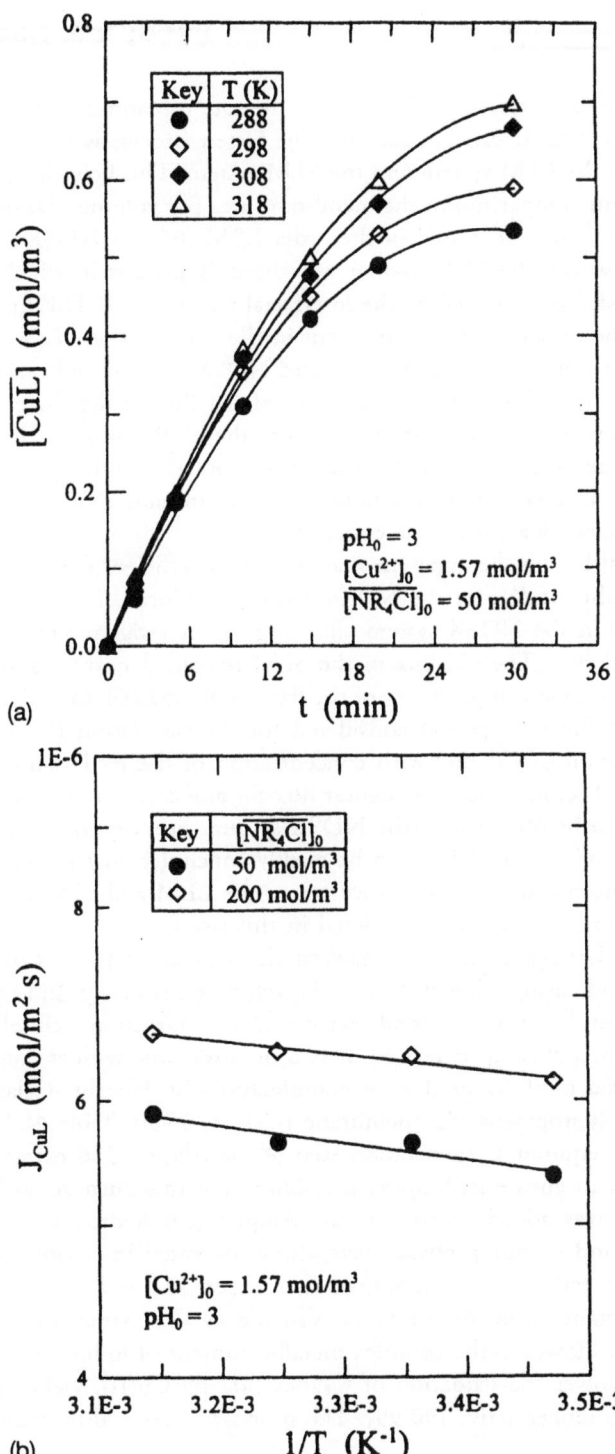

(a)

(b)

in the three experimental systems, representing the rate of cadmium removal, are shown in Figure 15a. It can be seen that the fastest process is the NDSX system, followed by the ELM system and the SLM system. This behavior could not be expected from comparison of the membrane area per volume of solution in the feed tanks (m^2/m^3), decreased in the order ELM (652) > NDSX (350) > SLM (34.7). However, the ELM system was the only process in which the kinetics of Cd transfer was affected by the interfacial reaction rate. This finding was attributed to the presence of the surfactant in the composition of the LM, a component that is not required in the SlM and NDSX systems and can alter the chemical reaction mechanism. The variation of Cd flux of the three contactors with time is presented in Figure 15b. Again, the NDSX process allowed the highest initial flux for a given initial concentration of Cd. This result derived from the advantageous geometry of the hollow-fiber membrane module, yielding the highest specific area within the contactor.

In addition, although the apparent mass transfer coefficient for the SLM was 10 times the value of K_{mc} for the NDSX, the driving force of the flux was considerably higher in the NDSX system, allowing the investigators to obtain higher values of Cd flux. The Cd flux in the SLM remained nearly constant, because of the disadvantageous geometry of the flat membrane cell, the variation of concentration in the time period considered for the simulation (1 h) was very low. The variation of Cd flux with concentration of the feed solution is presented in Figure 15c, in which the highest flux for any concentration of Cd in the feed is seen to be obtained in the NDSX system. From examination of Figure 15c together with Table 13, it can be readily concluded that the NDSX process offers optimum system performance versus the ELM and SLM configurations for the chemical reactions considered in this work.

Furthermore, Urtiaga et al. [94] analyzed the separation process of Cd from phosphoric acid, using Aliquat 336 as the selective extractant and water as the back-extractant, by means of nondispersive solvent extraction technology. The NDSX experimental setup consisted of two or three hollow-fiber contactors. The commercial modules used were manufactured by Hoechst-Celanese (Table 3) with a polypropylene HF membrane (Celgard X-30, Table 4). This continuous process required a regeneration step of the Aliquat 336 extractant to its chloride form to ensure its long-term stability. For this purpose, a third membrane module was added to convert the Aliquat 336 hydroxide to the chloride. The feed and stripping phases were always operated in a continuous mode (Fig. 16a). In earlier studies [88,95–98], Cyanex 302 and other phosphorus-based extractants were found to be suitable as Cd extractants from highly acidic media. However, the complex metallic content of industrial grade phosphoric acid involves the oxidation of Cyanex 302 by Cu(II), yielding an extractant that is no longer active [90,99]. The proposed chemical mechanism

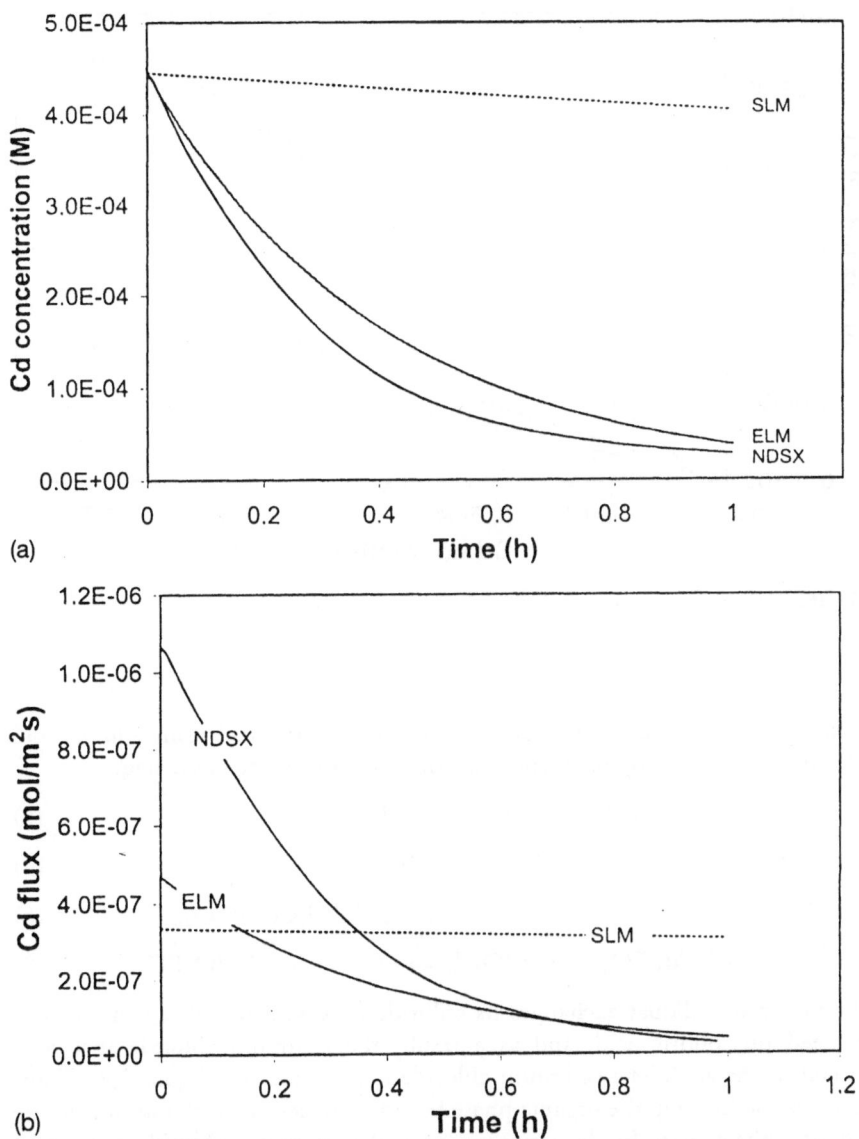

Figure 15 (a) Simulated Cd concentration in the respective feed tanks of the NDSX, ELM, and SLM systems; volume of the feed was 4 L. (From Ref. 90.) (b) Simulated evolution of the Cd flux with time in the NDSX, ELM, and SLM systems; volume of the feed was 4 L. (From Ref. 90.) (c) Simulated flux of Cd as a function of cadmium concentration in the respective feed tanks of the NDSX, ELM, and SLM systems. (From Ref. 90.)

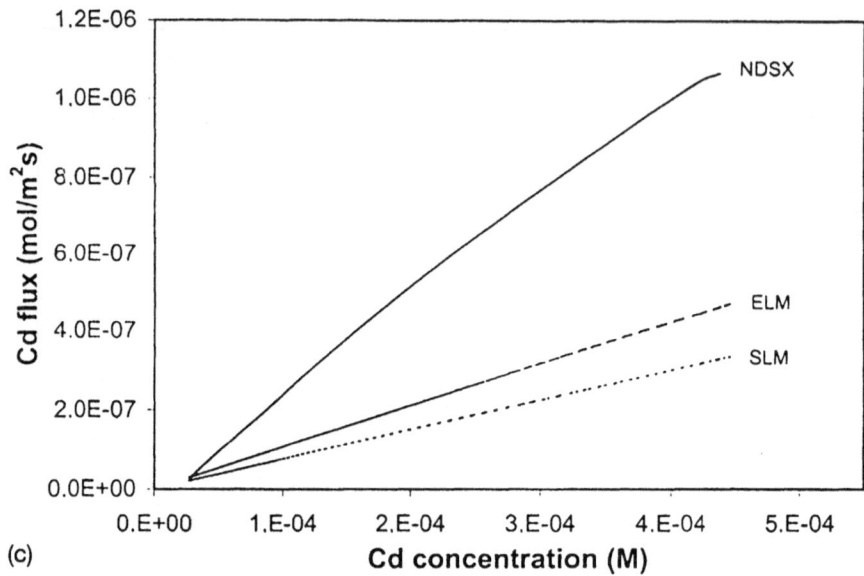

(c)

Figure 15   Continued

of the separation process of cadmium from phosphoric acid using Aliquat 336 (Q) and water as stripping agent is as follows. In the extraction stage:

$$H_3PO_4 + 3QCl_o \Leftrightarrow Q_3PO_{4o} + 3HCl \tag{6}$$

$$Cd_3(PO_4)_2 + 6HCl \Leftrightarrow 3CdCl_2 + 2H_3PO_4 \tag{7}$$

$$3CdCl_2 + 6QCl_o + Cd_3(PO_4)_2 \Leftrightarrow 3Q_2CdCl_4Cd_3(PO_4)_{2o} \tag{8}$$

$$3Q_2CdCl_4Cd_3(PO_4)_{2o} + 12QCl_o \Leftrightarrow 6Q_2CdCl_{4o} + 2Q_3PO_{4o} \tag{9}$$

The extractant Aliquat exchanges its chloride ions with the phosphate ions of the feed phosphoric acid, and as a result, the cadmium chloride species is formed in the acid. The cadmium chloride is extracted by Aliquat 336. Thus, in the extraction step the organic phase becomes loaded with cadmium chloride and phosphate ions. In the stripping step, the cadmium chloride is released into the water stream and the phosphate form of Aliquat 336 is transformed into its hydroxide form, releasing phosphoric acid into the aqueous phase, thereby reducing the pH value of the stripping solution. The addition of a third step for the regeneration by equilibration of Aliquat 336 converts Aliquat 336 to the chloride form, fulfilling the next requirement for continuing the Cd removal process.

Table 13 Equations of Cd Flux and Parameters in the Three LM Systems

System	Equation of flux	Parameters
NDSX	$N(z) = K_{mc}(\bar{C}_{oi} - \bar{C}_o)$ $N_{overall} = \frac{1}{L}\int_0^L N(z)\,dz$	$K_{mc} = 8.33 \times 10^{-8}$ m/s
ELM	$N = \frac{1}{a_s}\left(-\frac{dc}{dt}\right) = \frac{\beta}{a_s}\,C_0\exp(-\beta t) = \dfrac{1}{(1/k_z) + (a_z k' r)}\,C$	$a_s = 652$ m²/m³, $\beta = 6.89 \times 10^{-4}$ s⁻¹, $k_c = 2.35 \times 10^{-6}$ m/s, $k'_r = 1.25 \times 10^{-3}$ s⁻¹
SLM	$N = \frac{1}{(A/V)}\left(-\frac{dc}{dt}\right) = \frac{\alpha}{(A/V)}\,C_0\exp(-\alpha t) = PC$	$A/V = 34.7$ m²/m³, $a = 2.61 \times 10^{-5}$ s⁻¹, $P = 7.52 \times 10^{-7}$ ms⁻¹

Source: Ref. 90.

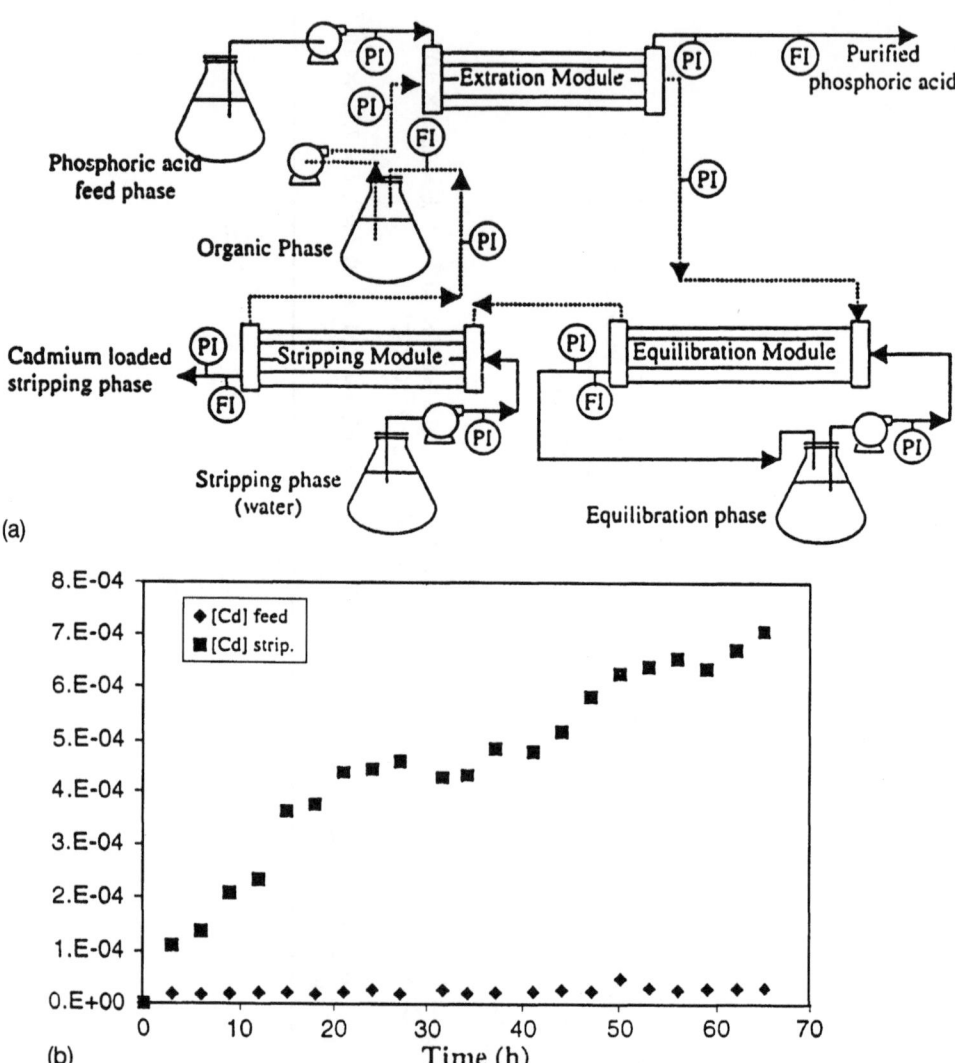

Figure 16 (a) Experimental setup with three hollow-fiber modules, including the equilibrium step of the organic phase. (From Ref. 94.) (b) Cadmium concentration in the feed and stripping phase at the outlets of the extraction and stripping modules, with the equilibrium step of the Aliquat 336 extractant. (From Ref. 94.)

In the stripping and equilibration stages, we write:

$$Q_2CdCl_4 + 2H_2O \Leftrightarrow 2QOH + CdCl_2 + 2HCl \tag{10}$$

$$Q_3PO_4 + 3H_2O \Leftrightarrow 3QOH + H_3PO_4 \tag{11}$$

$$QOH_o + HCl \Leftrightarrow QCl_o + H_2O \tag{12}$$

Co extraction of Zn and Fe was also observed. Figure 16b shows the evolution of cadmium concentration at the outlet of the extraction and stripping modules in a long-term experiment. It can be seen that the concentration of Cd was reduced from 50 mg/L at the inlet of the module to values below 2 mg/L at its outlet. The percentage of cadmium extraction remained constant over the whole period, proving that the efficiency of Aliquat 336 as an extractant remained unaltered. The concentration of Cd in the stripping phase followed a different pattern, showing an increase with time until a steady state was reached. The apparent reason was that mass transport rate of cadmium in the stripping step was lower than that of the extraction step, resulting in the accumulation of cadmium in the organic phase. Also, the flow rate of the stripping phase had no influence on the extraction percentage, whereas the variation of the flow rate of the feed phase introduced a remarkable effect that was due to the influence on the extraction process of the cadmium concentration in the organic phase. Analysis of the steady state results of Cd transport from the acidic feed phase to the stripping phase led to the membrane mass transport coefficient, $K_m = 1.76 \times 10^{-8}$ m/s.

The performance of a commercial-scale hollow-fiber extraction system was investigated by Seibert and Fair [101] to study the scale-up. A large-scale extraction/distillation system was used to extract hexanol from water into octanol. In membrane extractor studies, the octanol-rich phase was fed on the tube side, whereas in the packed-column studies, the octanol-rich phase was chosen as the dispersed phase. This chemical system was selected because of its high solute distribution coefficient. The membrane extractor studied in this work differed from the parallel flow module (Fig. 17a), consisting of a central perforated tube sealed in the middle. The shell-side fluid enters through the perforation on one side on the module, flows around the baffle, and leaves the module by reentering the perforations. Since the module is baffled, it may be considered to be a staged device. Thus each module may be represented as having two actual stages of separation. The characteristics of three modules are given in Table 14. The flow diagram of the process is shown in Figure 17b.

Since the membrane extractor is a staged device, mass transfer efficiency may be described in terms of an overall stage efficiency. The equivalent performance demonstrated how readily the membrane extraction process could be scaled up with larger flow rates. For comparison, mass transfer data were obtained from a 0.425 m diameter extraction column filled with 3.2 m beds of

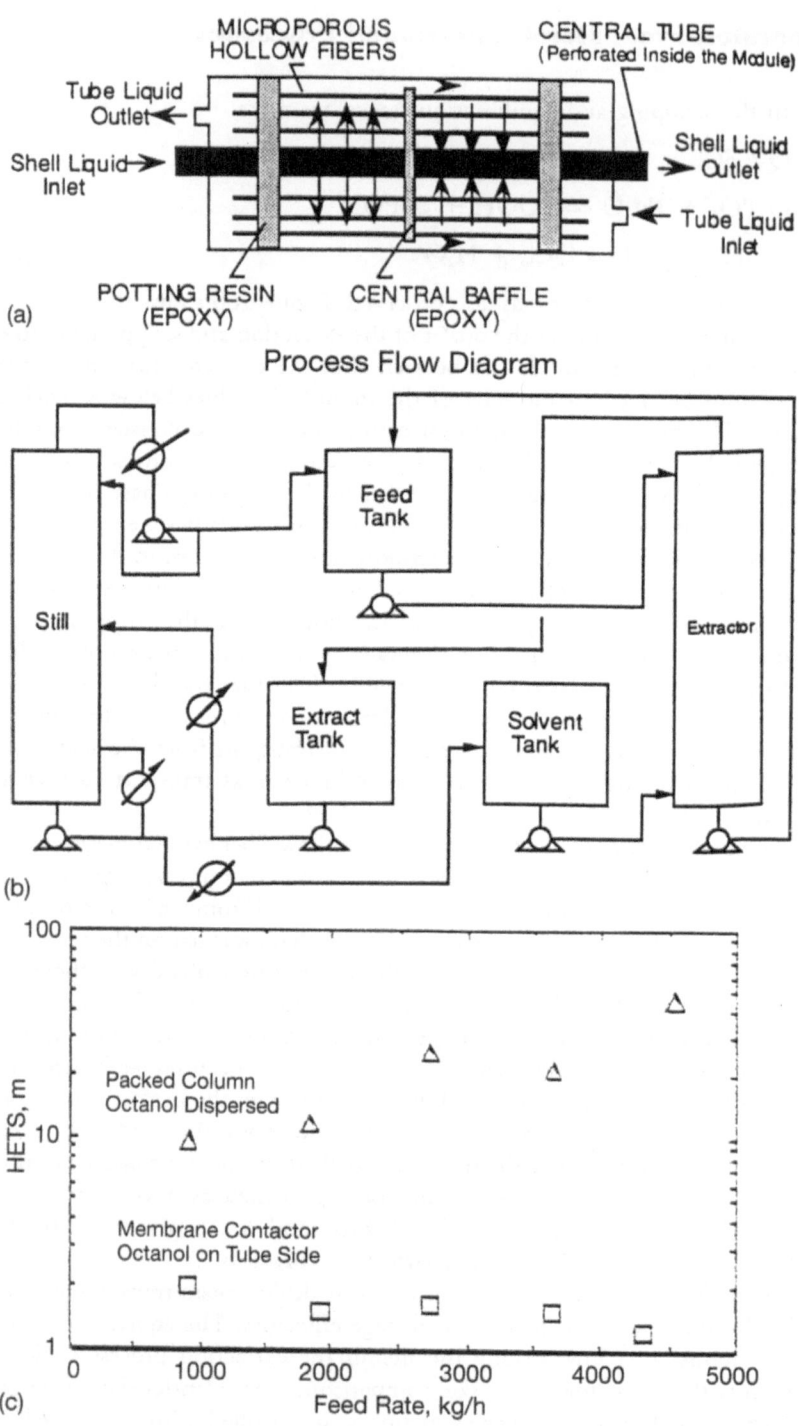

MICROPOROUS HOLLOW FIBERS

CENTRAL TUBE (Perforated Inside the Module)

Tube Liquid Outlet

Shell Liquid Inlet

Shell Liquid Outlet

Tube Liquid Inlet

POTTING RESIN (EPOXY)

CENTRAL BAFFLE (EPOXY)

(a)

Process Flow Diagram

Feed Tank

Still

Extractor

Extract Tank

Solvent Tank

(b)

Packed Column Octanol Dispersed

Membrane Contactor Octanol on Tube Side

HETS, m

Feed Rate, kg/h

(c)

Table 14   Characteristics of the Membrane Module

Module type	Liqui-Cel ExtraFlow
Membrane material	Polypropylene
Casing material	316 SS
Potting rasin	Epoxy
Modules per case	2
Baffles per modules	1
Stages per module	2
Module diameter	9.8 cm
Module length	71.0 cm
Effective fiber length	63.5 cm
Fiber outside diameter	300 μm
Fiber inside diameter	240 μm
Porosity	0.3
Number of fibers per module	30,000
Contact area per module	81,380 cm^2
Contact area per stage	40,690 cm^2
Interfacial area	27.0 cm^2/cm^3
Tortuosity	2.6

Source: Ref. 101.

a type 2 structured packing. As shown in Figure 17c, the mass transfer efficiency of the membrane extractor was superior, because of insufficient mass transfer area in the packed column. Visually, only a few drops per minute could be observed in the windows of the packed contactor. In the modeling of the baffled membrane extractor, the tube-side fluid is taken to be in plug flow while the shell-side fluid is completely mixed. The overall tube-side coefficient is calculated from the three resistances to mass transfer:

$$\frac{1}{K_{ot}} = \frac{1}{k_t} + \frac{1}{k_m} + \frac{m_{ts}}{k_s} \tag{13}$$

The Murphree efficiency for the tube side is given by $E_{m,t} = \alpha/(1 + \alpha/2)$, where $\alpha = K_{ot}A_i/q_t$. The derivation of the Murphree tube-side efficiency may be found

---

Figure 17   (a) Schematic view of the Liqui-Cel membrane extractor. (From Ref. 101.) (b) Process flow diagram of the experimental system. (From Ref. 101.) (c) Comparison of mass transfer in the membrane extractor with a type 2 structured packing for the *n*-octanol–*n*-hexanol/water system. Solvent flow rate was 45 kg/h. (From Ref. 101.)

elsewhere [102]. The overall stage efficiency may be calculated from the Murphree efficiency:

$$E_o = \frac{\ln[1 + E_{m,t}(\lambda - 1)]}{\ln \lambda} \tag{14}$$

The model was found to correlate well with the shell-side-controlled hexanol system, especially at moderate to high shell-side flow rates.

B. Analytical Applications of Microporous Membrane-Based Liquid–Liquid Extraction (MMLLE)

To design a suitable and complete analytical system, membrane-based extraction techniques should be coupled successfully with any of the important analyte determination systems such as AAS, mass spectrometry, gas chromatography, and capillary electrophoresis [35]. Membrane-based extraction techniques offer efficient alternatives to classical sample preparation techniques by making use of the advantages of liquid–liquid extraction (e.g., the possibility of tuning the selectivity by chemical means) and by avoiding the disadvantages (e.g., solvent consumption and the need for manual handling). These techniques permit high selectivity and high enrichment factors, as well as giving good possibilities for automation. Sample preparation methods based on liquid–liquid extraction are now gradually being replaced by either microporous membrane liquid–liquid extraction (MMLLE) or supported liquid membranes SLMs [31]. These techniques proved to be more compatible with requirements because of their simplicity. Several investigators, including Warshawsky and Pohlandt, proposed the use of hydrophobic polymers as a support for an organic phase [30]. A cylindrical hollow-fiber design has been used to construct extraction units with channel volumes of 1.3 μL [103]. The membrane is typically a porous PTFE membrane soaked in a suitable organic solvent with low water solubility. Thus a three-phase system, aqueous–organic–aqueous, is created. Typical membrane liquids are n-undecane, di-n-hexyl ether, and tri-n-octyl phosphate, either neat or containing additives. The time for use before the liquid has to be refilled varies from days to months. Generally, nonpolar membranes are more stable than the more polar ones. A comprehensive review by Jönsson and Mathiasson [35] deals with a number of examples from the fields of environmental and biomedical analysis using SLM or MMLLE techniques.

C. Membrane Cell and Theoretical Background

MMLLE is a two-phase extraction in which a hydrophobic membrane separates an aqueous phase and an organic phase in a flow system [36]. MMLLE can be performed in a typically designed membrane unit shown in Figure 18. In each block a groove is machined and liquid connections are provided at both ends. By clamping the blocks together with a membrane between, one flow-through channel is formed on each side of the membrane. The volume of each channel is typically in the range $10-1000$ μL. The acceptor phase is an organic solvent, which also fills the pores of the hydrophobic membrane. This forms a two-phase system, aqueous–organic; the organic phase is partly in the membrane pores and partly in the acceptor channel. The acceptor phase can be stagnant or flowing. If the acceptor phase is stagnant, the only driving force for the mass transfer is the attainment of a distribution equilibrium between the aqueous and organic phases, and efficiency will then be higher if the partition coefficient is large (i.e., the hydrophobicity of the analyte is large). The mass transfer can be improved if the acceptor phase is continuously or intermittently pumped, removing the extracted species/molecules successively from the acceptor. Equation (15) was derived for the extraction efficiency of an MMLLE system, assuming that the partition is in equilibrium and the flows in the channel are parallel:

$$E = \frac{1 - F_D}{F_D + F_A K} \tag{15}$$

where F_D and F_A are the volume flow rate of the donor phase and acceptor phase, respectively. Thus, compounds with large distribution coefficients K are more efficiently extracted, and the extraction efficiencies increase with acceptor flow rate. A high acceptor flow rate maintains low concentrations of the analytes in the acceptor phase and thus maximizes the concentration gradient, leading to a high mass transfer rates. However, owing to the dilution of analyte, the acceptor flow leads to a smaller concentration enrichment factor, E_e:

$$E_e = \frac{1}{1/K + F_A/F_D} \tag{16}$$

At small acceptor flow rates, the enrichment factors approach the maximum

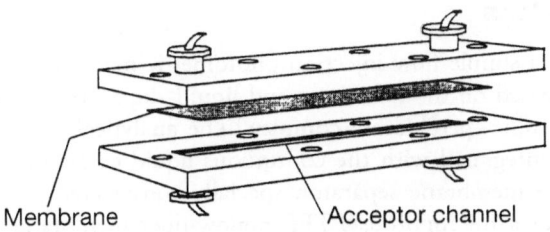

Membrane   Acceptor channel

Figure 18 Membrane unit cell used for MMLLE-based analytical applications. (From Refs. 20 and 35.)

value $E_{e,max} = K$. MMLLE is conceptually similar to continuous dialysis, and Eq. (16) is valid for that technique under the assumption of equilibrium between the phases and with the value of $K = 1$.

When the acceptor flow is in the opposite direction to the donor flow, the concentration gradient between the two phases is larger and the mass transfer will be somewhat more efficient. This situation is mathematically far more complex than the parallel flow assumption used earlier because it is not possible to assume equilibrium anywhere in the cell. Appropriate numerical solutions to the partial differential equations have been derived [104] for the corresponding problem in dialysis.

The long-term stability of the liquid membrane extraction system necessitates some consideration. The organic solvent must be selected to be insoluble in water and to have sufficiently low viscosity, which is readily accomplished with certain nonpolar liquids. The most commonly used liquid is n-undecane, which forms membranes that can be used continuously for several weeks without physical and chemical changes. When there is a need for more polar membranes, di-n-hexyl ether or mixtures of this solvent with n-undecane are often used, and such membranes are almost as stable as pure n-undecane membranes. To achieve special selectivity, various additives to the liquid membrane are required. This may compromise stability, and the use of additives that are as hydrophobic as possible is advantageous. For practical use, an SLM preparation should be stable for at least one working day. The material of the liquid membrane support seems to influence the stability somewhat, the most commonly used PTFE membranes being slightly better than polypropylene membranes. Also, pore size influences membrane stability [4]. The SLM can be regenerated in a few minutes by simply soaking the membrane support in the desired liquid, wiping, and reinstalling the SLM in the membrane holder. For hollow-fiber membranes, in situ regeneration has been shown to work well [105], and this should also be the case for flat membranes.

D. Extraction, Determination, and Speciation of Metals

Richter et al. [106] developed a simple flow injection photometric method for the determination of copper, based on the formation and liquid–liquid extraction of the analyte–bathocuproine–perchlorate complex. The analytical reaction of complex formation is integrated with the continuous liquid extraction procedure. A hollow-fiber-type membrane separator, specially constructed for the determination, consisted of a microporous PTFE hollow-fiber membrane (Japan GoreTex), 75 cm long with an inner diameter of 1 mm, an outer diameter of 1.8 mm, and 2 μm pore size, inserted into a helically coiled glass tube of 2.0 mm inner diameter. The coiled glass contained one inlet and one

Table 15   Analytical Features of the Method

Limit of detection	15 ng/mL
Determination range	48–3000 ng/mL
Reproducibility ($n = 11$)	RSD%
200	2.24 ng/mL
1000	1.62 ng/mL
Sample throughput	20

Source: Ref. 106.

acceptor system. The dichloroethane stream flowed as a segmented phase through the hollow-fiber membrane. The extractant used was 0.005 M 2,9-dimethyl-4,7-diphenyl-1,10-phenanthroline (bathocuproine). Table 15 shows the analytical features of the method. The effects of several foreign ions usually present in natural waters were examined by analysis of synthetic samples containing 300 ng/mL Zn, 300 ng/mL Cu, 300 ng/mL Fe, 50 μg/mL Ca, 50 μg/mL Mg, and 0.5% NaCl. The recovery was 104.3 ± 2.8%. Validation of the method was achieved by analysis of a certified reference material (Table 16). Determination of copper was then carried out in a river water sample (collected in April 1998, Los Almendros River, Chile). The concentration found was 1.01 ± 0.02 μg/mL, consistent with that determined by AAS (0.97 ± 0.04 μg/mL).

In the laboratory of one of the present authors, a flow probe optosensor for chromium(VI) was developed by Castillo et al. [107], who used flat-sheet liquid-supported membranes (FSLSM). This work presents a new strategy to promote selective transport of Cr(VI) based in the use of FSLSM as the interface of the flow probe sensor. Diffusion-limited transport of the target analytes across FSLSM represents a powerful novel tool and excels most of the conven-

Table 16   Determination of Copper in Certified Reference Materials

Certified Reference Material (μg/mL)	Concentration (μg/mL)	
	Determined[a]	Certified
Metal element in water GBW 08607[b]	1.03 ± 0.09	1.02 ± 0.01
HPS certified wastewater CWW − TMD[c]	1.04 ± 0.03	1.000 ± 0.005

[a] Mean of five determinations.
[b] From laboratory of the government chemist, LGC, UK.
[c] From U.S. high purity standards.
Source: Ref. 106.

tional approaches, since it combines the stages of extraction, stripping, and regeneration into a single step.

The combination of the transport of anionic chromium species through FSLSM and the subsequent reaction between Cr(VI) and diphenylcarbazide (DPC) in a sulfuric acid medium was evaluated. The first step of permeation is the partition of the Cr(VI) anions into the liquid anion-exchange supported membrane. This is a spontaneous process based on the reaction of $HCrO_4^-$ with the cationic ionophores (Aliquat 336 or Alamine 336) present in the liquid membrane to form ion pairs. The stripping of Cr(VI) from the membrane to the receiving solution includes a combination of a reduction step of $HCrO_4^-$ by DPC (DPC_{red}) to produce Cr(III), and the specific complexation of Cr(III) with the oxidized form of the resulting, DPC_{ox}, to $Cr(DPC_{ox})_3$, leads to an increase of both the membrane phase diffusivities and the concentration gradients of the targeted Cr(VI) anions. The scheme of the transport process of analytes from the sample to the probe cavity of the sensor head is shown in Figure 19a.

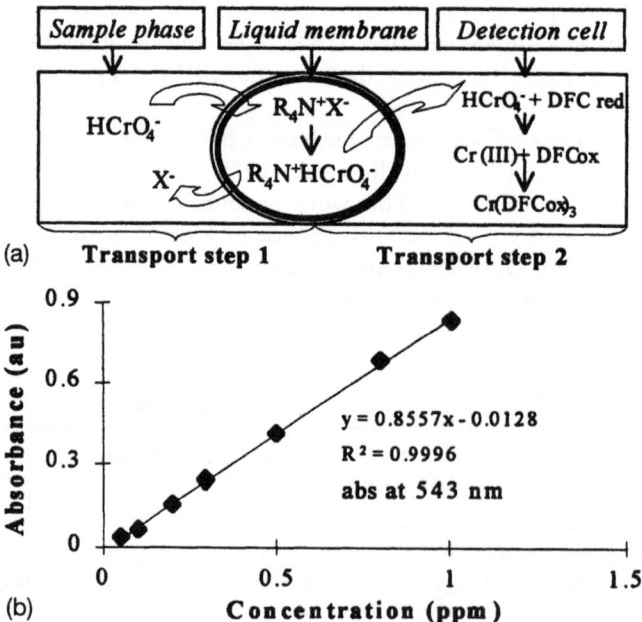

Figure 19 (a) Scheme of the transport process of analytes from the sample to the probe cavity of the sensor head. (From Ref. 107.) (b) Sensor response as a function of Cr(VI) concentration in the sample. (From Ref. 107.)

The Cr(III) complex is not retained by the supported extractants of the liquid membrane, resulting in increased diffusivities of $HCrO_4^-$, the concentration gradient of the targeted ions being kept high and the number of free ions capable of back-diffusion being strongly reduced by complexation. Contamination by undesirable transition metal ions is not favorable either in transport through the membrane or in complexation with DFC. Higher efficiency was obtained with Aliquat 336 than with Alamine 336 in the membrane, and thus the former was selected for further experiments (Fig. 19b). Recently, an FSSLM method for the speciation of chromium was developed by Djane and coworkers [108]. The method is based on the selective extraction and enrichment of anionic Cr(VI) and cationic Cr(III) species in two serially connected SLM units. Aliquat 336 and DEHPA, respectively, were used as the selective extractants in the membrane liquid. Graphite furnace atomic absorption spectrometry (GFAAS) served for the final determination. Optimized conditions for the DEHPA membrane was as follows: sample solution at pH 3, acceptor solution 0.1 M HNO_3, and 10% w/w carrier in kerosene. The corresponding values for the Aliquat membrane were pH 7, 0.75 M HNO_3, and 6% carrier in di-n-hexyl ether. This gave an extraction efficiency for Cr(III) and Cr(VI) of 90 and 40%, respectively. The method was used to measure the concentration of Cr(III) and Cr(VI) in surface water from an abandoned tannery site. Storage experiments at different pH values showed that preservation at a neutral pH gave almost constant values over a period of one month. At the acidic value (pH 3), the concentration of Cr(VI) decreased rapidly while the concentration of Cr(III) increased. The detection limit, expressed as three times the standard deviation of enriched blank samples, was 0.01 μg/L. The authors demonstrated that FSSLM can be used for chromium speciation studies of complex matrices such as natural water. By using different membrane liquids, soluble Cr(III) and Cr(VI) species and those complexes loosely bound to colloidal particles, which can be broken down during the residence time of the sample plug in the donor channels of the SLM unit, can be selectively enriched, avoiding the need for any chromatographic separation prior to final determination.

The FSSLM technique was also found to be useful in pretreatment for handling complex matrices such as wastewater and urine, where sample cleanup is necessary before the final analysis. Ndung'u et al. [109] developed a method for simultaneous trace determination of metal ions Zn, Ni, and Co at micrograms-per-liter levels in river water. They used ion pair chromatography (IPC) and performed the sample cleanup and preconcentration steps by using the FSSLM technique. The detection limits in reagent water for Zn, Ni, Co, Cd, and Mn after 120 min enrichment time were 0.15, 0, 15, 0.06, 0.3, and 0.12 μg/L, respectively.

To check the FSSLM technique for the enrichment of metals in a flow system with offline AAS, Papantoni and coworkers [110] reported the possibility of using this technique for the estimation of ultratrace level concentrations (i.e., at <ppb levels). With purification of reagents, the possibility of reaching detection limits at low parts-per-trillion levels seemed good. The stability of the membrane system made 24 h time-integrated field sampling possible.

In another study, Buffle and coworkers [29] developed a technique based on HFSLMs consisting of 1,10-didecyl-diaza-18-crown-6 (Kryptofix 22DD) plus dodecanoic acid in a 1:1 mixture of toluene and phenylhexane as diluent, for speciation of trace metals such as Cu, Pb, and Cd in natural waters. Two types of single hollow-fiber module (nonflow and flow systems) were used for free metal ion separation and preconcentration. The sample solution consisted of desired concentrations of Cu(II), Pb(II), or Cd(II) nitrate in 10^{-2} M morpholinoethanesulfonic acid (MES) adjusted to pH 6 with LiOH. The stripping solution contained 5×10^{-4} mol/L CDTA adjusted to pH 6.4 with NaOH. The carrier solution 0.1 M Kryptofix 22DD plus 0.1 M dodecanoic acid in the diluent and an Accurel PP q3/2 (Akzo) hydrophobic hollow fiber were used as the support for the liquid membrane. High preconcentration factors (100–3000) under natural water pH conditions were reported for Pb(II), Cu(II), and Cd(II). The time required for achieving this preconcentration ranged between 5 and 120 min when sample and receiver solution volumes were 250 mL and 60 μL, respectively. These membranes were stable for at least 7 days of continuous use. The authors claimed that by combining HFSLMs with sensitive analytical detectors (e.g., voltammetric ones), detection limits of metal ions below 10^{-10} mol/L could be achieved.

Djane and coworkers [111] presented a method for trace metal determination in complex matrices combining supported liquid membrane sample cleanup and enrichment with potentiometric stripping analysis (PSA) in a flow system. The membrane contained 40% DEHPA dissolved in kerosene. The SLM-PSA technique allowed the determination of lead in acidified urine samples at micrograms-per liter. The combination is potentially useful for the rapid determination of other trace metals in complex matrices in the presence of organic compounds that can cause electrode fouling. Likewise, an SLM extraction technique for the determination of Cr(VI) in occupational hygiene and environmental samples was developed by Ndung'u and others [112]. Samples containing Cr(VI) were extracted by using the SLM technique for 20 min, and Cr(VI) was determined online in a flow system by reaction with 1,5-diphenylcarbazide (DPC). The corresponding detection limits for an alternative approach, using electrothermal atomic absorption spectrometry (GFAAS) offline after 20 min SLM enrichment, were 0.04 and 0.03 μg/L when ammonium and phosphate buffers, respectively, were used. The SLM methodology for Cr(VI) determination was validated by using a new certified reference material, CRM 545, for Cr(VI) and total leachable chromium. The results of the validation are shown in Table 17. The values for total leachable Cr shown in Table

Table 17 Analytical Results for the Determination of Cr(VI) and Total Leachable Cr in a Certified Reference Material (CRM 545)[a]

Buffer[c]	Cr(VI) amount [g/kg (s)][b]			Total leachable Cr [g/kg (s)]	
	Certified value	SLM-GFAAS	SLM-DPC-FIA	Certified value	Direct GFAAS
Phosphate	40.16 (0.60)	40.02 (0.63)	40.22 (0.81)	39.47 (1.30)	40.87 (1.29)
Ammonium	40.16 (0.60)	40.23 (0.34)	40.13 (0.82)	39.47 (1.30)	42.56 (2.33)

[a] Welding dust was loaded onto a glass fiber filter. Each value is a mean of three replicate determinations using a 20 min SLM extraction.
[b] Calculated after subtraction of the blank values.
[c] Used during ultrasonification step.
Source: Ref. 111.

17 were determined by direct GFAAS injections (no SLM extraction) after ultrasonic extraction and appropriate dilution. Generally, good agreement was achieved between the certified values and those obtained by using both SLM-GFAAS and SLM-DPC-flow injection (FIA) techniques. Christensen et al. [112], who were involved in the certification of CRM 545, attributed the fact that the total leachable Cr content of the welding dust is somewhat lower than the Cr(VI) content to the uncertainty of the results used for calculation of the certified values. The author claimed that the developed method is adaptable for both field sampling and lab analyses.

IV. EMULSION AND BULK LIQUID MEMBRANES UTILIZING HOLLOW-FIBER CONTACTORS

Hollow-fiber contactors also find applications in other processes, where they are utilized either to stabilize the process or to improve the efficiency. The conventional emulsion liquid membrane extraction in a stirred contactor has two main disadvantages [113].

1. On prolonged contact with the feed stream, the emulsion swells with water, increasing the internal phase volume. Swelling and water uptake cause a reduction in the stripping reagent concentration in the internal phase, which in turn reduces the stripping efficiency. Furthermore, the solute that has been concentrated in the internal phase is also diluted, resulting in lower separation efficiency of the liquid membrane.

2.   Leakage of the internal phase contents into the feed stream may occur because of membrane rupture. Leakage, like swelling, also reduces the efficiency of separation. Making more stable emulsions with an optimal surfactant could minimize the leakage, but this makes the subsequent de-emulsification and product recovery steps more difficult. Lower shear rates would also minimize leakage, but mass transfer resistances could then become very significant.

To eliminate the problems just described, an ELM can be used to simultaneously extract and strip metal from an aqueous stream by dispersion-free contacting in a hollow-fiber contactor [113]. This arrangement combines the advantages of ELM separation (simultaneous extraction and stripping) and dispersion-free solvent extraction. The absence of a high shear rate, encountered in agitated dispersion, minimizes leakage of the internal stripping phase. In addition, internal aqueous droplets cannot directly contact the feed aqueous phase; thus the possibility of swelling is greatly diminished. Figure 20a shows the ELM extraction in a hydrophobic HFC. Hu and Wiencek [44] reported an extraction technique, using an ELM within a hollow-fiber contactor (24 cm long, shell i.d. = 1.5 cm, number of fibers = 85, pore size = 0.2 μm, fiber o.d. = 0.1 cm, fiber i.d. = 0.06 cm) for copper extraction using LIX 84 as extractant, 1 wt% ECA 5025 as surfactant, and tetradecane as the solvent. A pressure vessel delivered the organic phase, which flowed through the tube side in a once-through mode. The aqueous phase, delivered by a pump in total recycle mode, flowed through the shell side of the module. As in earlier studies, to prevent the organic phase from flowing out of membrane pores and mixing with the aqueous phase, a higher pressure (about 35 kPa difference) was applied to the aqueous phase. The equilibrium constant (K_{eq} = 3.9) and mass transfer correlations of the systems were obtained for the shell side as follows:

$$Sh = 0.245 Re^{0.6} Sc^{0.33} \tag{17}$$

for the tube side as follows:

$$Sh = 1.895 Pe^{0.33} \tag{18}$$

and for the membrane as

$$k_{mem} = 14.74D \tag{19}$$

where D is the diffusivity (cm^2/s). The relative resistances of extraction processes under various operating conditions are as follows:

$$\frac{1}{K} = \frac{1}{k_M^s} + \frac{1}{O_{sm}H_{sm}k_f} + \frac{1}{A_{lm}A_sK_pk_c^m} + \frac{1}{A_tA_sK_pk_C^t} \tag{20}$$

where A_{lm} and A_t are the log-mean membrane surface area and the tube-side membrane surface area, respectively, and k_f is the reaction rate constant. The pseudo–partition coefficient K_p is defined as follows:

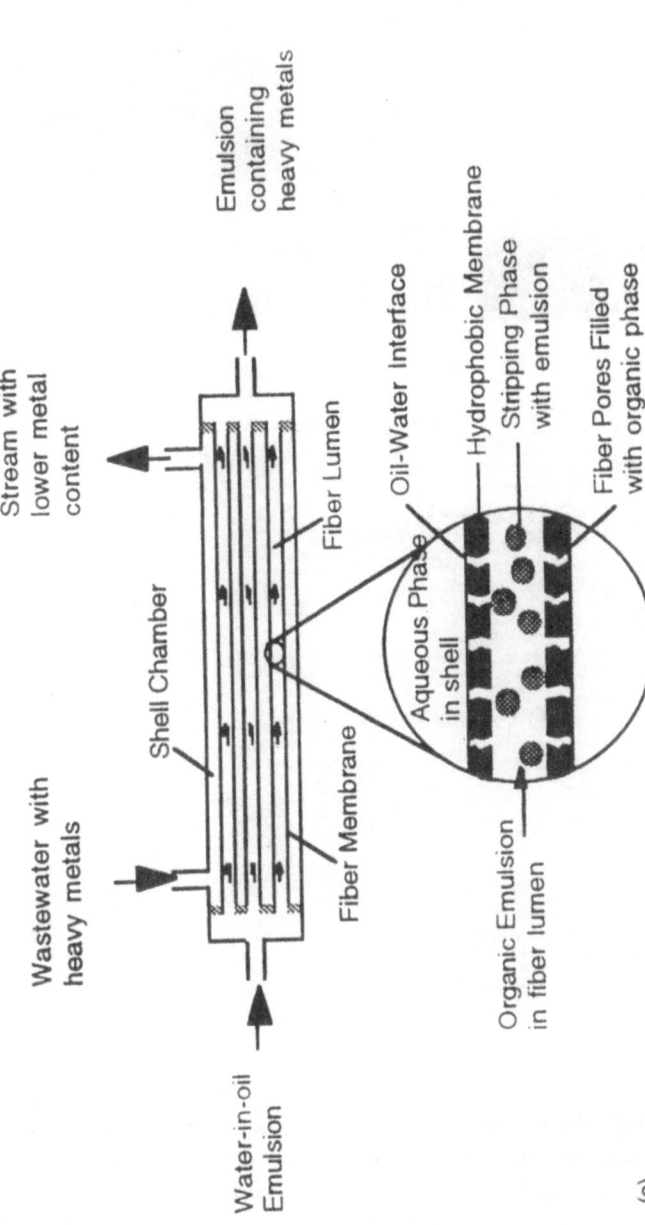

(a)

Figure 20 (a) ELM extraction in a hollow-fiber contactor. (From Ref. 44.) (b) Typical experimental setup for carrying out ELM extraction in HF contactors. (From Ref. 44.) (c–g) Comparison to overall extraction resistance in HFC solvent extraction, contributed by aqueous phase mass transfer, chelating reaction, membrane phase mass transfer at various conditions: (c) Effect of shell (aqueous) flow rate, $M_{sini} = 500$ mg/L, $O_{tini} = 20$ wt%, $Q_t = 1$ mL/min. (d) Effect of tube (organic) flow rate, $O_{tini} = 20$ wt%, $M_{sini} = 2000$ mg/L, $Q_s = 2$ mL/min. (e) Effect of initial copper concentration, $O_{tini} = 20$ wt%, $Q_s = 1$ mL/min, $Q_t = 128$ mL/min. (From Ref. 44.) (f) Effect of initial oxime concentration, $M_{sini} = 250$ mg/L, $Q_s = 2$ mL/min, $Q_t = 1$ mL/min. (g) Residence under "normal" conditions. $Q_{tini} = 5$ wt%, $Q_s = 8$ mL/min, $Q_t = 8$ mL/min. (From Ref. 44.) (h–i) Comparison of simple solvent extraction with ELM extraction in an HFC. SS and ELM represent simple solvent extraction and extraction using ELM. (h) $M_{sini} = 10,000$ mg/L, $Q_{tini} = 5$ wt%, $Q_t = 1$ mL/min. (i) $M_{sini} = 1000$ mg/L, $O_{tini} = 5$ wt%, $Q_t = 1$ mL/min. (j) Extraction in an "imaginary" HF contactor, $M_{sini} = 1000$ mg/L, $O_{tini} = 5$ wt%, $Q_s = Q_t = 4$ mL/min. (From Ref. 44.)

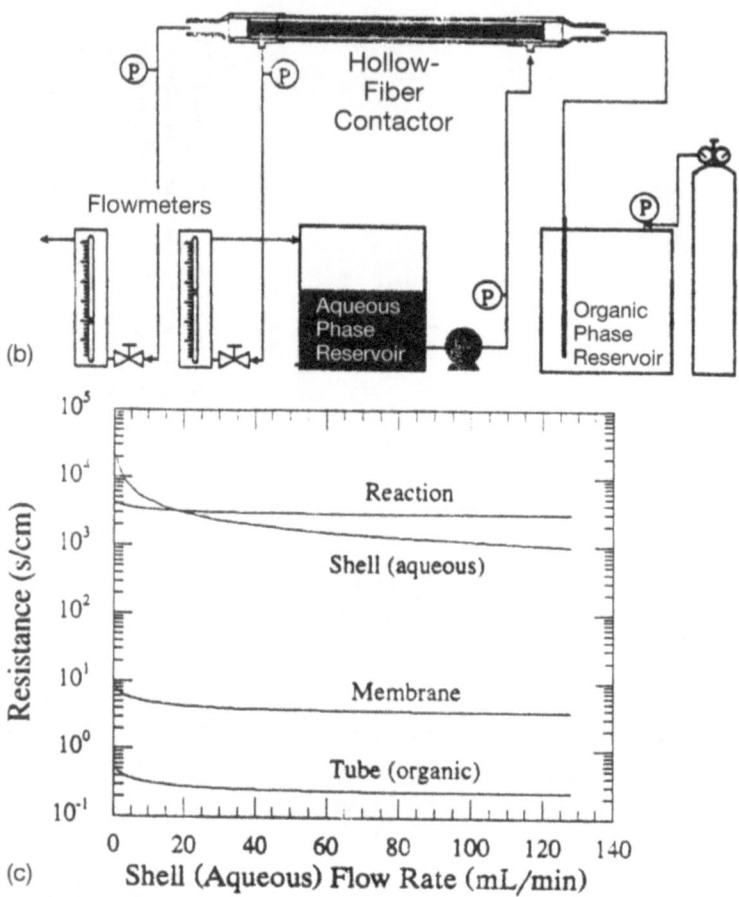

(b)

(c)

Figure 20   Continued

$$K_P = \frac{K_{eq} O_{sm}^2}{H_{sm}^2} \tag{21}$$

The four terms on the right-hand side of Eq. (20) are the resistances due to shell-side mass transfer, reaction, membrane phase mass transfer, and organic phase mass transfer, respectively. The results from the model prediction are plotted in Figure 20c–g. In most cases, resistance from the chelating reaction controls the rate of the extraction process. The membrane resistance is usually more than two orders of magnitude lower than the reaction resistance, and the organic phase resistance is even smaller.

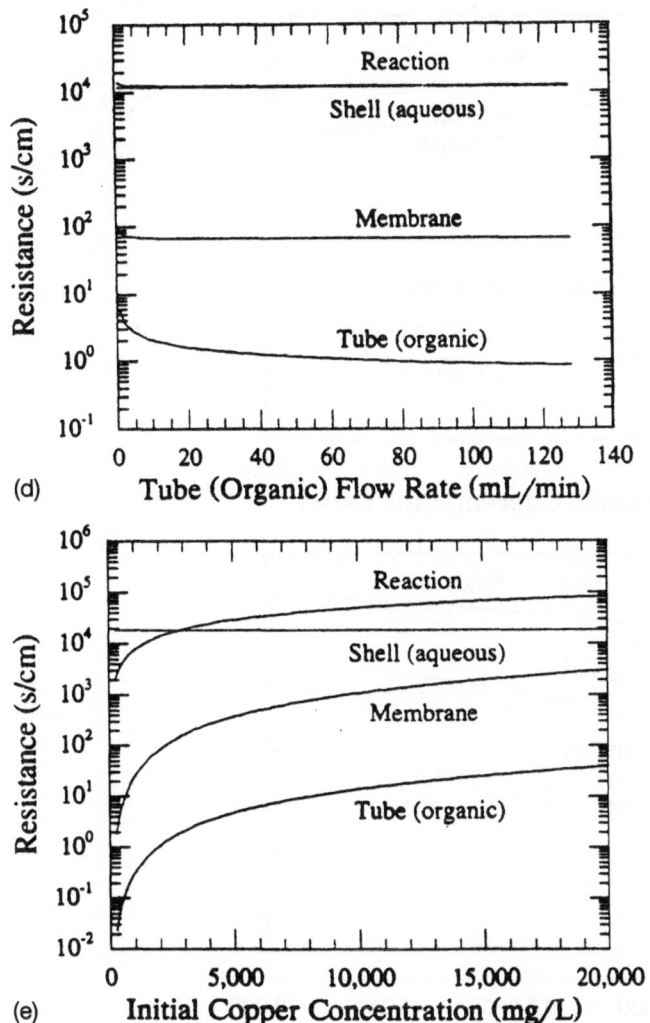

(d) Tube (Organic) Flow Rate (mL/min)

(e) Initial Copper Concentration (mg/L)

The data from simple solvent extraction (SS) and ELM extraction have been compared, and some representative data are shown in Figure 20h–i. The points in the plots are taken from experimental data and the lines are from interpolations (i.e., not from model simulations). It was observed that ELM extraction outperforms simple solvent extraction when the extraction is near the equilibrium limit, as shown in Figure 20h. This tends to happen when the

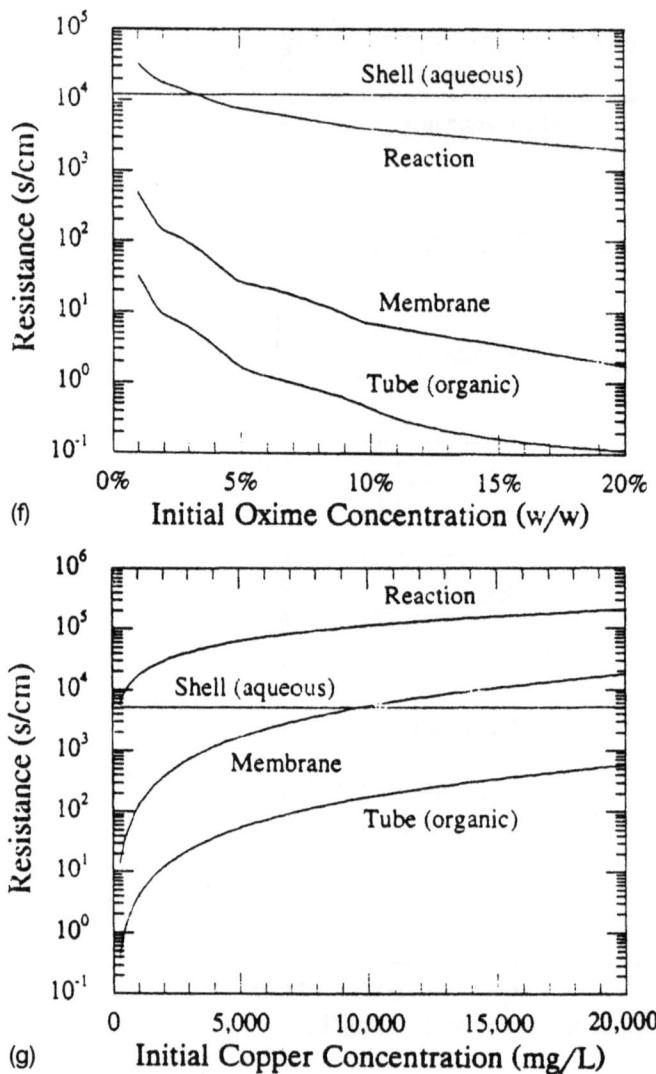

(f)

(g)

Figure 20 Continued

aqueous phase is concentrated, the aqueous flow rate is high, the organic phase flow rate is low, and the oxime concentration is low. In this particular case, when simple solvent extraction extracted about 13% of the copper in the aqueous phase, ELM extraction was able to achieve 50% extraction by eliminating equilibrium limitations. On the other hand, if the extraction is far from the

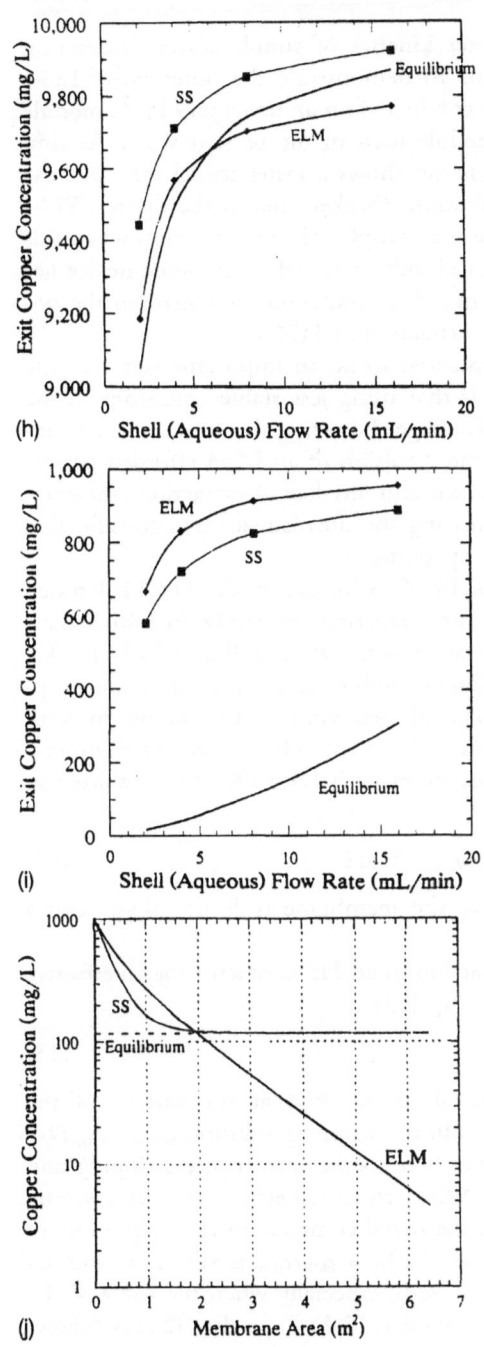

(h)

(i)

(j)

equilibrium limit (Fig. 20i), the faster kinetics of simple solvent extraction results in a higher level of extraction. To demonstrate the potential of ELM extraction in HFCs, simulations were conducted on an imaginary HFC module that is 100 times longer than the module used in the present work. As seen from Figure 20j, simple solvent extraction shows a faster initial rate, but the slope decreases as it approaches equilibrium (broken line in the figure). ELM extraction starts to have an advantage over simple solvent extraction when this module approaches 2 m^2. Since a typical industrial-scale membrane device has a surface area ranging from 3 to 10 m^2, this simulation demonstrates the potential industrial application of ELM extraction in HFCs.

Emulsion stability was not considered to be an important factor in the HFC. The present authors' opinion is that using less stable emulsions in the contactors would confer an added advantage. Lowering the surfactant concentration (even as far down as 0%) in the emulsion of an ELM afforded performance comparable to that of an HFC without any loss of extraction efficiency. This suggests the possibility of eliminating the downstream de-emulsification process for a substantial reduction in operating costs.

To improve the performance of BLM, Schlosser et al. [115,116] tested the use of hollow-fiber membranes. Their experimental study, including mass transfer characteristics of an HF pertractor with two bundles of hydrophobic microporous hollow fibers in pertraction (a combination of extraction and stripping with a single liquid membrane) of neodymium and holmium with D2EHPA as a carrier, was reported in 1999 [115]. The overall stoichiometry of the extraction equilibrium of neodymium with DEHPA can be written as follows [117–119]:

$$Nd^{3+} + 3(HA)_{2\,org} \Leftrightarrow NdA_3(HA)_{3\,org} + 3H^+ \tag{22}$$

where org denotes species dissolved in the membrane (solvent) phase, and a similar reaction pertains to holmium.

The flux density of Nd in pertraction in an HF contactor for an effective interfacial area $A_F \varepsilon$ can be defined by Eq. (23):

$$J = K_P A_F \varepsilon M^{-1}(c_F - c_{FRm})_{ls} \tag{23}$$

where c_F is the concentration of metal in the feed at the surface of the fiber corresponding to its concentration in the stripping solution $c_{FRm} = c_{Rm} D_{rR}/D_{rF}$, where D_{rR} and D_{rF} are the distribution coefficients on the stripping and the extraction interface, respectively. When stoichiometric excess of a strong acid is used in the stripping solution, the metal complex on the strip interface is effectively destroyed. Then the value of $D_{rR} = c_{MR}/c_{RM}$ is very low, and accordingly the value of c_{FRm} is practically zero, especially when the value of the distribution coefficient on the feed interface is high. From Eq. (23) it follows that

$$J = K_P A_F \varepsilon M^{-1} c_{Fls} \tag{24}$$

From the overall material balance of the feed and stripping solution streams in the steady state conditions, we can write

$$J = V_F' M^{-1}(c_{Fo} - c_F) = V_R' M^{-1}(c_R - c_{Ro}) \tag{25}$$

Combining Eqs. (24) and (25) and assuming a fresh stripping solution, when $c_{Ro} = 0$, the overall mass transfer coefficient in the pertractor can be calculated as follows:

$$K_P = \frac{V_R' c_R}{A_F \varepsilon c_{Fls}} \tag{26}$$

where the logarithmic mean concentration of metal in the feed stream is

$$c_{Fls} = \frac{c_{Fo} - c_F}{\ln(c_{Fo}/c_F)} \tag{27}$$

A new pertractor, PV-10, with two distributed bundles of hollow fibers, one for the feed and the second for the stripping solution, was built (Fig. 21a) [120]. These two U-shaped bundles with 20 mixed hollow fibers each, were inserted through the bottom ends of two vertical glass tubes. The same microporous polypropylene hollow fibers, Celgard X-10 (Hoechst-Celanese) were used in both bundles. The porosity of the fiber walls was 30%. the inner and outer diameters of the HF were 0.20 and 0.264 mm, respectively. The basic parameters of the PV-10 pertractor were as follows: effective length of fibers, 491.5 mm; inner geometrical surface area of one bundle, ~61.76 cm². The membrane phase was pulsed by a moving piston, with adjustable amplitude h and frequency of stroke ω_M, connected to the pertractor. The pulsation velocity was set to $\omega_M \cdot h = 6.6$ mm/s, a value found to be an optimum in earlier work. A steady state for the pertraction of Nd and Ho was achieved after about 7 h of operation at 25°C.

The flux of holmium through the LM is much lower, less than one-third, of that of neodymium (Fig. 21b). The much higher buildup of Ho in the membrane phase is a consequence of slower decomposition of the Ho/DEHPA complex than decomposition of Nd/DEHPA. Neodymium is effectively recovered in the HF pertractor with high yield (79–82%) even in short fibers of the contactor or for a short residence time. With increasing flow rate of the feed in the fiber lumen, u_F, the yield of Nd decreases owing to the lower residence time, but the overall mass transfer coefficient and flux density increase. This suggests that the resistance in the feed phase plays some role. It can be assumed that the value of the overall mass transfer coefficient, as well as flux, will approach a steady value at $u_F \sim 6$ cm/s, as it was observed in other systems such as Zn, phenol, and butyric acid pertraction [120,121]. With increasing

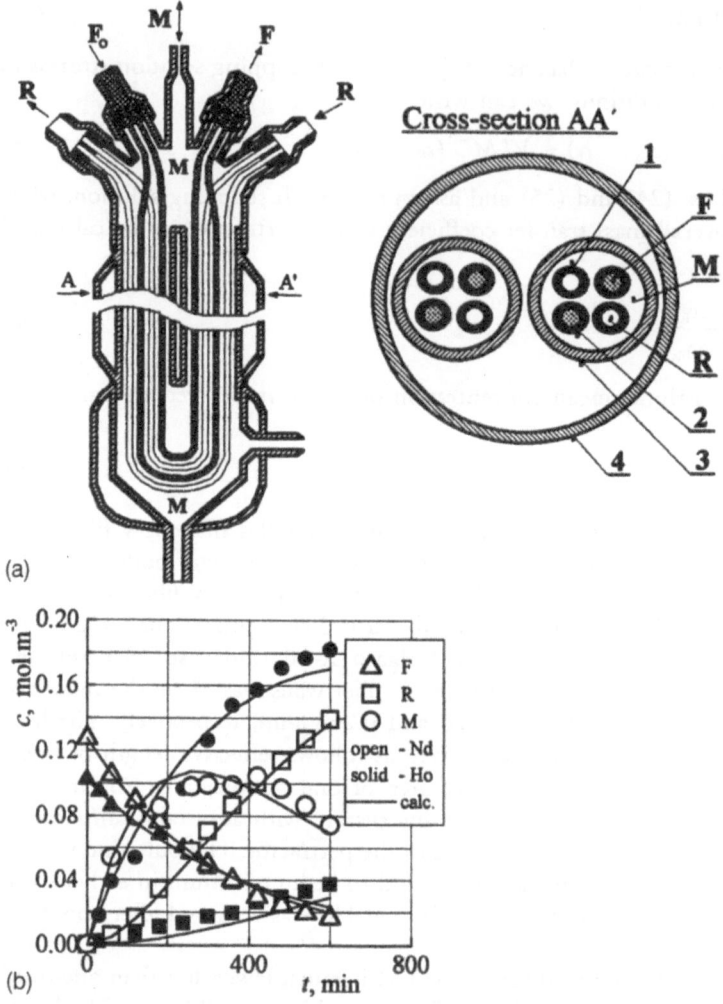

Figure 21 (a) Diagram of the hollow-fiber pertractor PV-10 with two distributed bundles of hollow fibers: F, feed; M, liquid membrane phase; R, stripping solution; 1, microporous HF for strip; 2, microporous HF for feed; 3, glass tube; 4, glass thermostat jacket. (From Ref. 115.) (b) Time dependence of metal concentrations in the three phases of the two-chamber contactor in pertractor of neodymium and holmium. (From Ref. 115.)

feed concentration the value of the overall mass transfer coefficient decreases while the flux into the stripping solution increases. As presented by Cara et al. [122], the extraction rate constant of Nd through the well-mixed layered BLM is practically independent of feed concentration. This indicates that the mass transfer resistance in an HF bundle is probably responsible for the decrease of K_p. Experimental values of the steady state mean concentration of Nd in the membrane phase are much lower than the equilibrium concentration when the distribution coefficient of around 100 for the actual experimental conditions is taken into account. The yield of Nd in the stripping solution decreases only slightly with the feed concentration in the concentration interval studied.

The overall mass transfer coefficient and flux of holmium were a little lower than for neodymium, but their differences were much smaller than in the pertraction through a layered LM. This set of discrepancies can be explained by a relatively large influence of the rate of complex decomposition on the transfer rate through the layered LM. In the BLM employing an HF contactor, the main mass transfer resistance of the membrane phase is in the wall pores and the relative contribution of the resistance due to the reaction kinetics is small.

Using the same pertractor described in Reference 115, pertraction was used to study zinc [3,115] and neodymium decontamination [5,117] with D2EHPA. The overall stoichiometry of the extraction equilibrium of zinc with D2EHPA can be written in a manner analogous to Eq. (22) as follows:

$$Zn^{2+} + 1.5(AH)_{2 \, or} \Leftrightarrow ZnA_2AH_{or} + 2H^+ \qquad (28)$$

Zinc and neodymium are efficiently recovered with high yield (82–99%) even in a short contactor or for short residence time. With increasing feed concentration, the value of the overall mass transfer coefficient of Zn and Nd decreased, while the flux into the stripping solution increased. The rate constants of the complex formation and destruction of Zn–D2EHPA and Nd–D2EPHA are practically independent of the feed concentration. Therefore, mass transfer resistance in an HF bundle is probably responsible for decrease of K_p. The steady state mean concentrations of Zn and Nd in the membrane phase are much below their equilibrium concentrations when the distribution coefficients of around 100 for the actual experimental conditions are considered.

V. MODELING OF HOLLOW-FIBER, MEMBRANE-BASED, LIQUID–LIQUID EXTRACTION (HFMLLE) AND KINETIC ANALYSIS

Two models have generally been suggested for the evaluation of mass transfer coefficients: the resistance in series model and the chemical equilibrium and kinetic model. The mathematical modeling of extraction membrane systems can be based on two different approaches. The first attempts to characterize the mass transfer process in terms of an overall or a local permeability coefficient, with average or local concentrations being considered along the membrane. The effects of hydrodynamic conditions are thus separated from the system properties. However, more accurate modeling can be achieved by considering the velocity and concentration profiles along the hollow fiber by means of appropriate mass conservation equations and, when the associated boundary conditions are applied at the fiber wall, it is often possible to derive analytical solutions to these equations. When this is not the case, numerical methods must be used.

Recently, the application of a hollow fiber membrane extractor has become popular for kinetic studies on solvent extraction and liquid membrane extraction [44,45,47,48,123–126], because the flow pattern of fluids that governs the diffusion process is an ideal laminar flow in the hollow fiber. Consequently, it is easy to exactly evaluate the effect of diffusion on the observed extraction rate, which differs from those in a vigorously stirred tank and a stirred transfer cell.

Along this line, Alam et al. [123] used an HF setup (Fig. 22a) to study the extraction kinetics of palladium from chloride media with bis(2-ethylhexyl)monothiophosphoric acid (MSP-8) in Exxsol d-80 aliphatic kerosene diluent at 303 K in a membrane extractor. For completeness, we show a schematic diagram (Fig. 22b), and an internal view of the hollow-fiber contactor (Fig. 22c). The extractor consists of a cylindrical glass tube in which a single hollow fiber is inserted. This fiber is made of hydrophobic PTFE. The aqueous solution was pumped into the lumen of the fiber, while the organic solution was fed into the annulus of the extractor. Both aqueous and organic phases flow cocurrently. At steady state, raffinates of the aqueous phase and loaded extracts of the organic phase were sampled and analyzed for palladium concentration. The extraction rate of palladium, J_M, was calculated by:

$$J_M = \frac{C_{Pd\,org} - C_{Pd}}{Q_{aq}/2\pi r_1 L} = \frac{C_{Pd\,org} Q_{org}}{2\pi r_1 L} \tag{29}$$

A loading test of palladium with MSP-8 indicated that the metal:reagent ratio was 1:2. The interfacial tension, γ, of the dimeric MSP-8 abruptly decreased with increasing MSP-8 concentration in its high concentration region, while the interfacial tension of the palladium complex was affected by its own concentration. Therefore, MSP-8 had interfacial activity, while the adsorption of the palladium complex at the interface is negligible. Figure 22d shows the relation between the extraction rate of palladium J_M and the MSP-8 concentration in the organic phase, and that between J_M and the chloride ion concentration in the aqueous phase. It was found that extraction rate was propor-

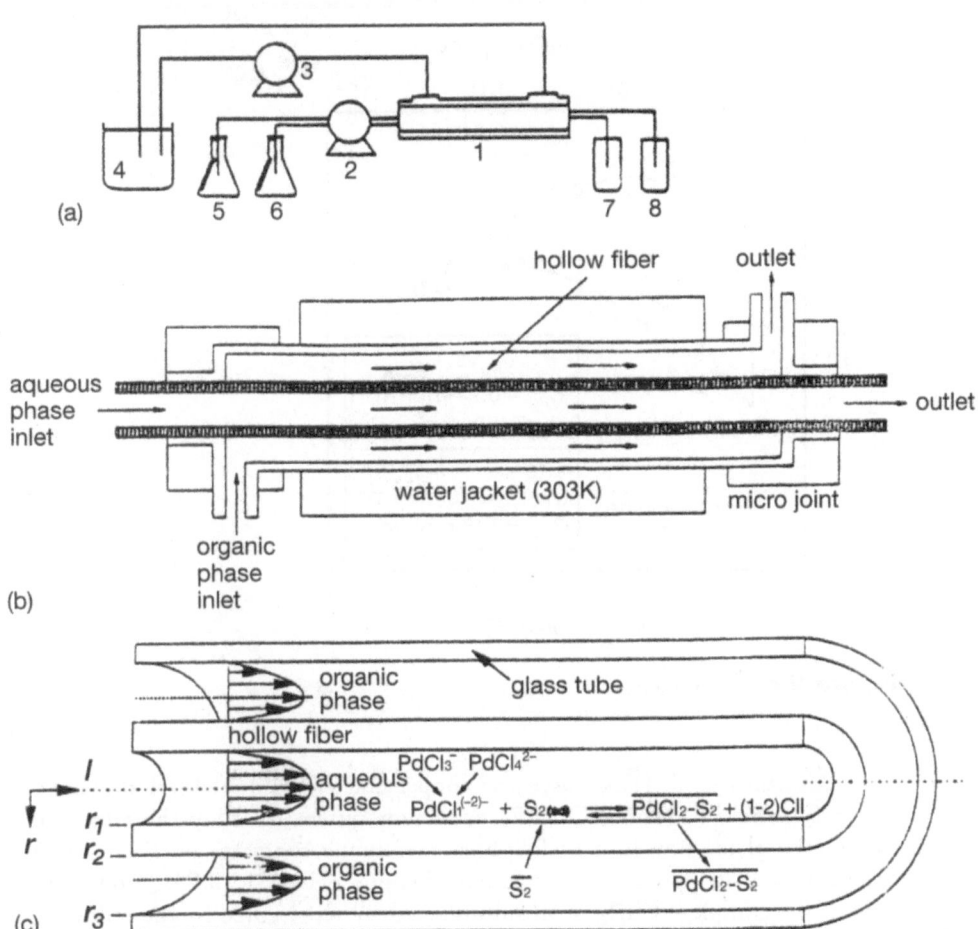

Figure 22 (a) Experimental setup. (b) Schematic diagram of the hollow-fiber membrane extractor. (c) Internal view of the hollow-fiber membrane extractor. (From Ref. 123.) (d) Relations between extraction rates J_M and concentration of dimeric MSP-8, C_{S2}, and chloride ion concentration C_{Cl}. (From Ref. 123.)

tional to the extractant concentration at low extractant concentrations. At high concentrations, the extraction rate deviates downward from the linear relationship, a result that is attributable to the adsorption of extractant as well as the diffusional effect of palladium. The extraction rate dependence on the chloride ion concentration at low concentrations is small, with a slope of $-n$ (where $n < 1$).

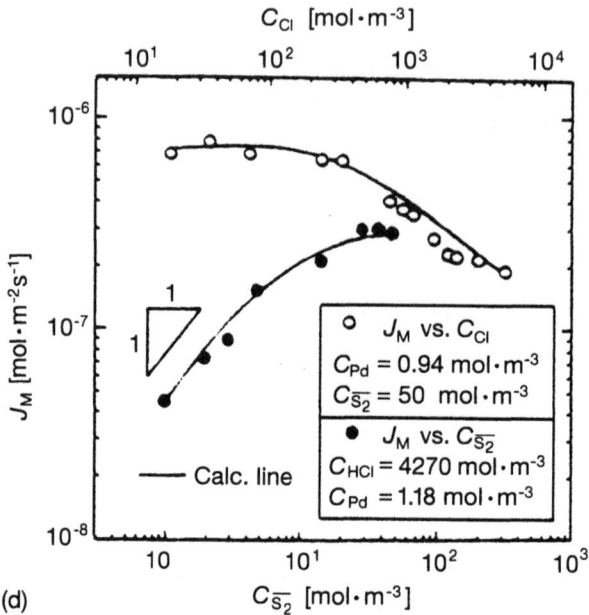

(d)

Figure 22   Continued

Yoshizuka et al. [124], used the copper and *N*-8-quinolyl-*p*-dodecylben-zenesulfonamide (C12phSAQ, HR) extraction system to study the effects of the interfacial reaction in a hollow-fiber membrane contactor on the rates of extraction and stripping. The extraction of copper with C12phSAQ and stripping of the complex (CuR$_2$) with sulfuric acid are represented by the following reaction:

$$Cu^{2+}_{aq} + 2HR_{org} \Leftrightarrow CuR_{2\,org} + 2H^+_{aq} K_{ex} \tag{30}$$

The reaction of Eq. (30) expresses the following interfacial reactions:

$$HR_{org} \Leftrightarrow HR_{ad} K_{ad} \tag{31}$$

$$Cu^{2+}_{aq} + HR_{ad} \Leftrightarrow CuR^+_{ad} + H^+_{aq} K_1 \tag{32}$$

$$Cu^{2+}_{aq} + 2HR_{ad} \Leftrightarrow CuR_{2\,org} + H^+_{aq} K_2 \tag{33}$$

If in reaction (33) k_2 and k'_2 are the forward and backward reaction rate constants, respectively, then $K_2 = k_2/k'_2$. Among the interfacial reaction steps, the reaction of Eq. (33) is the rate-determining type. When the Langmuir adsorption isotherm is taken to express the adsorption of HR and CuR$^+$ at the interface, the interfacial extraction rate of Eq. (30), *R*, is obtained as follows:

$$R = \frac{k_2 K_1 (K_{ad}/S_{HR})^2 (C_{Cu} C_{HR}^2/\alpha_H - C_{CuR2}\alpha_H/K_{ex})}{[1 + K_{ad}C_{HR} + K_{ad}K_1 C_{Cu} C_{HR}^2/\alpha_H]^2} \tag{34}$$

The values of the constants are given in Table 18.

When it is considered that aqueous solutions containing copper or sulfuric acid flow through the lumen of a hollow fiber, whereas organic solutions containing HR and CuR_2 flow cocurrently through the shell side of the contactor (Fig. 22a–c), the laminar flow patterns are represented as follows:

$$u_{aq}(r) = 2Q_{aq}(\pi r_1^2)^{-1}[1 - (r/r_1)^2] \tag{35}$$

$$u_{org}(r) = \frac{2Q_{org}(\pi r_1^2)^{-1}\{(r_3/r_1)^2 - (r/r_1)^2 + [(r_3/r_1)^2 - (r_2/r_1)^2]\ln(r/r_3)/\ln(r_3/r_2)\}}{(r_3/r_1)^4 - (r_2/r_1)^4 - [(r_3/r_1)^2 - (r_2/r_1)^2]/\ln(r_3/r_2)} \tag{36}$$

The interface between the aqueous and organic solutions is the inner surface of the hollow fiber, because the fiber is hydrophobic. Furthermore, it can be assumed that the activity of the hydrogen ion in the aqueous solution is kept constant along the lumen of the hollow fiber, and that in the denominator of Eq. (34) under extraction and stripping conditions, $1 + K_{ad}C_{HR} \gg K_{ad}K_1 C_{Cu}C_{HR}^2/\alpha_H$. Therefore, the interfacial extraction rate of copper and the interfacial stripping rate of CuR_2 are simplified to

$$R = \frac{k_2 K_1 (K_{ad}/S_{HR})^2 (C_{Cu} C_{HR}^2/\alpha_H - C_{CuR2}\alpha_H/K_{ex})}{[1 + K_{ad}C_{HR}]^2} \tag{37}$$

$$R' = \frac{k_2'(C_{CuR2}\alpha_{H0} - K_{ex}C_{Cu}C_{HR}^2/\alpha_{H0})}{[1 + K_{ad}C_{HR}]^2} \tag{38}$$

Figure 23 shows the relationship between β and τ_M for various values of ϕ and X_{RO} (defined in Fig. 23). The calculated curves for $\phi > 1.0$ approach the asymptote $\beta = 0.159\tau_M$ with increasing τ_M, indicating that E approaches the equilibrium value at the outlet of the membrane contactor. In this case, E

Table 18 Values of Constant in Eq. (34)

$k_2 = (1.7 \pm 0.5) \times 10^8 \ m^2/(mol \cdot s)$
$K_1 = (1.7 \pm 0.3) \times 10^{-2} \ —$
$K_{ad} = (2.0 \pm 0.3) \times 10^{-2} \ m^3/mol$
$S_{HR} = (3.2 \pm 0.2) \times 10^5 \ m^2/mol$
$K_{ex} = 3.3 \pm 0.3 \ —$

Source: Ref. 124.

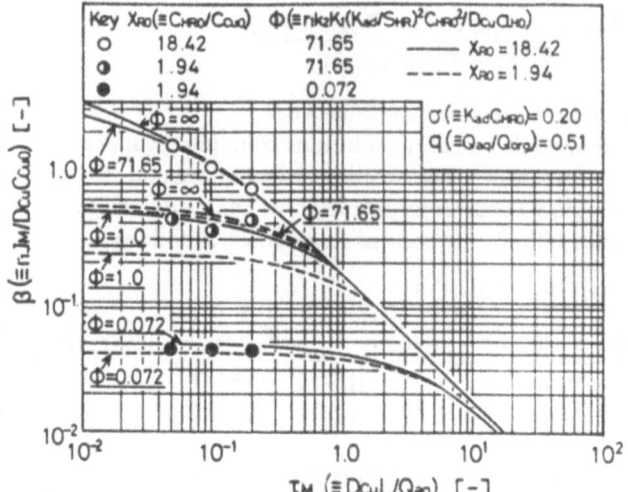

Figure 23 Relationship between β and τ_m. (From Ref. 124.)

approaches 1.0 with increasing contact time of each solution as can be found from the relation

$$\beta = \frac{E}{2\pi\tau_M} \tag{39}$$

The experimental results shown in Figure 23 were obtained for various flow rates, keeping the ratio of Q_{aq} to Q_{org} (i.e., the ratio τ_M to τ_R) constant. The effect of the interfacial reaction on the average extraction rate of copper was negligibly small when $\phi = 71.7$ (i.e., at pH 5.0 and $C_{HR\,org} = 10.0$ mol/m^3).

Kubota and coworkers [127] used a microporous hydrophobic hollow-fiber membrane extractor and a calix[4]arene carboxyl derivative [(25,26,27,28-tetrakis(carboxymethoxy)5,11,17,23-tetrakis(1,1,3,3-tetramethylbutyl)calix[4]-arene] as the extractant dissolved in toluene to study the extraction of three rare earth metals (Er, Ho, and Y). The HF contactor consisted of a PTFE (Japan Gore Tex). The HF unit was inserted in a cylindrical glass tube, as shown in Figure 24a. The permeation rate of metal ions through the membrane was

Figure 24 (a) Schematic diagram of membrane extractor and liquid flow pattern of each phase in the membrane extractor. (From Ref. 127.) (b) Relation between P_M and $C_{(H_4R)_2}$ in the presence of sodium ion: C_{Na}, 0.050 M; C_{Mi}, 0.1 mM; pH 2.5, for Ho (▲), Er (□), and Y (○). (From Ref. 127.)

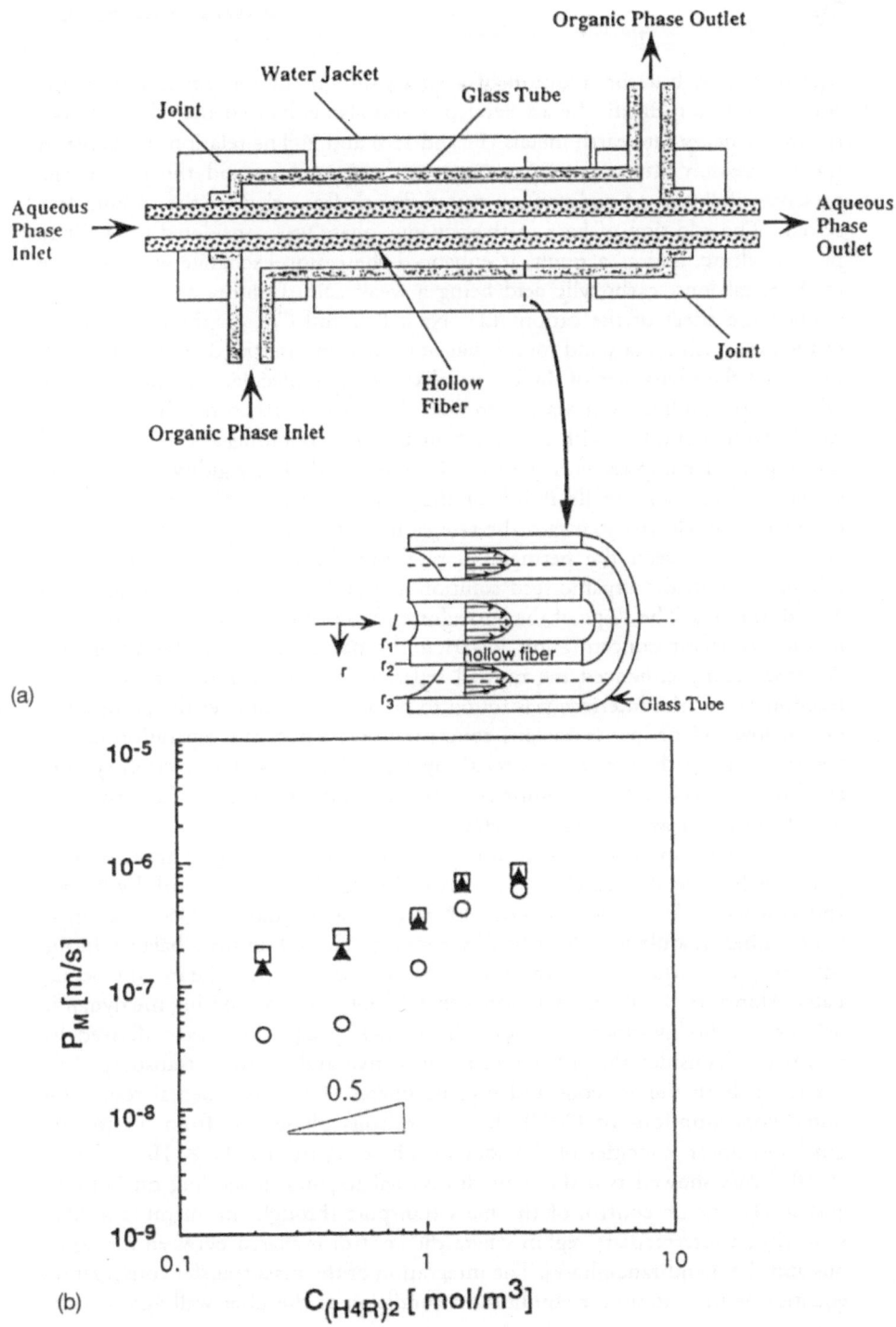

(a)

(b)

relatively slow, but the addition of a small amount of sodium ions into the aqueous solution drastically accelerated it and also enhanced the selectivity between the heavy rare earth metals (Er and Ho) and Y. The relationship between the permeation rate of the three rare earth metal ions and the pH in the presence of 0.050 M sodium ion was different from that in the sodium-free system. The addition of base to the aqueous phase was considered to facilitate proton release, and as a result, it enhanced the cation exchange with the rare earth metal ions, carboxylic acid being a weak acid. Oshima et al. [128] examined the effect of the cations Li^+, Na^+, K^+, and Cs^+ on the extractability of the rare earth metals and found that only sodium ion, the diameter of which just fitted the cavity size of the cyclic calix[4]arene molecule, was effective, The effect of the sodium ion was, therefore, due to the structural change in the calix[4]arene complex, with rare earth metals incorporating a sodium ion and forming a 1:1 complex with the calix[4]arene [128]. The rigidity of the cyclic structure decreased the flexibility of the functionalized carbonyl groups and enhanced the selectivity toward the rare earth metal ions. Figure 24b shows the relationship between the permeation rate and the concentration of dimeric calix[4]arene in the organic feed solution at pH 2.5 in the presence of 0.050 M sodium ions. The slope of the curve for each metal was almost 0.5, a decrease in the extractant concentration lowering the permeation rate. However, this decrease greatly enhanced the mutual separation of metal ions. The chemical reaction rate at the interface was found to be predominant over the permeation rate at low pH or low extractant concentrations, where the separation factors between rare earth metals were relatively high. This model extraction system, employing a hollow-fiber membrane extractor, was effective for the separation of Y from the heavy rare earth metals.

Different approaches were used to study the modeling of mass transfer processes by several researchers [15,17,40,129–132]. In earlier work by Alonso and coworkers [17], the extraction of Cr(VI) with Aliquat 336 in a single hollow-fiber module was described by a steady state solute mass balance in the aqueous phase, while the chemical reaction was considered to be instantaneous. Later, Alonso et al. [47] used a more general approach, considering the dynamic behavior of hollow-fiber modules. The modeling equations were derived by means of a consideration of the axial convective and diffusive transport of all species in both the aqueous and organic phases. The experimental results for initial concentrations of Cr(VI) in the aqueous phase, C_0, from 50 to 500 g/m^3 and linear velocities of the aqueous phase, v, from 2.95×10^{-3} to 1.18×10^{-8} m/s showed two different diffusional regimes depending on both C_0 and v: (1) kinetic control of the mass transport through the membrane fiber wall, (2) an intermediate region where the control is shared between the aqueous and the membrane phases. The integration of the mass transfer conservation equation with a nonlinear equilibrium condition at the fiber wall agreed satisfactorily with the results. An optimization of the parameters D, the solute diffusivity in the aqueous phase, and K_{eq}, the equilibrium constant of the extraction chemical reaction, yielded $D = 2.3 \times 10^{-9}$ m^2/s and values of K_{eq} dependent on C_0. For the analysis of the experimental system of interest, it was sufficient to solve for the extraction profile in a single fiber to predict the performance of the entire hollow-fiber extraction module. The steps in the extraction process were assumed to be as follows [17]:

Step 1.   The metal ion/metal species in the aqueous phase diffused from the bulk to the aqueous–organic interface (inside wall of fiber or support) through the boundary layer.

Step 2.   At the aqueous–organic interface, metal ion/metal species reacted with the extractant in the organic phase in the membrane pores to form the metal–organic complex.

Step 3.   Metal–organic complex diffused from the aqueous organic interface to the outside wall of flat-sheet support/fiber through the organic-filled membrane pore; free extractant diffused in the opposite direction from the organic phase/shell side into the pore.

Step 4.   Metal–organic complex diffused from the outside flat-sheet support/fiber wall to the organic phase bulk (or shell side, which was flowing cocurrent or countercurrent to the aqueous phase).

The system under study was the reversible extraction equilibrium of Cr(VI) with Aliquat 336 (Q):

$$CrO_4^{2-} + 2QCl_{org} \Leftrightarrow Q_2CrO4_{org} + 2Cl^- \tag{40}$$

The chemical equilibrium was described by the mass action law:

$$K_{eq} = \frac{[Q_2CrO4_{org}][Cl^-]^2}{[CrO_4^{2-}][QCl_{org}]^2} \tag{41}$$

Given the total carrier concentration in all forms (carrier and carrier–solute complex species), $C_{M\,org}$ being known, the solute–carrier complex equilibrium concentration at the fluid membrane interface, $C_{i\,org}$, could be obtained.

When mass transfer through the organic membrane governed the kinetic control, the expression for the solute flux was:

$$N = Ks(C_{T\,org} - C_{org}) \tag{42}$$

where $C_{T\,org} = 0.1$ M was the maximum solute–carrier complex concentration in the organic phase, $K = 8.08 \times 10^{-8}$ m/s was the membrane mass transfer coefficient, and $s = 1.305$ was the shape factor of the hollow fibers.

Kubaczka and coworkers [133] presented a method for solving the generalized Maxwell–Stefan equations in order to perform membrane-based solvent extraction in multicomponent systems—for example, the separation of aromatic from aliphatic hydrocarbons (ternary or higher systems) by means of tetraethylene glycol. Two general methods were developed for the calculation of mass fluxes in the membrane pores filled with the liquid phase. Since, these methods required knowledge of the mutual diffusivities D_{ij} (i and j referring to components in the multicomponent mixture) and the membrane diffusion coefficients D_{im}, a calculation procedure was proposed to evaluate them. The membrane resistance was calculated for both hindered and unhindered diffusion in the membrane pores. A system of kinetic and balance equations was given that described multicomponent mass transfer in membrane extractors, with the resistance on both sides of the membrane and within the membrane itself properly accounted for. Unlike other studies, the procedure for the determination of the shell-side resistance was based on the use of local transfer coefficients calculated by the method of Miyataki and Iwashita [134], assuming an analogy with heat and mass transfer. The average difference between the calculated outlet concentrations and the experimental results was 15%. In an extended study, Kubaczka and Burghardt [135] reported data analysis showing that when the norm of a matrix was used, the effect of individual mass transfer resistances in multicomponent extraction, expressed as reciprocals of the matrices of mass transfer coefficients, could be assessed correctly. If the resistances identified by the norm as negligible were omitted, the overall accuracy of the simulations remained unaffected. Since when the membrane resistance was neglected, the membrane still determined the mass fluxes in the stationary system of coordinates, data analysis showed that the omission of this effect led to an appreciable drop in the accuracy of the calculations. The omission of diffusivity dependencies also generated considerable errors. In conclusion, this method for the calculation of mass transfer resistances based on the evaluation of matrix norms was found to be quite useful in the extraction process investigated. Since the problem of assessing transport resistances in multicomponent systems is common to all mass transport processes, it is highly probable that the method will prove helpful in dealing with other processes of this type.

In the mid-1990s, membrane processes were simulated by using the gPROMS (general process modeling system) software package [136,137]. Both distributed (e.g., membrane modules) and lumped (e.g., storage tanks) unit operations could be described. The user of gPROMS is shielded from much of the mathematical complexity and can therefore concentrate on the underlying modeling issues. When a complete gPROMS model of the process is available, it is possible to carry out steady state and dynamic simulations by specifying the equipment being simulated, the values of the parameters and input variables, the initial state of the system, and any external action imposed on it. For distributed unit operations, one also needs to specify the approximation method to be applied to each of the spatial domains and the granularity and order of approximation [47]. Alonso et al. [88] developed the modeling of LM processes using this simulator to solve the mathematical equations arising from the mass balances in hollow-fiber devices. However, a database including thermodynamic properties (equilibrium and partition coefficients) and mass transfer parameters was not yet available. Recent studies for the modeling and mass transfer of Cr(VI) with Aliquat 336 [47] and Cd(II) with Cyanex 302 [88] with hollow fibers were performed utilizing gPROMS, and reasonable agreement between model predictions and experimental measurements was obtained.

Alonso and Pantelides [47] reported the modeling and simulation of the extraction and stripping of Cr(VI) by using Aliquat 336 in a HF contactor, the work performed earlier [17] being selected for model development including model parameters, using the gPROMS. Steady state simulations of a single extraction membrane module were carried out for both the cocurrent and countercurrent cases. The flow rates used were 0.275 cm³/s for the aqueous stream and 0.63 cm³/s for the organic stream. The authors obtained the radial profiles of the chromium and chloride concentrations in the aqueous phase as well as the complex and Aliquat 336 concentrations in the organic phase and the mean concentration of chromium and chloride in the aqueous phase as functions of the axial position were obtained. Some of the results are shown in Figure 25; it should be noted that in Figure 25b the aqueous concentrations for cocurrent and countercurrent operations practically coincide. Models were developed for both single-function and dual-function membrane modules, and these were combined to model the entire process carrying out simultaneous extraction and stripping. The integrated process models incorporated the possibility of partial recycling of both the extraction and the stripping product streams. Although these models were dynamic, they could be used for the direct prediction of steady state conditions. Some mathematical complications arising in this context was identified and resolved by the authors.

Yang and Cussler [48a] explored the competitive extraction of copper and nickel with kerosene solutions of D2EHPA in a hollow-fiber membrane module. The reaction changed from first order to second order as the reaction proceeded. The first-order behavior largely resulted from mass transfer resistances in the membrane and in the aqueous boundary layer. The second-order reaction occurring in dilute solutions was a consequence of an interfacial reaction. The equilibrium extraction and the extraction rates were fitted to the expected equations, but an unexpected and rich chemical complexity was found. The mass transfer coefficients were evaluated by Eqs. (49) and (50). Systematic deviations of the data in Figure 26 from the expected trend were found. These deviations for both copper and nickel suggested that the mass transfer coefficients increased at the lower concentrations occurring at longer times, a result that was the opposite of what the authors had expected. The following equation was therefore suggested to evaluate the overall mass transfer coefficient K across solvent-wet hollow fibers:

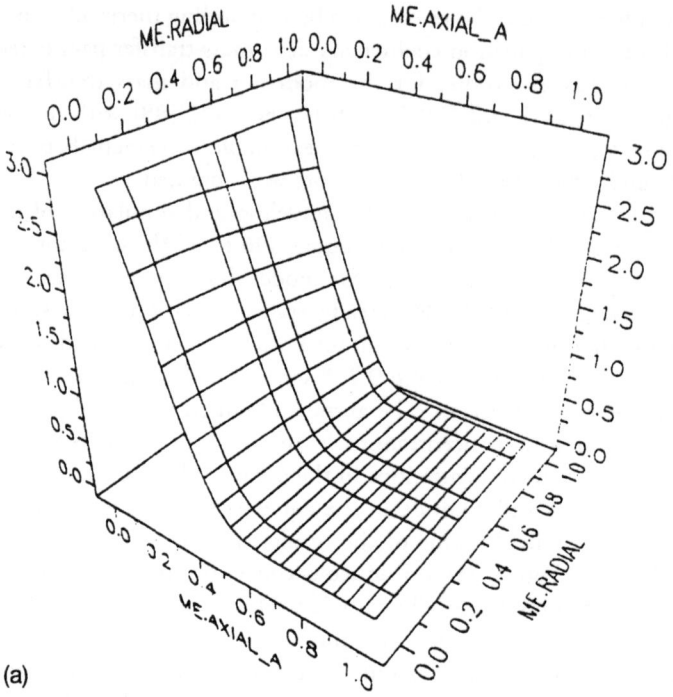

(a)

Figure 25 (a) Concentration profile of CrO_4^{2-} in the aqueous phase in cocurrent operations: ME.RADIAL, normalized radial position; ME.AXIAL_A, normalized axial position. (From Ref. 47.) (b) Average concentration of CrO_4^{2-} in the aqueous and organic phases in cocurrent and countercurrent operation. (From Ref. 47.)

$$\frac{1}{K} = \frac{1}{k_a} + \frac{1}{Hk_m} + \frac{1}{Hk_o} \tag{43}$$

where k_a, k_m, and k_o are the individual aqueous, membrane, and organic mass transfer coefficients, respectively, and H is the partition coefficient. The diffusion, rather than the chemistry, was responsible for the extraction kinetics. The kinetics of the ion exchange at the water–membrane interface could, however, be slow, influencing K, and so Eq. (43) must be supplemented by a new term:

$$\frac{1}{K} = \frac{1}{k_a} + \frac{1}{Hk_m} + \frac{1}{Hk_o} + \frac{1}{k_i} \tag{44}$$

where k_i is the effective rate of interfacial reaction at the surface. In other words, $1/k_i$ is the resistance to the mass transfer of this interfacial reaction. This new term accounted for the differences between an equilibrium interface assumed

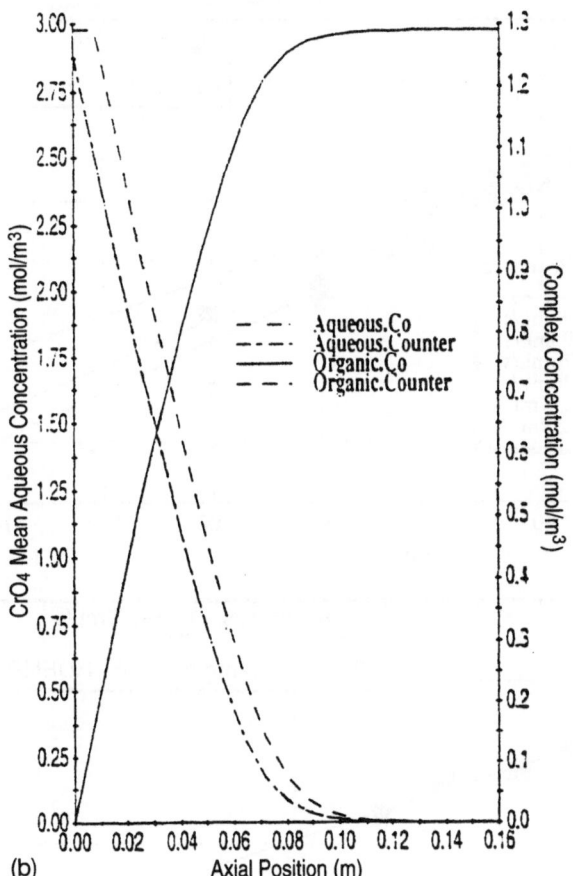

(b)

by Eq. (43) and a nonequilibrium interface due to the slow formation of complex.

The present authors showed that these results had implications for other metals subjected to hollow-fiber extraction. Based on the expertise gained over the last several years, they speculated about the extent to which other metal extractions in hollow-fiber contactors were practically attractive, as shown in Table 19. The following considerations were applied: the value of the metal, the availability of an extraction system, the extraction environment (especially the acidity), the ease of the extraction, and the selectivity of the extraction. For example, beryllium could be quantitatively extracted over a large pH range by many extractants, including β-diketones, 8-hydroxyquinoline, cupferron, and

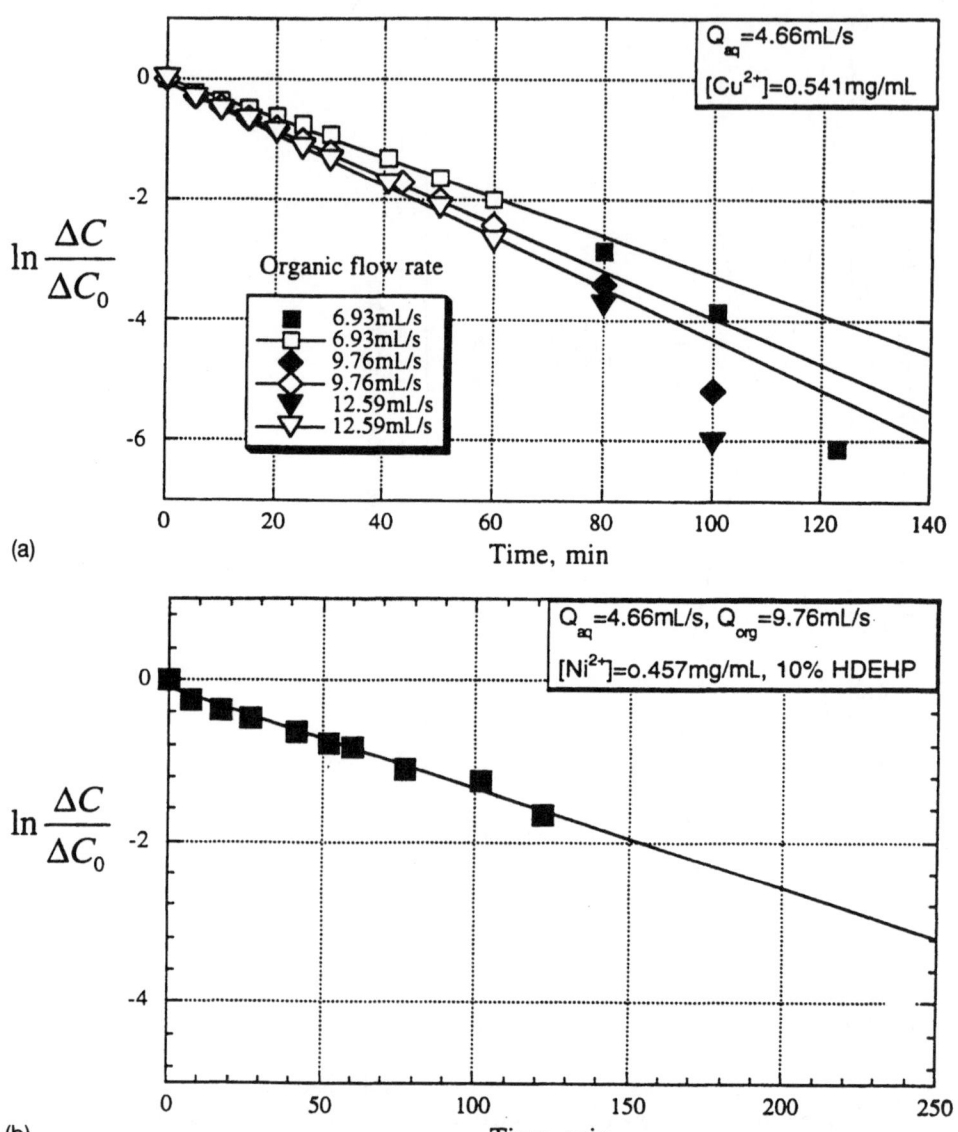

Figure 26 (a) Copper and (b) nickel concentrations vs time. The semilogarithmic behavior at small times is expected, but the systematic deviation at larger times is not. (From Ref. 48a.)

Table 19 Potentials of Metal Extractions Determined by Using Hollow-Fiber Contactors[a]

	I	II											III	IV	V	VI	VII
1	H (1)																
Rating																	
2	Li (3)	Be (4)											B (5)	C (6)	N (7)	O (8)	F (9)
Rating	G	G															
3	Na (11)	Mg (12)												Al (13)	Si (14)	P (15)	S (16)	Cl (17)
Rating	P	G												P				
4	K (19)	Ca (20)	Sc (21)	Ti (22)	V (23)	Cr (24)	Mn (25)	Fe (26)	Co (27)	Ni (28)	Cu (29)	Zn (30)	Ga (31)	Ge (32)	As (33)	Se (34)	Br (35)
Rating	P	F/P	G	G	G	G	G	P	G	G	G	G	F	F	G	F/P	
5	Rb (37)	Sr (38)	Y (39)	Zr (40)	Nb (41)	Mo (42)	Tc (43)	Ru (44)	Rh (45)	Pd (46)	Ag (47)	Cd (48)	In (49)	Sn (50)	Sb (51)	Te (52)	I (53)
Rating	F	G	G/F	G/F	F/P	G	F/P	G/F	$	G	G/F	G/F	F	G	G/F	F/P	
6	Cs (55)	Ba (56)	La (57)	Hf (72)	Ta (73)	W (74)	Re (75)	Os (76)	Ir (77)	Pt (78)	Au (79)	Hg (80)	Tl (81)	Pb (82)	Bi (83)	Po (84)	At (85)
Rating	G	P	G	F/P	F/P	G/F	P	?	?	F/P	G/F	G/F	F	G	G/F	F/P	
Lanthanoids	Ce (58)	Pr (59)	Nd (60)	Pm (61)	Sm (62)	Eu (63)	Gd (64)	Tb (65)	Dy (66)	Ho (67)	Er (68)	Tm (69)	Yb (70)	Lu (71)			
Rating	G	G	G	G	G	G	G	G	G	P	P	P	P	P			
Actinoides	Th (90)	Pa (91)	U (92)	Np (93)	Pu (94)	Am (95)	Cm (96)	Bk (97)	Cf (98)	Es (99)	Fm	Md	No	Lw			
Rating	F	P	F	P	P	F	G	G	G								

[a] G, good; F, fair; P, poor; ?, not known.
Source: Ref. 48a.

organophosphorus acids. Its selective extraction was also readily achieved via acetyl acetone and amine [138]. Therefore, the possibility of beryllium extraction in HF contactors was rated as good. The potential of the hollow-fiber separation of lithium was also rated as good because of its growing value in batteries and its availability in seawater and salt lakes, although the existence of a suitable extraction system was unclear. In contrast, for germanium only a few choices of organic solvents existed, and whereas selective extraction was extremely good when carbon tetrachloride was used, extraction required a high concentration (>8 M) of HCl. Because the membrane and the potting material could be severely challenged in this environment, germanium extraction was rated as poor. Similarly, iron being cheap, the merit of an HF-based separation of it was doubtful, even though several systems promised extraction. Extraction of some of the actinides (such as Th, U, Am) using an HF contactor was claimed to be fair, but the extractive separation of Pu and U was taken to be poor, although extraction of Pu(IV) was reported to be promising under the experimental conditions used (viz., 2 M HNO_3 and 30% TBP) [48b].

Nuclear fuel reprocessing plants generate a large amount of low-level radioactive wastes, such as condensate waste, which could be treated with the HFMLLE technique under suitable chemical conditions [48c]. Kathios et al. [22] conducted preliminary tests of a microporous hollow-fiber membrane module for the liquid–liquid extraction of actinides to evaluate the feasibility of recovering actinides or processing radioactive wastes. This work involved the analysis of the potential for using these modules for process-scale metal separations such as radioactive waste stream cleanup and environmental remediation. MHF modules were used to remove neodymium (a surrogate for americium) from 2 M nitric acid using DHDECMP (dihexyl-N,N-diethylcarbemoyl methylphosphonate) and CMPO (n-octylphenyl-N,N-diisobutylcarbamoyl methylphosphine oxide) as extractants in diisopropylbenzene. The modules were also used to concentrate the neodymium from the organic phase by back-extraction of it into 0.01 M nitric acid. The extractants removed trivalent metal ions from the aqueous phase by forming the complex $Nd(NO_3)_3 \cdot 3CMPO$ or $Nd(NO_3)_3 \cdot 3DHDECMP$. The results were used to determine the number of modules required to achieve one theoretical extraction stage, that is, one that first allowed mass transfer between the aqueous and the organic phases to reach equilibrium and then allowed complete separation of the two phases to occur before they exited the process. The number of modules per theoretical stage, N^{TS}, is

$$N^{TS} = \frac{A_i^{TS}}{A_i} = \left[\frac{C_{fi} - C_{fo}^{TS}}{C_{fi} - C_{fo}} \right] \left[\frac{\Delta C_{lm}}{\Delta C_{lm}^{TS}} \right] \tag{45}$$

where A_i is the total membrane surface area of the module based on the internal diameter, ΔC_{lm} is the logarithmic mean concentration difference, ΔC_{lm}^{TS} is the logarithmic mean concentration difference in a theoretical stage, A_i^{TS} is the total membrane surface area of all modules based on the internal diameter, C_{fi} is the solute concentration in the aqueous feed, C_{fo} is the solute concentration in the effluent streams, and C_{fo}^{TS} is the solute concentration in aqueous steam in a theoretical stage. The modules used in this analysis were small laboratory-scale versions specifically designed for experimental purposes. In an actual process application, pilot-scale (or larger) modules could be used, each one providing as much as 10 times the membrane area of a laboratory-scale version. An increase in surface area could also be achieved by connecting a number of modules in series. Earlier deficiencies in the design of HF contactors have been removed. Newer modules have an advanced fiber mesh that allows the organic liquid to flow more uniformly throughout the fiber bundle. Also, new shell-side baffle technology has been introduced that may significantly improve the performance.

In 1996 Valenzuela et al. [23,24] used a hollow-fiber membrane contactor to conduct application-oriented studies of copper recovery from a dilute solution produced in a mining process. This solution was an acid leach residual solution of a Chilean molybdenite concentrate that contained variable but significant concentrations of copper, as well as molybdenum, rhenium, some iron and arsenic, and other nonvaluable metals [139,140]. Extraction of copper was strongly dependent on a high molarity of carrier extractant in an HF-based extraction. A moderate increase in copper extraction was observed when the acidity of the feed solution was decreased. As shown in Table 20, extraction of copper was maximal in comparison to some other metal ions and provided evidence for the feasibility of copper separation from the main impurities of the leach solution, including molybdenum present at high concentrations. The concentration of Mo(VI) in the product phase took place only when the pH

Table 20 Extraction of Copper in the Presence of Other Metals as a Function of Initial pH for 1.0 M LIX 860: Cu, 0.57 g/L; Mo, 4.0 g/L; Re, 0.3 g/L; Fe, 0.30 g/L; As, 0.40 g/L

pH	Extraction (%)[a]				
	Cu	Mo	Re	Fe	As
0.77	63.1	26.2	n.d.	n.d.	n.d.
1.38	73.2	12.4	n.d.	n.d.	n.d.
1.90	81.4	3.10	n.d.	n.d.	n.d.
2.51	85.3	n.d.	n.d.	n.d.	n.d.

[a] n.d., not detected.
Source: Ref. 24.

of the solution exceeded 1.9, because no cationic species of this metal existed in aqueous solution maintained below this pH [139,140]. Likewise, no extraction of rhenium was detected because this metal existed only as an anionic perrhenate species, not extractable by LIX 84. Extraction of iron and arsenic was insignificant too.

The extraction rate of copper with 8-octanesulfonamidoquinoline (OSAQ) dissolved in toluene was measured in a cocurrent tubular membrane extractor, using a hollow fiber as solid support [141]. The results were analyzed by a heterogeneous interfacial reaction model that considered adsorption of OSAQ at the interface between the aqueous and organic phases. At low pH values a first-order dependence with respect to extractant concentration was observed, with a slope nearly equal to 1. On the other hand, the extraction rate was proportional to the first power of copper concentration between pH 1.6 and 5.6. However, in the high pH and high copper concentration ranges, all the curves tended to converge to a constant value of metal flux. The relation between the interfacial tension and the OSAQ concentration indicated that the extractant exhibited interfacial activity.

Removal of the toxic metals Cu(II) and Cr(VI) was achieved by Yun et al. [19], who used nondispersive microporous hydrophobic hollow-fiber membrane-based solvent extraction, with the aqueous waste feed flowing through the fiber bores and organic extractant flowing countercurrently on the shell side in a once-through mode. Figure 27b shows the flow pattern in the hollow-fiber contactor schematically; for a simplified view of the fiber wall and its surroundings, see Figure 27a. Figure 27c shows the effect of changing the shell-side organic phase flow rate, keeping the aqueous flow rate constant, for 10% v/v LIX 84 (anti-2-hydroxy-5-nonylacetophenone oxime) in n-heptane. It was noted that beyond an organic flow rate of 0.045 mL/s there was hardly any effect, but below this flow rate, the module extraction efficiency changed significantly. In the entire extraction process, the interfacial reaction was considered to be very important. Hence, most of the copper ions could be easily extracted when the concentration of the oxime was in excess due to an increase in the organic phase flow rate, since the aqueous inlet concentration was only 500 mg/L and the aqueous stream flow rate was low (0.07 mL/s). This connection was confirmed by a theoretical prediction of k_e, the forward reaction rate constant, using two cases of shell-side mass transfer. Either the shell-side correlation mentioned earlier in the published literature was used (long dashes for $k_e = 9.0 \times 10^{-6}$ cm/s and dots for $k_e = 3.0 \times 10^{-6}$ cm/s) or an infinite mass transfer coefficient in the shell side was used (short dashes and dash-dots for $k_e = 9.0 \times 10^{-6}$ and 3.0×10^{-6} cm/s). At low aqueous flow rates, the exit concentration of copper was extremely low (as low as 1 mg/L). Furthermore, the predicted behavior using $\beta = 17.4$ was significantly better than that with $\beta = 5.8$ (dots). Using the same HF setup for Cr(VI) removal, the exit concentration of Cr(VI) in the aqueous phase at a fixed flow rate of the organic phase (0.031 mL/s) for a 100 mg/L Cr(VI)-containing feed aqueous phase was studied as a function of the aqueous phase flow rate. A very low concentration of Cr(VI) could be achieved in the exit aqueous phase [<1 mg/L Cr(VI)] for an aqueous rate flow rate up to 0.6 mL/s, 20 times higher than the organic flow rate. The mathematical model for solvent extraction of copper developed in this study predicted the experimental data well.

Sirkar and coworkers [142] developed a separation technique using HF-based NDSX (under the name "hollow fiber contained liquid membrane": HFCLM) for the separation of two cations [e.g., Cu(II) and Zn(II)], a cation and an anion [e.g., Cu(II) and Cr(VI)], and two cations and one anion [e.g., Cu(II), Zn(II), and Cr(VI)]. The extraction selectivity of Cu(II) and Zn(II) by LIX 84 and D2EHPA, respectively, in a two-fiber-set HFM extractor (one for extraction and another for stripping) was significantly enhanced owing to competitive extraction. The efficiencies of extraction of Cu(II) and Cr(VI) by LIX 84 and tri-*n*-octylamine (TOA), respectively, were reported to increase owing to self-control of the aqueous feed pH (the protons released from extraction of cations were partly/totally consumed by the extraction of anions). In HFM modules, the extraction rates of Zn(II) and Cu(II) by D2EHPA and LIX 84, respectively, were controlled by the organic boundary layer resistances as well as by the interfacial reaction resistances. In subsequent studies using the same HFCLM module containing multiple sets of fibers, Sirkar et al. [143] reported the simultaneous extraction of the heavy metal species Cu(II), Cr(VI), Zn(II), and Hg(II) into an appropriate mixed solvent, followed by recovery and concentration of these heavy metals into an aqueous stripping solution. In earlier studies [142], these investigators had used two separate organic reagents to extract cations and anions, but these systems were considered to be excessively complicated because of the typical arrangement of HFCLM fibers for them. Therefore, to simplify the process, a mixture of the chelating extractant LIX 84 and the basic extractant TOA in heptane diluent simultaneously extracted Cu(II) and Cr(VI) with synergistic effect. A three-fiber set of microporous hollow fiber devices was then fabricated and used to remove, separate, concentrate, and recover Cr(VI) and Cu(II) simultaneously from synthetic wastewaters into separate basic and acidic stripping solutions, respectively.

The setup of the novel synergistic membrane-based extractor is shown in Figure 28a,b. In the hollow-fiber membrane device, LIX 84 in a diluent flowed through the bore of one set of fibers and D2EHPA in a diluent flowed through the bore of the other set of fibers as the aqueous feed containing Cu(II) and Zn(II) flowed in the shell side. Figure 28c shows that the copper extraction rate and the mass transfer coefficient were strong functions of the aqueous feed flow rate; however, the variation of copper concentration in the aqueous feed had little effect on the mass transfer coefficient. The aqueous boundary layer

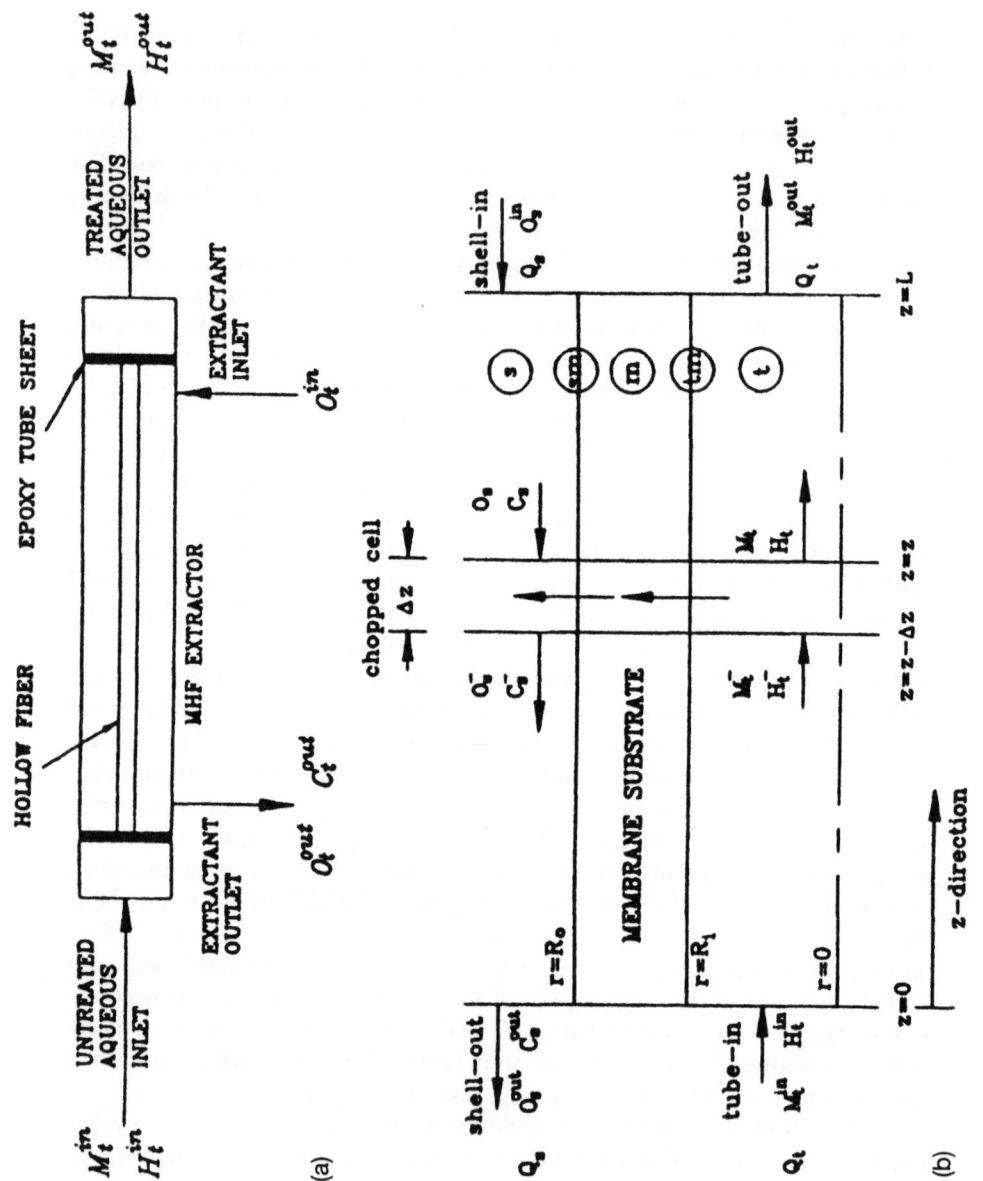

(a)

(b)

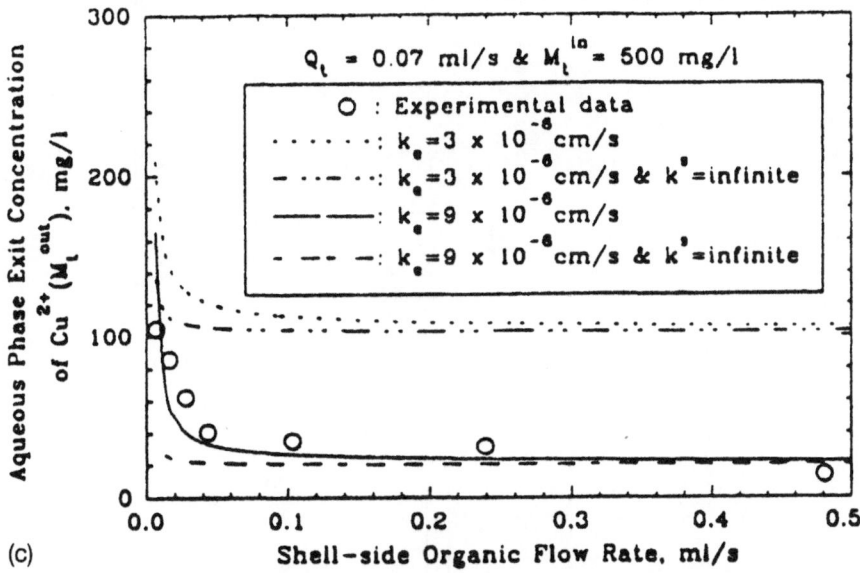

Figure 27 (a) Once-through flow configuration of a MHF module. (b) Schematic for modeling of copper extraction in a hollow-fiber module (a and b on previous page). (From Ref. 19.) (c) Exit concentration of the aqueous phase of copper from the hollow-fiber module as a function of the organic phase flow rate when 10% v/v LIX 84 in *n*-heptane was used. See text and Ref. 19 for more details. (From Ref. 19.)

resistance in the HFM extractor, the interfacial reaction resistance, and the organic boundary layer resistance, of which the first one plays the major role, controlled the copper extraction rate. On the other hand, an increase in flow rate of D2EPHA in kerosene significantly increased the zinc extraction rate and reduced zinc concentration in the aqueous outlet solution when the concentration of D2EPHA was lower than 0.60 M. This result implied that at lower D2EHPA concentrations the interfacial reaction should not be neglected. The increased zinc extraction rate and the mass transfer coefficient with an increase in the aqueous feed flow rate implied that the aqueous boundary layer resistance was an important contributor to the total mass transfer resistance. However, its effect was not as strong as that of the flow rate of the D2EHPA in kerosene.

The authors [143] studied the separation of mixed cations like Cu(II) and Zn(II) while the feed stream contained also $HCrO_4^-$. The aqueous phase was fed into the shell side of module 1 while two organic extracts were passed through the two individual fiber sets. Aqueous raffinate from this experiment was sent to module 2, where only one extractant (LIX 84 in *n*-heptane) was used to drastically reduce the Cu(II) content. In run 3, a synthetic solution of

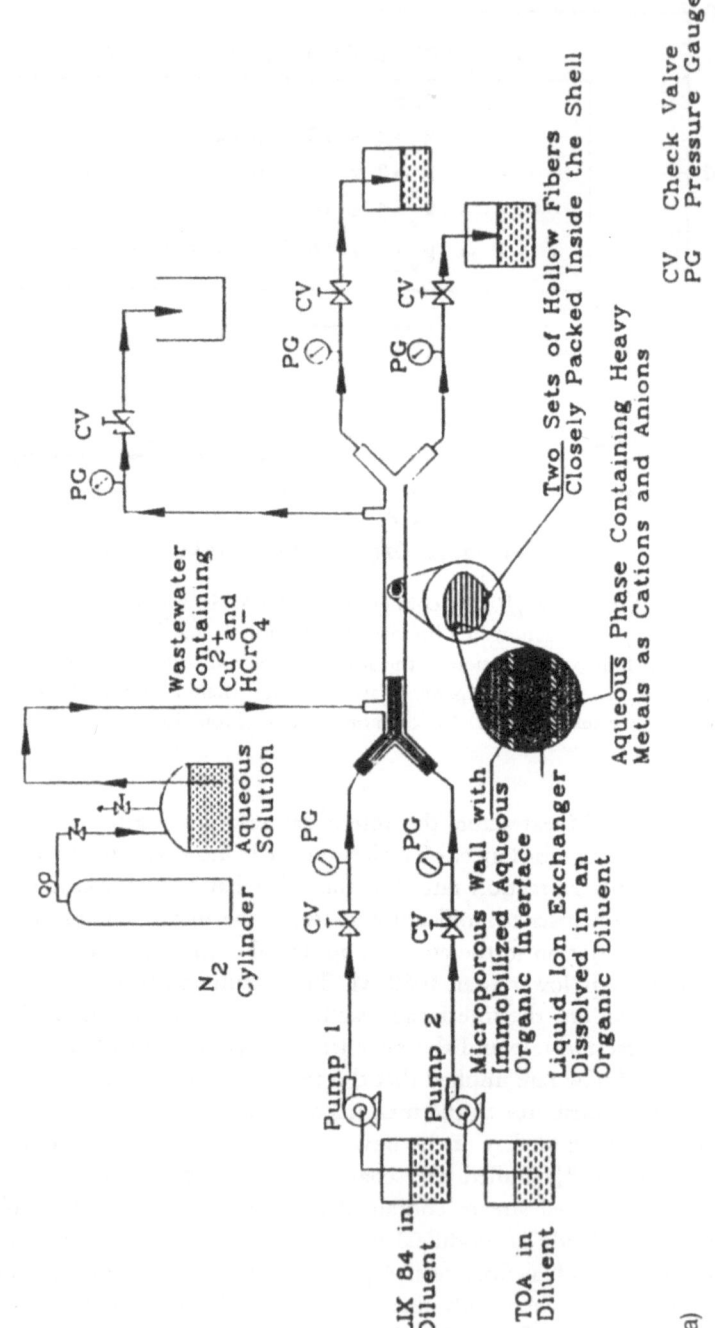

(a)

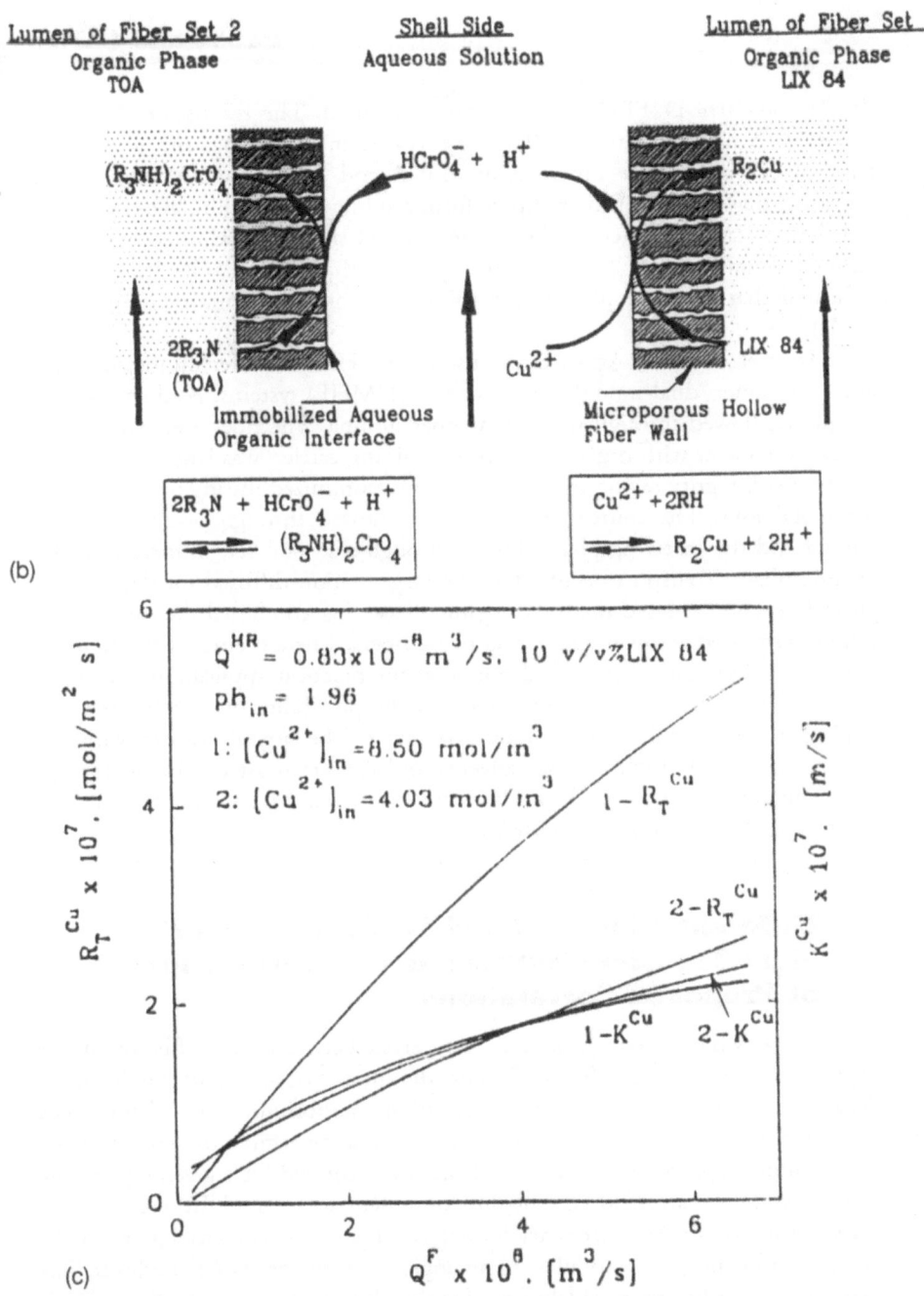

Figure 28 (a) Setup of novel synergistic membrane-based extractor for the removal of heavy metals from wastewater (a on previous page). (From Ref. 143.) (b) Schematic of the synergistic extraction Cu(II) and Cr(VI) in a module containing two sets of hollow fibers. (From Ref. 115.) (c) Effect of aqueous phase flow rate on the mass transfer rate of copper extraction in an HFM extractor. (From Ref. 143.)

Zn(II) and 20% D2EPHA in *n*-heptane was used. The results are shown in Table 21. A high extent of Cu(II) and Cr(VI) removal along with a very low removal of Zn(II) were obtained in runs 1 and 2. The results from run 2 suggested that the pH of the raffinate from module 1 was still in favor of Cu(II) rather than Zn(II) extraction. Therefore, the pH should be adjusted before the raffinate from module 1 is sent for Zn extraction. The results from run 3 indicated that a higher pH of feed solution favored a significant removal of Zn(II).

Izatt et al. [144] reported the use of the HFNDSX technique, under a different name: "dual-module hollow-fiber" (DMHF) systems. Feed and receiving phases flowed through the hollow-fiber lumina. The fiber area, which was always in contact with organic-solvent-containing carrier, was large. The loss of membrane integrity owing to solvent loss was minimized during permeation of the metal ions. The source phase generally flowed through one set of fiber lumina and the receiving phase flowed through a second set, immersed in the organic phase. Transport occurred as the target cation diffused through source phase fibers and entered the bulk organic phase that contained the carrier. The cation permeated across the receiving phase fibers, where it entered the aqueous receiving phase. This system had potential for practical applications owing to improved stability, but a disadvantage was the possibility of membrane pores fouling up as a result of continuous contact of the organic carrier with the membrane. Highly hydrophobic solvents and carriers must be chosen for optimal membrane function. A similar setup was used recently for the preconcentration of metal ions and assay [29].

A. Different Methods Used for Evaluation of Mass Transfer Coefficients and Verification of Proposed Correlations

The rate at which a component is transferred between two different phases depends on the mass transfer coefficient, the interfacial area, and the degree of departure of the component from its partition equilibrium. Evaluation of mass transfer coefficients is of importance because these determine the rate at which equilibrium is approached and control the time required for a given separation, and therefore the size and cost of the equipment to be used. To estimate the mass transfer coefficients necessary for engineering analysis and design, researchers resort to conceptual modules, to analogies, and to correlative methods. The two-film model has been widely used for decades, the Chilton–Colburn [145] analogy between momentum, heat, and mass transfer having been one of the most extensively and frequently applied, and a large variety of correlations for different geometries and hydrodynamic conditions have been proposed. Good

Table 21 Extraction and Separation of Cr(VI) and Cu(II) Via a Three-Fiber-Set HFCLM Device[a]

Organic membrane composition[b]	Feed flow rate (mL/min)	Run time (h)	pH_{out}	0.1 M NaOH flow rate (mL/min)	Cr(VI) in NaOH (ppm)	2 M H_2SO_4 flow rate (mL/min)	Cu(II) in H_2SO_4 (ppm)
Extractant 1	2.33	20	3.82	0.28	58.0	0.12	69.0
Extractant 2	1.43	24	3.47	0.36	150.0	0.10	228.0
Extractant 3	1.83	30	3.45	0.49	93.0	0.29	62.0

[a] Feeds: Cr(VI), 220 ppm; Cu(II), 1000 ppm; pH 4.19.
[b] Extractant 1: 5% v/v LIX 84 and 3.6% v/v TOA in kerosene. Extractant 2: 25% v/v TOA in kerosene. Extract 3: 50% v/v LIX 84 and 25% v/v TOA in kerosene.
Source: Ref. 143.

reviews can be found in the literature [49,146–148]. Various methods as presented in this section were adopted to evaluate mass transfer coefficients.

The results of metal concentration measurements allowed the calculation of the overall mass transfer coefficient, K_f, defined by

$$J = K_f \left(C_f - \frac{C_{e/s}}{H} \right) \tag{46}$$

where J is the solute flux, C_f is the concentration in the feed solution, $C_{e/s}$ is the concentration in the extract or strip solution, and H is the distribution ratio of Au(I). Thus, K_f is based on a feed-driving force. As derived by D'Elia, Cussler, et al. [149–151] for cocurrent flow, the key equation for the calculation of K_f is

$$\frac{1/Q_f + 1/Q_{e/s}H}{1/V_A + 1/V_{e/s}H} \ln \left[\frac{C_{e/R}^0 H - C_f^0}{(C_{e/s}^0 - C_f^0) + (V_f/HV_{e/s})(C_f^0 - C_f)} \right]$$
$$= t \left\{ 1 - \exp \left[\frac{-4K_f V_m}{d(1/Q_f + 1/Q_{e/s}H)} \right] \right\} \tag{47}$$

Similarly, for countercurrent the equation is

$$\left\{ \frac{[1/Q_f - 1/Q_{e/s} \exp[-(4K_f V_m/d)(1/Q_f + 1/Q_{e/s})]]}{(1/V_w + 1/V_s H)} \right\}$$
$$\ln \left[\frac{C_{e/s}H - C_f^0}{(C_{e/s}^0/H - C_f^0) + (V_f/HV_{e/s})(C_f^0 - C_f)} \right]$$
$$= t \left\{ 1 - \exp \left[\frac{-4K_A V_m}{d(1/Q_f - 1/Q_{e/s}H)} \right] \right\} \tag{48}$$

In these equations Q_f and $Q_{e/s}$ are the feed and extract/strip flows, V_f and $V_{e/s}$ are the feed and extract/strip volumes, C_f^0 and $C_{e/s}^0$ are the concentrations of the solute in the feed and in the extract/strip solutions at time zero, C_f is the concentration of the solute in the feed at time t, V_m is the volume of all the hollow fibers, and d is the diameter of one fiber.

In mass transfer, the local mass transfer coefficient (as the Sherwood number) has generally been correlated with the Reynolds and Schmidt numbers according to a power law. To estimate mass transfer coefficients, Wilson [152] proposed an analogy based on a method of calculating film coefficients and dealing with vapors that has been universally used in the last decades [153,154]. It consists of plotting the overall resistance to mass transfer (inverse of the mass transfer coefficient) as a function of the inverse of the fluid velocity, which enables the determination of the addition of other resistances, as well as their dependence on the flux velocity.

In nondispersive solvent extraction systems, overall mass transfer coefficients are a weighted average of the individual mass transfer coefficients in the aqueous feed phase, across the membrane, and in the organic phase. Often, one of the three individual coefficients is much smaller than the others, its inverse dominating the overall mass transfer coefficient. Recent work assumes that the main resistance to the solute transport lies in the membrane. Among those who deal with ionic species in the extraction step, it is also widely assumed that chemical reactions are fast enough to be considered to be instantaneous; then the reacting species are present in equilibrium concentrations at the interface everywhere.

When a resistance-in-series model is assumed, the reciprocal of the overall mass transfer coefficient (the total resistance to mass transfer) can be described as the sum of the mass transfer resistances inside the fiber (feed phase), across the fiber wall (membrane resistance), and outside the fiber (organic phase). Therefore, the expression for overall mass transfer coefficient K_M can be written as follows [49,54]:

$$\frac{1}{K_M} = \frac{1}{k_i} + \frac{1}{k_f} + \frac{d_i/d_{lm}}{k_m H} + \frac{d_i/D_h}{k_s H} \tag{49}$$

The first term on the right-hand side of Eq. (49) represents the resistance due to interfacial reaction, the second term indicates the mass transfer resistance in the aqueous phase, the third term is the membrane resistance, and the fourth term is the resistance of the organic extracting solvent. The overall mass transfer coefficient can be calculated from the individual transfer coefficients k_i, k_f, k_m, and k_s. The following expression takes into account the empirical relations for k_f and k_s, which depend on the hydrodynamics of the feed and organic phases, respectively:

$$\frac{1}{K_M} = \frac{1}{k_i} + (D_t/d_i)^{-1} \left(\frac{d_i^2 \nu_t}{LD_t}\right)^{-1/3} + \frac{(d_i/d_{lm})\tau t_m}{k_m D_m \varepsilon}$$
$$+ \frac{d_i/D_h}{H\beta L(\nu_s D_h/\eta_s)^{0/6}} \left(\frac{D_s}{\eta_s}\right)^{-0/33} \left(\frac{D_s}{D_h}\right) \tag{50}$$

Yang and Cussler (48a) presented the relationship in terms of flux and metal concentration for establishing relevant equations, similar to Eq. (51), for Co and Ni extraction. They developed the following relation for gold cyanide extraction with LIX 79 (R_{org}) in alkaline cyanide media at pH 10.5:

$$\frac{1}{k_i} = \frac{1}{k_e[H^+][R_{org}]} \tag{51}$$

and setting $H = K_{ex}[H^+][R_{org}]$, hence $[H^+][R_{org}] = H/K_{ex}$, so that upon suitable substitution in Eq. (51), one would have $1/k_i = 1/(k_e[H/K_{ex}])$.

In Eqs. (49) and (50), k_i is the effective rate of interfacial reaction at the surface; k_f, k_m, and k_s are the mass transfer coefficients in the aqueous feed, membrane, and the organic solvent, respectively; and d_i, D_h, and d_{lm} are the internal, external, and logarithmic mean fiber diameters, respectively. The value of k_i can be determined by methods published in the literature [19,48,143]. As described in earlier work, correlations for mass transfer in hollow fibers or small tubes were established for the tube side and the shell side by Rasad and Sirkar [155] and Dahuron and Cussler [150], respectively. For the tube-side mass transfer coefficient, the following correlation was given:

$$Sh = \frac{k_f d_i}{D_t} = 1.64 \left(\frac{D_t}{d_i}\right)\left(\frac{d_i^2 v_t}{L D_t}\right)^{1/3} \tag{52}$$

In general, this correlation predicts mass transfer coefficients with reasonable accuracy for $N_{Gz} > 4$ but overestimates them for $N_{Gz} < 4$ [156]. In the recent studies by Kumar et al. [78,79], N_{Gz} ranged from 5.5 to 8.2. For the shell-side mass transfer coefficient, the following correlation was given [19,157,158]:

$$Sh = \frac{k_s D_h}{D_s} = \beta\,\frac{D_h(1 - \phi)}{L}\,Re^{0.6}Sc^{0.33} \tag{53}$$

where D_h is the hydraulic diameter, D_t is the diffusion coefficient of the solute in the tube side, d_i is the inner fiber diameter, L is the fiber length, D_s is the diffusion coefficient of the solute in the shell side, v_t and v_s are the velocities of the liquid inside the fiber and shell side, respectively, and ϕ is the packing fraction. The Reynolds number Re is $v_s D_h/\eta_s$, the Schmidt number Sc is η_s/D_s, and $\beta = 5.85$ for hydrophobic membranes and $0 < Re < 500$ and $0.04 < \phi < 0.4$. In the recent studies [78,79] $3.0 < Re < 11.0$ and $\phi = 0.35$. The membrane mass transfer coefficient can be determined from $k_m = D_m \varepsilon/t_m \tau$ [159,160].

Membrane characteristics such as thickness, porosity, tortuosity, and pore size are generally given by the membrane manufacturer/supplier (Table 4). The membrane tortuosity was also determined by the Wakao–Smith relation, expressed as the inverse of the membrane porosity that almost matches with the value suggested by Celgard GmbH [161].

Kim and Stroeve [162,163] considered the steady state for the fluid phase containing the solute initially and analyzed theoretically the mass transfer of a solute, describing the velocity and concentration profiles along a hollow fiber by means of the continuity mass conservation equation and associated boundary conditions for the solute of the inner fluid.

This chapter discusses two approaches for evaluating mass transfer coefficients. The first [17,164] was explained in Section 5, on modeling. The second is based on macroscopic mass balances of the metal ions in the fluid phases in HF modules and in homogenized stirred tanks, dealing with kinetic analysis of the non-steady state system [165]. A widely applicable methodology was used for the kinetic analysis. The reader is referred to the original publication [165] for details.

The resulting set of first-order coupled differential equations was integrated by means of a Runge–Kutta method, and the mass transfer parameters were obtained from a comparison of the simulated results and the experimental data, using the standard deviation as the criterion in the optimization of the parameters. The kinetic results demonstrating the evolution of chromium(VI) concentration with time in the feed and stripping tanks are shown in Figure 29, where concentrations are presented as dimensionless variables.

For liquid–liquid extraction systems, where the chemical equilibrium is defined by means of a constant distribution coefficient, the experimental data have been usually fitted to a logarithmic expression [166,167].

The dimensionless Cr(VI) concentration in the organic solution at equilibrium data fit the expression $R = H_r Y$, and a constant value of H_r was used in evaluating mass transfer coefficients. After resolution of the set of differential equations, K_R was considered to be the optimization parameter in the comparison between the experimental and simulated curves. The optimization was

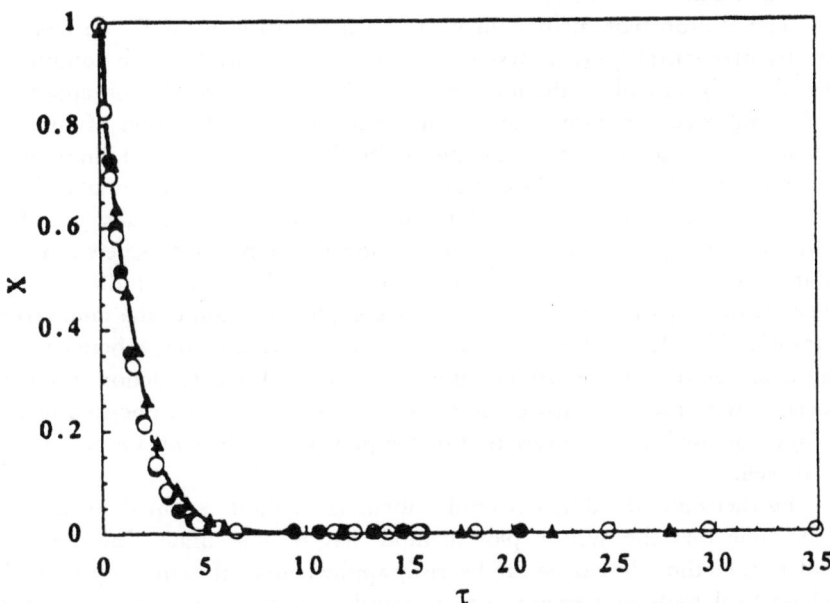

Figure 29 Experimental dimensionless Cr(VI) concentration, X, in the feed aqueous phases vs dimensionless time τ, for each cycle of experiments. (From Ref. 165.)

done on the basis of the minimal weighted standard deviation for the dimensionless concentration of Cr(VI) in the BEX solution, similar to the earlier method published by Alonso et al. [17].

The optimal values of the overall mass transfer coefficients ranged between 2.84×10^{-4} and 1.12×10^{-4} cm/s for extraction and 1.90×10^{-6} to 2.68×10^{-6} cm/s for back-extraction. Liquid–liquid extraction systems that incorporate a carrier in the formulation of the organic phase require nonlinear expressions for the description of the chemical equilibrium at conditions of high loading capacities of the organic carrier. The mass transfer resistance of the membrane in the extraction process is a more complicated expression than that generally given ($1/k_m H$). Kinetic analysis and determination of the mass transfer parameters would require integration of the set of coupled differential equations with the incorporation of nonlinear equilibrium expressions to separate the different contributions to the extraction mass transfer parameter.

As a further refinement of the methods for calculating mass transfer coefficients, Viegas et al. [168a] considered the use of a constant (case a) or a variable (case b) partition coefficient. The extraction experiments provided very regular and smooth solute concentration profiles. In Figure 30 the solute concentration evaluation in the tube-side phase is plotted as a function of time, the tube-side Reynolds number is varied, and the shell-side Reynolds number is kept constant ($Re_s = 1.31$).

The Wilson plot methodology can be used satisfactorily for developing mass transfer correlations in systems operating under steady state conditions when the only variable is the fluid velocity. Nevertheless the Wilson approach can be improved for systems in which a parameter is a function of another variable or for transient state operations. In these cases its major limitations stem from the fact that it involves a two-step calculation, two sequential fittings, leading to unnecessary loss of information and incoherent results. Another drawback of the procedure is that no variation in the parameters between and within enforcement can be taken into account. In the system studied by the present authors, partition coefficients are a complex function of the solute concentration. Therefore, when this relation is taken into account, a better modeling is achieved. For the system studied, a nearly 20-fold reduction in errors associated with the estimated parameters was observed when a one-step methodology was used, and the variation of the partition coefficient was accounted for, as well.

Furthermore, the different results obtained by these two methods lead us to conclude that the Wilson plot method may be inaccurate. The one-step calculation method developed can be easily applied, given the currently available mathematical tools that enable the analytical manipulation of equations and fittings with appropriate expressions. The empirical correlations that best describe the system studied are for the tube and shell side, respectively:

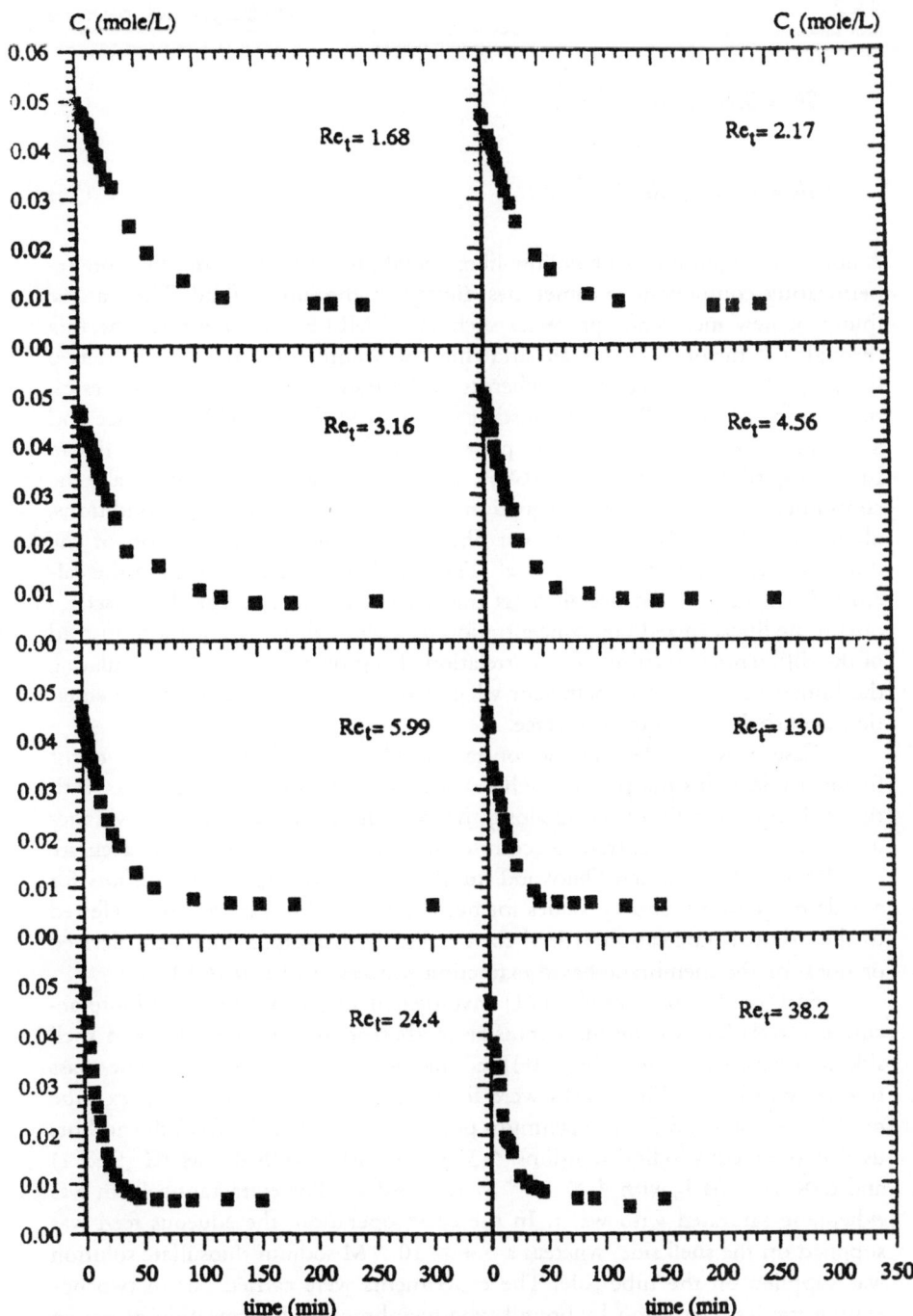

Figure 30 Solute concentration in the aqueous phase as a function of time at a constant $Re_s = 1.31$ and various Re_t. (From Ref. 168a.)

$$Sh_t = 0.20 Sc_t^{1/3} Re_t^{1.01} \left(\frac{d_i}{l}\right)^{1/3} \tag{54}$$

$$Sh_s = 8.71 Sc_s^{1/3} Re_s^{0.74} \left(\frac{d_h}{l}\right)^{1} \tag{55}$$

Among the applications of hollow-fiber membrane processes, the membrane-permeating component is sometimes diluted in the lumen fluid. This can be found in new membrane processes such as HFMLLE. In these cases the flux through the membrane is small, and thus the volumetric flow rate of the two fluids can be taken as constant, whereas in the traditional membrane processes, this cannot be done. When a constant external (wall and shell) resistance and a constant shell concentration are present, the boundary conditions are linear. In this regard, Qin and Cabral [168b], who attempted to solve numerically the continuity mass conservation equation with the linear boundary conditions, obtained and correlated the lumen Sherwood number as a function of the dimensionless parameters Sh_w and z'. The correlation greatly improved the calculated accuracy of the overall mass transfer coefficient and could be used to obtain the linear mixed-cup concentration by a set of algebraic equations instead of the differential equation. The correlation also provided a basis for calculating the lumen mass transfer coefficient when a more complicated boundary equation occurred at the lumen surface.

Case a, which demands a constant shell concentration, does not occur frequently in industrial practice, whereas case b deals with a constant Sh_w, with the shell concentration varying along the axial direction, according to whether operation is in the cocurrent or countercurrent mode. Empirical but accurate correlations of the lumen Sherwood number were provided by the authors for a wide range of Sh_w and G_z values for both cases a and b. The reader is referred to the original publication [168b] for details. Similar expressions were derived or used for the membrane-based extraction process [149,150,169,170].

Further, Takeuchi et al. [171] investigated means of obtaining more accurate correlations of the mass transfer coefficients on the tube side and shell side in a single oil-containing MHF contactor, as well as the membrane mass transfer coefficient. The MHFs were four PTFE hollow fibers (GoreTex tube of 50% porosity and 2 μm maximum pore diameter). For the feed, the authors used two aqueous iodine solutions: 5.3×10^{-3} M I_2 with 0.1 M KI (feed 1) and 6×10^{-3} M I_2 with 4×10^{-3} M KI (feed 2). The extractant solvent was n-heptane saturated with water. In the SLM operation, the aqueous feed was supplied on the shell side, whereas a 5.4×10^{-3} M sodium thiosulfate solution was supplied on the tube side. The experiments were carried out in two operating modes: extraction by bound-type membrane and permeation across an SLM. The tube-side mass transfer coefficients, which were comparable to those of the Levêque solution and to earlier correlations for laminar flow inside tubes, were examined in terms of the dimensionless Sh, Re, and Sc numbers in general correlations. The values of Sh and Re on the tube side were calculated based on the inner diameter of the MHF, d_i, as the characteristic length. The results are plotted as Sh vs Re in Figure 31a. It was found that the Re dependence of Sh is 1/3 at $Re < 700$. The solid lines represent Eq. (57) given shortly, but the experimental results for $Re > 700$ deviate remarkably from each line. This suggested that a transition from laminar to turbulent flow arose by $Re \sim 700-$

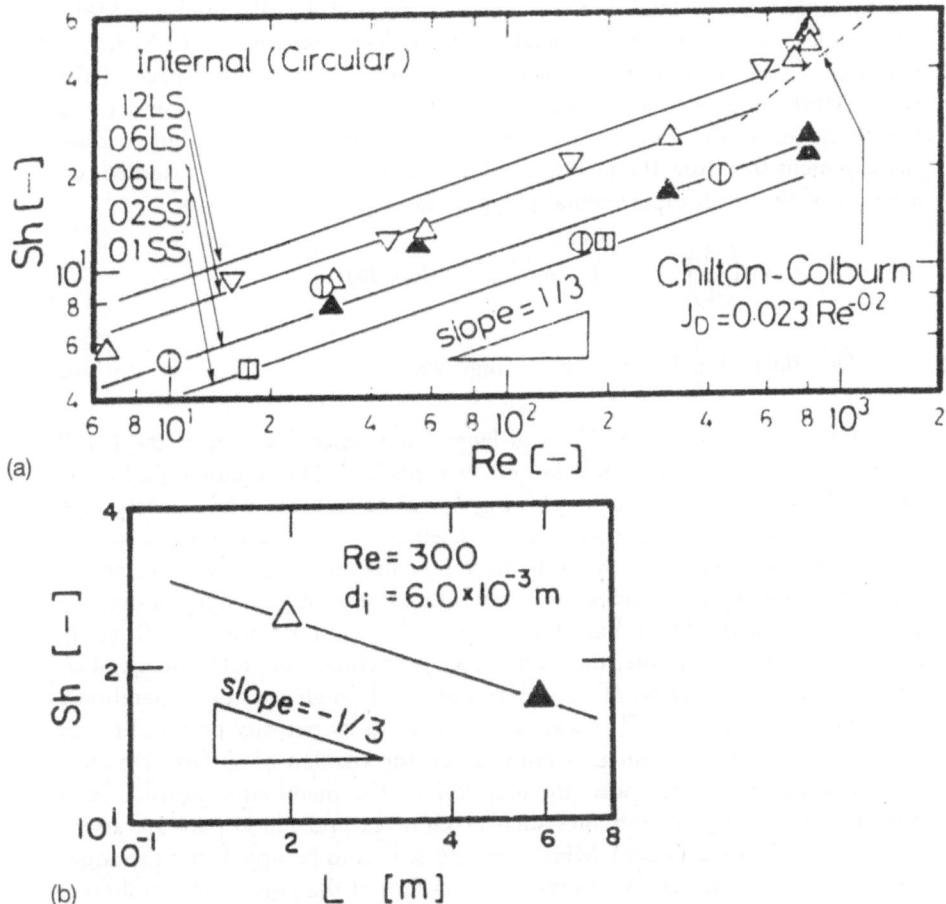

(a)

(b)

Figure 31   (a) Effect of Reynolds number on Sherwood number for tube-side flow. (From Ref. 171.) (b) Hollow-fiber length dependence of the tube-side mass transfer coefficient. (From Ref. 171.)

800 for the aqueous flow inside the oil-containing MHF, unlike the case of fluid flow inside a solid tube. Thus, a comparison was made between the present results and the Chilton–Colburn [145] equation, as shown by the dashed line in Figure 31a.

$$\frac{k_t d_i}{D} = 1.4 \left(\frac{d_i}{L}\right)^{1/3} \left(\frac{d_i u_t}{\nu}\right)^{1/3} \left(\frac{\nu}{D}\right)^{1/3} \tag{56}$$

The film mass transfer coefficient for the aqueous flow on the shell side was obtained in the contacting mode, where the direction of the solute transfer was from the annular fluid into the external surface of the oil-containing MHF. The results were examined in general dimensionless correlations. The hydraulic diameter d_e was used as the characteristic length in the annular space of the single MHF module. For laminar flow in the shell side, the 1/3 power of Re was found to be the same as that on the tube side, while for turbulent flow, the exponent 0.8 from the analogy of Chilton and Colburn was assumed, on account of the small experimental range of Re covered in this work.

$$Sh = 0.85 \left(\frac{d_s}{d_o}\right)^{0.45} \left(\frac{d_e}{L}\right)^{1/4} Re^{1/3} Sc^{1/3} \quad \text{(low } Re\text{)} \tag{57}$$

$$Sh = 0.017 \left(\frac{d_s}{d_o}\right)^{0.57} Re^{0.8} Sc^{1/3} \quad \text{(high } Re\text{)} \tag{58}$$

Figure 31b shows the effects of length and inner diameter of the MHF on Sh, respectively, giving a relation of $Sh \propto (d_i/L)^{1/3}$. The results of k_m for the various MHFs are presented by plotting $k_m/(D\varepsilon/\delta)$ against d_i/d_o, where $D\varepsilon/\delta$ is a theoretical membrane mass transfer coefficient across a homogeneous oil layer of thickness δ; hence the ordinate represents the reciprocal of the tortuosity of the microporous membrane, $1/\tau$. The value of k_m depends not only on the geometry of the MHF but also its mode. If the MHFs used in this study had uniform pore structure, the value of k_m or τ should be independent of d_i/d_o as well as the direction of the solute transfer through the liquid membrane layer. Thus, the GoreTex fiber was not uniform in its porosity profile through the wall thickness. According to Pons [172], the fraction of surface free area, ξ, of microporous sheets was not identical to the membrane porosity. Also microporous (PTFE) membranes such as GoreTex and Fluoropore are asymmetrical [173] in the present MHF contactors. It is to be noted that the aqueous–organic interface is not always at the surface of the polymeric membrane. The thickness of the stagnant organic sublayer on the surface varies with the hydrophobicity of the polymeric support and the operating conditions, such as the pressure difference between the tube side and the shell side in the MHF module. This leads to uncertainty in the path length of the solute diffusion across the oil-containing membrane. Also, the pore entrance, and exit effects become more significant with an increase in thickness of the oil sublayer, as suggested by Malone and Anderson [174a]. When flowing on the tube side or the shell side, the oil could be partly interchanged with the membrane phase. This caused an apparent enhancement of k_m for the bound membrane. In conclusion, the membrane coefficient depended on the direction of solute transfer as well as the operating mode, tending to drop with an increase in the ratio of inner to outer diameter of the MHF. Thus, great care must be taken in the application of the tortuosity factor for the interpretation of the mass transfer across the wall of oil-containing MHF membranes.

B. Deviations in the Evaluation of Mass Transfer Coefficients

The increase in extractant flow rate was expected to increase extraction efficiency, thus lowering outlet raffinate metal ion concentrations. However, the outlet concentrations did not change a great deal. This could be attributed to the low packing density of the fibers in the working module, which resulted in channeling of the aqueous phase flow. That is, a significant amount of the aqueous phase and the metal ions passed through the module without coming in contact with the fibers carrying the organic extractant. It was suggested that flow bypassing and improper distribution on the shell side might be prevented by carrying out studies with more highly packed modules, or with cross-flow modules with or without baffles.

Mass transfer studies of a commercial-scale nonbaffled membrane extractor module indicated performance that was lower than expected from results obtained in smaller scale modules [174]. The toluene–acetone–water system was chosen for this study because of the availability of mass transfer data for commercial extraction contactors. Water was used to extract acetone from toluene with an initial feed composition of approximately 5 wt% acetone. The comparison of theoretical mass transfer with 100% utilization of the fiber surface area is shown in Figure 32a. This deviation was ascribed to a significant portion of fibers being bypassed by the shell-side fluid, with the result that only a fraction of the total fiber surface area was utilized. The hydraulics of membrane extraction was investigated to verify that bypassing had occurred. The results from the dye tracer studies of a module with 7500 fibers indicated that the actual residence time was three to four times smaller than the ideal residence time, giving evidence that significant bypassing had occurred in the commercial-scale membrane module. A typical residence time distribution is shown in Figure 32b. The ideal residence time was calculated from

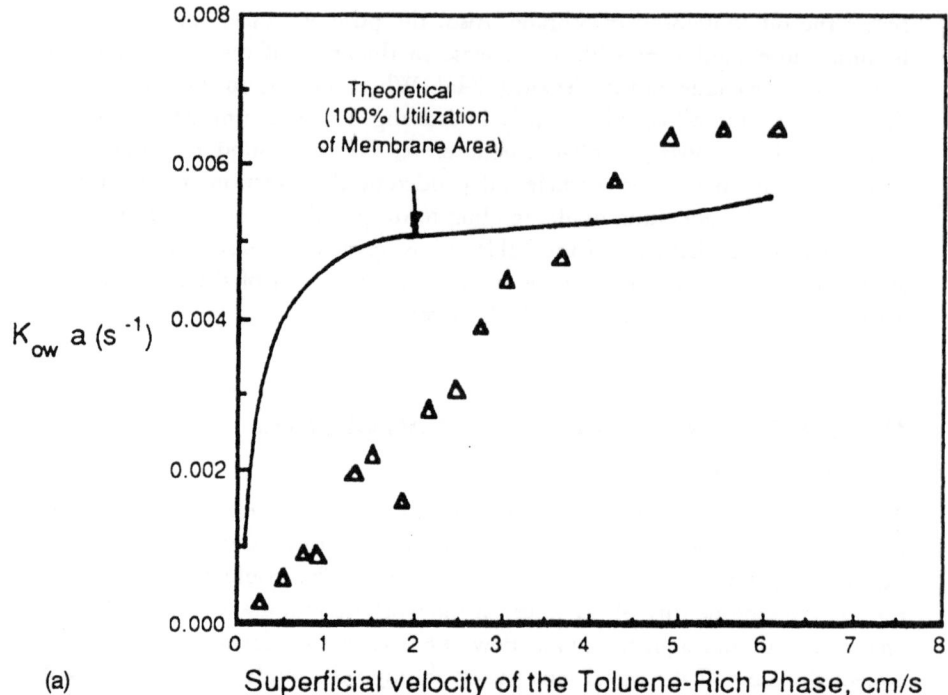

(a)   Superficial velocity of the Toluene-Rich Phase, cm/s

Figure 32 (a) Mass transfer efficiency of the membrane extractor for the toluene (shell)–acetone–water (tube) system: aqueous phase flow rate, 0.22 cm/s; direction of mass transfer, toluene → water; number of fibers, 7500. (From Ref. 174b.) (b) Comparison of experimental residence time with that predicted if all fiber area were utilized (conditions as in a). (From Ref. 174b.) (c) Typical residence time distribution on the shell-side of a commercial size membrane extractor studied for this work. (From Ref. 174b.)

$$t_{res} = \frac{\text{shell-side void volume}}{\text{shell-side flow rate}} \tag{59}$$

The bypassing parameter α was observed to decrease with shell-side flow rate to an asymptotic value of zero. A model for predicting the fraction of bypassed flow was developed, assuming the mass transfer model of Prasad and Sirkar [157,160] to be valid, and was determined from the experimental data of this work. The empirical correlation for predicting this parameter was $\alpha = \exp(-CU_s)$, where C and U_s are an empirical correlation constant and the shell-side superficial velocity, respectively.

Some studies assumed ordered packing [175] and with one exception [176] did not address the randomness of the hollow-fiber membrane module. Chen and Hlavacek [176] discussed the use of Voronoi tessellations and the

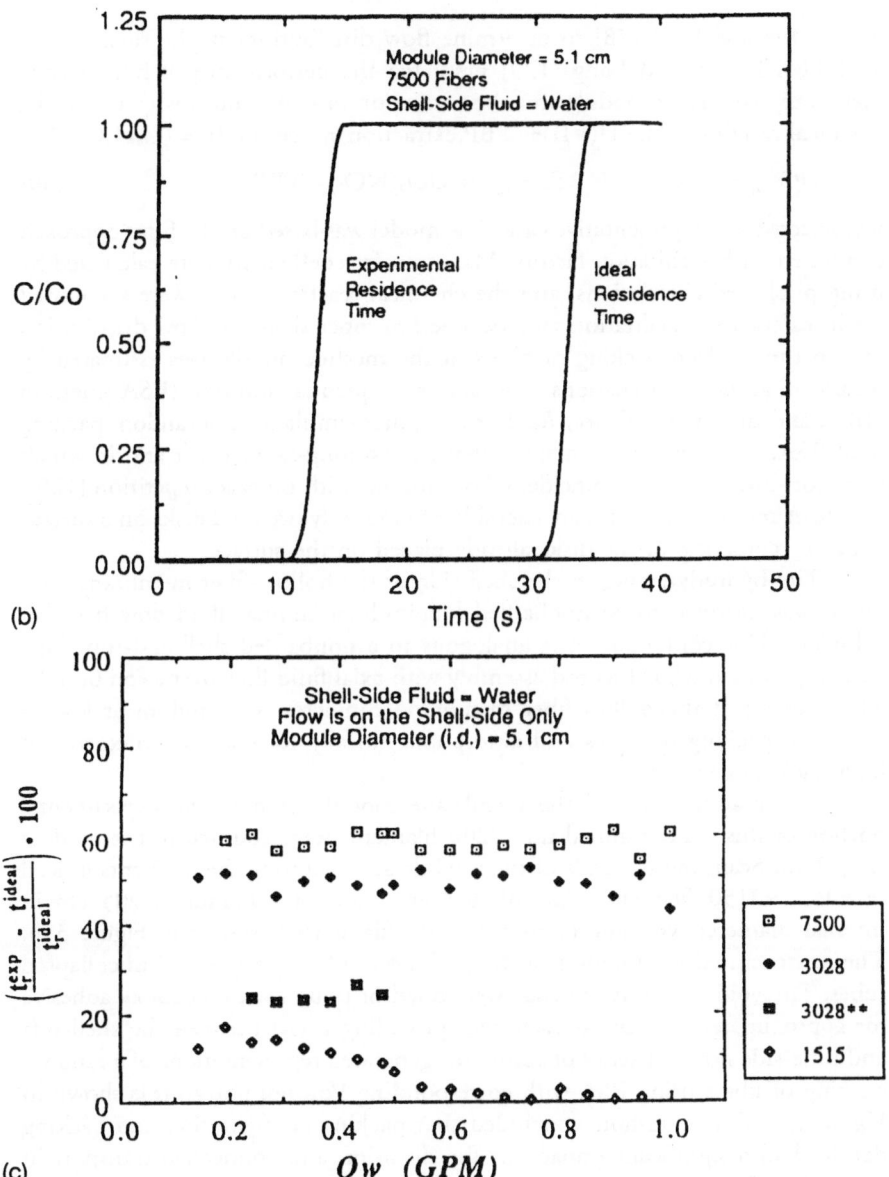

(b)

(c)

Rehme method [177,178] to determine flow distributions in the shell side of an HFM. Rogers and Long [179] predicted the performance of hollow-fiber membrane contactor modules for the extraction of compounds with interfacial chemical reactions. The U(VI) + TBP extraction system [TBP = $(C_4H_9O)_3PO$]

$$UO_{2\,aq}^{2+} + 2NO_{3\,aq}^{-} + 2TBP_{org} \Leftrightarrow UO_2(NO_3)_2 \cdot TBP_{2\,org} \qquad (60)$$

was selected as a representative case. The model was based on the films approach combined with facilitation factors. Mass transfer coefficients were calculated by using published correlations, and the chemical reaction effects were accounted for in calculating facilitation factors. The improper shell-side flow distribution due to the random packing of fibers in the module bundle was estimated by means of Voronoi tessellations, the random sequential addition (RSA) method [181], and the friction factor, Re. The computer simulations of random packing used RSA, a configuration in space that has no long-range order and in which any short-range order is coincidental or random with no exact repetition [180]. The RSA method generates an assembly of randomly oriented disks on a surface with no reordering of the disks already placed on the surface.

The hydrodynamics in the shell side of the hollow-fiber membrane contactor was assumed to be nonbaffled longitudinal laminar fluid flow between cylinders. This configuration is analogous to a nonbaffled shell-and-tube heat exchanger or a nuclear fuel rod assembly with axial fluid flow over a rod bundle. The packing of the hollow-fiber membrane contactor was random at low to moderate packing densities and amorphous (locally to globally structured) at high packing densities.

The manufacturer of the membrane module used in the experimental portion of this study claimed that 2200 filaments were counted in the module (Fig. 33a). Scanning the ends of the module and counting the filaments indeed revealed ~2150 filaments. An enlarged end view of a module 1.905 cm in internal diameter with approximately 2200 filaments is shown in Figure 33b. The unstructured randomness of the packing can be noted as well as collapsed tubes. The void spaces at the ends were filled or potted with an epoxy adhesive for approximately 1.5 cm at each end, providing a seal between the shell-side and tube-side flows. A series of computer-generated representations of a random packing of fibers using RSA with corresponding Voronoi polygons is shown in Figure 33c–f. The authors concluded that packing configuration and packing density had a significant impact on the diffusion and connective transport in the hollow-fiber membrane.

Cowder and Cussler [182] reported the effects of polydisperse hollow fibers on the performance of hollow-fiber modules. They proposed a new theoretical approach to the estimation of the average mass transfer coefficient in a hollow-fiber module from the average properties of the hollow fibers. The analysis was performed assuming that each solute diffused independently and that

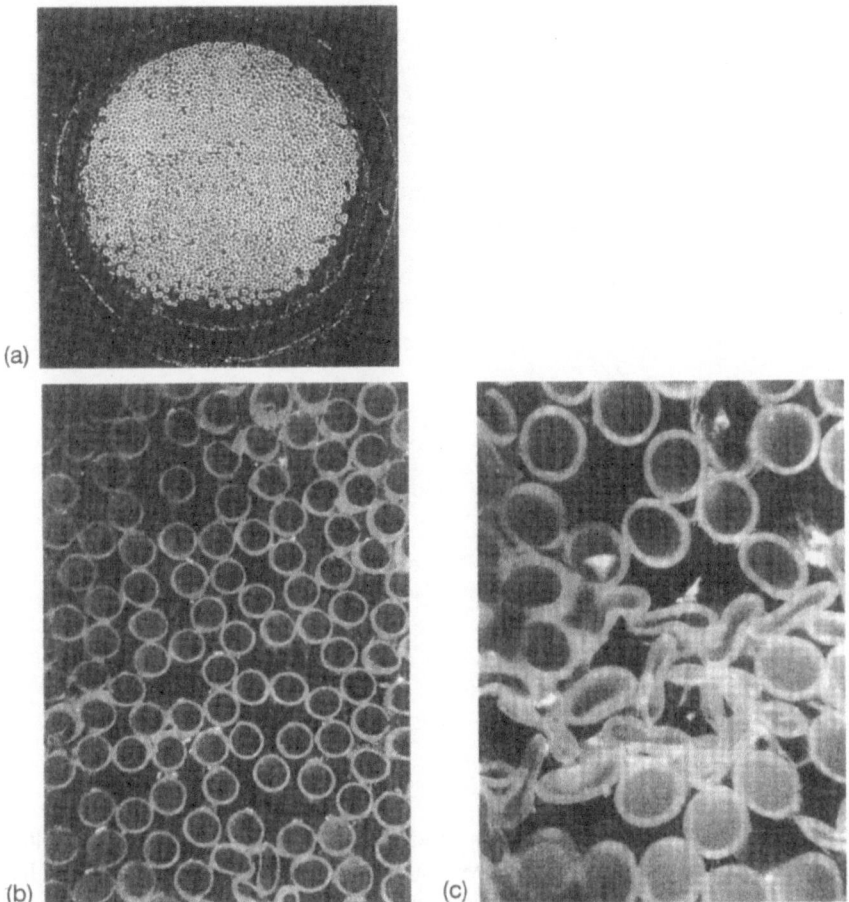

Figure 33 (a) Hollow-fiber module showing the tube-side opening having 1.905 cm i.d. with approximately 2200 fibers of 1.50 μm radii. (From Ref. 179.) (b) Sevenfold magnification of cross-sectional view of membrane module. Note the randomness, void spaces, and collapsed tubes. (From Ref. 179.) (c) End view of HF module, 7× magnification module tubes consisting of inconsistent structure and collapsed tubes. (From Ref. 179.) (d–g) Computer-generated cross sections of HFM with corresponding Voronoi polygons: (d) 500, (e) 1000, (f) 2200, and (g) 3380 filaments; 2.54 cm shell i.d., 1.50 μm fiber radius. (From Ref. 179.)

solute transport was unaffected by mass transfer resistances on the permeate side of the membrane. For polydisperse fiber diameters, polydispersity reduced the average mass transfer coefficient by roughly 40% for radii having a standard deviation of ca. 20%; for polydisperse fiber wall thicknesses, the average coefficient could be increased or decreased, but the changes were smaller. The fol-

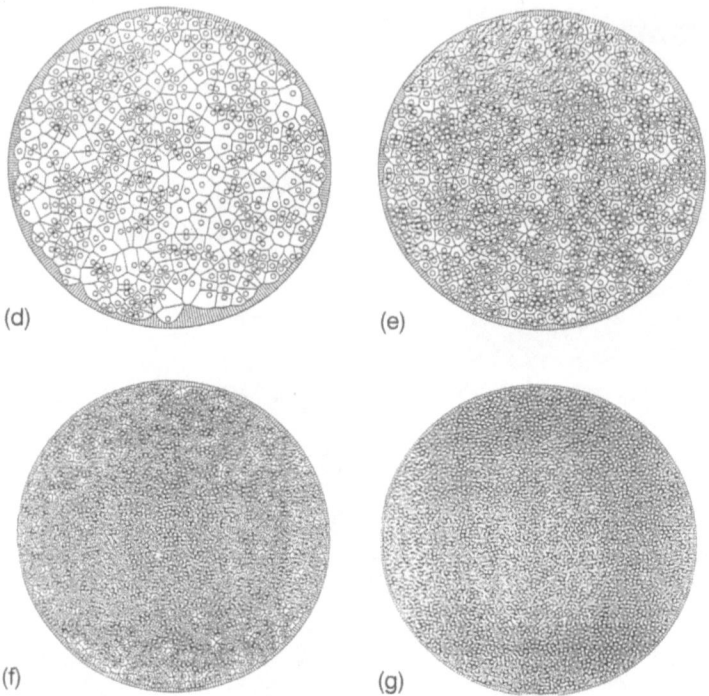

(d)   (e)

(f)   (g)

Figure 33   Continued

lowing equation was proposed for a module with n different batches of hollow fibers. The radii of a given batch were all equal, but the different batches had different radii. Thus the distribution was not nearly Gaussian, but rather a group of Dirac delta functions. In this case, the average overall mass transfer coefficient $\langle K \rangle$ could be calculated from

$$\frac{\langle K \rangle}{P/\delta} = [\Sigma \phi \sigma_i^4][\Sigma \phi \sigma_i^2]1 + \ln \left(\frac{c_{10}}{c_{i1\langle R \rangle}} \right) \frac{\ln[\Sigma \phi \sigma_i^4]}{\Sigma \phi \sigma_i^4 \exp[-\ln(c_{10}/c_{i1\langle R \rangle})(\sigma_i^{-3} - 1)]}$$

(61)

where $\sigma_i = R_i/\langle R \rangle$ denoted the dimensionless radius of batch i, R the radius of a fiber, and $\langle R \rangle$ the average fiber radius; $c_{i1\langle R \rangle}$ was the outlet concentration of a fiber of average radius, c_{10} the inlet concentration, ϕ the fraction of mono-disperse fibers in batch i, P the permeability of the wall, and δ the wall thickness of a fiber.

The deviation observed between the theoretical and the experimental val-ues of the mass transfer coefficient could have had several causes, as explained by a number of researchers [54,155,157,171,175,179,183–185]. In a hollow-fiber contactor, the tube-side flow regime was likely to be laminar owing to the very small lumen diameter. Otherwise, the pressure drop along the fiber would be exceedingly high. In any such correlation, the effects of any improper distribution of flow on the tube side resulting from particular design of inlet and outlet headers would be hidden [186] and the effects of fiber inlet crimping or plugging at the tube sheet would be unknown. On the other hand, hollow-fiber modules usually exhibited channeling, and stagnant regions on the shell side occurred because of nonuniform distribution of the fibers and their elongation due to use of organic solvents [155,187,188]. The hydrodynamics of the organic phase circulating in the shell side would be affected by these deviations from ideal flow, causing an incorrect evaluation of the mass transfer coefficient. The discrepancy between theoretical and experimental values could also be due to the inaccurate diffusion coefficient of the solute in the organic phase [183], directly linked to Eqs. (52) and (53).

This theoretical/experimental discrepancy was reported to be more pronounced in stripping experiments, stripping being more affected by the organic flow rate in the shell side. This could be a result of irregular shell-side flow, caused by stagnant zones, preferential pathways, and deficient mixing due to a nonuniform distribution of fibers, and of their possible deformation by action of the organic solvent [155,185,186]. These effects were important mainly when commercial modules were used, with nonuniformly spaced fibers. Improper shell-side flow distribution due to random packing of fibers in the module bundle probably also contributed to the inaccuracy of shell-side mass transfer coefficients, as observed by Rogers and Long [179]. In a single-fiber module contactor that did not present such improper distribution, Takeuchi et al. [171] obtained a shell-side Reynolds number exponent of 1/3. Therefore, any correlation was a result of the specific system and the conditions used. This nonuniform fiber spacing, affecting the mass transfer coefficients mainly when low flows were used [54,186], showed that for hand-made modules, where the fibers were carefully spaced, the correlations were similar to the theoretical predictions, whereas those obtained for commercial modules diverged greatly.

C. Performance Evaluation and Long-Term Stability of HF Contactors

To accomplish high mass transfer in a small space, Yeh and Hsu [189a] reported that it was desirable to utilize multiple passes of fluids as shown in Figure 34. Further, with a specified total number of hollow fibers, increasing the fiber passes as well as decreasing the total cross-sectional area of flowing channels inside the hollow fibers should increase the fluid velocity, thereby leading to improved mass transfer. Hence, Yeh and Hsu investigated the effect of fiber

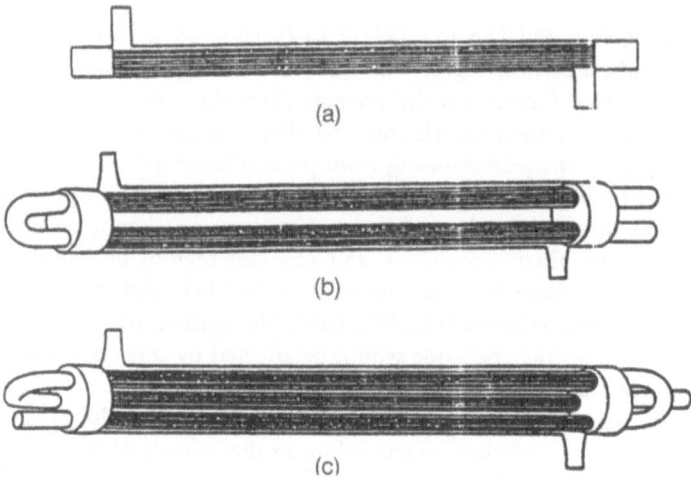

Figure 34 Microporous hollow-fiber (MHF) modules: (a) one-fiber-pass system; (b) two-fiber-pass system, and (c) three-fiber-pass system. (From Ref. 189.)

passes on solvent extraction performance in a module with multiple fiber passes and one shell pass, with fixed total number of hollow fibers. For a specific system, the mass transfer coefficient, k_a in the fiber side was a function of the fluid velocity u_a in that side only, increasing with u_a. For a device with m fiber passes and one shell-side pass with total fiber number N, the fluid velocities in the fiber side and the shell side and the equivalent diameter d_e of the module chamber were, respectively:

$$u_a = \frac{Q_a}{(n/m)(\pi d_i^2/4)} \tag{62}$$

$$u_b = \frac{Q_b}{(\pi/4)(d_i^2 - Nd_o^2)} \tag{63}$$

$$d_e = \frac{d_s^2 - Nd_o^2}{d_s - Nd_o} \tag{64}$$

where Q is the flow rate and d_s is the shell diameter, and d_i and d_o are the inside and outside diameters of the fiber. Figure 35 shows that the overall mass transfer coefficient k_i and the total mass transfer rate W increased not only when Q_a increased but also when m did. A considerable improvement in performance was obtained when a multiple-fiber-pass module with N/m fibers in each pass was used, instead of a one-fiber-pass module with the same total number N of hollow fibers.

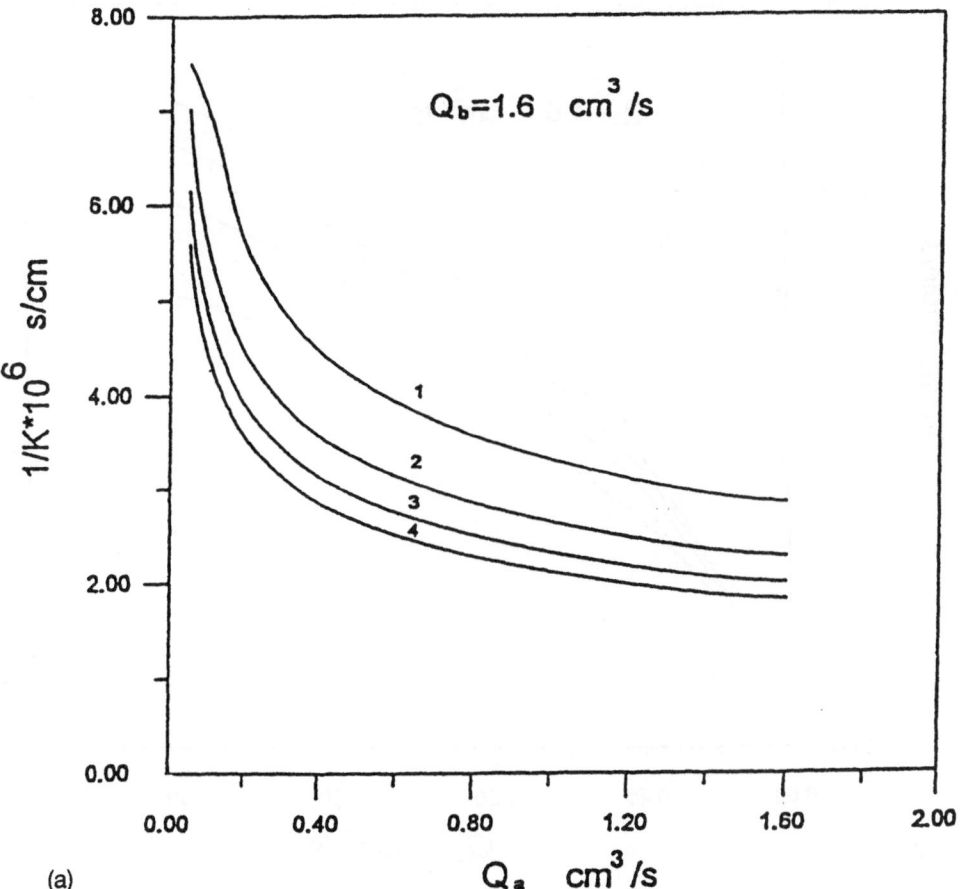

$Q_b = 1.6 \; cm^3/s$

(a)

$Q_a \; cm^3/s$

Figure 35 (a) Relation between mass transfer coefficient K and aqueous flow rate Q_a; numbers on curves designate number of passes. (From Ref. 189.) (b) Relation between the mass transfer rate W and Q_a, the numbers on the curves designating the number of passes. (From Ref. 189.)

Prasad and Sirkar [190] presented the results of a comprehensive study of multisolute extraction to establish long-term uses of hollow-fiber contactors. Using a variety of highly packed Celgard hydrophobic polypropylene hollow fibers in Liqui-Cel modules, they dealt with the recovery of organic compounds, such as 4-methylthiazole (MT) and 4-cyanothiazole (CNT). Recoveries exceeding 99% were obtained as functions of aqueous and organic flow rates, hollow-fiber diameters, and fiber packings for single or series-connected MHF devices. The overall mass transfer coefficient, height of transfer units (HTUs), and num-

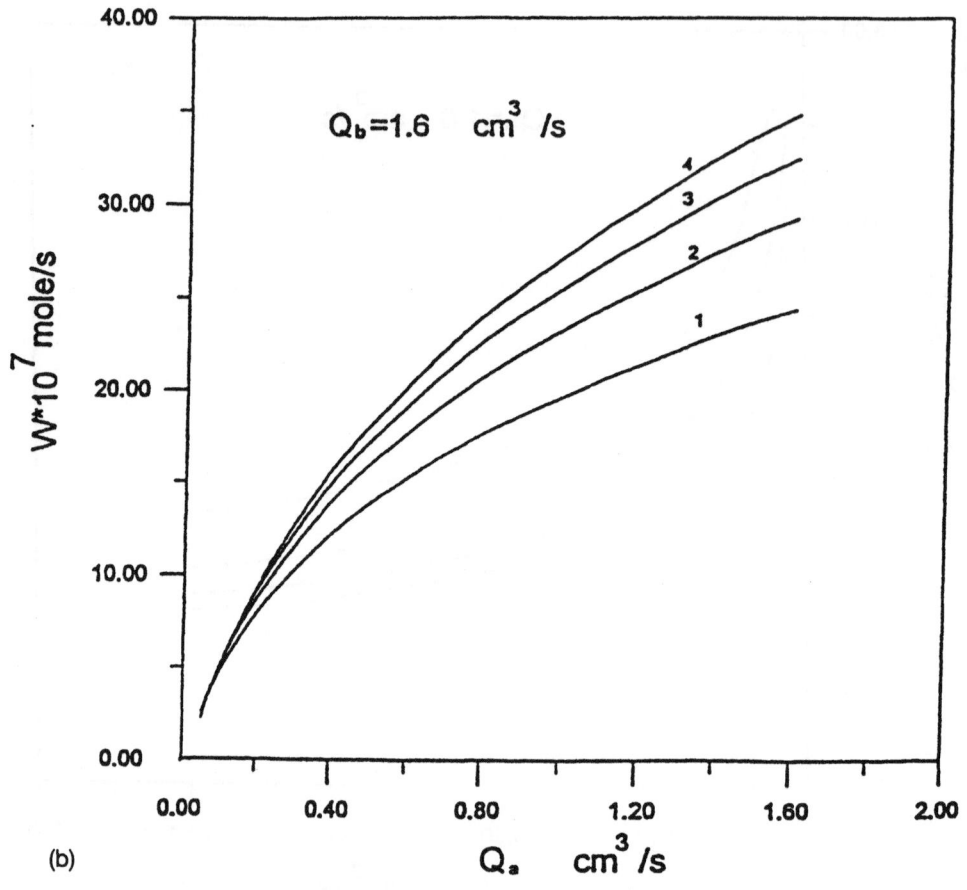

(b)

Figure 35 Continued

ber of transfer units (*NTU*s) obtained in these devices were calculated and correlated with the phase flow rates, hollow-fiber diameter, and other physical parameters. It was shown that for the aqueous phase on the tube side and once-through operation, the required module length, $L = HTU \times NTU$ could be obtained from

$$HTU = \frac{Q_{ab}}{K_w \alpha_i} \tag{65}$$

$$NTU = -\ln \frac{[(C_{fo} - C_{fo}^*)/(C_{ft} - C_{ft}^*)]}{1 - Q_{aq}/mQ_{org}} \tag{66}$$

where $\alpha_i = 4d_i N / d_s^2$. This study utilized six type 1 modules (with the following specifications: shell i.d., 1.9 cm; number of fibers, 600; fiber material, X-20 polypropylene; fiber i.d., 400 μm; wall thickness, 25 μm; fiber porosity, 40%; pore size, 0.03 μm; contact area, 0.11 m^2; length of module, 15 cm), one type 2 module (number of fibers, 800; contact area, 0.15 m^2; other characteristics the same as type 1), three type 3 modules (number of fibers, 1800; contact area, 0.21 m^2; other characteristics the same as type 1), and two type 4 modules (number of fibers, 4000; contact area, 0.193 m^2; fiber material, X-10 polypropylene; fiber i.d., 100 μm; wall thickness, 25 μm; fiber porosity, 20%; pore size, 0.03 μm; other characteristics the same as type 1). It was important to verify the performance of each module to see whether the various modules with the same number and type of fibers yielded similar results. The recoveries obtained with various modules of a given type under similar flow conditions were indeed quite close. A change of the solvent from benzene to toluene did not introduce any module-to-module variation.

Figure 36 shows the results obtained with one and two type 3 modules connected in series. Recoveries of both solute species were higher with two modules in series than with only one module. Similar results were obtained with two type 1 modules as well as with two type 4 modules connected in series. A simple explanation of these results was that having two modules in series doubled the contact area between the two phases, but additional complexities had to be considered. When two hollow-fiber modules were connected in series, the length of the module was doubled and the mass transfer coefficient therefore should have been lower than that obtained with one module only. According to Eq. (67), Sh should be inversely proportional to a fractional power of L:

$$ Sh = 1.62 \left(d^2 \frac{\sigma}{LD} \right)^{1/3} \tag{67} $$

That this was not the case, as seen in Figure 36b, could be explained by noting that the two modules were connected with tubing having a 6.25 mm inner diameter. The concentration profile started to develop in the first module but underwent a rapid mixing when the flow entered the connecting tubing. The flow then entered the second module and the velocity and concentration profile developed again. The foregoing reasoning would explain the high mass transfer coefficient obtained with two modules connected in series. From a mass transfer point of view, it was thus better to have two modules in series rather than one long module. When solvent extraction was carried out with the aqueous phase on the shell side of the type-3 MHF module, the recoveries of the two solutes remained fairly constant and were lower than those obtained with the aqueous flow on the tube side. This result was most likely due to the significant backmixing on the shell side. Although these MHF modules had a high surface area

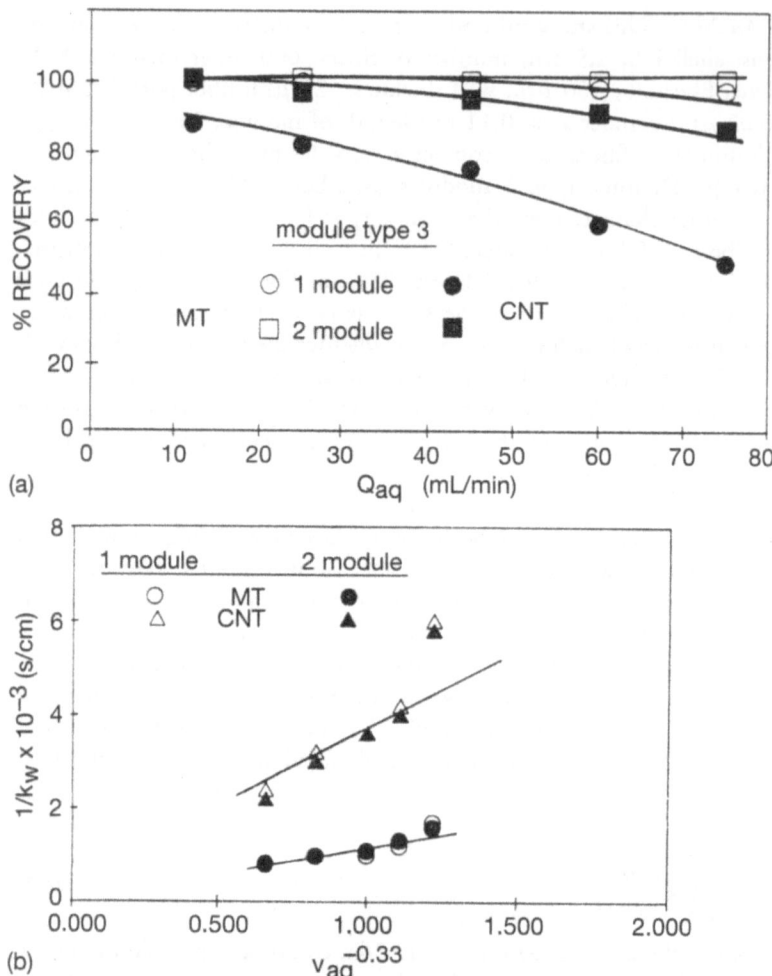

Figure 36 (a) Recoveries of MT and CNT as a function of aqueous flow rate for one and two type 3 modules, respectively, connected in series: Q_{or} = 30 mL/min. (From Ref. 190.) (b) Overall aqueous phase based mass transfer resistance for MT and CNT as a function of aqueous phase flow rate for one and two type 3 modules connected in series: Q_{or} = 30 mL/min. (From Ref. 190.)

per volume, the void volume in the shell side was still fairly large, leading to bypassing and secondary flows in the shell at the higher flow rates. The module orientation (i.e., vertical or horizontal) might affect the mass transfer, and in the vertical configuration, the flow direction could be from top to bottom or from bottom to top. The performance variation between the three configurations was small, however.

To check long-term behavior, solvent extraction of MT and CNT from aqueous solutions was carried out over 64 days with a type 3 hollow-fiber module and benzene as the solvent. The experiment was run in a once-through mode on weekdays and in a batch recirculation mode over weekends and holidays. The solute recoveries and mass transfer coefficients obtained in this manner were fairly constant and close to those obtained earlier in short-term studies. The module was subsequently cut open and the Celgard X-20 hollow-fiber membranes were studied by means of scanning electron micrography. The micrographs showed that long-term and continuous exposure to the solutes at fairly high pH (~10) and the benzene solvent did not affect the pore structure and dimensions of the polypropylene membrane. The developed relations were used to design a large MHF extractor for MT and CNT recovery from a large stream. This involved connection of a known number of modules in series and then calculation of the number of such series arrangement needed in parallel for a desired feed flow rate. Table 22 presents the results of calculations for such a modular assembly for CNT recovery. The total number of modules needed for the desired recovery of 98.3% of CNT dropped rapidly as more and more modules were connected in series. An increase in the number of modules in series increased the ability to handle a higher flow rate in each parallel limb for a specified recovery. This higher flow rate led to higher *HTUs* in each limb, but the increased total length, due to an increase in the number

Table 22 Modules Needed for 98.3% Recovery from Aqueous Stream

Modules in series, SM	Parallel configuration, PM	Total modules	Pressure drop (kPa)
4	34	136	144
5	24	120	256
6	19	114	388
7	15	105	572
8	12	96	820
9	11	99	1008
10	9	90	1364
15	5	75	3684

Source: Ref. 190.

of modules in series, was sufficient for 98.3% recovery. However, this increased flow through each module led to a higher pressure drop in the modular assembly. It was noted that the aqueous phase pressure was always to be maintained higher than the organic phase (shell-side) pressure. These pressure drops are also presented in Table 22. Because of pressure drop considerations, the practical number of modules needed was not necessarily the predicted minimum.

Lemanski and Lipscomb [191] reported a theoretical analysis of shell-side flows and their influence on mass transfer. Fluid flow within the shell of a hollow-fiber bundle was analyzed, based on volume averaging of the relevant conservation of mass and momentum equations. The ratio of the effective mass transfer coefficient to the intrinsic mass transfer coefficient of the fiber depends on two dimensionless groups: module geometry number, MG = $(k_{rr}/k_{zz})(L/R)^2$, and the intrinsic mass transfer coefficient number, MT = $k_a \tau$. The values of k_{rr} and k_{zz} were determined as a function of fiber packing geometry and fiber packing fraction. For small MT, the best performance was found for small values of MG. However, upon increasing MT, performance decreased dramatically for small MG, while little change occurred for large MG. Therefore, the best overall performance was found for large MG. The performance changes were interpreted in terms of the residence time distribution, θ. For small MT, the large-θ behavior of F (the cumulative residence time) controlled the performance, while the small-θ behavior controlled performance for large MT. Values of MG for many commercial hollow-fiber membrane modules appeared to fall in the range 10–100. Typical commercial devices possessed an aspect ratio of 5–10 and a fiber packing fraction of roughly 0.4–0.6. For these parameter ranges, MG varied from approximately 6 to 60. In this range, module performance was high but not too sensitive to fluid flow rate or MT. Thus, this represented a desirable design. In addition to the papers described in this section, a few papers described earlier should be consulted [18–21a,22,37,44,78,79,174b] for the details on performance evaluation and long-term uses of HF contactors.

VI. SPECIAL USES OF HOLLOW-FIBER MEMBRANES AND CONTACTORS

A. Semiconductors

Membrane contactors are used in the production of ultrapure water for semiconductor manufacturing, where remarkably low levels of contaminating gases must be ensured. For example, oxygen concentration must be reduced to the ppb level to avoid uncontrolled native silica oxide growth in water immersion systems. Unlike conventional deoxygenation approaches such as nitrogen bubbling and vacuum degasification, membrane contactors provide a uniform water dispersion and are insensitive to changes in flow rate. Nitrogen bubbling has another disadvantage: even though it removes oxygen, it saturates the water with another troublesome gas, nitrogen. Membrane contactors are also used to add CO_2 to water to increase the aggressiveness of a rinse step [192]. In addition, membrane contactors are used commercially for ozonization of water in silicon wafer cleaning applications. Specific uses include removal of organic and metal contamination, control of native oxide growth, and photoresist stripping. Unlike conventional approaches, this method of ozonization is bubble free, an important feature in the semiconductor industry because bubbles attract particles that later form troublesome deposits on wafer surfaces. Membrane contactors also allow manufacturers to enjoy the advantages of ozone in lieu of conventional cleaning chemicals like sulfuric acid and peroxides. That is, ozone is milder and more environmentally friendly, reduces water usage, and requires less floor space to handle. Membrane contactors have become the technology of choice for ozonization in Japan [2,192].

B. Beverage Industry

Among the few reported commercial applications of HFs, a Pepsi bottling plant in West Virginia has operated a bubble-free membrane-based carbonation line since December 1993. The plant processes ~400 L/min of beverage with 26.4 × 71.1 cm² Liqui-Cel Extra Flow modules having a total interfacial area of 193 m². The advantages over conventional carbonation methods include reduced foaming, improved yields, lower CO_2 pressures, and minimal drop in filler speed when filling at elevated temperatures [2,192,193]. Liqui-Cel contactors are also used in beer production. Several plants are using the technology for CO_2 removal followed by nitrogenation to obtain a dense foam head, and others are employing membranes to remove oxygen from beer to preserve flavor. Some breweries use Liqui-Cel equipment to strip oxygen from water and then using the deoxygenated water to dilute beer containing 9–10% alcohol. The high alcohol content is obtained from high gravity brewing, a process that offers the brewer increased capacity [2].

C. Human Gills

One of the most interesting applications of hollow-fiber membranes was reported in 1989 by Yang and Cussler [151,194] for developing artificial gills that later helped in the design of human gills. The HF module shown in Figure 37 used the hollow-fiber fabric as a vane mounted diagonally in an open-ended rectangular box. Gas flowed through each hollow-fiber lumen; the gas entered and left through tubular manifolds mounted along the diagonal as shown. The entire box was towed through the liquid, so that liquid flowed both through and around the box. This type of module had occasionally been suggested as

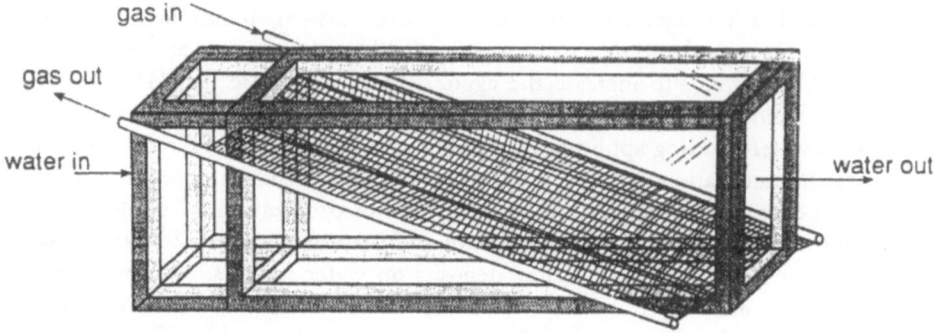

Figure 37   Vane HF module geometry used in artificial gill applications. The vane module is an open-ended box with a vane of hollow-fiber fabric mounted diagonally inside. (From Ref. 194.)

an artificial gill, either for a human diver or for a small nonnuclear submarine [195]. The purpose of this module was to harvest dissolved oxygen for the diver's breathing or for the fuel cells of the submarine.

D. De-emulsification of Water/Oil/Solid Emulsions

Typical de-emulsification methods found in literature are additions of de-emulsifying agents, pH adjustment, gravity or centrifugal settling, filter coalescers, heater treaters, electrostatic coalescers, and membranes [195]. There are advantages and disadvantages to each of these de-emulsification techniques. In processes using additives, there are additional problems of disposal of adsorbent or additive, or contamination of the recovered oil or water with the additive. The adjustment of the pH can sometimes be used to break oil-in-water emulsions, but effects of pH and salts are not significant for water-in-oil emulsions. Gravity settling, centrifuging, or heating can break some emulsions, but centrifuges are capital intensive and expensive to run and maintain. Electric field methods have been used to de-emulsify water-in-oil emulsions, and electrostatic coalescers are widely used in the petroleum industry [196]. The electric field method is also employed for the breaking of emulsions used in liquid membrane technology [196]. Extremely high voltages (10–20 kV) are required to cause droplet coalescence. Hsu and Li [197] showed that using insulated electrodes with a non-water-wettable coating prevented sparking, thus making it possible to use voltages as high as 20 kV. Larson et al. [198] also used a high-voltage cell with insulated electrodes for de-emulsification of coarse and micro emulsions. In view of these difficulties, membrane modules are found to be more appropriate for de-emulsification because they are more compact and can be run continuously. Compared with other processes, they offer the following advantages of splitting emulsions: low energy cost, especially with microfiltration membranes that operate at low pressures; no moving parts; no degradation due to heating; and no extra safety considerations as in high-voltage de-emulsification.

In an interesting application, hollow-fiber membranes were utilized successfully by Tirmizi and coworkers [199] for de-emulsification of water/oil/solid emulsions. The separation due to preferential wetting was studied, implying that the wetting phase permeated through the pores while the nonwetting phase would be retained. A hydrophobic membrane having pore sizes from 0.02 to 0.2 μm allowed the permeation of an oil phase at almost zero pressure and retained the water phase, even though the molar mass of water (18 g/mol) is much smaller than that of the oil (198 g/mol for the tetradecane used). Permeation rates were compared with batch gravity settling rates at the same temperature. For emulsions with a high surfactant content, a combination of hydrophobic membranes, hydrophilic membranes, and electric field separation was investigated to get complete separation and to recover both phases. A membrane de-emulsification unit using hydrophobic membranes alone and a hydrophobic membrane unit operated simultaneously with the electrostatic cell are shown in Fig. 38. A clarified oil phase was collected as permeate from the membrane unit, and the aqueous phase was collected from the bottom of the electrostatic cell. Table 23 provides details of the membrane modules tested, such as membrane area, pore size, and tube internal diameter as well as performance data [199], such as pure component breakthrough pressure and permeate purity.

The de-emulsification with hydrophobic membranes indicated that the permeating tetradecane had a very high purity for all tests (<500 ppm residual water). Gravity settling (the conventional method) was applied to water-in-oil emulsions containing 5, 10, 20, and 30 kg/m³ ECA5025 (Parafimins/Exxon Chemicals) in tetradecane with 80% (v/v) oil and 20% (v/v) water. The emulsions were allowed to settle for 3 weeks. After settling, the emulsions separated into three layers: the top, clear, oil layer containing less than 280 ppm water; a middle layer containing more than 1.0% water; and a bottom thick emulsion layer containing 58–95% water. A clear water phase was not obtained after gravity settling or centrifugation. Table 24 compares the gravity settling and membrane separation results in module D (polypropylene). The percentage recovery of a clear oil phase, as a function of surfactant content, can be used to compare the two methods. As the surfactant concentration increased from 5 kg/m³ to 30 kg/m³, the stability of the emulsion greatly improved, and the recovery by gravity settling dropped from 60% to 20%. However, the hydrophobic membrane was able to recover 85–90% tetradecane even from stable 30 kg/m³ ECA 5025 emulsions. Emulsions with as low as 0.5 kg/m³ surfactant

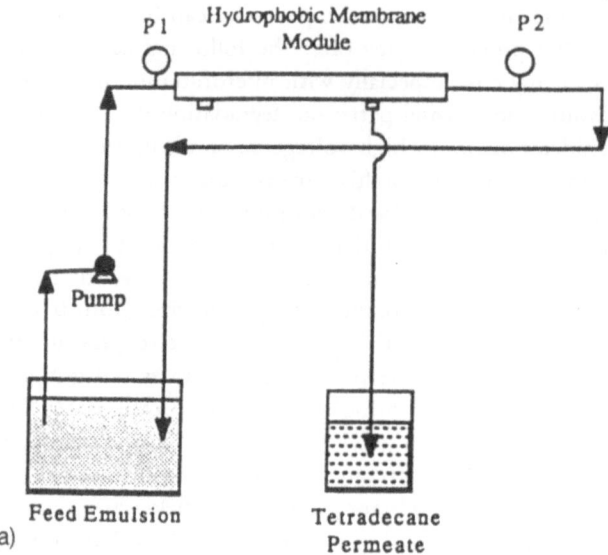

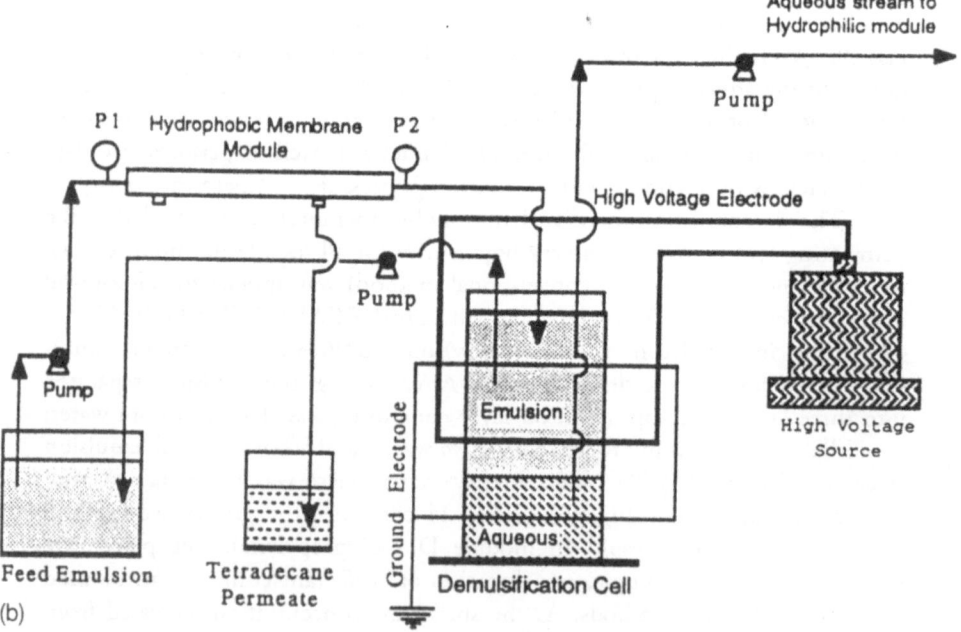

Figure 38 Experimental setup for de-emulsification of emulsion with MHF. (a) Membrane de-emulsification where the membrane module is either a hydrophobic module alone or hydrophilic and hydrophobic modules in series for the combined separation case. (b) Membrane separation with electricity field coalescence. (From Ref. 199.)

Table 23 Evaluation of Hydrophobic Hollow-Fiber Membrane Modules for Oil–Water Separation[a]

Module no.	Product, manufacturer	Material	Area (cm²)	Pore size (μm)	Tube i.d. (cm)	Breakthrough pressure of pure component (kPa)		Water content permeate for tetradecane (ppm)[*,b]
						Tetradecane–air*	Water–air*	
A	PC-240/5C Hoechst Celanese	Polypropylene	4000	0.05	0.024	0	>207	49
B	LM1P06 Enka	Polypropylene	400	0.1	0.06	0	>207	46
C	Ceramic (single tube) (U.S. Filter)	Ceramic	55	0.02	0.7	0	>138	28
D	LM2P18 (Enka) 2 Modules	Polypropylene	200	0.2	0.18	0	>138	31

[a] All characteristics of membranes supplied by respective manufacturers except those denoted by *, which were measured as part of this work.
[b] Separation of pure tetradecane ($C_{14}H_{30}$) from water, no surfactant.
Source: Ref. 199.

Table 24 Tetradecane Recovery by Gravity Settling vs Hydrophobic Membrane Separation

Surfactant content in tetradecane (kg/m^3)[a]	Recovery of tetradecane (%)	
	From upper clear layer after gravity settling[b]	By hydrophobic membrane[c]
5.0	60	92.2
10.0	40	84.5
20.0	45	93.0
30.0	20	89.8

[a]Composition: 80% v/v water; surfactant, ECA 5025.
[b]Of 25 mL samples after 3 weeks.
[c]Model D emulsion (see Table 23).
Source: Ref. 199.

after 48 h of gravity settling separated out only the oil, with 48 ppm water, a middle cloudy layer with 0.71% water, and the bottom layer with 66% water.

Emulsions with surfactant concentrations higher than 2 kg/m^3 were de-emulsified with a combination of electrostatic and membrane separation techniques, using the apparatus shown in Figure 38b. When the electric cell with membrane was operated with 1000 mL of emulsion for 60 minutes, 702 mL of oil phase containing 830 ppm water, and 154 mL containing 3.2 ppm tetradecane were recovered. The emulsion loss (144 mL) was due to the holdup volume of apparatus. If the electric cell had not been used for this emulsion (i.e., if only membrane separation had been applied), the aqueous phase could not have been recovered.

VII. CONCLUDING REMARKS AND FUTURE TRENDS

Dispersion-free membrane extraction (DFME) using microporous hollow-fiber extractors is a refreshingly new separation technique that has gained profound importance owing to its versatile potential and industrial applicability. This chapter dealt with investigations and developments of dispersion-free membrane extraction technology over the last 10 years in the environmental and hydro-metallurgy fields. Although a number of membrane module geometries are possible, hollow-fiber modules have received the most attention. Extensive research work has been taken up by several researchers to develop the hollow-fiber membrane liquid–liquid extraction (HFMLLE) technique aimed at extracting multicomponents from diverse aqueous streams by using a variety of well-packed hydrophobic hollow fibers. Very high recoveries are possible with small well-packed Liqui-Cel extractors depending on the flow conditions used. The few pilot plant studies reported to date indicate that the change of scale does not impose negative effects on the DFME process. For example, the experimental test work with wastewater from a galvanic process showed good performance for a pilot plant [21a]. This technique may be especially useful in environmental engineering because of nondispersive pore contact and the extremely large specific interfacial area. Small *HTU* values pointed to HFMLLE applications in hydrometallurgy to resolve difficult tasks—for example, the recovery of gold from alkaline cyanide media in the presence of base metals [78,79] and the separation of Cd in the presence of bulk quantities of Ni [37]. In spite of the small dimension of hollow fibers, a specific flow rate that is large in comparison to that from conventional extraction columns could be obtained. One of the studies indicated that it was possible to achieve similar extraction rates in a short, densely packed HF module rather than the conventional extraction column of 6 m length. This finding can be related to the high interfacial area and short diffusion paths that prevailed over the disadvantages of laminar flow.

Analytical applications of HFMLLE showed this to be a very promising speciation-sensitive technique for the in situ preconcentration and separation of trace compounds in water. More promising results have been achieved when applied to separation and preconcentration steps in analytical schemes pertaining to transduction mechanisms in chemical sensors based on transport of the active component through the membrane between sensors and sample.

A wide range of steady state and dynamic simulations was carried out to demonstrate the flexibility of the models describing the effects of various operating conditions on the performance of HFMLLE. Models have been developed for the single-function and dual-function modules, and these have been combined to model entire separation processes carrying out simultaneous extraction and stripping of metallic species from aqueous solutions. The integrated process models incorporate the possibility of partially recycling both the extraction and stripping product streams. Furthermore, although these models are dynamic, they can also be used for the direct prediction of steady state conditions. Other methods for the determination of mass transfer coefficients also were discussed. The method suggested by D'Elia, Cussler, and Dahuron [149,150,169] for mass transfer coefficient evaluation is quite often used by other researchers.

The important uses of HF contactors in the field of semiconductor and beverage industries have claimed success. The concept of applying hollow-fiber membranes in designing a synthetic human gill is interesting, and further work should be taken up in this direction. The emulsion liquid membrane (ELM) extraction of metal ions using an HF contactor is a promising technique that overcomes the problems associated with conventional ELMs. De-emulsification of emulsions by the use of an HF contactor could be useful to make the ELM process more feasible for industrial applications.

Future research is necessary to collect comprehensive pilot plant experiences to permit us to comment on integrated membrane process dealing with hydrometallurgical and environmental problems. HFMLLE scaling up has reached the barrier of commercialization, which might be overcome by arriving at an understanding of the efficiency parameters that would lead to insights into the separation characteristics of a given system. This chapter should help to design future work in this direction by providing the readers with the type of data representation required, allowing intercomparison between the different systems and covering, as far as possible, the permeability, mass transfer, selectivity, and stability properties. Furthermore, in the field of radioactive waste treatment, the possibility of membrane systems that perform clean, specific, low-energy separations of radionuclides calls for the development of additional commercially available selective extractants.

The implementation of new approaches directed toward the scrutiny of the interfacial phenomenon and issues within the membrane phase by modern physicochemical techniques will give important information for the understanding and the correct application of theory in these systems.

If future investigations are carried out by selecting critical separation problems based on their typical kinetics as described in References 70, 78, and 79, major breakthroughs in the performance of DFME systems using HF contactors can be expected. To obtain the best results with HFMLLE, the characteristics of hollow-fiber contactors for providing efficient mass transfer are most important. Therefore, to facilitate further commercialization of HF contactors, the challenges that need to be addressed include better understanding of the shell-side flow, leading to more reliable scale-up procedures, and improved materials of construction, offering a wide range of solvent compatibility.

Designing and operating HFMLLE systems involve many decisions related to the design of individual modules (i.e., their type, diameter, length, and number of fibers in them), the network structure (i.e., the number and connectivity of the various units), and the operating conditions (flow rates, recirculation ratios, etc.). Thus a more systematic approach based on the use of rigorous mathematical optimization appears to be essential.

ACKNOWLEDGMENTS

This work was supported by the Spanish Ministry of Science and Culture, CICYT (QUI 99-0749). Dr. Anil Kumar Pabby acknowledges financial support from the Comisión Interministerial de Ciencia y Tecnologia, Spain for awarding a Visiting Scientist fellowship. Thanks are also due to K. Balu, Director of the Nuclear Recycle Group, BARC, India, and D. D. Bajpai, Head of PREFRE, BARC, India, P. K. Dey, Chief Superintendent, PREFRE, BARC, and R. K. Singh, PREFRE, BARC, for their valuable suggestions during the preparation of this manuscript. Special thanks are due to Dr. J. P. Shukla, Radiochemistry Division, BARC, Mumbai, for his useful suggestions and critiques.

TRADE NAMES, ABBREVIATIONS, AND SYMBOLS

Acorga P-500	Commercial aldoxime
Alamine 336	Mixture of trioctyl and tridecyl amine
Aliquat 336	Tridodecylammonium chloride
Bathocuproine	4,7-Diphenyl-2,9-dimethyl-1,10-phenanthroline
Bathophenanthroline	4,7-Diphenyl-1,10-phenanthroline
B21C7	Benzo-21-crown-7
calix[4]arene carboxyl derivative	[25,26,27,28-tetrakis(carboxymethoxy)-5,11,17,-23-tetrakis 1,1,3,3-tetramethyl] butyl calix[4] arene
CDTA	Cyclohexanediaminetetraacetic acid
C12phSAQ	N-8-Quinolyl-p-dodecylbenzenesulfonamide
CMPO	Octyl(phenyl)-N,N-diisobutylcarbamoylmethyl phosphine oxide
CNT	4-Cyanothiazole
CTA	Cellulose triacetate
Cyanex 302	Bis-(2,4,4-trimethylpentyl)monothiophosphinic acid
Cyanex 272	Bis-(2,4,4-trimethylpentyl)phosphinic acid
DA18C6	Dicyclohexano-18-crown-6
DC18C6	1,10-Didecyldiaza-18-crown-6
DtBuC18C6	Di-$tert$-butylcyclohexano-18-crown-6
DHDECMP	Dihexyl-N,N-diethylcarbamoylmethylenephosphonate
NDecB21C7	n-Decylbenzo-21-crown-7
D2EHPA	Di-(2-diethylhexyl)phosphoric acid
DEPA	Di-(2-diethylhexyl)phosphate
DPC	Diphenylcarbazide
DTMPPA	Di-(2,4,4-trimethylpentyl)phosphinic acid
DTPA	Di-(2-ethylhexyl)dithiophosphoric acid
ECA 5025	Parafimins
EDTA	Ethylenediaminetetraacetic acid
FIA	Flow injection analysis

GFAAS	Graphite furnace atomic absorption spectrometry
Kelex 100	96% 7-(4-Ethyl-1-methoxyoctyl)-8-quinolinol
Kryptofix 22D	1,10-Didecyldiaza-18-crown-6
LIX 54	40% β-hydroxyoxime in kerosene
LIX 63	Commercial hydroxyoxime
LIX 65 N	2-Hydroxy-5-nonylbenzophenone oxime
LIX 79	N,N'-Bis(2-ethylhexyl)guanidine
LIX 84	anti-2 Hydroxy-5-nonylacetophenone oxime
LIX 860	β-Hydroxyoxime
MES	Morpholinoethanesulfonic acid
MSP-8 or MTPA	Di-(2-ethylhexyl)monothiophosphoric acid
MT	4-Methylthiazole
OSAQ	8-Octylsulfonamidoquinoline
PP	Polypropylene
PVC	Polyvinyl chloride
PTFE	Polytetrafluoroethylene
NPOE	o-Nitrophenyl octyl ether
'Oct[4]CH$_2$COOH	p-t-Octylcalix[4]arene[25,26,27,28-tetrakis(carboxy-methoxy)-5,11,17,23-tetrakis(1,1,3,3,tetramethylbutyl)-calix[4]arene]
PC-88 A	2-Ethylhexylphosphonic acid mono-2-ethylhexyl ester
TBP	Tri-n-butyl phosphate
TOA	Tri-n-octylamine
TOMAC	Trioctylmethylammonium chloride
TOPO	Tri-n-octyl phosphine oxide
a	specific extraction area, m^{-1}
a'	specific stripping area, m^{-1}
A_i	contact area per stage, m^2
c_{io}, c_{iw}	organic and aqueous phase concentrations of species i, mol/cm^3
C	metal concentration (g/cm^3)
$C_0(0)$	initial concentration of the component in the lumen side reservoir, M
$C_0(t)$	concentration of the component in the lumen side reservoir at time t, M
C_A	solute A concentration in the tank
C_{Ai}	initial solute A concentration in the tank
C_A^m	solute A concentration in the extraction module
$C_A^{mE^*}$	equilibrium solute A concentration in the extraction module
C^*	dimensionless solute concentration at the wall

$\bar{C}_M$	total carrier concentration in all forms (carrier and carrier–solute complex), mol/m^3
C_s	shell bulk concentration of the component when the concentration is constant
$C_{s,0}$	inlet concentration of the component in the shell or the bulk concentration of the component in the shell-side reservoir, M
$C_{s,0}(t)$	concentration of the component in the shell-side reservoir at time t, M
$C_{s,0}$	initial concentration of the component in the shell-side reservoir, M
$\bar{C}_T$	maximum solute–carrier complex concentration in the organic phase, mol/m^3
C_0	initial solute concentration in the aqueous phase, mol/m^3
$\bar{C}_i$	solute–carrier complex equilibrium concentration at the fluid membrane interface, mol/m^3
d	diameter of one fiber, cm
D_s	diffusion coefficient of solute in the shell side
D_m	diffusion coefficient of metal complex in the membrane, cm
D_h	hydraulic diameter, cm
D_t	diffusion coefficient of solute in the tube side, cm^2/s
d_{ti}, d_i or R_i	hollow-fiber inside diameter, μm
d_{to}, d_o or R_o	hollow-fiber outside diameter, μm
E_{mt}	Murphree tube-side efficiency
E	extent of copper extraction defined in Eq. (39) (see Ref. 124)
h	amplitude of the membrane phase pulsation, m
H or H_e	partition coefficient of extraction as expressed in Eqs. (46–48)
H or H_r	partition coefficient of stripping as expressed in Eqs. (46–48)
H	concentration of hydrogen ion in the aqueous phase as expressed in Eqs. (20), (21)
HETS	height equivalent to a theoretical state
HTU	height of transfer unit
HFM or MHF	hollow fiber microporous membrane contactor
i.d.	inner diameter
J	flux density, g/m^2/s
J	mass transfer flux, mol/m^2/s

k_e	forward reaction constant of Eq. (49), cm/s
k_i	rate of interfacial reaction defined in Eq. (51)
k_m or $\bar{K}$	membrane mass transfer coefficient, cm/s
k_s	organic mass transfer coefficient, cm/s
k_{st} or K_s	mass transfer coefficient for stripping, cm/s
k_t	tube-side mass transfer coefficient, m/s
k_w	interfacial mass transfer coefficient for extraction, m/s
k_w'	interfacial mass transfer coefficient for stripping, m/s
k_{im}	membrane mass transfer coefficient for species i, m/s
k_{io}	local mass transfer coefficient of species i in organic phase, m/s
k_{iw}	local mass transfer coefficient of species i in aqueous phase, m/s
k_{be}	break-up rate constant of ELM at critical concentration of surfactant concentration, s^{-1}
k_m	pore or membrane mass transfer coefficient, m/s
k_2	rate constant for forward reaction in Eq. (37), $m^2/mol/s$
k_2'	rate constant for backward reaction in Eq. (38), $m^{-1} s^{-1}$
K_1	equilibrium constant of Eq. (32), m^2
K_2	equilibrium constant of Eq. (33), m^2
K_{ad}	equilibrium concentration of Eq. (31), m^3/mol
$K_{avg,z}$	average lumen mass transfer coefficient from the entrance to z; m/s
$K_{avg,z,0}$	average overall mass transfer coefficient of the module, m/s
K_E	overall mass transfer coefficient in extraction module, m/s
K_o	mass transfer coefficient based on organic phase, m/s
K_o'	mass transfer coefficient for the organic phase in BEX module, m/s
K_p	overall mass transfer coefficient in HF pertractor, m/s
K_{ot}	overall tube-side mass transfer coefficient, cm/s
K_w	overall mass transfer coefficient based on aqueous phase, m/s
K_{me} or K_m	membrane mass transfer coefficient, m/s
K_f or k_f	mass transfer coefficient of the aqueous feed, cm/s
K_R	overall mass transfer coefficient in back-extraction module, m/s
K_{Au}^E, K_{Au}^S	coefficients for extraction and stripping, respectively, cm/s
K_{ms}	mass transfer coefficient in stripping modules, m/s
K_e'	equilibrium constant of extraction reaction, $mol^{-2} L^{-2}$

l	planar diffusion path length, m
L	hollow-fiber length, m
m_i	distribution coefficient for species i
n	number of metal ions bound to each carrier molecule
NTU	number of transfer units
N_{Gz}	Graetz number = $d^2\, v/dL$
N_{Re}	vd/η
N_{Sc}	η/D
N_{Sn}	kd/D
O	concentration of organic extractant as expressed in Eqs. (20), (21)
q_t	total volumetric flow rate on tube side per module, cm^3/s
Q	flow rate of feed solution, m^3/s
Q'	flow rate of stripping solution, m^3/s
r	hollow-fiber radius, cm
r	radial diameter of extractor, m
r_1 and r_2 or r_o	internal and external diameters, m
r_3	inner radius of membrane extractor, m
R	dimensionless solute concentration in the BEX solution
$R_m^o, R_f^o, R_s^o, R_i^o, R_{st}^o$	fractional resistances due to membrane, feed, solvent, interfacial reaction and stripping, respectively
S	cross-sectional area of the extraction cell, m^2
s	section of the feed aqueous phase in module
s_{Oe}	section of the organic phsae in the BEX module
s_{Oi}	section of the stripping phase in the BEX module
s_R	section of the feed aqueous phase in BEX module
t_m	thickness of the fiber membrane, cm
T	temperature, K
t	time, s
u, v	linear velocity in the fiber lumen, m/s
V	volume of the aqueous feed solution, m^3
V_{wi}^o	initial content of water at time t, m^3
V_{wi}	content of water at time, m^3
V_e or V_O	organic tank volume, m^3
V_f or V_A	feed tank volume, m^3
V_s or V_R	stripping tank volume, m^3
V_m	volume of hollow fibers, cm^3
v_A	linear velocity in the feed phase
v_O	linear velocity in the organic phase
v_R	linear velocity in the feed phase
$\dot{V}$	volumetric flow rate, m^3/s

X	dimensionless solute concentration in the feed solution
Y	dimensionless solute concentration in the organic solution
z	axial distance in the module
Z	dimensionless distance
α_{HO}	activity of hydrogen ion at initial stage
δ	wall thickness of a hollow fiber
ε	membrane porosity
λ	porosity of perforated baffle plate
v	linear flow velocity (m/s) expressed in N_{Re} and N_{Gz}
ρ	separation selectivity expressed by Eq. (4)
γ	interfacial tension of solvent
η	viscosity of a solvent as expressed in Eq. (50)
ξ	fraction of surface free area
σ_i	$R_i/\langle R \rangle$ dimensionless radius of batch i (see Eq. (61))
θ	residence time distribution
μ	viscosity, cP
τ	dimensionless time
ω_M	frequency of pulsation of the membrane phase, s^{-1}
ν	linear velocity of aqueous phase as expressed in Eq. (52)

Subscripts

A	solute in the feed solution
O	solute in the organic solution
R	solute in the BEX solution
e	extraction
f	inlet
i	interface between two phases, species i in hollow-fiber section
m	membrane
o	organic, outside surface
p	outlet
s	stripping
s	shell side
t	tube side
w	aqueous film

Superscripts

*	equilibrium value
mE	extraction module
mR	back-extraction module
T	tank

REFERENCES

1. KK Sirkar. Chem Eng Commun 157:145–184, 1997.
2. A Gabelman, ST Hwang. J Membr Sci 159:61–106, 1999.
3. RA Bartsch, JD Way. In: RA Bartsch and JD Way, eds. Chemical Separation with Liquid Membranes. ACS Symp Ser No 642. Washington, DC: American Chemical Society, 1996, pp 1–10.
4. AM Sastre, A Kumar, JP Shukla, RK Singh. Sep Purif Methods 27:213–298, 1998.
5. JA Jonsson, L Mathiasson. Trends Anal Chem 11:106, 1992.
6. RD Noble, JD Way. In: RD Noble and JD Way, eds. Liquid Membrane: Theory and Applications. ACS Symp Ser No 347. Washington, DC: American Chemical Society, 1987, pp 1–26, 111–121.
7. WS Winston Ho, KK Sirkar. Membrane Handbook. New York: Van Nostrand Reinhold, 1992.
8. J Gyves, ER San Miguel. Ind Eng Chem Res 38:2182, 1999.
9. M Ruppert, J Draxler, R Marr. Sep Sci Technol 23:1659, 1988.
10. GT Frank, KK Sirkar. Biotechnol Bioeng Symp 17:303, 1986.
11. RN Paunovic, ZZ Zavargo, MM Tekic. Chem Eng Sci 48:1069, 1993.
12. R Basu, KK Sirkar. Solvent Extr Ion Exch 10:119, 1992.
13. CH Yun, R Prasad, KK Sirkar. Ind Eng Chem Res 31:1709, 1992.
14. PR Alexander, RW Callahan. J Membr Sci 35:571, 1987.
15. (a) Y Sato, K Kondo, F Nakashio. J Chem Eng Jpn 23:23, 1990. (b) Y Sato, K Kondo, F Nakashio. J Chem Eng Jpn 22:200, 1989.
16. (a) M Kubota, M Goto, F Nakashio, T Hano. Sep Sci Technol 30:777, 1995. (b) M Matsumoto, Y Tsutsumi, K Kondo, F Nakashio. J Chem Eng Jpn 23:233, 1990.
17. AI Alonso, AM Urtiga, A Irabien, I Ortiz. Chem Eng Sci 49:901, 1994.
18. UA Daiminger, AG Geist, W Nitsch, PK Plucinski. Ind Eng Chem Res 35:184, 1996.
19. CH Yun, R Prasad, AK Guha, KK Sirkar. Ind Eng Chem Res 32:1186–1195, 1993.
20. BW Reed, R Klassen, AE Jansen, JJ Akkerhuis, BA Bult, FIHM Oesterholt. Removal of hydrocarbons from waste water by membrane extraction. Paper presented at the AIChE Spring National Meeting, Atlanta, April 27–21, 1994.
21. (a) AI Alonso, B Galán, M Gonzalez, I Ortiz. Ind Eng Chem Res 38:1666, 1999. (b) I Ortiz, B Galán, A Irabien. J Membr Sci 118:213, 1996. (c) AI Alonso, B Galán, A Irabien, I Ortiz. Sep Sci Technol 32:1543, 1997.
22. DJ Kathios, GD Jarvinen, SL Yarbro, BF Smith. J Membr Sci 97:251, 1994.
23. FR Valenzuela, C Basualto, J Sapag, C Tapia. Miner Eng 10:1421, 1997.
24. FR Valenzuela, C Basualto, J Sapag, C Tapia. Miner Eng 9:15, 1996.
25. FR Valenzuela, C Basualto, J Sapag, C Tapia. Bol Soc Chil Quim 41:57, 1996.
26. A Kumar, AM Sastre. Ind Eng Chem Res 39:146, 2000.
27. A Kumar, AM Sastre. Spanish Patent 98-01736 (1998).
28. A Rossel, C Palet, M Valiente. Anal Chim Acta 349:171, 1997.

29. N Parthasarthy, M Pelletier, J Buffle. Anal Chim Acta 350:183, 1997.
30. MJC Taylor, DE Barnes, GD Marshall. Anal Chim Acta 265:71, 1992.
31. NK Djane, K Ndung'u, F Malcus, G Johansson, L Mathiasson. Fresenius J Anal Chem 358:822, 1997.
32. NK Djane, IA Bergdahl, K Ndung'u, A Schütz, G Johansson, L Mathiasson. Analyst 122:1073, 1997.
33. J Noguerol, C Palet, M Valiente. J Membr Sci 134:261, 1997.
34. RC Johnson, K Koch, RG Cooks. Ind Eng Chem Res 38:343, 1999.
35. JA Jönsson, L Mathiasson. Trends Anal Chem 18:318, 325, 1999.
36. Y Shen, JA Jönsson, L Mathiasson. Anal Chem 70:946, 1998.
37. B Galan, FS Roman, A Irabien, I Ortiz. Chem Eng J 70:237, 1998.
38. D Kralj, GRM Breembroek, GT Witkamp, GM van Rosmalen, L Brecevic. Solvent Extr Ion Exch 14:705, 1996.
39. AA Alhusseini. Sep Sci Technol 35:825, 2000.
40. A Sengupta, R Basu, KK Sirkar. AIChE J 34:1698, 1988.
41. S Majumdar, KK Sirkar, A Sengupta. In: W Ho and KK Sirkar, eds. Membrane Handbook. New York: Van Nostrand Reinhold, 1992, pp 764–808.
42. KK Sirkar. In: RA Bartsch and JD Way, eds. Chemical Separation with Liquid Membranes. ACS Symp Ser No 642. Washington, DC: American Chemical Society, 1996, pp 222–238.
43. Š Schlosser, I Rothova. Sep Sci Technol 29:765, 1994.
44. SYB Hu, JM Wiencek. AIChE J 44:570–581, 1999.
45. ZF Yang, AK Guha, KK Sirkar. Ind Eng Chem Res 35:1383, 1994.
46. R Basu, K Sirkar. J Membr Sci 75:131, 1992.
47. A Alonso, CC Pantelides. J Membr Sci 110:151, 1996.
48. (a) C Yang, EL Cussler. J Membr Sci 166:229–238, 2000. (b) NS Rathore, A Kumar, JV Sonawane, RK Singh, JP Shukla, DD Bajpai. J Membr Sci 2001. (c) AB de Haan, PV Bartels, J de Graauw. J Membr Sci 45:281, 1989.
49. EL Cussler. Diffusion: Mass Transfer in Fluid Systems. Cambridge: Cambridge University Press, 1984.
50. T Araki, H Tsukube, eds. Liquid Membranes: Chemical Applications. Boca Raton, FL: CRC Press, 1990.
51. RW Baker, EL Cussler, W Eycamp, WJ Koros, RL Riley, H Strathmann. Membrane Separation Systems. Recent Development and Future Directions, Park Ridge, NJ: Noyes Data Corp, 1991.
52. M Mulder. Basic Principles of Membrane Technology. Dordrecht: Kluwer Academic Publishers, 1991.
53. I Osada, T Nakagawa. Membrane Science and Technology. New York: Dekker, 1992.
54. EL Cussler. Hollow fiber contactors. In: JG Crespo and KW Boddeker, eds. Membrane Processes in Separation and Purification. Dordrecht: Kluwer Academic Publishers, 1995.
55. HK Lonsdale. J Membr Sci 10:81, 1982.
56. R Marr, A Kopp. Int Chem Eng 22:44, 1982.
57. PR Danesi. Sep Sci Technol 19:857, 1984–1985.

58. RD Noble, JD Way, A Bunge. In: JA Marinsky and Y Marcus, eds. Ion Exchange and Solvent Extraction, A Series of Advances, vol 10. New York: Dekker, 1988.

59. RD Noble, CA Coval, JJ Pellegrino. Chem Eng Prog 58, March 1989.

60. R Baker, I Blume. Coupled transport membranes. In: MC Porter, ed. Handbook of Industrial Membrane Technology. Park Ridge, NJ: Noyes Data Corp, 1990.

61. R Kopunec and TN Mahn. J Radioanal Nucl Chem 183:181, 1994.

62. HC Visser, DN Reinhoudt, F de Jong. Chem Soc Rev 23:75, 1994.

63. JP Shukla, A Kumar, RK Singh, RH Iyer. In: RA Bartsch and JD Way, eds. Chemical Separation with Liquid Membranes. ACS Symp Ser No 642. Washington, DC: American Chemical Society, 1996.

64. AJB Kemperman, D Bargeman, T Van Den Boomgaard, H Strathmann. Sep Sci Technol 31:2732, 1996.

65. AA Elhassadi, DD Do. Sep Sci Technol 34:305–329, 461, 1999.

66. BW Reed, MJ Semmens, EL Cussler. In: RD Noble and SA Stern, eds. Amsterdam: Elsevier, 1995.

67. SS Kulkarni, EW Funk, NN Li. Membranes. In: WSW Ho and KK Sarkir, eds. Membrane Handbook. New York: Van Nostrand Reinhold, 1992.

68. Y Tsujita. The physical chemistry of membranes. In: Y Osada and T Nakagawa, eds. Membrane Science and Technology. New York: Dekker, 1992.

69. X Feng, RYM Huang. Ind Eng Chem Res 36:1048, 1997.

70. ER San Miguel, J Gyves. In: M Cox and M Valiente, eds. Proceedings of the International Solvent Extraction Conference, Solvent Extraction for the 21st Century, Barcelona, July 11–16, 1999.

71. K Akiba, M Ito, S Nakamura. J Membr Sci 129:9, 1997.

72. ER San Miguel, J Gyves, JC Aguilar, ML Balinas, MTJ Rodriguez. Manuscript in preparation (2000).

73. J Gyves, JC Aguilar, ER San Miguel. In: M Cox and M Valiente, eds. Proceedings of the International Solvent Extraction Conference, Solvent Extraction for the 21st Century, Barcelona, July 11–16, 1999, p 1005.

74. A Geist, J Kim, P Plucinsky, W Nitsch. In: M Cox and M. Valiente, eds. Proceedings of the International Solvent Extraction Conference, Solvent Extraction for the 21st Century, Barcelona, July 11–16, 1999, p 1489.

75. Y Zhu, J Chen, R Jialo. Solvent Extr Ion Exch 14:61–68, 1998.

76. (a) P Schöner, W Nitsch, P Plucinski, A Fedorov. In: M Cox and M Valiente, eds. Proceedings of the International Solvent Extraction Conference, Solvent Extraction for the 21st Century, Barcelona, July 11–16, 1999, p. 1495. (b) P Schöner, P Plucinski, W Nitsch, U Daiminger. Chem Eng Sci 53:2319, 1998.

77. I Ortiz, AI Alonso, AM Urtiaga, M Demircioglu, N Kocarcik, N Kabay. Ind Eng Chem Res 36:1048, 1999.

78. A Kumar, AM Sastre. Ind Eng Chem Res 2001 (in press).

79. A Kumar, AM Sastre. AIChE J 47:328, 2001.

80. A Kumar, AM Sastre. In: M Cox and M Valiente, eds. Proceedings of the International Solvent Extraction Conference, Solvent Extraction for the 21st Century, Barcelona, July 11–16, 1999.

81. A Kumar, AM Sastre. Removal of toxic heavy metals from waste water using hollow fiber non-despersive membrane extraction technique. Second Interna-

tional Conference on Advanced Wastewater Treatment, Recycling and Reuse, Milan, September 14–16, 1998.

82. R Haddad, A Kumar, AM Sastre. J Membr Sci 2000 (submitted).
83. MB Mooiman, JD Miller. The solvent extraction of gold from aurocyanide solutions. Proceedings of the AIChE International Solvent Extraction Conference, New York, 1983, p 530.
84. MB Mooiman, JD Miller. Miner Metall Process 153, August 1984.
85. R Haddad, A Kumar, AM Sastre. Chem Eng J 2000 (submitted).
86. L Hartinger. Handbook of Wastewater and Recycling Technology in the Metalworking Industry, Munich: Hanser, 1991.
87. WRA Vauck, HA Müller, Operations in Chemical Engineering. Leipzig: VEB Deutscher Verlag für Grundstoffindustrie, 1989.
88. AI Alonso, AM Urtiaga, S Zamacona, A Irabien, I Ortiz. J Membr Sci 130: 193–203, 1997.
89. RS Juang, IP Huang. Ind Eng Chem Res 39:1409, 2000.
90. AM Urtiaga, A Alonso, I Ortiz, JA Daoud, SA El-Reefy. J Membr Sci 164:229, 2000.
91. I Ortiz, F San Román, B Galán, AM Urtiaga. In: M Cox and M Valiente, eds., Proceedings of the International Solvent Extraction Conference, Solvent Extraction for the 21st Century, Barcelona, July 11–16, 1999, p 543.
92. SA Daoud, SA El-Reefy, HF Aly. Sep Sci Technol 33:537, 1998.
93. T Gallego, ES Perez de Ortiz. In: DM Maron, ed. International Symposium on Liquid–Liquid Two Phase Flow and Transport Phenomena, Anatalya, Turkey, November 3–7, 1997. New York: Begell House, 1998, pp 301–310.
94. AM Urtiaga, S Zamacona, MI Ortiz. Sep Sci Technol 34:3279, 1999.
95. Hoechst AK. US Patent 4,503,016 (1985).
96. American Cynamide. US Patent 5,028,401 (1991).
97. SA Ercros. Spanish Patent ES2,020,138 (1991).
98. AM Sastre, A Kumar, JP Shukla, RK Singh. Sep Purif Methods 27:223, 1998.
99. R Prasad, KK Sirkar. In: WSW Ho and KK Sirkar, eds. Membrane Handbook. New York: Van Nostrand Reinhold, 1992.
100. C Cote, D Bauer. Rev Inorg Chem 10:121, 1989.
101. AF Seibert, JR Fair. Sep Sci Technol 32:573, 1997.
102. A Sengupta, BW Reed, AF Seibert. Paper presented at the San Francisco AIChE meeting, November 16, 1994.
103. E Thordarson, P Palmarsdottir, L Mathiasson, JA Jönsson. Anal Chem 68:2559, 1996.
104. B Berhardsson, E Martins, G Johansson. Anal Chim Acta 167:111, 1985.
105. S Palmarsdottir, E THordarson, LF Edholm, JA Jonsson, L Mathiasson. Anal Chem 69:127, 1997.
106. P Richter, MI Toral, R Manriquez. Anal Lett 32:601, 1999.
107. E Castillo, E Castillo, M Granados, MD Prat, AM Sastre, A Kumar, JL Cortina. In: M Cox and M Valiente, eds. Proceedings of the International Solvent Extraction Conference, Solvent Extraction for the 21st Century, Barcelona, July 11–16, 1999, p 47.

108. NK Djane, K Ndung'u, C Johnsson, H Sartz, T Tornstorm, L Mathiasson. Talanta 48:1121, 1999.
109. K Ndung'u, NK Djane, JA Jonsson, L Mathiasson. Analyst 120:1471, 1995.
110. M Papantoni, NK Djane, K Ndung'u, JA Jonsson, L Mathiasson. Analyst 120: 1471, 1995.
111. NK Djane, A Armalis, K Ndung'u, G Johansson, L Mathiasson. Analyst 123: 393, 1998.
112. K Ndung'u, NK Djane, F Malcus, L Mathiasson. Analyst 124:1367, 1999.
113. JM Christensen, K Brytriasen, K Vercoutere, R Ornells, P Quevauviller. Fresenius J Anal Chem 363:28, 1999.
114. B Raghuraman, J Wiencek. AIChE J 39:1885, 1993.
115. C Gabriel, Š Schlosser, M Muñoz, M Valiente. In: M Cox and M Valiente, eds. Proceedings of the International Solvent Extraction Conference, Solvent Extraction for the 21st Century, Barcelona, July 11–16, 1999, p 1015.
116. Š Schlosser, I Rothova, H Frianova. J Membr Sci 80:99, 1993.
117. K Yoshizuka, Y Sakamoto, Y Baba. Proc ISEC 90 805, 1990.
118. K Yoshizuka, Y Sakamoto, Y Baba. Ind Eng Chem Res 31:1372, 1992.
119. JS Preston, AC du Preez. Solvent Extr Ion Exch 9:237, 1991.
120. E Forgová, Š Schlosser. First European Congress on Chemical Engineering, Florencia, 1997, p 1321.
121. E Sabolová, Š Schlosser. Proceedings of the 13th International Congress, CHISA 98, Prague, 1998, p 12.
122. G Cara, Š Schlosser, M Muñoz, M Valiente. Paper 1524, Proceedings of the 11th International Congress, CHISA 96, Prague 1996, p 6.
123. MS Alam, K Inoue, K Yoshizuka. Ind Eng Chem Res 35:3899–3906, 1996.
124. K Yoshizuka, K Kondo, F Nakashio. J Chem Eng Jpn 19:312–316, 1986.
125. K Yoshizuka, Y Sakamoto, Y Baba, K Inoue, F Nakashio. Ind Eng Chem Res 31:1372–1378, 1992.
126. M Goto, F Nakashio, K Yoshizuka, K Inoue. J Membr Sci 74:207–214, 1992.
127. F Kubota, T Kakoi, M Goto, S Furusaki, F Nakashio, T Hano. J Membr Sci 165:149–158, 2000.
128. T Oshima, T Kakoi, P Kubota, K Ohto, M Goto, F Nakashio. Sep Sci Technol 33:1905, 1998.
129. R Basu, KK Sirkar. Solvent Extr Ion Exch 10:119, 1992.
130. J Kim, P Stroeve. J Membr Sci 45:99, 1989.
131. M Teramoto, H Tanimoto. Sep Sci Technol 18:871, 1983.
132. K Yoshizuka, R Yashukawa, M Koba, K Inoue. J Chem Eng Jpn 19:312–316, 1986.
133. A Kubaczka, A Burghardt, and Mokrosz. Chem Eng Sci 53:899–917, 1998.
134. O Miyataki, H Iwashita. Int J Heat Mass Transfer 34:322, 1991.
135. A Kubaczka, A Burghardt. Chem Eng Sci 55:2907–2916, 2000.
136. P Barton, C Pentelides. AIChE J 40:966, 1994.
137. M Oh, C Pentelides. Comput Chem Eng 20:611, 1996.
138. AK De, SM Khopkar, RA Chalmers. Solvent Extraction of Metals, London: Van Nostrand Reinhold, 1970.

139. F Valenzuela, S Poblete, J Spag, C Tapia, C Basualto. Bol Soc Chil Quim 40: 25, 1995.
140. F Valenzuela, J Andrade, J Spag, C Tapia, C Basualto. Miner Eng 8:893, 1995.
141. F Valenzuela. Latin Am Appl Res 24:33, 1994.
142. AK Guha, CH Yun, R Basu, KK Sirkar. AIChE J 40:1223, 1994.
143. ZF Yang, AK Guha, KK Sirkar. Ind Eng Chem Res 35:1383, 4214, 1996.
144. RM Izatt, JD Lamb, RL Bruening. Sep Sci Technol 23:1645, 1988.
145. TH Chilton, AP Colburn. Ind Eng Chem 26:1183, 1934.
146. RH Perry, DW Green, eds. Perry's Chemical Engineers' Handbook, 6th ed. New York: McGraw-Hill, 1984.
147. RW Rousseau, ed. Handbook of Separation Process Technology, 1st ed. New York: Wiley, 1987.
148. V Gekas, B Hallstrom. J Membr Sci 30:153, 1987.
149. NA D'Elia, MS thesis, University of Minnesota, Minneapolis, 1985.
150. L Dahuron, EL Cussler. AIChE J 34:130, 1988.
151. MC Yang, EL Cussler. AIChE J 32:1910, 1986.
152. EE Wilson. ASME 37:47, 1915.
153. AS Foust, LA Wenzel, CW Clump, L Maus, LB Anderson. Principles of Unit Operation, 2nd ed. New York: Wiley, 1980.
154. LM Coulson, JF Richardson. Chem Eng., vol 1, 3rd ed. Oxford: Pergamon Press, 1977.
155. R Prasad, KK Sirkar. J Membr Sci 50:153–175, 1990.
156. X Lévêque. Ann Mines 13:201, 1928.
157. R Prasad, KK Sirkar. AIChE J 34:177–188, 1988.
158. R Basu, R Prasad, KK Sirkar. AIChE J 36:450, 1990.
159. A Kiani, RR Bhave, KK Sirkar. J Membr Sci 20:125, 1984.
160. R Prasad, A Kiani, RR Bhave, KK Sirkar. J Membr Sci 26:79–97, 1986.
161. SB Iversion, VK Bhatia, K Dam-Johansen, G Jönsson. Characterisation of microporous membranes for use in membrane contactors, Proceedings of the Seventh International Symposium on Synthetic Membranes in Science and Industry, Decchema, Frankfurt am Main 1994, pp 22–23.
162. J Kim, P Stroeve. Chem Eng Sci 41:247, 1988.
163. J Kim, P Stroeve. Chem Eng Sci 44:1101, 1988.
164. AI Alonso, A Irabien, MI Ortiz. Sep Sci Technol 31:271, 1996.
165. I Ortiz, B Galán, A Irabien. Ind Eng Chem Res 35:1369, 1996.
166. AM Urtiaga, MI Ortiz, E Salazar, A Irabien. Ind Eng Chem Res 31:877, 1992.
167. JC Hutter, GF Vandegrift, L Nunez, DH Redfield. Environ Eng 40:166, 1994.
168. (a) RMC Viegas, M Rodriguez, S Luque, JR Alvarez, IM Coelhoso, JPSG Crespo. J Membr Sci 145:129, 1998. (b) Y Qin, JMS Cabral. AIChE J 43:1975, 1997.
169. NA D'Elia, L Dahuron, EL Cussler. J Membr Sci 29:309, 1986.
170. HM Yeh, CM Huang. J Membr Sci 103:135, 1995.
171. H Takeuchi, K Takahashi, M Nakano. Ind Eng Chem Res 29:1471, 1990.
172. MN Pons. Chem Eng J 35:201, 1987.
173. H Takeuchi, K Takahashi, W Goto. J Membr Sci 34:19, 1987.

174. (a) DM Malone, JL Anderson. AIChE J 23:177, 1977. (b) AF Seibert, X Py, M Mshewa, JR Fair. Sep Sci Technol 28:343, 1993.
175. SR Wickramasinghe, J Michael, J Semmens. J Membr Sci 62:371, 1991.
176. V Chen, M Hlavacek. AIChE J 40:606, 1994.
177. RK Shah, AL London. Advances in Heat Transfer: Laminar Flow Forced Convection in Ducts. New York: Academic Press, 1987.
178. S Kakac, RK Shah, AE Bergles. Low Reynolds Number Flow Heat Exchangers. Washington, DC: Hemisphere, 1983.
179. JD Rogers, RL Long Jr. J Membr Sci 134:1, 1997.
180. EL Hinrichsen, J Feder, T Jossang. J Stat Phys 44:793, 1986.
181. J Lemaitre, A Gervois, JP Troadec, N Rivier, L Ammi, L Oger, D Bideau. Philos Mag 67:347, 1993.
182. RO Cowder, EL Cussler. J Membr Sci 134:235, 1997.
183. PR Alexander, RW Callahan. J Membr Sci 35:57–71, 1987.
184. JK Park, HN Chang. AIChE J 32:1937–1947, 1986.
185. MJ Costello, AG Fane, PG Hogan, RW Schofield. J Membr Sci 80:11, 1993.
186. SR Wickramasinghe, MJ Semmens, EL Cussler. J Membr Sci 69:235, 1992.
187. M Dax. Semiconductor International. December 1996.
188. TM Bush, SJ Hardwick, MJ Wikol. Overcoming the barriers to cleaning with bubble free ozonated deionised water. IEEE/SEMI Advanced Semiconductor Manufacturing Conference and Workshop, Boston, September 23–25, 1998.
189. HM Yeh, YS Hsu. Sep Sci Technol 33:757, 1998.
190. R Prasad, KK Sirkar. J Membr Sci 50:153, 1990.
191. J Lemanski, GG Lipscomb. AIChE J 41:2322, 1995.
192. JL Humphrey, GE Keller II. Separation Process Technology. New York: McGraw-Hill, 1997, p 74.
193. KK Sirkar. Membrane separations: Newer concepts and applications for the food industry. In: RK Singh and SSH Rizvi, eds. Bioseparation Processes in Foods. New York: Dekker, 1995, pp 353–356.
194. MC Yang, EL Cussler. J Membr Sci Artif Gills 42:273, 1989.
195. KJ Lissant. Demulsification Industrial Applications, Surfactant Science Ser, vol 13. New York: Dekker, 1983, p 94.
196. SE Taylor. Colloids Surf 29:29, 1988.
197. EC Hsu, NN Li. Sep Sci Technol 29:115, 1985.
198. K Larson, B Raghuraman, J Wiencek. J Membr Sci 91:231, 1994.
199. NP Tirmizi, R Bhavani, J Wiencek. AIChE J 42:1263, 1996.

Index

471